Fundamental Aspects of
Heterogeneous Catalysis
Studied by Particle Beams

NATO ASI Series

Advanced Science Institutes Series

A series presenting the results of activities sponsored by the NATO Science Committee, which aims at the dissemination of advanced scientific and technological knowledge, with a view to strengthening links between scientific communities.

The series is published by an international board of publishers in conjunction with the NATO Scientific Affairs Division

A	**Life Sciences**	Plenum Publishing Corporation
B	**Physics**	New York and London
C	**Mathematical and Physical Sciences**	Kluwer Academic Publishers
D	**Behavioral and Social Sciences**	Dordrecht, Boston, and London
E	**Applied Sciences**	
F	**Computer and Systems Sciences**	Springer-Verlag
G	**Ecological Sciences**	Berlin, Heidelberg, New York, London,
H	**Cell Biology**	Paris, Tokyo, Hong Kong, and Barcelona
I	**Global Environmental Change**	

Series B: Physics

Fundamental Aspects of Heterogeneous Catalysis Studied by Particle Beams

Edited by

H. H. Brongersma and R. A. van Santen

Eindhoven University of Technology
Eindhoven, The Netherlands

Plenum Press
New York and London
Published in cooperation with NATO Scientific Affairs Division

Proceedings of a NATO Advanced Study Institute on
Fundamental Aspects of Heterogeneous Catalysis Studied
by Particle Beams,
held September 3–14, 1990,
in Alicante, Spain

Library of Congress Cataloging-in-Publication Data

NATO Advanced Study Institute on Fundamental Aspects of Heterogeneous
  Catalysis Studied by Particle Beams (1990 : Alicante, Spain)
   Fundamental aspects of heterogeneous catalysis studied by particle
  beams / edited by H.H. Brongersma and R.A. van Santen.
       p.   cm. -- (NATO ASI series. Series B, Physics ; v. 265)
   "Proceedings of a NATO Advanced Study Institute on Fundamental
  Aspects of Heterogeneous Catalysis Studied by Particle Beams, held
  September 3-14, 1990, in Alicante, Spain"--T.p. verso.
   "Published in cooperation with NATO Scientific Affairs Division.
   Includes bibliographical references and index.

   1. Heterogeneous catalysis--Congresses.   I. Brongersma, H. H.
  (Hidde Herman), 1940-   . II. Santen, R. A. van (Rutger A.)
  III. North Atlantic Treaty Organization. Scientific Affairs
  Division. IV. Title. V. Series.
  QD505.N3645 1990
  541.3'95--dc20   91-20199
                                                        CIP

ISBN-13: 978-1-4684-5966-1   e-ISBN-13: 978-1-4684-5964-7
DOI: 10.1007/978-1-4684-5964-7

© 1991 Plenum Press, New York
Softcover reprint of the hardcover 1st edition 1991

A Division of Plenum Publishing Corporation
233 Spring Street, New York, N.Y. 10013

PREFACE

Present day heterogeneous catalysis is rapidly being transformed from a technical art into a science-based technology. A major contribution to this important change is the advance of surface spectroscopic techniques able to characterize the complex surfaces of the heterogeneous catalytic system.

The Advanced Study Institute (on which the current proceedings is based) has as its primary aim the bringing together of a variety of lecturers, outstanding in those fields of experience, to enable a broad coverage of different relevant approaches. Not only catalyst characterization but also catalytic reactivity had to be covered in order to relate catalyst properties with catalyst performance. Since modern catalysis relates catalytic performance to microscopic molecular catalyst features, theoretical electronic aspects also had to be included.

The Advanced Study Institute had a unique feature in that it brought together physicists, catalytic chemists and chemical engineers whom rarely directly interact. From physics especially new experimental possibilities of beams were emphasized. At present it is possible to obtain very detailed information on model catalysts, whilst the applications to practical catalysts are gaining rapidly in sophistication.

Apart from the plenary lectures, the Institute included "hot topics" to highlight special developments and offered participants the opportunity to present contributed papers (either orally or as a poster). These contributions formed an integral part of the summer school and significantly enhanced the interaction between participants. Inclusion of the hot topics and contributed papers in these proceedings give them an added topical value.

This book is organized in three different sections: General Catalysis, Surface Reactivity and Catalyst Characterization.

Our very sincere thanks are due to the North Atlantic Treaty Organization for sponsoring our Advanced Study Institute and for the support received from the Dirección General de Investigatión Cientifica y Técnica del Ministerio de Education y Ciencia, Real Sociedad Española de Fisica, Air Force Office of Scientific Research, Universitat d'Alacant and the Ajuntament d'Alacant.

We should like to take this opportunity to express our deep appreciation to Alberto Gras-Marti, co-director of this Institute, for invaluable help as chairman of the local organizing committee.

Hidde H. Brongersma
Rutger A. van Santen

CONTENTS

GENERAL CATALYSIS

REACTIVITY

CATALYST CHARACTERIZATION

Nuclear Magnetic Resonance

Photons

Electrons

Neutrons

Ions

INTRODUCTION TO THE USE OF PARTICLE BEAM TECHNIQUES IN

HETEROGENEOUS CATALYSIS

Richard W. Joyner

Levehulme Centre for Innovative Catalysis
Department of Chemistry, University of Liverpool
PO Box 147, Liverpool, L69 3BX, UK

Heterogeneous catalysis finds very widespread application and is of profound academic interest as well as major industrial importance. This brief chapter provides an introduction to some of the concepts of interest in contemporary catalysis, while it is left to individual presentations to introduce the wide range of techniques under consideration.

Catalysts are used to accelerate the rates of desired chemical reactions. Elementary textbooks suggest that the catalyst is unchanged in use and this may sometimes be the case. All useful catalysts have acceptable combinations of <u>activity</u>, <u>selectivity</u> and <u>lifetime</u>. The activity is a measure of the rate of the catalysed reaction. An academic may express this as a turnover number (TN):-

$$TN = \text{molecules produced / active site / s} \tag{1}$$

whereas the industrialist will often wish to know the space time yield, (STY):-

$$STY = \text{mol of desired product / l of catalyst / h} \tag{2}$$

To determine the turnover number we need to know the number density of active sites, which is not easy to measure and can be a controversial quantity.

The aim in many catalytic reactions is to reach a local thermodynamic minimum, ie. to catalyse only one of a number of possible reactions. An example would be:

$$C_2H_4 + \tfrac{1}{2}O_2 \quad \xrightarrow[\text{reaction}]{\text{desired}} \quad \begin{array}{c} C_2H_4O \\ \text{ethene oxide} \end{array}$$

$$\xrightarrow[\text{reaction}]{\text{unwanted}} \quad \begin{array}{c} CO_2/H_2O \\ \text{(incomplete conversion)} \end{array}$$

In oxidations of this sort conversion to CO_2/H_2O is more favourable

Fundamental Aspects of Heterogeneous Catalysis Studied by Particle Beams,
Edited by H.H. Brongersma and R.A. van Santen, Plenum Press, New York, 1991

1

thermodynamically than formation of ethene oxide, but it is clearly economically very undesirable.

Acceptable lifetime is difficult to define, since due to the ingenuity of reactor design engineers, the range is from < 1s to many years. Hydrocarbon cracking catalysts, as discussed by Knozinger (page 7), are active for only a few seconds and must then be reactivated. Promoted iron catalysts for ammonia synthesis, however can remain active for more than ten years.

All three aspects of catalytic behaviour are important, but efforts to develop or improve present catalysts have in recent years centred on selectivity. Relatively low activities and short lifetimes may be accommodated, but the consequence of poor selectivity is wasted feedstock and this is often economically prohibitive. If we are entering an era of higher oil prices, the emphasis on selectivity will become even more marked.

The papers collected in this book are mainly concerned with the application of a range of particle beam techniques and spectroscopies to characterise catalysts and improve our knowledge of catalytic phenomena. Three broad classes of catalytic materials are important, oxides, metals and sulphides. The many varieties of oxide are described by Knozinger, (page 7). Oxides are widely used as catalyst for oxidation reactions. They also have an important role as catalyst supports, where high area is necessary to sustain an expensive metal in small particle form, (eg. platinum or rhodium). Zeolites, or molecular sieves form an important special class of oxides. These materials, which contain channels and cavities of molecular dimension penetrating right through the crystal structure, are of immense interest and value. They are particularly amenable to study by nuclear magnetic resonance and neutron methods and are considered in some detail by Pfeiffer, (page 155) and Jobic, (page 259). Sulphides show many similar characteristics to oxides and are also considered by Knozinger.

Metal catalysts are important both in unsupported form and supported on refractory oxides, as already mentioned. Only a restricted number of the available metallic elements can be used as metal catalysts. Many are easily oxidised by traces of water, and only the group VIII and group Ib metals can be prepared and maintained in metallic form, even in the most strongly obtainable reducing environments, $(P(H_2)/P(H_2O) = 10^6)$. Much current interest centres on the role of promoters such as alkali metals, which improve catalytic performance when added to a catalyst, but are inactive on their own. There is also a considerable interest in bimetallic and alloy catalysts. It is sometimes possible to improve the performance of a metal catalyst by using two metals which are not capable of bulk alloy formation but which can coexist in small particles, (diameter < ca 2nm). Such catalysts are referred to as bimetallics.

Catalysts are so widely used in the chemical, petroleum and pharmaceutical industries that it is not useful to attempt a catalogue of catalysed reactions. Before moving to consider the role of particle beams in catalytic studies, it is worth noting, however, that the fastest growing catalytic application of the last decade has been in pollution control. Automotive exhaust treatment and selective reduction of nitrogen oxides, (NO_x), in power station stack gases are making a major contribution to a cleaner environment.

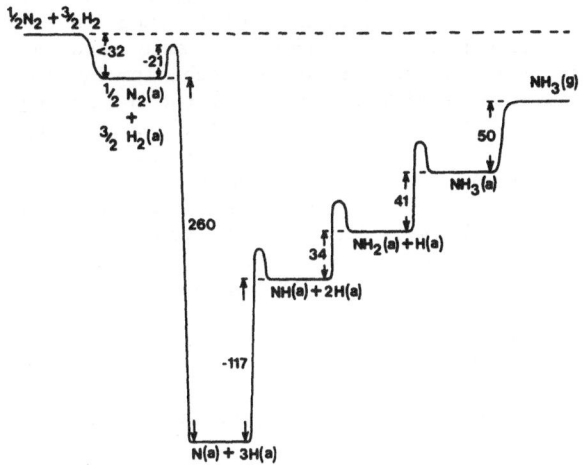

Fig. 1. The one-dimensional potential energy diagram for the synthesis of ammonia at a potassium promoted iron surface, (Ertl, 1983). Energies are in kJ mol⁻1 and the heat of formation of ammonia is -46kJ mol⁻1.

A growing number of particle beams are useful in catalytic studies and these form the main focus of this ASI. Table 1 indicates where the main discussion of each technique can be found.

At this stage it is worth asking what the information obtained from these and other characterisation techniques is used for - why is catalyst characterisation performed? In most academic and many industrial laboratories the aim is to understand the mechanism by which the catalyst operates. Much has been achieved and one of the best examples is shown in Fig. 1. This is a one-dimensional potential energy representation of the way in which ammonia is synthesised from a nitrogen/ hydrogen mixture over the classic promoted iron, Haber catalyst, and deduced by Ertl, (1983). Stolze and Norskov, (1985,1988) have used this diagram and other experimental data to calculate the rate of production of ammonia in a commercial plant, achieving an answer very close to the true value and thus representing a major validation of the mechanism.

Catalyst characterisation is also useful in optimisation of a new catalyst formulation and this is the aim in many industrial studies. A detailed description of the catalyst will assist in determining the best composition and the most suitable preparation method for the new catalyst.

Characterisation evidence can also be of immense value where it is necessary to diagnose catalyst failure. Sometimes a catalyst prepared according to a well tried method is found to be inactive, or to decay rapidly in use, and characterisation studies can explain what has happened.

The range of techniques applied in catalyst characterisation is extremely wide, (see eg. Deviney and Gland, 1985). One rule and one guideline which have been found to be reliable in practise may be mentioned. The simple rule is:

NEVER RELY ON ONLY ONE TECHNIQUE

It is important to note this in a volume with a strong base in technique, but it is of general applicability. Each technique imposes its own perspective on the problem at hand and this can be misleading in some situations, so careful choice is important. The information in this volume should be of material help in many cases.

The guideline is to choose in-situ techniques where possible. A number of useful approaches have been developed and are discussed here, including infra-red and Raman spectroscopy, X-ray adsorption and the range of neutron methods. Nuclear magnetic resonance is also being applied in a quasi in-situ mode, which is already yielding exciting results on the mechanism of methanol conversion to hydrocarbons over the zeolite ZSM-5, (Anderson and Klinowski, 1989,1990) and X-ray diffraction is also of value.

The highest aim of catalyst characterisation is to establish the relationship between the electronic and geometric structure of the catalyst and its performance. This may involve in-situ or ex-situ studies of catalytic materials, or studies of model catalysts. Surface science studies, usually employing highly characterised single crystal surfaces, are often used. Surface science contributes to our understanding of catalysis in many ways, (Joyner, 1990), and several important surface techniques are reviewed in this volume. A number of these are concerned with determination of surface structure and it cannot be overstated how important accurate structural data is for developing our detailed understanding of catalysis (MacLaren et al, 1986, van Santen, page 85.)

Modern heterogeneous catalysts may contain up to ten components, so that when reactant and product molecules are included the result is very complex indeed. Much contemporary catalytic research seeks to break down the complexity by studying pairwise interactions, for example those involving eg:

> metal and support
> catalyst and promoter
> oxide and oxide
> sulphide and oxide
>
> or proton and the zeolite framework.

As this volume demonstrates, much progress has already been made and the wide range of powerful techniques now available guarantees that the next decade will be extremely exciting.

REFERENCES

Anderson, M.W., and Klinowski, J., (1989), Nature (London), 339, 200.
Anderson, M.W., and Klinowski, J., (1990), J. Amer. Chem. Soc.,　; 1.
Deviney M. L., and Gland, J.L., (1985), "Catalyst Characterisation Science," American Chemical Society Symposium Series No. 288, Washington.
Ertl, G., (1983), "Catalysis, Science and Technology," Ed. Anderson, J. R., and Boudart, M., Springer Verlag, Berlin, Vol. 4, pp 273.
Joyner, R.W., (1990), JCS Faraday Trans., 86; 2675.
MacLaren, J. M., Pendry, J. B., Joyner, R. W., and Meehan, P., (1986) Surface Sci., 175; 263.
Stolze, P., and Norskov, J., (1985), Phys. Rev. Lett., 55; 2502.
Stolze, P., and Norskov, J., (1985), Surface Sci., 197, L230.

HETEROGENEOUS CATALYSTS AND CATALYTIC PROCESSES

Helmut Knözinger

Institut für Physikalische Chemie
Universität München
Sophienstrasse 11, 8000 München 2, FRG

INTRODUCTION

Heterogeneous catalysts play an extraordinary role in chemical and petrochemical industry and in environmental control. About 90% of all processes in chemistry and petrochemistry are employing catalysts. The catalyst consumption worldwide was approximately US $ 2.800 million in 1984 excluding car exhaust catalysts and was estimated to be about US $ 5.000 million in 1985.[1] These numbers clearly emphasize the economic importance of catalysts and catalytic processes. In this context it must be noted that the value of the catalyst applied in a given process to produce a particular product under consideration is typically less than 1 % of the total cost. Hence, the value added by the catalyst is to be considered as very high as compared to the other cost generating factors.

The application of catalysts leads to savings of raw materials and energy consumption and to reductions in atmospheric pollution. They also play a major role in environmental control. Chemical transformations are always accompanied by energy conversions. An example for the saving of raw materials combined with an improved utilization of the liberated heat of reaction is the catalytic synthesis of phthalic anhydride. The non-catalytic oxidation of naphthaline in the liquid phase gave yields of only 5 - 15 %. In contrast, the catalytic gas phase oxidation with air over supported vanadium oxide catalysts now-a-days yields approximately 90 % of the desired product. In addition the liberated heat of conversion is used to produce overheated water vapour so that the energetic efficiency of the process is significantly improved.

Nitric oxide as an intermediate in nitric acid production was obtained in the early days via the conversion of N_2 and O_2 in an electric discharge at temperatures above 3000 K. This process required 60 000 kWh per ton N_2 consumed. Now-a-days nitric oxide is produced via the catalytic combustion of ammonia over Pt grids at about 1000 K with a total energy consumption (including the NH_3 synthesis) of about 10.000 kWh per ton of N_2 consumed. The modern catalytic NO production thus only requires

Fundamental Aspects of Heterogeneous Catalysis Studied by Particle Beams,
Edited by H.H. Brongersma and R.A. van Santen, Plenum Press, New York, 1991

7

17 % of the energy consumed in the early non-catalytic process.

Side-products in chemical processes may lead to environmental pollution. The application of selective catalysts can dramatically reduce the formation of these undesired side-products or catalysts can be used for their conversion into tolerable compounds. The best-known example for the latter application certainly is the car exhaust catalyst. Another example is the hydrogenolytic removal of sulfur compounds (hydrodesulfurization) from petroleum distillates using CoMo-catalysts. Sulfur dioxide would otherwise be emitted when combusting the sulfur-containing petroleum fractions. The H_2S produced via hydrodesulfurization can either be removed by reaction with metal oxides or washed out from the H_2 with suitable solvents. It can then be transformed catalytically into elemental sulfur (Claus process) which on catalytic oxidation yields SO_2 and finally sulfuric acid. This series of processes nicely demonstrates how the application of catalysts can be combined in pollution control and production of valuable chemicals.

FORMS OF HETEROGENEOUS CATALYSTS

The specific properties of a heterogeneous catalyst are determined by its chemical formulation and, in the case of supported catalysts, the distribution and dispersion of the active components. However, technical catalysts have to meet a number of additional requirements which are dictated by the process proper in which they are applied. Hence, properties such as pore structure and distribution, mechanical strength, thermal stability and resistance toward sintering, particle and grain sizes etc. must be optimized for a given process. Moreover, the build-up of a pressure drop across the catalytic bed should be minimized for fixed-bed reactors, while catalyst powders consisting of small (and hence light) particles have to be used in fluidized bed operation. Besides powders having particle sizes in the range 50 - 200μm for the latter application, various forms of coarse catalyst particles such as pellets, spheres, tablets, granules and extrudates with typical dimensions of 0.2 - 0.5 cm have been developed for their use in fixed bed reactors. In addition, still larger specimens such as those shown in Fig. 1 a are in use. As support materials they typically consist of oxides (e.g. Al_2O_3) or ceramic materials. The biggest units

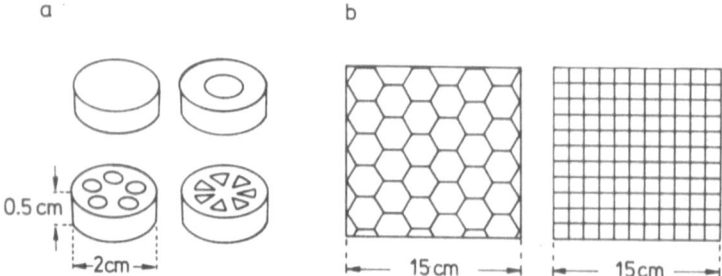

Fig. 1. a) Shapes of catalyst support tablets
b) Possible channel geometrics in monolithic supports.

are the so-called monoliths; cross-sections of monoliths are shown in Fig. 1 b, indicating typical channel structures. The dimensions of these units can vary from the order of 10 - 20 cm (e.g. car exhaust catalysts) up to 15 x 15 x 100 cm (e.g. selective catalytic reduction SCR of NO by NH_3). When used as catalyst supports the monolith may consist of steel or ceramic materials such as cordierite which has an extremely low thermal expansion coefficient. Both materials are applied for car exhaust catalysts. The monolithic material is covered with a so-called wash-coat consisting of alumina which serves as the actual support for the catalytically active material (e.g. noble metal). In other applications the monolith is directly formed from the catalytically active mass (e.g. $(V_2O_5-MoO_3)/TiO_2$ for SCR).

In the following, a few typical classes of heterogeneous catalysts of technological importance are briefly described and for each case their application will be demonstrated on the basis of one selected process. Obviously this cannot be done here in any detail. Relevant references will therefore be included which will provide the interested reader with an in-depth treatment of the various processes in consideration.

BINARY OXIDES

Simple binary oxides, such as SiO_2, MgO and the transition aluminas, are terminated by a surface hydroxyl layer for energetic reasons.[2,3] Thermal treatment at sufficiently high temperatures in vacuum leads to partial dehydroxylation via condensation of adjacent hydroxyl groups. The surface of the oxides is then constituted by hydroxyl groups, oxygen ions and coordinatively unsaturated (cus) exposed cations,[2,3] as schematically indicated in Fig. 2. The individual properties of these surface groups and atoms are determined by the bonding character in the particular oxide (electronegativity of the metal constituent).

Fig.2, Schematic representation of the dehydroxylation process of (a) alumina, (b) silica, and (c) magnesia surfaces.

The hydroxyl groups can act as H-bond donor or acceptor sites, as basic sites or as proton (Bronsted) acidic sites. Oxygens function as H-bond acceptor and basic sites and cus metal cations are electron pair acceptor sites, i.e. Lewis acidic sites. The overall surface properties of an oxide are dependent on the relative contributions of the three types of constituents and on the degree of hydroxylation. Thus, the MgO surface can be classified as basic. [3] Due to the high ionic bond character and the small charge / radius ratio of Mg^{2+}, the basic properties of O^{2-} and OH^- ions are dominating the surface properties here. Nevertheless, $Mg^{2+}O^{2-}$ pairs can act as acid-base pair sites which are able to dissociate e.g. C-H bonds. [4] Even dihydrogen is heterolytically dissociated on dehydroxylated MgO surfaces. [5]

$\eta-$ and $\gamma-Al_2O_3$ surfaces are amphoteric in nature when they are partially dehydroxylated. The relatively strong Lewis acidity brought about by cus Al^{3+} cations is compensated for by the basic function of O^{2-} and OH^- species. Protonic acidity is not developed. Acid-base pair sites, however, play an important role also on alumina surfaces. [6]

The Si-O bond in SiO_2 has a strong covalent character. [2] As a consequence, the surface behaves almost neutral, although surface silanol groups do act as H-bond donors, their properties being similar to alcoholic OH groups. On dehydroxylation, siloxane bonds are formed as indicated in Fig. 2b in which the Si atoms remain 4-coordinated. When fully symmetric siloxane bridges are developed, they are relatively inert. Pure silicas therefore develop hardly any catalytic activity.

The acid-base properties of binary oxides determine their catalytic behaviour. It should be noted that simple binary oxides typically do not develop strong protonic acidity, although mixed oxides, particularly amorphous silica-aluminas and crystalline alumino silicates (zeolites) do function as solid proton acids as will be discussed in section 2. Since proton acidity plays an important role in catalytic mechanisms, SiO_2-Al_2O_3 and zeolites will be included in this discussion.

The dehydration of alcohols is catalyzed by oxides, the detailed mechanism of this elimination reaction and hence, the product distribution being determined by the particular surface properties of the oxide catalyst. [7,8] Three principal cases can be distinguished:

<u>Case A:</u> Proton acidity is dominating as in SiO_2-Al_2O_3. In this case the dehydration reaction follows a so-called E1 elimination which can be described by the following sequence of steps in which oxonium ions ROH^+_2 are formed by proton addition and carbenium ions R^+ act as intermediates:

$$ROH + H^+ \longrightarrow ROH_2^+ \tag{1}$$

$$ROH_2^+ \longrightarrow R^+ + H_2O \tag{2}$$

$$R^+ \longrightarrow Alkene + H^+ \tag{3}$$

<u>Case B:</u> Basic properties are dominating as on MgO. As described above, MgO surfaces provide strong basic sites (denoted B below). These can abstract protons from hydrocarbon molecules [4]

$$-\overset{|}{\underset{|}{C}} - H + B \longrightarrow -\overset{|}{\underset{|}{C}}{}^- + BH^+ \tag{4}$$

and thus form carbanions as intermediates. Alcohols are dehydrated on MgO via this route which is denoted E1cB elimination mechnism. [7]

Case C: Amphoteric surface properties as on aluminas. Alcohol molecules on aluminas are adsorbed via H-bond systems.[8,9] Proton fluctuations in these H-bond systems may lead to polarization of the adsorbed alcohol molecule. These dynamic polarizations will lead to enhancements of the acidity of ß-hydrogen atoms (see Scheme 1) which can thus undergo acid-base interactions with surface basic sites. This leads to a concerted rupture of the C_α —OH and C_β — H bonds with formation of the alkene product.[9] This mechanism is denoted an E2 elimination

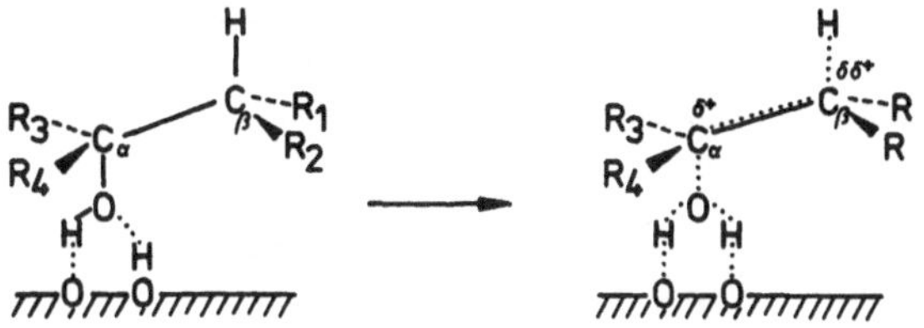

Scheme 1

It is interesting to note that pure SiO_2 is not active in alcohol dehydration due to the lack of proton acidity *and* of surface basic sites.

SILICA-ALUMINAS AND CRYSTALLINE ALUMINO-SILICATES (ZEOLITES)

Amorphous silica-aluminas can be prepared in a variety of ways.[10] In these materials Al^{3+} is incorporated at a tetrahedral Si-position within the poorly ordered framework of SiO_4 tetrahedra. The negative charge thus produced within the network is usually compensated by a proton

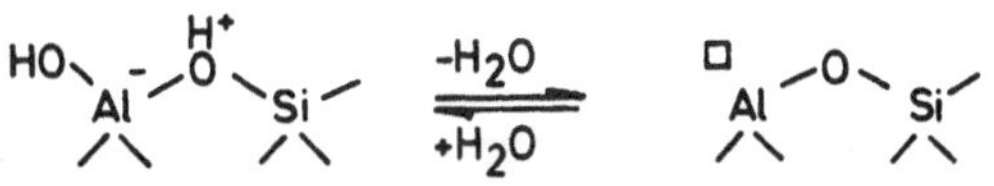

Scheme 2

which is thought to be associated with an oxygen atom bridging the Al and a Si atom. This charge compensating proton provides the protonic or Bronsted acidity of these materials.[2] As indicated in scheme 2, water can be removed on thermal treatment. This dehydroxylation would create a vacancy at the Al atom, thus exposing a tricoordinate Al^{3+} which acts as a Lewis acid site.[10] Scheme 2 describes the interconversion of Bronsted and Lewis acid sites by a dehydroxylation/rehydroxylation process.

SiO_4 and $[AlO_4]^{-1}$ tetrahedra are also the primary building blocks of crystalline alumino-silicates. These can be interconnected in a multitude of ways so as to provide a large variety of zeolite families.[10] As examples, the structures of zeolite A and of the faujasite (zeolites X and Y) are shown in Fig. 3. The secondary building block here is a cubooctahedron (also known as "sodalite" cage), in which the tetrahedral Si or Al atoms are located at the corners of polygons with the oxygen atoms approximately halfway between them. Zeolite A results when these cubooctahedra are interconnected through double four-membered rings to form a regular three-dimensional structure. If the sodalite cages are linked together via double six-membered rings, the resulting material is called faujasite. As can be seen in Fig. 3, large cages are being formed in this way which are accessible through windows of well-defined diameter depending on the mode of interconnection of the sodalite cages. The faujasite zeolite Y has an aperture of 0.74 nm (large pore zeolite). These structures are responsible for the molecular sieve properties of zeolites.

Many other structures are known today, among which the silica-rich so-called silicalites play a major role as catalysts. [10-12] The most prominent structure here is the zeolite ZSM-5. This material contains a pore structure consisting of two linear, intersecting pore systems running in [010] and [100] directions. The diameter of the wider pores ([010] direction) is close to 0.55 nm (medium pore zeolite).

The charge-compensating ions are initially Na^+ ions which can be exchanged for NH^+_4.[10] These may then be decomposed thermally so as to yield protons as charge-compensating ions which would act as Bronsted sites as indicated in scheme 3:

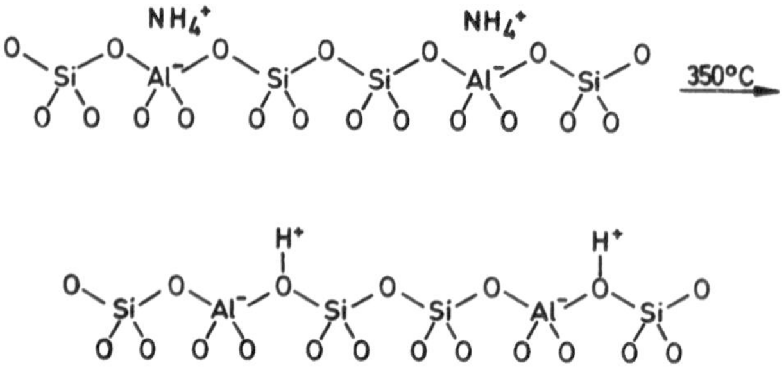

$$\text{350°C}$$

Scheme 3

Again the acidic protons are thought to be associated with oxygens bridging an Al and a Si atom. Their density can obviously be controlled by the Si/Al ratio. Moreover it is now possible to also control the acid strength distribution by incorporation of other trivalent cations such as e.g. Ga^{3+} or Fe^{3+} into the zeolite framework, [13,14] e.g. the ZSM-5 structure. Thus, the Bronsted acid strength of ZSM-5 type zeolite has been shown [14] to decrease in the sequence: [Al] - HZSM - 5 > [Ga] - HZSM - 5 > [Fe] - ZSM - 5. The synthesis of titanium silicates led to an additional new family of zeolites providing catalytic activity for oxidation reactions with H_2O_2 as oxygen donor. [15,16]

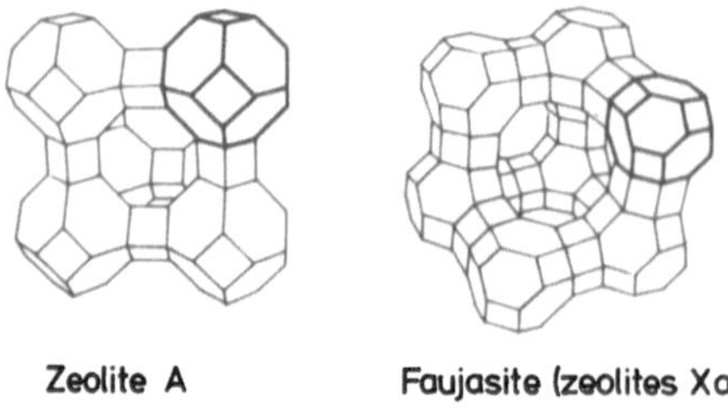

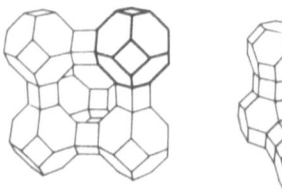

Zeolite A   Faujasite (zeolites X and Y)

Fig. 3. Structures of zeolite A and of faujasites.

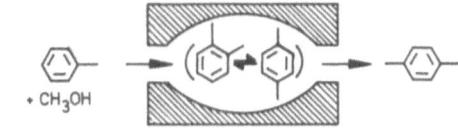

Restricted Transition State Selectivity

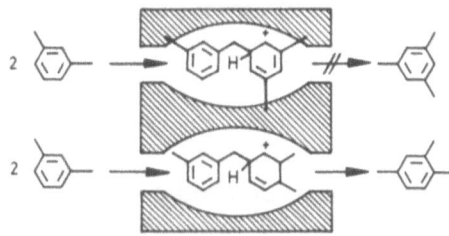

Fig. 4. Schematic representation of various forms of shape
selectivity.

The channel structure of zeolite materials is responsible
for their unique shape-selectivity in catalytic conversions.
According to Csicsery [17,18] three different modes of shape-selectivity can be distinguished, which are schematically represented in Fig. 4. These depend on the channel/aperture dimensions of the zeolite relative to sizes and shapes of reactant
and/or product molecules and of the transition state, respectively. Reactant shape selectivity is operative when in a mixture
of reactants one is excluded from entering the zeolite channels
while a second is not (e.g. branched vs straight-chain hydrocarbon, see Fig. 4). If from a mixture of products formed within
zeolite cages, only one can escape through the apertures, the
product distribution is controlled by product shape selectivity.
For example, toluene alkylation with ethylene to form methyl
ethyl benzenes is an acid catalyzed reaction, which when carried
out with $AlCl_3$ leads to a mixture containing typically 50 % m-,
25 % o-, and only 25 % p-methyl ethyl benzene. In contrast, when
a zeolite catalyst is used, the least bulky p-product can be
obtained with 95 % selectivity. Finally, certain reaction routes
may be suppressed by the restricted transition state selectivity
if bulky intermediates, e.g. in bimolecular reactions (e.g.
transalkylations, see Fig. 4), were involved.

An interesting example of shape selectivity is the simultaneous catalytic cracking of n-hexane and 3-methyl pentane [11]
as shown in Fig. 5. On a large pore LaY zeolite the predicted
higher reactivity of the branched hydrocarbon is obvious from
its significantly greater conversion. In contrast, n-hexane
conversions are greater on the medium pore HZSM - 5, probably

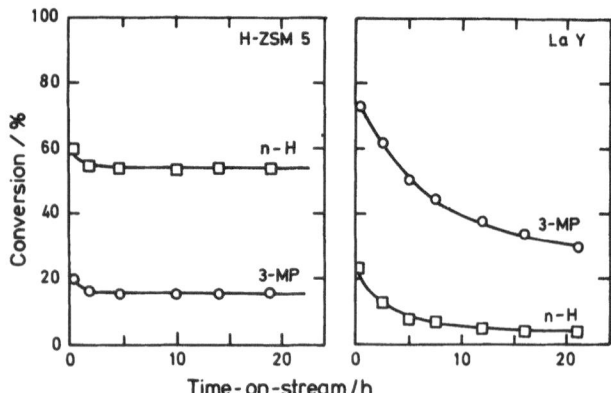

Fig. 5. Cracking of mixtures of n-hexane (n-H) and 3-me-
thyl-pentane (3-HP) over HZSM-5 and LaY zeolite
catalysts under comparable experimental conditions.

indicating the effect of reactant shape selectivity. It is also
obvious that the large pore zeolite LaY deactivates much faster
than the medium pore ZSM - 5. Deactivation is due to coke formation. It has been argued [19] that coke is deposited within the
channel system of large pore zeolites, whereas it only forms on
the external surface of medium pore zeolites thus not affecting
their catalytic performance. Although the mechanism of coke
formation is certainly not entirely understood, the low tendency
toward coke formation in medium pore zeolites is probably due to
the restricted transition state selectivity within narrow pores.
[20]

Catalytic cracking is a technologically extremely important
process in petroleum industries. [10,21,22] Petroleum distillates
typically provide an excess of high-boiling hydrocarbons which
can be transformed into valuable products, e.g. gasoline, by catalytic cracking. Cracking, i.e. the scission of C-C bonds, is
a proton catalyzed reaction where the formation of carbenium
ion intermediates is the first and most important step. [10] Amorphous silica-aluminas have therefore been used as catalysts
which, however, deactivate extremely fast due to coke formation
(life time in the order of seconds). Modern FCC (fluid catalytic
cracking) catalysts contain high percentages of zeolites (e.g.
LaY) which not only are significantly less susceptible to coking
but also yield improved product compositions, e.g. greater
naphtha yields with reduced formation of gases and C_4-hydrocar-
bons. [10]

MIXED METAL OXIDES

Many mixed metal oxides are characterized by the fact that
they can release oxygen from their bulk very readily. Ternary
oxides providing this property are particularly those containing
Bi and Mo (so-called bismuth molybdates), Fe and Sb, U and Sb,
and Sn and Sb. These materials typically undergo rapid isotopic
exchange with $^{18}O_2$ from the gas phase throughout their bulk
suggesting high oxygen mobility in addition to their tolerance
of oxygen deficiency. Many defect perovskites of general formula

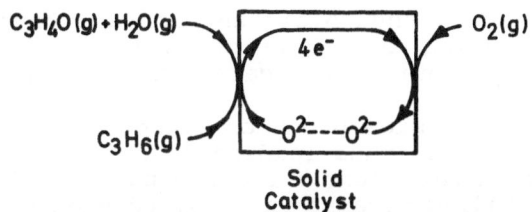

Fig. 6. Redox processes in a solid oxidation catalyst.

ABO_{3-x} (A ≡ alkaline earth ions, B ≡ first row transition metal ions, with $0 \leq x \leq 0.5$) also have a remarkable flexibility in their oxygen non-stoichiometry. [23-26] The structural frameworks of the metal cations in perovskites are retained in oxygen-deficient phases, thus accounting for the reversibility of production and annihilation of oxygen vacancies.

These properties of the ternary oxides mentioned are responsible for their activity as oxidation catalysts. [10,27] It has been demonstrated beyond doubt by isotopic [18]O labelling that the solid catalyst releases structural oxygen in the catalytic oxidation reaction, the lattice oxygen being incorporated into the product molecules. Oxygen vacancies thus produced can be replenished by incorporation of O_2 from the gas phase. This process is facilitated by the electrons being released in the catalytic oxidation step. Fig. 6 demonstrates this reduction – reoxidation cycle of the solid catalyst schematically for the formation of acrolein C_3H_4O from propene C_3H_6. This selective oxidation is catalyzed e.g. by bismuth molybdates. [10] The reaction is thought to proceed via the catalytic cycle shown in Fig. 7. [28] An allyl intermediate at a Mo site is formed by hydrogen abstraction from propene followed by oxygen transfer from the catalyst. A second hydrogen atom is abstracted from the hydrocarbon molecule in the desorption step of the product acro-

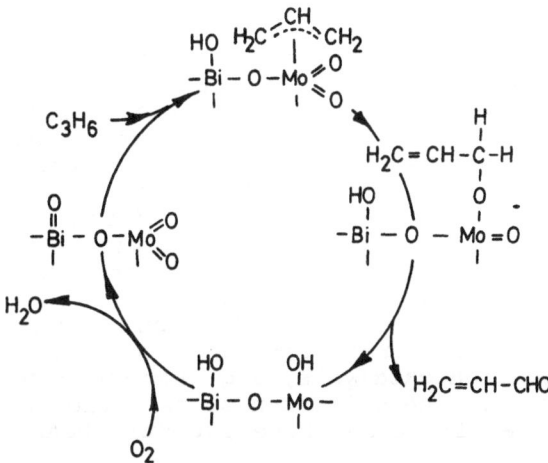

Fig. 7. Catalytic cycle for propene oxidation to acrolein.

lein. The catalyst is then hydrogen-rich and oxygen-deficient. Reoxidation by gaseous O_2 and condensation of hydroxyl groups (abstraction of H_2O) brings the catalyst back into its original state.

The formation of side products cannot easily be avoided in such complex processes, thus leading to unfavourable selectivities. The selective catalytic oxidation to valuable chemicals is particularly demanding because the desired products are far less favoured than the thermodynamically stable endproduct CO_2. Therefore, selectivities of industrial catalysts are improved by promoters and additives. The modern technical catalyst for acrolein production contains nine elements, namely besides Bi and Mo also Fe, Ni, Co, P, W, Si and K. With this catalyst acrolein selectivities $\geq 90\%$ at propene conversions ≥ 95 % can be achieved at 1,5 bar in the temperature range 620 - 720 K. Many other valuable products can be obtained with high selectivities from propene when only the catalyst is designed properly. Thus, ternary oxides containing Sn and Mo would give acetone, those containing Bi and Sb benzene, while those containing Cu and Cr would lead to total oxidation to CO_2.

The kinetics of selective oxidation reactions is generally based on the Mars- van Krevelen redox mechanism [10] which involves the reoxidation of the catalyst according to equ. (5):

$$O_2 + 2 \; [\;] + 4e^- \longrightarrow 2 \; [O^{2-}] \qquad (5)$$

where [ ] denotes an oxygen vacancy. This process is highly improbable as a single step as it involves the simultaneous transfer of four electrons. A stepwise reduction of dioxygen is much more plausible. It can be envisaged as follows:

$$O_2(gas) + e^- \longrightarrow O_2^-(surf.)$$

$$O_2^-(surf.) + e^- \longrightarrow O_2^{2-}(surf.)$$

$$\text{(6a-e)}$$

$$O_2^{2-}(surf.) \longrightarrow 2 \; O^-(surf.)$$

$$O^-(surf.) + e^- \longrightarrow O^{2-}(surf.)$$

$$O^{2-}(surf.) + [\;] \longrightarrow [O^{2-}](bulk)$$

As discussed above, lattice oxygen $[O^{2-}]$(bulk) is assumed to be involved in selective oxidation. O^{2-} species are in fact much less reactive than the anionic molecular surface species which when present in high density will lead to total oxidation. It therefore becomes evident that a variety of factors must be optimally tuned so as to make an oxidation catalyst highly selective. These factors include high bulk oxygen mobility, high electron mobility, optimized metal-oxygen bond strength, low residence time on the surface of the oxidized precursor of the desired product, etc. High selectivities with a tendency toward total oxidation will occur when the residence times of hydrocarbon molecules on the surface are long and when simultaneously the stationary concentrations of O_2^-(surf.) and/or O_2^{2-}(surf.) species are high. These requirements have to be met for efficient total oxidation catalysts, e.g. for environmental protection. Guidelines for the selection of oxidation catalysts for a wide variety of applications have recently been given by Spivey.[29]

SUPPORTED OXIDES

Supported oxides are frequently referred to as monolayer-type catalysts suggesting that the active supported oxide phase is spread into a monolayer on the surface of a supporting oxide such as e.g. SiO_2, Al_2O_3, TiO_2, etc. Typical active oxides are oxides of the group V b, VI b and VII b elements, namely V, Cr, Mo, W, and Re. They are anchored onto the support surface in the form of oxo-anions. These can be monomeric or dimeric as e.g. chromate and dichromate species supported on silica. [2] In most supported oxide materials, however, the active phase is built up by polyanionic species such as polyvanadates on TiO_2 (anatase), and polymolybdates and polytungstates on Al_2O_3 and TiO_2 (anatase). It is interesting to note that these supported materials develop Bronsted acidities which are typically higher than those of the support. WO_x on ZrO_2 e.g. has been reported to develop superacid properties. [30]

These materials are active catalysts in a variety of processes in their oxide form, while some of them act as precursors for an active catalyst. Mo- and W-oxides supported on γ-Al_2O_3 are the precursors of the active sulfided form which is used in the petroleum industry in hydrodesulfurization (vide infra) and related processes. [31] Most technical catalysts of this class of supported oxides also contain promoters and additives to optimize their properties. Table 1 summarizes several oxide combinations and their uses in technologically important processes.

Table 1. Uses of Supported Oxide Catalysts

Active Oxide	Support Oxide	Uses	Comments
Mo, W	γ - Al_2O_3	Hydrotreating Hydrodesulfurization Hydrodenitrogenation Methanation Water Gas Shift	Precursor
Cr	SiO_2, SiO_2-Al_2O_3	Polymerization	
V	SiO_2, Al_2O_3	Selective Oxidation	
V, Mo, W	TiO_2 (anatase)	Selective Catalytic Reduction	
Re	Al_2O_3	Metathesis	

As an example for the application of supported oxide catalysts, the so-called selective catalytic reduction (SCR) of nitrogen oxides in the flue gases from coal-driven power plants by ammonia in the presence of excess oxygen shall briefly be described. [32,33] The catalytic process operates at 570 - 670 K and converts NO into N_2 and H_2O:

$$4NO + 4NH_3 + O_2 \longrightarrow 4N_2 + 6H_2O. \qquad (7)$$

The catalyst is based on TiO_2(anatase) as a support and contains oxides of vanadium, molybdenum or tungsten, or, in practice, combinations of two of these oxides, namely of vanadium and molybdenum or tungsten. The materials can be produced by mixing powders of TiO_2, MoO_3 (or WO_3) and V_2O_5 in the required proportions by intensive wet milling. Binders are added, followed by kneading. The resulting paste is then used to form plate-type catalytic converter elements or honeycomb-type elements by extrusion. These elements are then dried and calcined at typically 770 K and finally brought together into modules of about 3 m^3. In this case, the plate or honeycomb-type elements consist entirely of the catalytically active mass.

Mechanistic studies of this redox-type reaction have been reported for vanadium-based catalysts. [34,35] Little is known, however, about the reaction path on the more complex technical multicomponent catalysts. Experimental evidence indicates that the catalytic activity correlates [36] with the catalysts' acidity and that oxo-groups of the supported oxide are involved in the catalytic act. A major problem with the SCR process still is catalyst deactivation by arsenious oxide and pore blocking by ash and dust particles. The legislation on atmospheric pollution control will require for many other factories and plants, such as e.g. garbage incineration facilities, catalytic reduction of nitrogen oxides and oxidation of carbon monoxide, hydrocarbons and abatement of industrial odours. This field of catalytic pollution control will therefore continue to play an important role in research and development.

SUPPORTED SULFIDES

As mentioned above, Mo and W oxides supported on γ-Al_2O_3 are precursors for sulfided catalysts. [31] It has been demonstrated beyond doubt by many physical techniques that MoS_2 platelets in high dispersion are being formed during the sulfidation process, whereby the polyanionic precursor seems to play a determining role. The technical catalysts contain Co or Ni as promoters. These are located in three different phases, namely within the support in the form of a surface spinel, Co - or Ni-sulfides, and MoS_2 platelets which are decorated along the edges by Co- or Ni-atoms. [31,37-39] The latter phases are considered to be the active phases in catalytic hydrodesulfurization, and are denoted CoMoS or NiMoS phases. [31,37-39] A schematic representation of the cobalt-promoted catalyst [38] is shown in Fig. 8. The exact coordination and position of the

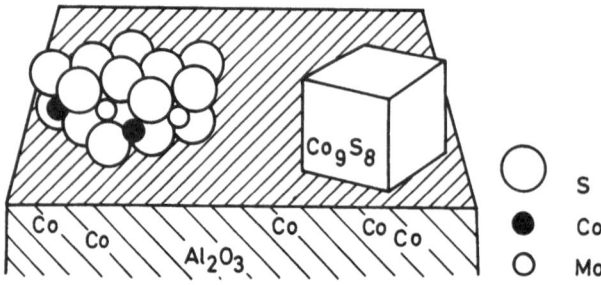

Fig. 8. Schematic representation of a sulfided CoMo catalyst.

18

promoter atoms is still not clear although recent Ni EXAFS studies seem to shed some light on this problem. Niemann et al. [40] proposed a possible change in the local environment of the promoter atoms at the (10$\bar{1}$0) edge plane of MoS$_2$ from a four fold square coordination to a five fold coordinated site during a reaction cycle. Louwers and Prins [41] suggested a five fold square pyramidal coordination for the Ni promoter atoms. Also, the nature of active sites and of the promoter action is still a matter of controversy. [31,39] Special sites in which Mo and promoter atoms undergo electronic interactions are proposed, while others consider the MoS$_2$ platelets simply as a secondary support which guarantees high dispersion and the appropriate coordination for the Co or Ni atoms, these being the active sites. [39] It should also be mentioned that stoichiometric MoS$_2$ certainly does not exist under catalytic conditions at elevated H$_2$ pressures. MoS$_2$ can take up considerable amounts of hydrogen under such conditions so that the Mo sulfide phase should better be described as an H$_x$MoS$_2$ phase rather then MoS$_2$. [31,42,43]

Catalytic hydrodesulfurization HDS consists in the hydrogenolytic scission of C-S bonds in organic molecules with formation of H$_2$S. It is a major process in petroleum industry, [10,44] which consumes large quantities of the above described Mo- or W-based catalysts. HDS is necessary since petroleum distillates contain sulfur-containing compounds which would lead to SO$_2$ emission on direct combustion and which would poison acidic and noble metal catalysts in subsequent refining processes. The sulfur-containing compounds present in petroleum include the families of thiophenes, benzo- and dibenzothiophenes, benzo-naphthothiophenes etc. [10] The sulfur content of the petroleum depends on its provenance. Typically, naphthas contain 400 ppm sulfur, while middle distillates may contain up to 1 % sulfur. These concentrations must be brought down to < 0.5 ppm and < 0.2 ppm, respectively, by catalytic HDS. The technical process is carried out at temperatures between 570 and 670 K and in the pressure range 20 - 60 bar depending on feedstock quality. The reaction sequence for dibenzothiophene as a typical example

(8)

consists of initial C-S bond scission followed by slower hydrogenation steps. [10]

METALS

Bulk metals are used in only a few processes as catalysts where a reasonably high surface area is stabilized by so-called structural promoters. The most prominent example of this class of catalysts is undoubtedly the NH$_3$ synthesis catalyst [45] (vide infra). Since metals, in particular noble metals, have high surface free energies, they tend to aggregate unless small particles can be stabilized by placing them on the surface of suitable supports. This class of supported metal catalysts is technologically extremely important. They are efficiently used in selective hydrogenations, [46,47] the control of motor-vehicle exhaust gases (three-way catalyst)[48] and in catalytic reforming. [10,49,50]

Ammonia Synthesis Catalyst
The catalyst is prepared by fusing magnetite with the

optimized amounts of promoters. A typical technical catalyst contains about 1 % K_2O, 2 % CaO, 2,5 % Al_2O_3, and small amounts of the oxides of silicon, magnesium, titanium, zirconium and vanadium. Alumina plays an important role (together with CaO) as so-called structural promoter during reduction and catalysis. It covers parts of the surface of iron crystallites and thus stabilizes them against sintering. Surface areas of the order of magnitude of 20 m^2/g can thus be obtained. Potassium oxide is known to act as an electronic promoter which assists in the dissociative chemisorption of dinitrogen. [51]

The ammonia synthesis reaction from N_2 and H_2:

$$N_2 + 3\ H_2 \longrightarrow 2NH_3 \qquad (9)$$

is probably the best understood catalytic reaction today. It can be described [51] microscopically by the sequence of elementary steps (10 a-g):

$$H_2 + 2s \rightleftharpoons 2H_{ad}$$

$$N_2 + s \rightleftharpoons N_{2,ad}$$

$$N_{2,ad} \rightleftharpoons 2N_{ad}$$

$$N_{ad} + H_{ad} \rightleftharpoons NH_{ad} \qquad (10\ a\text{-}g)$$

$$NH_{ad} + H_{ad} \rightleftharpoons NH_{2,ad}$$

$$NH_{2,ad} + H_{ad} \rightleftharpoons NH_{3,ad}$$

$$NH_{3,ad} \rightleftharpoons NH_3 + s$$

Here s denotes a surface site.

Ammonia is produced on a very large scale. Its technological and economic importance is due to the need of nitric acid and fertilizers. Fig. 9 illustrates the scheme of an ammonia synthesis plant which consists of a sequence of catalytic processes. The raw materials for the ammonia synthesis are natural gas, water and air. In the first step sulfur-containing compounds have to be removed from the natural gas by HDS. In a

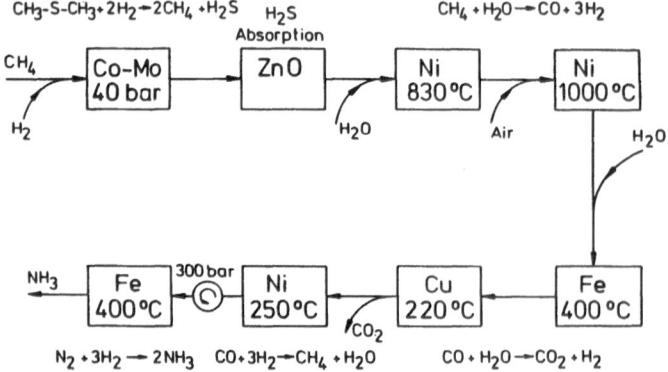

Fig. 9. Catalytic processes in ammonia synthesis.

subsequent steam reforming process, methane is converted on a Ni catalyst into CO and H_2, thus providing the hydrogen for the synthesis reaction. Remaining methane is then converted with air again on a Ni catalyst. Dinitrogen for the synthesis of ammonia is introduced with the air. CO - a catalyst poison - is transformed in further steps over Fe- and Cu - catalysts into CO_2 which is subsequently washed out. Remaining traces of the catalyst poison CO are finally hydrogenated to methane on a Ni catalyst prior to pressurizing the synthesis gas to 300 bar. The gas is then preheated to 670 K and fed into the synthesis reactor. Typically, equilibrium conversions of about 50 % can be achieved under such conditions.

Supported Metal Catalysts - Catalytic Reforming

High metal surface areas can be obtained when small metal particles are stabilized on support surfaces in high dispersion. [52,53] Dispersion is defined as the fraction of metal atoms being exposed. For example, a Pt particle with a diameter of about 1 nm contains less than 50 atoms and has a dispersion close to 100 %. This is particularly important for the catalytically highly active, but very expensive noble metals. For economic reasons one has to achieve optimal catalytic efficiency per unit mass of noble metal, and hence, small particles providing high dispersion are demanded. Platinum is in fact one of the most widely used noble metals in catalysis. It finds application in car exhaust catalysts, [48] in hydrogenations [46,47] and catalytic reforming. [10,49,50] The Pt content of the catalysts is usually rather small, typically less than 1 % wt and average metal particle sizes are around 1-2 nm. Particles may grow with concomitant reduction in dispersion under working conditions. The catalysts must then be regenerated in oxidizing atmospheres which leads to a redispersion of the metal in the form of oxides. [54] The regenerated material is then followed by reduction. The life time of metal catalysts used in hydrocarbon reactions can also be limited by coke deposition. It has been shown that this catalyst fouling can be significantly reduced by addition of a second component such as e.g. rhenium, germanium, tin or lead.

A typical example for a supported bimetallic catalyst is the reforming catalyst which combines the characteristic metal functions (hydrogenation, dehydrogenation) with the acidic functions of an appropriate acidic support (carbocation chemistry: isomerization, ring closure, ring enlargement). The reforming catalyst is thus a bifunctional catalyst. A technical reforming catalyst contains typically 0.3 wt % Pt and 0.3 wt %Re supported on an η-Al_2O_3 support the surface of which has chlorine incorporated to increase its acidity.

Catalytic reforming [10,49,50] has been introduced because the products obtained from catalytic cracking do not have sufficiently high octane rating. High quality gasoline should preferentially contain branched alkanes and aromatics. The alicyclic compounds and n-alkanes which are abundant among the cracking products, must be converted into the desired products. In the reforming process (typical conditions: 10-20 bar, H_2, 770 K) alkanes and cycloalkanes are dehydrogenated on the Pt particles. The resulting alkenes can then migrate onto the acidic support surface where they form carbenium ions and undergo skeletal rearrangements and ring closures and enlargements. Proton abstraction would then produce a new olefinic compound which can again be hydrogenated or further dehydrogenated by the metal particles to yield the desired branched alkanes and aromatics, respectively.

CONCLUDING REMARKS

A large variety of highly sophisticated catalytic processes have been developed largely by empirical approaches. The basic mechanistic understanding on a microscopic scale of a process under consideration including the detailed function of the catalytic surface is hardly available to date with perhaps one exception, namely ammonia synthesis. The final goal of scientific investigations into catalytic mechanisms would be an understanding of the dynamics of the working surface. Only very few studies on model catalysts such as single crystal surfaces have been published in this area up to now. The coming years will surely confront us with fascinating results which will become available by the development of more and more sophisticated new methods and techniques, such as time- and spatially resolved spectroscopies.

REFERENCES

1. K. H. Schmidt, Chem. Ind. 35: 69 (1983), and H. Kral, Chem. Ind. 41:44 (1989).

2. H. P. Boehm and H. Knözinger, In: "Catalysis-Science and Technology", J. R. Anderson and M. Boudart, eds., Springer, Berlin, Heidelberg, New York, Vol. 4, p. 39, 1983.

3. H. H. Lamb, B. C. Gates and H. Knözinger, Angew. Chem. Int. Ed. Engl. 27: 1127 (1988).

4. E. Garrone and F. S. Stone, Proc. 8th Intern. Congr. Catal., Berlin, 1984, Vol. 3 : 441 (1984).

5. S. Coluccia, F. Boccuzzi, G. Ghiotti and C. Mirra, Z. Phys. Chem. (Frankfurt) 121 : 141 (1980).

6. H. Knözinger and P. Ratnasamy, Catal. Rev.-Sci. Eng. 17 : 31 (1978).

7. H. Knözinger, In: "The Chemistry of the Hydroxyl Group", S. Patai, ed., p. 641, Interscience, London, New York, Sydney, Toronto, 1971.

8. H. Knözinger, Angew. Chem. Int. Ed. Engl. 7 : 791 (1968).

9. H. Knözinger, H. Bühl and K. Kochloefl, J. Catal. 24 : 57 (1972).

10. B. C. Gates, J. R. Katzer and G. C. A. Schuit, "Chemistry of Catalytic Processes", Mc Graw, New York, 1979.

11. J. Weitkamp, S. Ernst, H. Dauns and E. Gallei, Chem. Ing. Tech. 58 : 623 (1986).

12. J. W. Ward, in: "Applied Industrial Catalysis", Vol. 3, p. 271, B. E. Leach, ed., Academic Press, New York, London, 1984.

13. R. M. Barrer, "Hydrothermal Chemistry of Zeolites", Academic Press, New York, London, 1982.

14. J. Weitkamp, personal communication.

15. G. Perego, G. Belusi, C. Corus, M. Taramasso, F. Buonomo and A. Esposito, in: "Studies in Surface Science and Catalysis", Vol. 28, p. 129, Elsevier, Amsterdam, 1986.

16. B. Notari, in: "Studies in Surface Science and Catalysis", Vol. 37, p. 413, Elsevier, Amsterdam, 1988.

17. S. Csicsery, Pure Appl. Chem. 58 : 841 (1986).

18. S. Csicsery, in: Proc. Symp. "Shape Selective Catalysis, Route to Chemical Fuels", ACS Meeting, Seattle, USA, March 20-25, 1983, p. 116.

19. P. Dejaifve, A. Auroux, P. C. Gravelle, J. C. Vedrine, Z. Gabelica and E. G. Derouane, J. Catal. 70 : 123 (1981).

20. D. E. Walsh and L. D. Rollmann, J. Catal. 49 : 369 (1977).

21. J. Scherzer, Catal. Rev.-Sci. Eng. 31 : 215 (1989).

22. H. Pines, "The Chemistry of Catalytic Hydrocarbon Conversions", Academic Press, New York, London, Toronto, Sydney, San Francisco, 1981.

23. A. Reller and T. Williams, Chem. Britain 25 : 1227 (1989).

24. J. M. Thomas, D. A. Jefferson and G. R. Millward, Joel News 23 E : 7 (1985).

25. R. J. H. Vorhoeve, in "Advanced Materials in Catalysis", J. J. Burton and R. L. Garten, eds., Academic Press, New York, San Francisco, London, 1977, pp. 129.

26. L. G. Tejuca, J. L. G. Fierro and J. M. D. Tascon, Advan. Catal. 36 : 237 (1989).

27. G. I. Golodets, "Heterogeneous Catalytic Reactions Involving Molecular Oxygen", Studies in Surface Science and Catalysis, Vol. 15, Elsevier Publ., Amsterdam (1983).

28. R. K. Grasselli and J. D. Burrington, Ind. Eng. Chem. Prod. Res. Dev. 23 : 293 (1984).

29. J. J. Spivey, in: "Catalysis", Specialist Periodical Reports, Vol. 8, p. 157, Roy. Soc. Chem., London (1989).

30. K. Arata and M. Hino, Proc. Intern. Congr. Catal., Calgary 1988, Vol. 4, p. 1727, M. J. Phillips and M. Ternan, eds., The Chemical Institute of Canada, Ottawa, 1988.

31. H. Knözinger, Proc. Intern. Congr. Catal., Calgary, 1988, Vol. 5, p. 20, M. J. Phillips and M. Ternan, eds., The Chemical Institute of Canada, Ottawa, 1989.

32. G. Baumbach, Staub - Reinhalt. Luft 44 : 285 (1984).

33. M. H. Javad, _Brennst.-Wärme-Kraft_ 37 : 39 (1985).

34. A. Miyamoto, Y. Yamazaki, M. Inomata and Y. Murakami, _J. Phys. Chem._ 85 : 2366 (1981).

35. F. J. J. G. Janssen, F. M. G. van den Kerkhof, H. Bosch and J. R. H. Ross, _J. Phys. Chem._ 91 : 5921 and 6633 (1987).

36. F. Hilbrig, Dissertation, Universität München, 1989.

37. P. Ratnasamy and S. Sivasanker, _Catal. Rev.- Sci. Eng._ 22 : 401 (1980).

38. H. Topsoe, B. S. Clausen, N.-Y. Topsoe and E. Pedersen, _I & EC Fundamentals_ 25 : 25 (1986).

39. R. Prins, V. H. J. de Beer and G. A. Somorjai, _Catal. Rev.-Sci.Eng._ 31:1 (1989).

40. W. Niemann, B. S. Clausen and H. Topsoe, _Catal. Letters:_ 4 : 355 (1990).

41. S. P. A. Louwers and R. Prins, personal communication.

42. J. Polz, H. Zeilinger, B. Müller and H. Knözinger, J. Catal. 120 : 22 (1989.

43. A. B. Anderson, Z. Y. Al-Saigh and W. K. Hall, _J. Phys. Chem._ 92 : 803 (1988).

44. D. C. McCulloch, in: "Applied Industrial Catalysis", Vol. 1, p. 69, B. E. Leach, ed., Academic Press, New York, London, 1983.

45. J. S. Merriam and K. Atwood, see Ref. 44, Vol. 3, p. 113.

46. R. L. Augustine, "Catalytic Hydrogenation", M. Dekker, New York, 1965.

47. G. Webb, in: "Catalysis", Specialist Periodical Reports, Vol. 2, p. 145, Roy. Soc. Chem., London, 1978.

48. W. S. Briggs, see Ref. 44, Vol. 3, p. 241.

49. M. D. Edgar, see Ref. 44, Vol. 1, p. 124.

50. D. A. Dowden, in: "Catalysis", Specialist Periodical Reports, Vol. 2, p. 1, Roy. Soc. Chem., London, 1978.

51. G. Ertl, in: "Catalytic Ammonia Synthesis", J. R. Jennings, ed., Plenum Press, New York, in print.

52. J. R. Anderson, "Structure of Metallic Catalysts", Academic Press, New York, 1975.

53. K. Foger, in: "Catalysis-Science and Technology", J. R. Anderson and M. Boudart, eds., Springer, Berlin, Heidelberg, New York, Vol. 6, p. 227, 1984.

54. S. A. Stevenson, J. A. Dumesic, R. T. K. Baker and E. Ruckenstein, "Metal-Support Interactions in Catalysis, Sintering and Redispersion", Chapman and Hall, London, 1987.

PRODUCTION OF INDUSTRIAL CATALYSTS

J.W. Geus

Department of Inorganic Chemistry
University at Utrecht, The Netherlands

SUMMARY

AFTER AN INTRODUCTION DEALING WITH SOME BASICS OF CATALYTIC REACTIONS, IT IS ARGUED THAT SOLID CATALYSTS CAN BE SEPARATED RELATIVELY EASILY FROM THE REACTION PRODUCTS. THE CONSTRAINTS IMPOSED ON CATALYSTS TO BE USED IN INDUSTRIAL FIXED BED REACTORS ARE SUBSEQUENTLY DEALT WITH. THE REQUIREMENTS OF RESTRICTED PRESSURE DROP AND A SUFFICIENTLY EXTENSIVE ACTIVE SURFACE AREA PER UNIT VOLUME CALL FOR THE USE OF (HIGHLY) POROUS CATALYST BODIES. SINCE MOST CATALYTICALLY ACTIVE MATERIALS SINTER RAPIDLY UNDER THE CONDITIONS OF THE THERMAL PRETREATMENT OR THE CATALYTIC REACTION, SMALL PARTICLES OF THE ACTIVE COMPONENT MUST BE STABILIZED BY APPLICATION ONTO A THERMOSTABLE SUPPORT.

THE DIFFERENT TYPES OF SUPPORTED CATALYSTS ARE DISCUSSED AS WELL AS DIFFERENT PROCEDURES TO PRODUCE SUPPORTED CATALYSTS. REACTION OF THE SUPPORT WITH THE ACTIVE PRECURSOR DURING DEPOSITION AS WELL AS DURING THERMAL PRETREATMENT IS DEALT WITH.

INTRODUCTION

The course of a chemical reaction can be visualised by a diagram in which the potential energy has been plotted as a function of the relative positions of the atoms involved in the reaction. For a very limited number of reactions only, the relative orientations of the atoms can be unambiguously indicated by a two-dimensional plot. If the energy is plotted in the third direction, a three-dimensional plot results. However, more than two dimensions are usually required to define the positions of the atoms. Consequently, multidimensional plots generally represent chemical reactions.

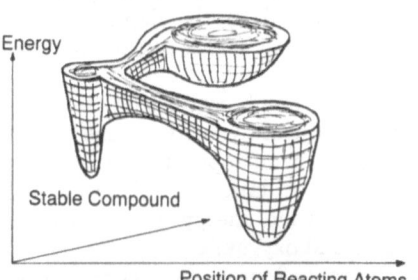

FIGURE 1   POTENTIAL ENERGY AS A FUNCTION OF RELATIVE POSITIONS OF ATOMS OF REACTING COMPOUNDS

Fundamental Aspects of Heterogeneous Catalysis Studied by Particle Beams,
Edited by H.H. Brongersma and R.A. van Santen, Plenum Press, New York, 1991

In the diagram states corresponding to stable compounds are indicated by pits. There are reactions where a pit is connected to a deeper pit via a trough the level of which continuously decreases from the higher to the lower pit. Such reactions proceed already as soon as the reactants are mixed. Precipitation of silver chloride upon mixing of silver nitrate and sodium chloride solutions is an instance of a spontaneously proceeding reaction. With many other reactions the troughs connecting the pits are exhibiting maxima. Reactions corresponding to such diagrams do not occur spontaneously, but ask for activation to proceed. Usually activation involves raising of the thermal energy. If the thermal energy is of the order of 0.1 of the height of the energy barrier, the reaction proceeds with a significant rate, which rises exponentially with the temperature. An instance of a reaction that calls for thermal activation is the oxidation of sulfur dioxide to sulfur trioxide. The reaction of carbon monoxide with molecular oxygen to carbon dioxide requires a very high thermal activation, in spite of carbon dioxide being substantially more stable than carbon monoxide. The above reactions correspond to a diagram containing two pits, one corresponding to sulfur dioxide and oxygen or carbon monoxide and oxygen, which is at a higher level, and the other corresponding to sulfur trioxide or carbon dioxide, which is at a lower level.

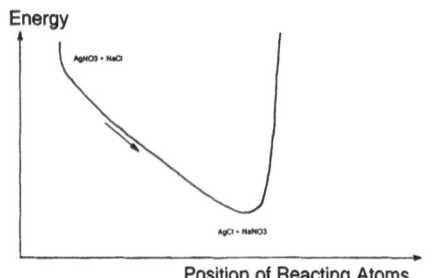

FIGURE 2

CROSS SECTION OF ENERGY DIAGRAM FOR REACTION PROCEEDING SPONTANEOUSLY AT ROOM TEMPERATURE.

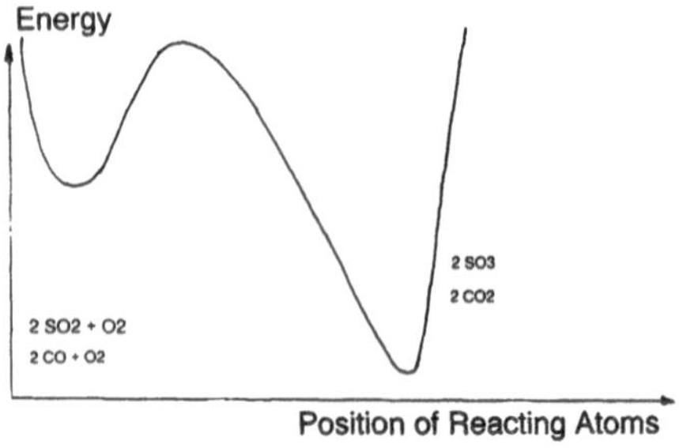

FIGURE 3

CROSS SECTION OF ENERGY DIAGRAM FOR REACTION ASKING FOR THERMAL ACTIVATION.

To bring about chemical reactions by thermal activation often meets with difficulties caused by the entropy. In our diagram the entropy can be very roughly associated with the width of the pits corresponding to the stable compounds. We can represent the number of atoms present in the different stable states by balls present in the pits and the thermal energy by vibrating our potential energy diagram in a vertical direction. At high temperatures and, hence, at high levels of vibration, the distribution of the balls in our model will be determined more by the width than by the depth of the different pits. The majority of the balls will be present in the pit that has the maximum width, which indicates that the chemical equilibrium is situated at the compound having the maximum entropy. At low levels of vibration, and, thus, at low temperatures, the distribution of the balls is dominated by the depth of the pits. Now the chemical equilibrium will be situated at the most stable compound corresponding to the deepest pit.

Often the maximum of the trough connecting two pits is thus high that at the temperature level required to sufficiently activate the molecules, the equilibrium is shifted to the state having a higher entropy and not to the state exhibiting a lower energy. An instance is the above oxidation of sulfur dioxide. At temperatures high enough to pass through the activation energy barrier, the equilibrium is at the side corresponding to sulfur dioxide and oxygen.

A catalyst enhances the rate of one or more chemical reactions. The acceleration of the chemical reaction is due to the catalyst chemically interacting with the molecules to be activated, and thus leading to a completely different potential energy diagram.

Owing to the chemical interaction with the catalyst, the pits corresponding to the initial and final state of the desired reaction are connected via a trough containing a barrier much lower than the trough in the diagram corresponding to the reaction without a catalyst. Since the catalyst is chemically binding one or more reactants more weakly than the reacting atoms in the final stable compound do, the chemical bond with the catalyst is rapidly broken. As a result, a catalyst is not affecting the chemical equilibrium, but is only increasing the rate at which the equilibrium is being established. A very small amount of catalyst can therefore cause reaction of an essentially infinite amount of reactants. To perform reactions where the unfavorable entropy prohibits the use of mere thermal activation, a catalyst has to be employed. An instance is the oxidation of sulfur dioxide, which proceeds only in the presence of a suitable catalyst.

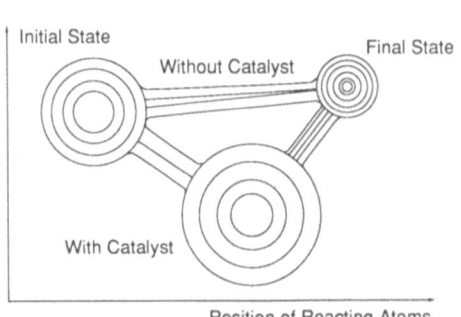 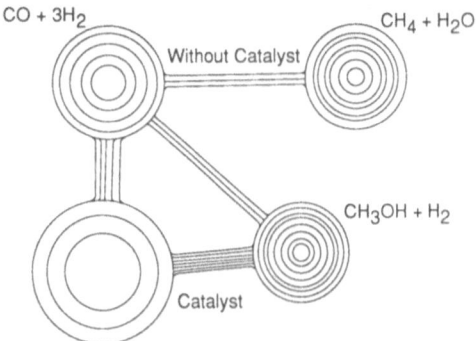

FIGURE 4

POTENTIAL ENERGY DIAGRAM WITH AND WITHOUT A CATALYST. CURVES ARE DRAWN AT EQUAL ENERGIES. THE CATALYST IS PROVIDING A LOWER PATH TO THE FINAL STATE.

FIGURE 5

SELECTIVITY INDUCED BY A CATALYST. THOUGH THE ENERGY OF METHANE LOWER, THE CATALYST PRODUCES METHANOL.

The first task of a catalyst therefore is to speed up the establishment of chemical equilibrium. However, there are many instances where more than one reaction is thermodynamically possible. Usually, only one of the reactions is favorable, and the other reactions must be suppressed as much as possible. Instances are the oxidation of ethylene to ethylene oxide and the reaction of carbon monoxide and hydrogen to methanol. With the oxidation of ethylene reaction to carbon dioxide and water is thermodynamically more favorable. The economic value of the thermal energy thus generated is, however, much less than that of ethylene oxide. Analogously carbon monoxide and hydrogen can react to methane and water, which are much less attractive than methanol. A suitable catalyst therefore only provides an easy pathway to the desired reaction products and blocks reaction to the unfavorable reaction products.

Summarizing the above, a catalyst accelerates and/or directs the course of chemical reactions. To allow a catalyst to convert a large amount of reactants, the reactants must intimately contact the catalyst, since a chemical bond between at least one of the reactants and the catalyst must be established. Moreover, the catalyst must be separated rapidly and effectively from the reaction products and subsequently be contacted with other reactants. The contact of reactants with homogeneous catalysts, in which the catalyst and the reactants are in the same phase, is effected most smoothly. However, subsequent separation of the catalyst from the reaction products is difficult. With heterogeneous catalysts contact of the reactants with the catalyst is established more difficult, but separation from the reaction products is fairly easy. Most favorable is the combination of gaseous reactants and a solid catalyst. Liquid reactants and a solid catalyst is also a combination much used.

SOLID CATALYSTS

With solid catalysts the catalytic action is due to formation of chemical bonds of one or more of the reactants with the surface atoms of the solid (chemisorption). An example is the dissociative adsorption of hydrogen on metal surfaces; the adsorbed hydrogen atoms are displaying a high activity in hydrogenation reactions. Another instance is the dissociative adsorption of molecular nitrogen on the iron catalyst used in the ammonia synthesis. The resulting adsorbed nitrogen atoms react smoothly with adsorbed hydrogen atoms to ammonia. Dissociative adsorption of dioxygen on silver leads to adsorbed oxygen atoms selectively oxidizing ethylene to ethylene oxide.

The heat of adsorption, which reflects the energy with which the active species is bonded to the surface, dominates the catalytic activity. Often a "volcano-shaped" curve is exhibited. At a low surface-bond strength, the stability of the adsorbed species is too low, whereas at a high bond strength the adsorbed species is too stable to bring about a rapid reaction. An intermediate bond strength, neither too low nor too high, leads to a maximum activity.

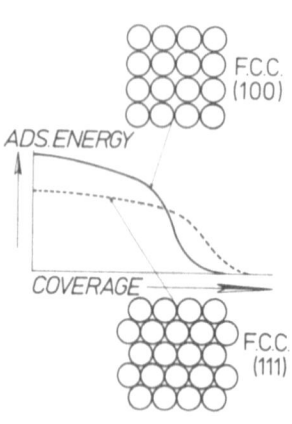

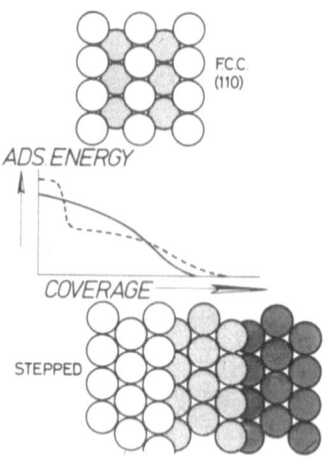

FIGURE 6

HEAT OF ADSORPTION AS A FUNCTION OF SURFACE COVERAGE FOR THE (111) AND (100) PLANE OF A F.C.C. METAL.

FIGURE 7

HEAT OF ADSORPTION AS A FUNCTION OF SURFACE COVERAGE FOR A (110) AND FOR A STEPPED SURFACE OF A F.C.C. METAL.

Most metal surfaces are displaying a high density of adsorption sites. The number of adsorption sites is of the same order of magnitude as the number of metal surface atoms. Usually the adsorption energy drops with the surface coverage. The decrease in the heat of adsorption can be due to the presence of sites of a different adsorption energy ("a priori heterogeneity"). The adsorption energy can depend on the surface structure. A (111) surface, e.g., exhibits another chemisorption energy than a (110) surface. The adsorption energy at steps on the surface is often different from that on flat surfaces. Interaction between the adsorbed species can also lead to a decrease in adsorption energy with coverage. Since the adsorption energy determines the catalytic activity, the activity can depend on the crystallographic metal surface exposed (structure-sensitive reactions). When the catalytic activity does not vary considerably with the surface structure, the catalytic activity is said to be structure-insensitive. Especially with oxides, the density of active sites is much lower than the number of surface atoms, though also with metals instances of a limited number of active sites present at, e.g., lattice defects are known. Taylor developed his well known theory of active sites, in which a very limited number of active sites is participating in the catalytic reaction, from results obtained on metal oxides.

High Number Of Sites of
Uniform Activity.

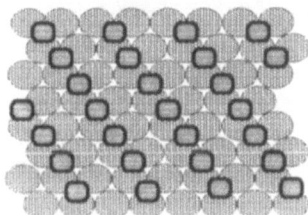

Activity Confined to Small
Fraction of Surface.

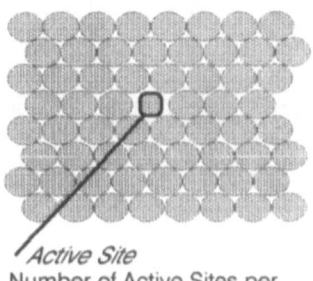

Active Site
Number of Active Sites per
Unit Surface Area (very)
Small.

FIGURE 8
HIGH DENSITY OF UNIFORMLY BINDING
SITES ON ACTIVE SURFACE.

FIGURE 9
ONLY A VERY SMALL FRACTION OF THE
SURFACE SITES IS ACTIVE. ACTIVE SITES
ARE GENERALLY CONFINED TO LATTICE
DEFECTS.

CONSTRAINTS ON SOLID CATALYSTS

Gaseous reactants are brought into contact with solid catalysts within either fixed bed or fluidized bed reactors. With a fixed bed reactor a flow of the reactants is passed through a bed of relatively large catalyst bodies. When a fluidized bed reactor is used, the catalyst particles are much smaller (generally about 100 µm). The friction of the gas flow with the catalyst particles is sufficiently high to lift the catalyst particles and to bring the particles in a vigorous motion. With a fluidized bed a very efficient heat transfer can be established. Also liquid reactants can be passed through a fixed catalyst bed. A special case with a liquid and gaseous reactant is the trickle flow reactor, in which a fluid is passed together with a gaseous reactant downwards through a fixed catalyst bed. In the catalytic desulfurization of heavier oil fractions hydrogen reacts with sulfur chemically bonded in the oil to hydrogen sulfide. This catalytic desulfurization is often carried out within a trickle flow reactor. Often the catalytic reaction of a gas and a liquid reactant is being performed with a solid catalyst suspended in the liquid. A flow of the gaseous reactant is passed through the catalyst slurry or dispersed by vigorous agitation into the slurry. In this survey we will concentrate on fixed bed catalysts used in gas-phase reactions.

The most important constraint on fixed bed catalysts is the pressure drop over the reactor. The pressure drop cannot be high, since this leads to the catalyst to be blown out of the reactor. Moreover, a high pressure drop asks for much mechanical energy to sufficiently raise the pressure before the reactor. The minimum size of the bodies to be used in fixed bed reactors is therefore of the order of 0.5 mm. With smaller particles the pressure drop is usually too high, unless very low linear gas velocities are used. Since the catalyst bodies must not disintegrate during loading of the reactor and during the reaction, catalyst bodies of a high mechanical strength are required. Furthermore catalysts have to exhibit a sufficiently high rate of production per unit volume, which is determined by the number of active site per unit surface area of the active component(s), the extent of the active surface area per unit volume, and the accessibility of the active surface within the catalyst. The accessibility of the active surface is determining the rate of the physical transport to and from the active surface.

FIGURE 10

PALLADIUM GAUZE USED IN THE SELECTIVE OXIDATION OF AMMONIA TO NITROGEN OXIDE TO PRODUCE NITRIC ACID. TOP LEFT-HAND SIDE : FRESH GAUZE. TOP RIGHT-HAND SIDE : GAUZE AFTER USE. BOTTOM : GAUZE AFTER USE AT HIGH MAGNIFICATION; NOTE RECRYSTALLIZATION DURING USE TO SMALL FACETTED CRYSTALS.

SPECIFIC SURFACE AREA OF SOLID CATALYSTS

The effect of the active surface area per unit volume on the activity is obvious. However, if the reaction can be carried out at high temperature levels, the active surface area of the catalyst is less important. An instance of a catalytic reaction where the active surface area is less important is the selective oxidation of ammonia to nitrogen oxide to produce nitric acid. As invented by Ostwald around 1900, Pt and Pd gauze are catalyzing the selective reaction to nitrogen oxide without substantial formation of molecular nitrogen and nitrous oxide. At the high reaction temperatures used with the oxidation of ammonia, a high active surface area is not required. The low surface area of the gauze is therefore sufficient to cause a virtually complete conversion of the ammonia. The pressure drop over the gauze is of course very low. In the figure an instance is given of an ammonia oxidation catalyst containing a small amount of gold present in palladium wires. Due to the reaction, the precious metal is reversibly oxidized, which leads to a strong change in the structure. Evidently intermediate oxidation causes facetting of the palladium, which leads to disintegration of the initially smooth metal surfaces.

In contrast to the oxidation of ammonia, an extended active surface area per unit volume is usually required to obtain a technically acceptable extent of conversion of the reactants. To produce, e.g., 10,000 ton per year a reactor of 20 m^3 is favorable. If the size of the reactor must, however, be 1,000 m^3 the process is technically hardly viable. A

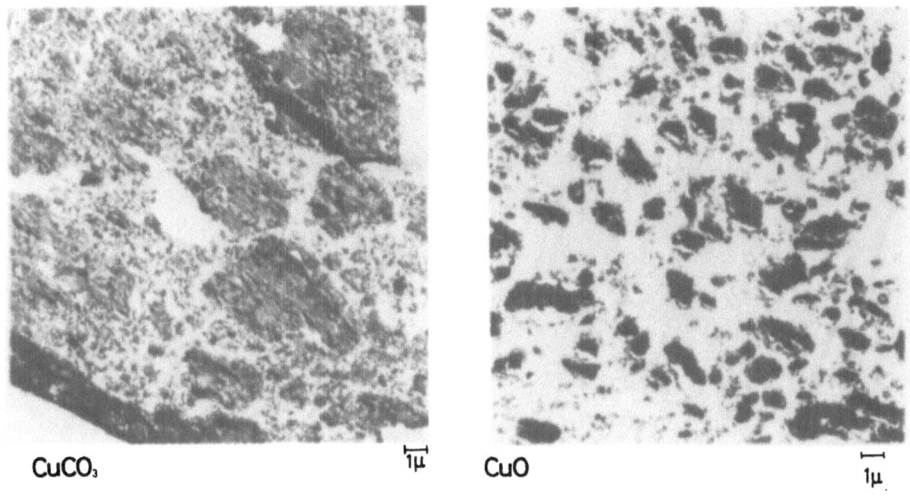

CuCO₃   1μ   CuO   1μ

FIGURE 11

LEFT-HAND SIDE : COPPER OXIDE(II) PREPARED BY DECOMPOSITION OF COPPER(II) CARBONATE.
RIGHT-HAND SIDE : THE SAME COPPER(II) OXIDE AFTER HYDROGEN-REDUCTION AND REOXIDATION.
REDUCTION HAS LED TO A CONSIDERABLE GROWTH IN THE PARTICLE SIZE AND THUS TO AN
APPRECIABLE DROP OF THE SURFACE AREA.

sufficiently high rate of production per unit volume calls for a high active surface area per unit volume. The effect of the active surface area is exemplified by the oxidation of carbon monoxide to carbon dioxide over copper oxide. Decomposition of copper carbonate leads to finely divided copper oxide, as represented in figure 11, which shows a micrograph of the copper oxide. Reduction with hydrogen to metallic copper causes an appreciable sintering of the copper particles. The sintering is due to the high mobility of copper atoms over metallic copper surfaces. Reoxidation of the metallic copper leads to copper oxide particles of a much larger size and, hence, a much lower surface area. The catalyst bed contained initially small copper oxide particles, which were after measurement of the catalytic activity reduced and subsequently reoxidized. The catalytic activity presented in figure 12 is hence measured on the same amount of copper oxide. Figure 12 indicates that the activation energy of the reaction has not changed by the reduction and reoxidation. The drop in activity is caused by the much lower pre-exponential factor, which is due to a sharp decrease in active sites caused by the increase of the size of the copper(II) oxide particles.

Since the minimum size of the catalyst bodies to be used in technical fixed bed reactors is about 0.5 mm, we will calculate the surface area of non-porous bodies of 0.5 mm. Assuming a void fraction of the catalyst bed of 50%, we arrive at a surface area of about 70 m^2 per m^3 of the reactor. We will next calculate the active surface area required to produce 10,000 tons per year within a reactor of 30 m^3. For the molecular weight of the product we assume 35. The catalytic activity can be expressed as the number of molecules reacting per site per second, which is known as the turn-over frequency. Usually the turn-over frequency is between 10^{-2} and 10^2 per site per second; here we will assume a turn-over frequency of 1 per site per second. A number of active sites of 10^{19} per m^2 is of the order of the maximum number of active sites. Consequently 1 m^2 of active surface area is producing 5×10^{-5} tons per day. Usually the number of days on stream of a technical reactor is of the order of 330 days per year. A rate of production

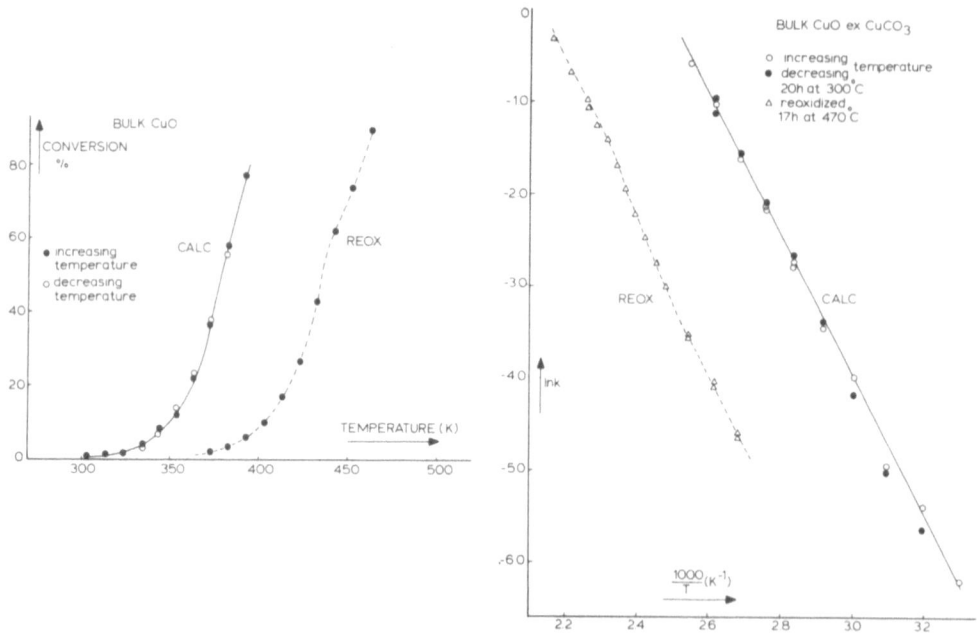

FIGURE 12

OXIDATION OF CARBON MONOXIDE OVER COPPER(II) OXIDE. IN THE ARRHENIUS PLOT (RIGHT-HAND SIDE) THE RATE CONSTANT HAS BEEN PLOTTED AS A FUNCTION OF THE RECIPROCAL ABSOLUTE TEMPERATURE. THE CURVE MARKED CALC REFERS TO THE CATALYST FRESHLY PREPARED BY DECOMPOSITION OF COPPER(II) CARBONATE; THE CURVE MARKED REOX TO THE CATALYST AFTER HYDROGEN-REDUCTION AND REOXIDATION. THE PARTICLE SIZE HAS INCREASED SHARPLY BY REDUCTION. THE RESULTING DROP IN SURFACE AREA AND THUS IN THE NUMBER OF ACTIVE SITES LEADS TO A CONSIDERABLY LOWER PRE-EXPONENTIAL FACTOR, WHICH BRINGS ABOUT A MUCH LOWER CONVERSION AT A FIXED TEMPERATURE.

of 1 ton per m^3 leads with 330 days on stream per year to a yearly production of 10,000 tons per year. To produce 1 ton per m^3 per day calls for 2 x 10^4 m^2 of active surface area. We have assumed the maximum number of active sites per unit surface area. It is therefore obvious that unless a catalytic reaction exhibiting a very high turn-over frequency is used, the surface area of non-porous catalyst bodies is much too small to provide a technically required rate of production. Consequently, highly porous catalyst bodies of at least 0.5 mm exhibiting a very high internal surface area have to be used in fixed bed catalytic reactors.

Transport through the narrow pores in the catalyst bodies can limit the rate of the reaction. Usually a trade-off between the pressure drop and diffusion limitation must be found. A low pressure drop calls for large catalyst bodies, whereas prevention of transport limitations inside the catalyst bodies requires short pores and, hence, small catalyst bodies. The limited pressure drop over the catalyst bed, and thus the minimum size of the catalyst bodies causes the main requirement on industrial catalysts to be a sufficiently high mechanical strength. Disintegration of catalyst bodies during loading into the reactor, thermal pretreatment, or utilization of the catalyst leads to severe problems. As a result, the elementary particles within the porous catalyst bodies must be tightly connected. If the elementary particles have a tendency to sinter, which brings about a loss in porosity and active surface area, the intimate contact between the elementary particles will cause rapid sintering.

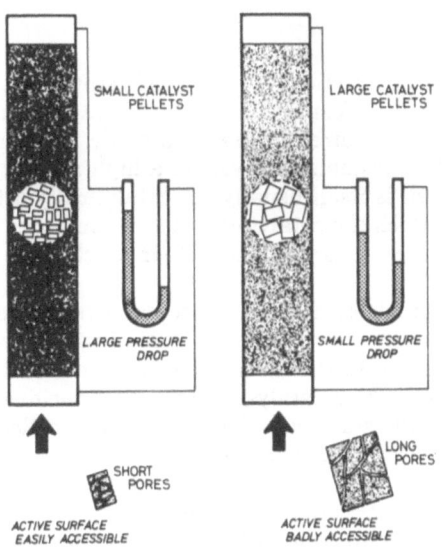

FIGURE 13

TRADE-OFF BETWEEN PRESSURE DROP AND TRANSPORT LIMITATIONS WITHIN CATALYST BODIES.
LEFT-HAND SIDE : SMALL CATALYST BODIES, HIGH PRESSURE DROP, LOW TRANSPORT RESISTANCE
WITHIN BODIES.
RIGHT-HAND SIDE : LARGE CATALYST BODIES, LOW PRESSURE DROP, HIGHER TRANSPORT RESISTANCE
WITHIN BODIES.

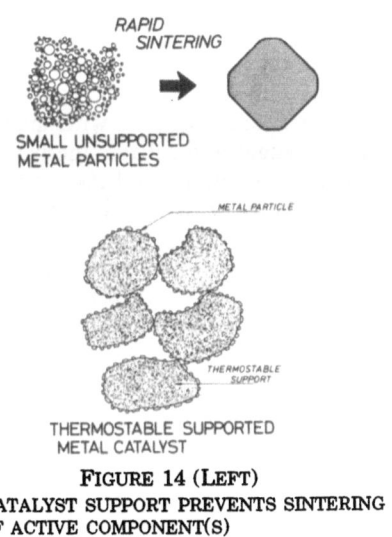

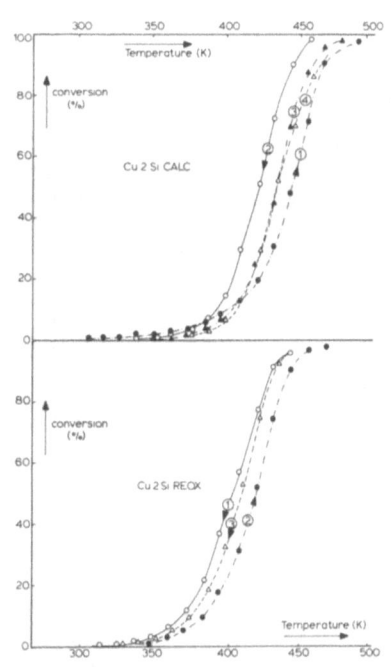

FIGURE 14 (LEFT)
CATALYST SUPPORT PREVENTS SINTERING
OF ACTIVE COMPONENT(S)

FIGURE 15 (RIGHT)
OXIDATION OF CARBON MONOXIDE OVER SILICA-SUPPORTED COPPER(II) OXIDE. REDUCTION AND
REOXIDATION NOT AFFECTING CATALYTIC PERFORMANCE. SLIGHT VARIATIONS IN CONVERSION CAUSED
BY VARYING WATER CONTENT IN FEED.

SUPPORTED CATALYSTS

Unfortunately, most catalytically active solids sinter fast at the temperatures of thermal pretreatment and use. Most metals sinter very rapidly provided the surfaces are not contaminated, which is required in many catalytic reactions. The generally used procedure to arrive at thermostable catalysts of a high activity is to use a support or carrier. The support, which is itself usually not catalytically active, is a thermostable, highly porous solid onto which the catalytically active component has been applied. The most frequently utilized carrier is alumina. Other much used carriers are silica and active carbon.

When the active particles are well dispersed over the surface of the support, contact between the active particles and, hence, sintering is prevented. To illustrate the effect of a support, we return to the oxidation of carbon monoxide over copper oxide. Figure 16 shows the activity of a catalyst in which copper oxide has been applied on silica. The surface area of the silica was about 200 m^2 per g. Reduction and reoxidation appear to affect the activity of the catalyst not significantly. The support apparently prevents sintering of the metallic copper particles resulting from the reduction.

Especially with selective catalytic reactions, the porous structure of the support bodies is highly important. With support bodies having narrow long pores, subsequent reaction(s) of the desired product is more likely, since the residence time of the initial product of the reaction is long. The selective oxidation of ethylene to ethylene oxide and of hydrogen sulfide to elemental sulfur is therefore performed on support bodies of a relatively low surface area (less than about 10 m^2 per g) having wide pores. Also the thermal conductance of catalyst beds is important. With highly exothermic or endothermic reactions, a sufficiently rapid transport of thermal energy out of or into the reactor presents difficulties. Generally several stages with intermediate heat exchangers are therefore used.

With most catalytic reactions, however, a high activity and, hence, a high active surface area per unit volume of catalyst is desired. If the active component contains more than one constituent, small particles of the required composition must be applied onto the support. With structure-sensitive reactions, application of very small active particles on the support may lead to unexpected results. The shape of the particles is determining the nature of the crystallographic surfaces exposed. If the equilibrium shape of the active component is containing flat areas with rounded-off corners and edges, as with many metals, the fraction of atomically more rough planes at the edges and corners increases when the particles are smaller (figure 16). Penetration of foreign atoms in between the surface atoms of an active component proceeds more readily with small particles. Penetration of foreign atoms into the surface of active metal particles generally affects the activity adversely. Often the aim is to raise the activity of a catalyst by decreasing the size of the supported active particles. However, exposition of unfavorable crystallographic planes or penetration of foreign atoms into the surface of the active

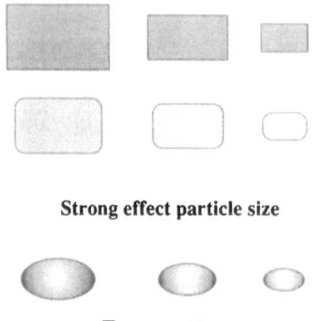

Strong effect particle size

FIGURE 16

EFFECT OF EQUILIBRIUM SHAPE OF ACTIVE PARTICLES ON NATURE OF EXPOSED LATTICE PLANES. EQUILIBRIUM SHAPE INDICATED IN THE CENTER LEADS TO MOST STRONG EFFECT ON THE STRUCTURE OF PLANES EXPOSED.

particles often bring about an activity smaller than that of larger active particles exposing a lower active surface area.

To unambiguously establish an effect of the surface structure on the activity, surface science techniques are indispensable. Different catalytic activities of surfaces exhibiting different LEED patterns indicates structure-sensitivity. However, many catalytic reactions proceed on a limited number of (defect) sites at the surface. The defect sites are not clearly evident in LEED patterns. Beam techniques can provide more accurate information about the atomic structure of solid surfaces. The technique that most workers in the field believe to be the most promising is scanning tunneling microscopy.

TYPES OF SUPPORTED CATALYSTS

Depending upon the costs of the reactor and that of the catalyst, the nature of supported catalysts is different. With precious metals, the costs of the catalyst are usually dominating over that of the catalytic reactor. It is therefore profitable to have a maximum fraction of metal atoms at the surface, and thus a maximum dispersion. With precious metal catalysts, consequently, very low loadings (usually lower than 1 wt.%) are used leading to very small metal particles (about 1nm) exhibiting dispersions of about 100%. Though the activity per unit volume of catalyst is relatively low, the activity per g of precious metal is at a maximum.

With base metal or metal oxide catalysts, the costs of the catalyst are much less important. Now the size of the reactor dominates, which calls for a maximum activity per unit volume, which is usually obtained at high loadings of the support. Base metal (oxide) catalysts therefore often show loadings of 30 to 50 wt.%. It is difficult to apply the active component(s) thus to avoid formation of clusters of small precursor particles rapidly sintering to large particles of a low surface area.

PRODUCTION OF (SUPPORTED) CATALYSTS

The procedures to produce industrial catalysts can be divided into two main groups, viz.,

i) selective removal of some constituents out of a non-porous precursor containing (a precursor of) the active component(s) and the support;

ii) application of (a precursor of) the active component(s) onto a separately produced support.

With the low loadings of precious metal catalysts, procedure (ii) is generally used, whereas the high loadings of base metal (oxide) catalysts often cause procedure (i) to be attractive.

a. Selective Removal

An important instance of selective removal is the ammonia synthesis catalyst. The starting material of this catalyst is a melt of magnetite, alumina, and potassium oxide produced in an electrical furnace in a non-oxidizing atmosphere. The alumina and potassium oxide content is low, viz., less than 2 to 5 wt.%. After processing the solidified melt to particles of the desired dimensions, the oxygen of the iron oxide is selectively removed by reduction with hydrogen. The small amount of alumina is sufficient to prevent sintering of the metallic iron particles. The shrinkage of the catalyst bodies during reduction is very limited and highly porous bodies are obtained. After it was established that the alumina merely prevents sintering of the iron particles, alumina was referred to as a structural promotor. The function of the potassium (oxide) promotor is still being debated.

Another instance of selective removal is the production of Raney nickel. With

Raney nickel, the starting material is a nickel-aluminum alloy, which is brittle. It is thus easy to process the melt of the aluminum alloy into smaller lumps that are contacted with a strongly alkaline solution, generally sodium hydroxide. The sodium hydroxide selectively dissolved the aluminum leaving a slurry of small nickel particles. The rapid settling of the relatively heavy nickel particles is favorable with liquid phase catalytic reactions. Separation of the catalyst is therefore easy.

Also the technical copper-zinc oxide methanol synthesis and carbon monoxide shift catalyst are produced by selective removal of water, carbon dioxide and oxygen from a coprecipitate of copper, zinc and some aluminum ions. Nickel-on-alumina catalysts are analogously produced by coprecipitation of nickel and aluminum ions, as well as copper/chromia catalysts from chromium and copper coprecipitates.

b. Application onto a Separately Produced Support

Application of an active precursor onto a separately produced support can be performed starting from either shaped bodies of the support or a powdered support. With a powdered support a separate shaping step is required, in which the loaded support is processed into the desired shape by extrusion or pelletizing. Starting from commercially available preshaped bodies of the support is attractive, since the desired chemical composition, shape, pore size distribution, and mechanical strength can be selected from the variety of support offered on the market. However, application of an active precursor homogeneously throughout preshaped bodies of a support is more difficult than to apply a precursor onto a powdered support.

With preshaped bodies of a support, impregnation with a solution of the active precursor and subsequent drying is usually carried out. Impregnation can be done with an amount of the solution of the precursor just sufficient to fill the pore volume (incipient wetness impregnation). The maximum loading that can be applied with incipient wetness impregnation is determined by the solubility of the active precursor and the pore volume. Multiple impregnation can lead to higher loadings, but the procedure is cumbersome. When impregnation is effected using a larger volume of the solution than can be taken by the pores of the support, higher loadings can be obtained. However, deposition of a substantial fraction of the precursor at the external edge of the body of the support often results.

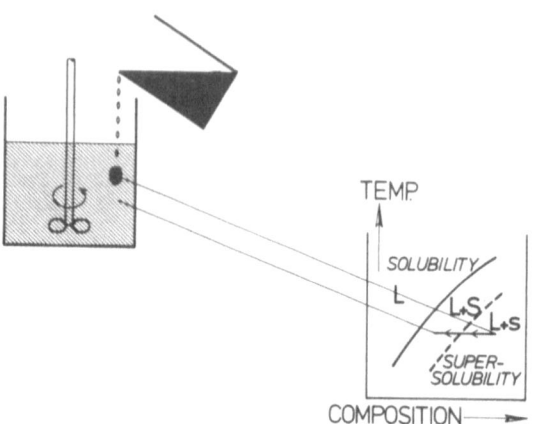

FIGURE 17

USUAL ADDITION OF PRECIPITANT LEADING TO INHOMOGENEOUS DISTRIBUTION OF ACTIVE PRECURSOR OVER SUPPORT.

PRECIPITATION FROM
HOMOGENEOUS SOLUTION

I. MIXING TO HOMOGENEOUS SOLUTION AT LOW TEMPERATURE; REACTION AT HIGH TEMPERATURE.

e.g.

$$CO(NH_2)_2 + 2H_2O \longrightarrow CO_2 + 2NH_4^+ + 2OH^-$$

II. INJECTION BELOW LEVEL OF LIQUID.

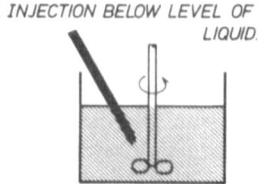

FIGURE 18

DEPOSITION-PRECIPITATION ONTO A SUSPENDED SUPPORT. PRECIPITATION FROM A HOMOGENEOUS SOLUTION BRINGS ABOUT NUCLEATION TO PROCEED EXCLUSIVELY ON THE SUPPORT.

Active material can only be lost by volatilization during the subsequent thermal treatment with impregnation and drying. No waste water is being produced. Precious metals, that are almost invariably used at low loadings, are therefore preferably applied by incipient wetness impregnation and drying.

With powdered supports, impregnation and drying is also used frequently, but crystallization of the precursor often leads to large particles. Precipitation of an active precursor in the presence of a suspended support is therefore often performed. With a slow nucleation, relatively large crystallites of the active precursor will be obtained. When nucleation of the precipitate proceeds rapidly, many small nuclei will result. The electrostatic charge on the nuclei of the precursor and on the support will determine whether the small nuclei will adhere to the support. When the mixing of the precipitant within the suspension is not very rapid, the temporarily locally high concentration will lead to flocculation of the small nuclei. Therefore relatively large clusters of small particles of the active precursor in between the particles of the support result.

To establish a very uniform distribution of small active particles over a support, the procedure of deposition-precipitation was developed. With deposition-precipitation a support facilitating nucleation of an active precursor is suspended in a solution of a suitable compound of the active component. Subsequently the concentration of the active precursor is raised homogeneously. Since nucleation of the active precursor at the surface of the support proceeds at lower concentrations than that in the bulk of the solution, precipitation exclusively on the surface of the support can be achieved. This procedure has been used with much success to produce a number of different catalysts. Figure 18 illustrates the basics of the method.

Often the interaction with the suspended support is more extensive. The support is reacting with the precipitating precursor to a different chemical compound. Instances are the reaction of suspending silica with precipitating copper and nickel ions. To obtain highly dispersed copper oxide on silica, precipitation from a homogeneous solution can be successfully used. Reaction to copper hydrosilicate is evident from temperature-programmed reduction. As shown in figure 19 the freshly dried catalyst exhibits reduction in two peaks, one due to copper(II) (hydr)oxide and the other peak displayed at a higher temperature due to copper(II) hydrosilicate. Reoxidation of the metallic copper particles leads to copper(II) oxide. After reoxidation, subsequent reduction proceeds therefore in one step. The water resulting from the reduction does not produce significant amounts of copper hydrosilicate with this catalyst.

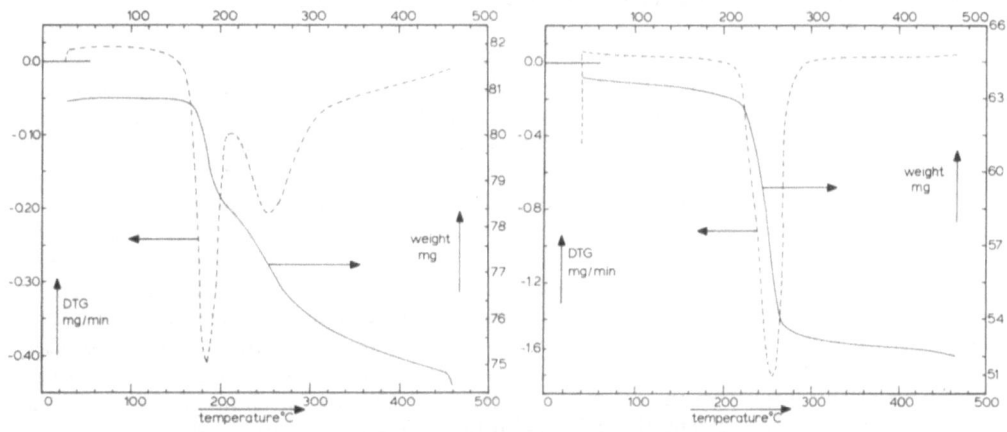

FIGURE 19

REDUCTION OF COPPER(II) (HYDR)OXIDE DEPOSITION-PRECIPITATED ON SILICA. LEFT-HAND SIDE : FIRST REDUCTION, TWO PEAKS ONE FROM COPPER(II) HYDROSILICATE. RIGHT-HAND SIDE : SECOND REDUCTION AFTER REOXIDATION. ONE PEAK DUE TO COMPLETE CONVERSION TO COPPER(II) OXIDE

The activity of the thus prepared catalyst in the oxidation of carbon monoxide is represented in figure 20. It can be seen that the activity of the fresh catalyst before reduction is disappointingly low. In spite of the very good dispersion and thus the high surface area, the conversion at e.g. 180 °C is much lower than that of the unsupported copper oxide and the less well dispersed copper oxide.

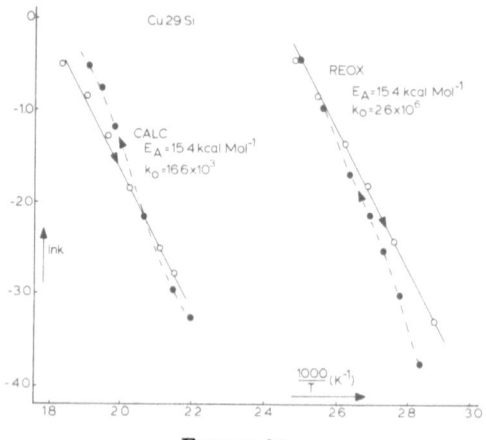

FIGURE 20

ARRHENIUS PLOT OF ACTIVITY IN CO OXIDATION OF 29 WT.% COPPER(II) OXIDE PREPARED BY DEPOSITION-PRECIPITATION ON SILICA. AFTER REDUCTION AND REOXIDATION THE ACTIVITY RISES STEEPLY DUE TO AN INCREASE IN THE NUMBER OF ACTIVE SITES

However, reduction and reoxidation now leads to a much higher activity, as apparent in figure 20. The low activity of the catalyst before reduction is due to the copper ions reacting during application with the silica to copper hydrosilicate. The activity of copper ions within copper hydrosilicate is much lower than in copper oxide. Reduction of the copper(II) ions of the hydrosilicate leads to small metallic copper particles that react to small copper(II) oxide particles on oxidation. The high surface area of the copper oxide particles results in a high catalytic activity.

THERMAL PRETREATMENT OF SUPPORTED CATALYSTS

Application of an active precursor usually leads to deposition of a hydrated salt or a hydroxide of the active component on the support. If the desired active component is the oxide of the metal compound deposited, calcination, i.e., keeping the catalyst at an elevated temperature in air, usually leads to the oxide. If a metal is the catalytically active component, reduction of the material deposited on the support is required. Since the reduced catalysts are generally pyrophoric, reduction is usually performed in situ in the reactor in which the catalyst has to be used. The catalyst is mostly calcined before being loaded into the reactor. First of all to diminish the amount of water being released during the reduction. However, the considerable increase in the mechanical strength of the catalyst caused by calcination renders previous calcination also attractive. Reduction of the freshly dried loaded support still containing the salt or hydroxide deposited often leads to smaller active metal particles than reduction of the previously calcined catalyst. In spite of the large amount of water set free and the low mechanical strength, the freshly dried catalyst is sometimes loaded into the reactor and reduced.

If reduction to the metal proceeds slowly and/or calls for high temperatures, the reduction is often performed in a special reactor (by the catalyst manufacturer). The reduced catalyst is passivated, that is the surface of the catalyst is reacted with oxygen by exposure to a gas flow containing e.g. 1 vol.% of oxygen. After loading the passivated catalyst into the final reactor, a relatively short reduction period at a relatively moderate temperature suffices to activate the catalyst.

The interfacial surface area of a catalytically active precursor with the support is large with well dispersed precursors. If the active component is liable to react with the support to a compound of a negligible or lower activity, a uniform distribution of very small particles can lead to a catalyst of a remarkably low activity. Especially with alumina, reaction to spinels of a low or negligible activity proceeds readily. Cobalt(II), iron(II), and nickel(II) react relatively rapidly and extensively with alumina to the corresponding aluminates. Calcination of cobalt and nickel on alumina catalyst often lead to catalysts of a low activity. Since formation of a spinel does not proceed with iron(III), only reducing conditions cause formation of iron(II) aluminate. Recrystallization of the support to another crystallographic phase of a very small surface area can also be induced by a well dispersed active component. An instance is vanadium(V) oxide on alumina. The vanadium oxide is accelerating the reaction to α-alumina of a surface area less than 10 m^2 per g.

CONCLUSION

Modern techniques and automation of well established procedures, such as X-ray diffraction and BET surface area determination, enable us to characterize solid catalysts and their precursors in detail. The effects of the detailed preparation procedure therefore can be evaluated presently very well. As a result the structure and chemical composition of inorganic materials of dimensions from 1 to 100 nm can be controlled to an increasing extent. Metal and alloy particles of about 1 nm, that is slightly larger than organic molecules, can be produced with a narrow particle size distribution. Small supported alloy particles often exhibit a remarkably small variation in chemical composition.

Broekhoff and de Boer have substantially developed the theory of capillary condensation of nitrogen within narrow pores. The resulting theory together with the development of the measuring apparatus has made possible an evaluation of transport effects of the rate and selectivity of catalytic reactions.

Beam techniques are valuable in assessing the effect of the structure and chemical composition of the surface of a catalytically active material on the activity and selectivity. Rutherford back scattering has proven very valuable in the elucidation of the interaction of active oxides or active precursors with supports. The latter experiments have been performed with model catalysts. The high surface sensitivity of LEIS renders this technique indispensable in the determination of the chemical composition of the surface layer. The technique as developed by Brongersma has revealed the surface composition of a number of technically important catalysts. The surface sensitivity is considerably higher than that of XPS, which is the technique generally used to assess the surface composition of supported catalysts.

REFERENCES

Only general references will be presented.
1.   "Preparation of Catalysts IV. Scientific Bases for the Preparation of Heterogeneous Catalysts" (B. Delmon, P. Grange, P.A. Jacobs, and G. Poncelet, eds.) Elsevier, Amsterdam (1987)
      Also the previous volumes of this Symposium Series are very useful.
2.   Charles L. Thomas "Catalytic Processes and Proven Catalysts" Academic Press, New York and London (1970)
3.   Martyn Twigg "Catalyst Handbook" Second Edition Wolfe Publishing Ltd England (1989)
4.   J.F. le Page "Catalyse de Contact. Conception, preparation et mise en oeuvre des

catalyseurs industriels" Editions Technip Paris (1978) (Also an English translation has been published)

5. J.R. Anderson "Structure of Metallic Catalysts" Academic Press New York and London (1975)

6. "Characterization of Catalysts" (J.M. Thomas and R.M. Lambert eds.) John Wiley & Sons Chichester (1980)

7. "Characterization of Heterogeneous Catalysts" (Francis Delannay ed.) Marcel Dekker New York (1984)

8. G.A. Somorjai "Chemistry in Two Dimensions: Surfaces" Cornell University Press Ithaca (1981)

IMPORTANCE OF SURFACE CHEMISTRY/CATALYSIS IN THE PROCESSING OF SEMICONDUCTORS

J. Enrique Ortega and Rodolfo Miranda

Departmento de Física de la Materia Condensada
Universidad Autónoma de Madrid. Cantoblanco
28049 Madrid

ABSTRACT

Thin layers of alkali metals deposited on a GaAs(110) surface can enhance by several orders of magnitude the oxidation kinetics of the substrate. Pure alkali oxides of different stoichiometry have been grown at 150 K on the GaAs(110) substrate, which does not react with the alkali oxides at this temperature. The spectral features of these different alkali oxides have been used to identify the oxygen species responsible for the Negative Electron Affinity activation of a Cs/O/GaAs photocathode, as well as the reaction products of the alkali-promoted oxidation of the GaAs(110) surface.

1. INTRODUCTION

Surface and interface chemistry of semiconducting materials is a field of paramount importance. Chemical reactions, such as, silicide formation, corrosion, nitridation and, above all, oxidation are commonly used in the semiconductor industry to produce metallic contacts, diffusion barriers, dielectric layers or gate oxides [1]. In some of these reactions, a catalytic process is used to lower the reaction temperature by decreasing some of the intermediate activation barriers. In others, the catalytic process is undesired, since it produces deleterious effects. An example is the corrosion of Al contacts by trace contamination of Cl^- ions, which dissolve the protective Al_2O_3 layer by a catalytic process [1].

In the fabrication of some MOS devices, it is necessary to grow uniform SiO_2 layers of a thickness of 300 Å. Thermal dry oxidation can produce thin oxide layers with the dielectric properties of thicker SiO_2 films (dielectric strength, 10^7 V·cm^{-1}, resistivity, 10^{16} Ω^2·cm). In order to reach superscaled MOSFETs with gate oxides only 100 Å thick, the temperature of oxidation of Si must be lowered, while controlling the thickness of the oxide with almost atomic precision and keeping satisfactory dielectric properties. The temperatures used today (1000 ^{0}C) limit the size reduction that can be achieved. A catalyst for the low-temperature oxidation of Si is then required.

Fundamental Aspects of Heterogeneous Catalysis Studied by Particle Beams,
Edited by H.H. Brongersma and R.A. van Santen, Plenum Press, New York, 1991

43

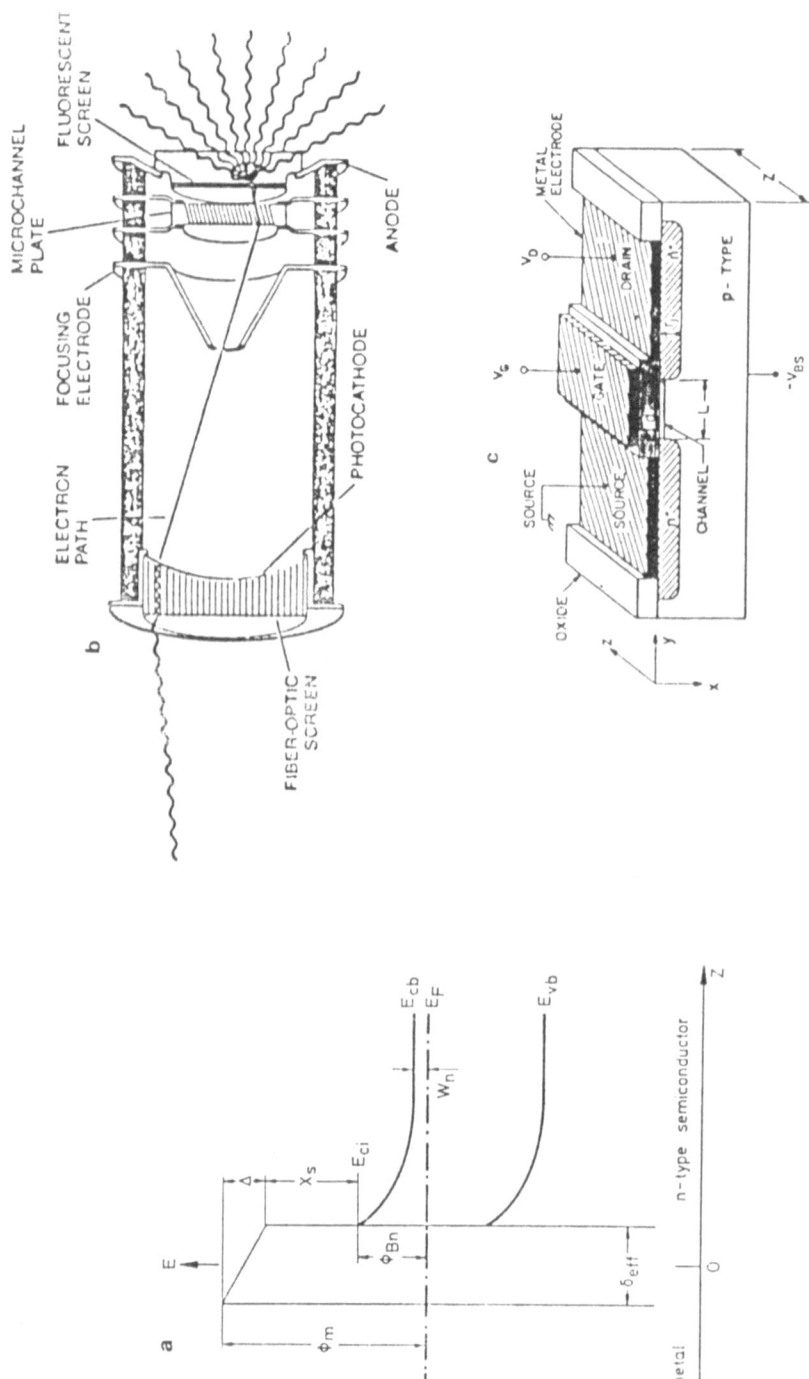

Fig. 1. Major applications of Alkali Metal/Semiconductor interfaces:
(a) Basic studies concerning the Schottky barrier formation,
(b) NEA photocathodes for image intensifiers, and
(c) Gate oxide in superscales MOSFET's produced by alkali metalpromoted oxidation of Si.

Since long, dopants and impurities in a Si crystal are known to promote the oxidation rate of Si. For example, P segregated at the SiO_2/Si interface enhances the oxidation rate [1]. Noble and transition metals deposited on the surface also promote the oxidation of Si [2]. These "catalysts", however, can not be removed from the oxide layer after the reaction. They can thus trap charges precisely where it is desired least: at the SiO_2/Si interface.

The properties of Alkali Metals (AM) deposited on the surface of semiconductors have been studied extensively in the past few years [2-4]. As illustrated in Fig. 1, there are three major reasons for this interest. AM/semiconductors interfaces are ideal model systems to study the formation of the ubiquitous metal/semiconductor junctions, addressing issues, such as, charge transfer, metallization, and origin of the Schottky barrier [3]. The second reason is that the properties of the interface between GaAs and a gas phase after deposition of Cs and reaction with oxygen are crucially affecting the performance of the Negative Electron Affinity (NEA) [4] photocathodes, which are at the heart of image intensifiers and night vision devices. These two reasons will not be discussed any further.

The third reason is that the AM's promote the oxidation of Si at room temperature, and that they can be removed by moderate thermal treatment (at 600 ^{0}C) producing a clean SiO_2/Si interface [2]. The AM-promoted oxidation has recently received widespread interest. In spite of much work, doubts were still present concerning both academic questions, e.g., the details of the oxidation mechanism, its local or non-local character, and industrial applications, e.g., can the alkali be totally eliminated from the SiO_2 film? The latter question justifies the initial scepticism of the industry: if some alkali ions remain trapped in the SiO_2 layer, increasing the number of mobile charges, they can cause serious instabilities in MOS devices.

Here we will summarize our recent findings [4] concerning the extension of the AM-promoted oxidation to GaAs. In this case the practical use of the method is doubtful, since standard thermal oxidation of GaAs results in a non-stoichiometric mixture of Ga_2O_3 and As_2O_3 with poor (breakdown field of 5×10^5 V·cm^{-1}, resistivity of 10^{14} Ω·cm) dielectric properties and a very high (10^{12} cm^{-2}) density of surface states. Oxides of this quality are not suitable for use as gate contacts on GaAs field effect transistors. On GaAs substrates, however, it is possible to isolate the various AM oxides and identify their crucial role in the microscopic mechanism. In the case of Si, the local character of the reaction has been directly visualized by Scanning Tunneling Microscopy (STM). The low residual level of alkali contamination on the SiO_2 film, as detected by SIMS, has stimulated the practical use of the AM–promoted oxidation of Si.

2. EXPERIMENTAL

The experiments have been carried out in a UHV system with base pressure in the low 10^{-10} Torr range. The techniques employed include Auger Electron Spectroscopy (AES), Low Energy Electron Diffraction (LEED), Work Function (WF) changes, and Photoelectron Spectroscopy (PES). PES was performed using Synchrotron Radiation from BESSY dispersed by the SX-700 and HE-PGM-2 monochromators. The energy resolution was 0.1-0.2 eV. Binding Energies (B.E.) are referred to the Fermi level (E_F) of the sample holder. Mirror-like GaAs(110) surfaces were obtained by cleaving in-situ GaAs bars (n or p-type, 3×10^{17} cm^{-3}) or by Ion Bombardment at 600 eV and Annealing at 800 K (IBA). Cleaved surfaces show no band bending, whereas surfaces prepared by IBA display in initial band bending with E_F fixed in the center of the band-gap. Alkali metals were evaporated from SAES Getters sources. A monolayer (ML) of alkali is defined as the first complete layer of alkali atoms in contact with the substrate. The purity of the oxygen was controlled with a quadrupole mass spectrometer.

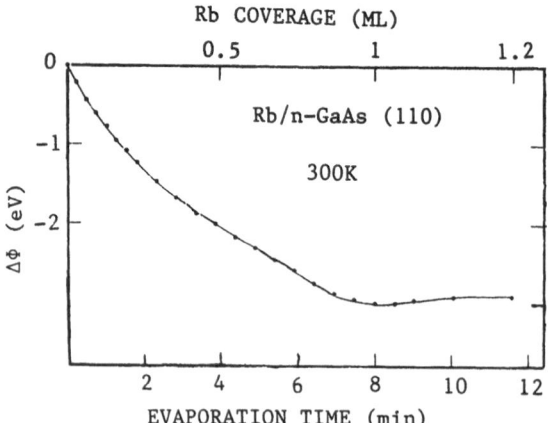

Rb COVERAGE (ML)

Rb/n–GaAs (110)

300K

EVAPORATION TIME (min)

0.25 ML Cs/GaAs (110)

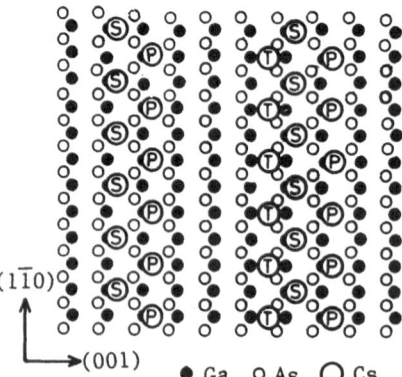

(1̄10)

(001)   ● Ga ○ As ◯ Cs

Fig. 2.   (a) Work function changes during deposition of Rb on n-GaAs(110). (b) Model for the geometry of Cs on GaAs(1̄10).

3. RESULTS AND DISCUSSION

3.1. Alkali metal adsorption

The alkali metals grow layer by layer on GaAs(110) at 300 K; a second layer does not start to grow until the first layer has been completed. The first layer which contacts the substrate always tends to be a hexagonal closely packed layer similar to the (110) face of the bcc structure typical for the bulk alkali metal [1]. The bonding between AM and GaAs is covalent, with a small transfer of charge of the order of 0.2 e⁻/atom. It is

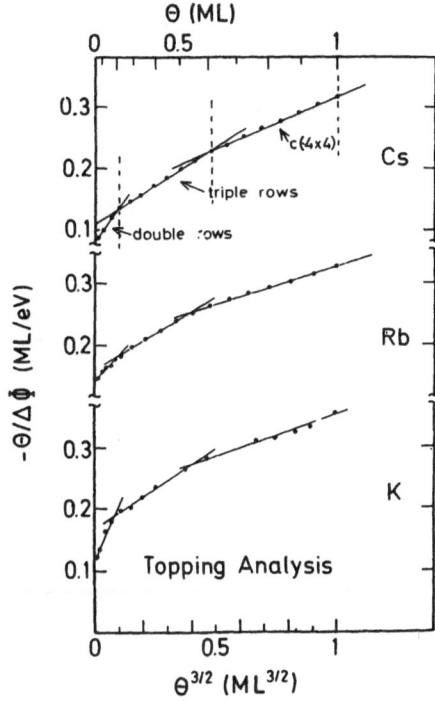

Fig. 2.   (c) Topping analysis [12] of the work function changes for adsorption of K, Rb, and Cs on n-GaAs(110)

thus appropriate to use the covalent radius of the AM to evaluate the number density of atoms in the ML. One obtains densities of 6.4, 5.8, and 4.8×10^{14} at/cm² for K, Rb, and Cs, respectively. The GaAs(110) surface has 4.43×10^{14} unit cells/cm². The adsorption geometry of AM on GaAs(110) is not well known. In the case of Cs, LEED [5] and STM [6] studies have shown that below 0.5 ML, the AM adatoms form zig-zag chains comprised of two (and above 0.12 ML, of three) rows of atoms along the [1\bar{1}0] direction.

This geometry is shown is Fig. 2b. The adsorption site is the quasi-threefold site associated with one Ga and two As atoms [6]. The bonding is primarily with the Ga dangling bond. At saturation at 300 K, the observed c(4×4) LEED pattern is consistent with a (110) plane for the Cs overlayer. The Cs/GaAs(110) interface, however, has been claimed not to be metallic under these conditions [7,8].

The saturation coverage at 300 K is 2, 1.2, and 1 ML for K, Rb and Cs, respectively, following the trend imposed by their vapor pressure at 300 K (1.17×10^{-8}, 4.82×10^{-7}, and 2.4×10^{-6} Torr). Larger coverages can be obtained by cooling the sample. At a coverage of about 2 ML of Cs, metallic behaviour is observed. The metallic character is deduced from the existence of a Fermi cutoff in UPS spectra, the fixing of the Fermi level in the same position for n and p-doped samples, and the presence of plasma losses [8]. The metallic character could be important for the promoted oxidation.

Adsorption of AM on semiconductor substrates produces a strong decrease in the work function (WF). Fig. 2a reproduces the evolution of the WF of a n-type GaAs(110) surface as a function of the coverage of Rb. The WF passes through a shallow minimum, $\Delta\phi_{min}$, and increases to a saturation value close to the WF of bulk crystals of alkali metals [9]. Similar data are available for the other alkali metals [4]. This behaviour is qualitatively the same as that observed for AM adsorption on metals [10]. One significant difference, however, is that $\Delta\phi_{min}$ appears at submonolayer coverages for metals and only close to completion of the first monolayer in semiconductors [11].

The Topping analysis [12] can provide values of the average dipole moments, μ, of adsorbed AM atoms from these WF measurements. Fig. 2c shows the result of such an analysis for Cs, Rb, and K on GaAs(110). Three linear regions appear corresponding to the formation of double rows, triple rows and the structure related to the c(4×4) LEED pattern. In all cases, the dipole moments decrease with the AM coverage.
For Cs, with 0, 0.2, and 0.5 ML, the numerical values of μ are 6.83, 5.07, and 3.76 debyes, respectively. The formation of triple rows above 0.1–0.2 ML of Cs produces the appearance of a shifted Cs-5p core–level emission in UPS [13].

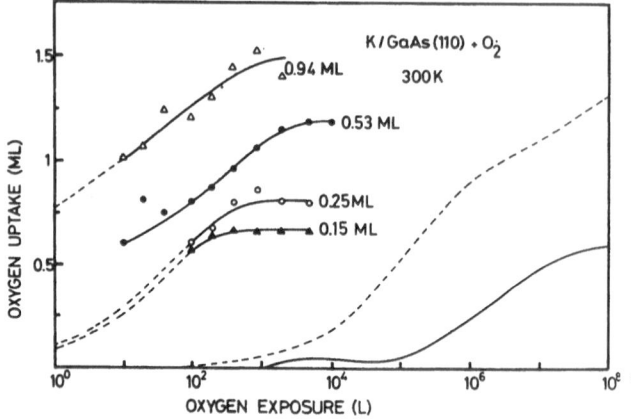

Fig. 3.   Oxygen uptake of GaAs(110) surfaces cleaved, illuminated with light from a Xe lamp and covered with various coverages of K. The exposure to O_2 was carried out at 300 K. The data have been taken with AES and translated to oxygen coverages as described in [15].

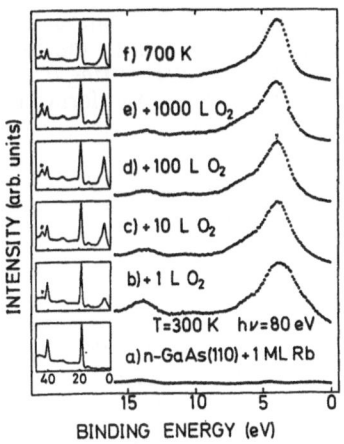

Fig. 4.   Photoemission spectra of the valence band and shallow core levels (in the inset) for GaAs(110) covered with 1 ML of Rb and exposed to O_2 at 300 K.

3.2. Oxygen adsorption on alkali–covered GaAs(110)

Spicer et al. [14] obtained the first indications that a Cs overlayer could enhance the oxidation rate of GaAs in the course of an investigation aimed at understanding the NEA process. Fig. 3 reproduces the first quantitative evaluation of the AM effect on the kinetics of oxygen adsorption as reported by Ortega et al. [15]. The continuous curve shows the adsorption rate of oxygen on a clean GaAs(110) surface at 300 K. The dashed curve reproduces data measured by Mönch and coworkers [16] illuminating the surface with light of an energy larger than the band gap. The uptake of oxygen is two orders of magnitude faster. Predeposition of K at submonolayer coverages increases the kinetics of oxygen adsorption up to eight orders of magnitude [15].

Photoemission spectroscopy reveals that this accelerated uptake of oxygen produces a substantial oxidation of the substrate [4, 15, 17]. Fig. 4 shows the valence band and the core level spectra for a representative example of 1 ML of Rb/GaAs(110) exposed to O_2 at 300 K. Upon oxygen exposure, satellites shifted to larger B.E. appear in the Ga–3d and As–3d core levels. They correspond to Ga and As atoms chemically bonded to 1 to 3 oxygen atoms and indicate the formation of Ga_2O_3 and As_2O_3, respectively [18].The atomic species produced by AM–promoted oxidation of GaAs are not different from those produced by thermal oxidation. The valence–band spectra reflect the oxidation of the substrate. Comparison with data measured during oxidation of clean GaAs shows that the ML of Rb enhances the rate of oxidation by a factor of 10^9 [4]. No sign of alkali oxides is seen in the valence band. Since the AM adlayer accelerates the oxygen uptakes and seems not to be affected by the oxidation process, one might think that the alkali metal atoms play the role of a classical catalyst. We will see below that this is not true. Incidentally, this experiment indicates that metallic character is not essential for the promoted oxidation to proceed.

There have been claims that the enhanced oxidation is due to a non–local, average, reduction of the WF produced by AM adsorption [19]. This non–local mechanism has been challenged by data showing that the WF decrease has no simple, quantitative correlation with the sticking coefficient for O_2 [20]. The sticking coefficient increases linearly with the AM coverage, while ϕ passes through a minimum. These data are consistent with early suggestions [2] of a local mechanism for the AM–promoted reaction. Furthermore, the evolution of the WF during oxygen exposure is not consistent with a non–local mechanism. It rather suggests the formation of AM–oxygen compounds which could be intermediates in the reaction.

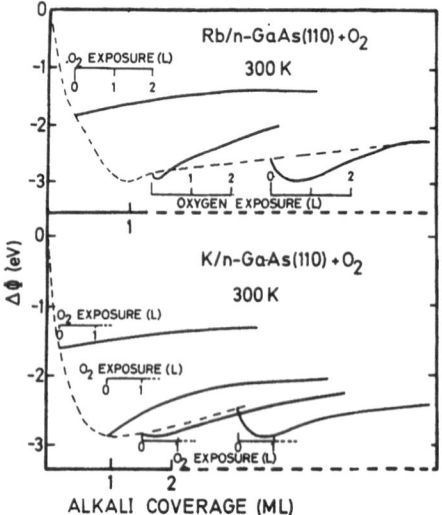

Fig. 5.   Variation of the work function of a n-GaAs(110) surface partially covered with K and Rb as a function of O_2 exposure at 300 K. The abscissa are in units of the alkali saturation coverage. The continuous curve shows the evolution of the work function with O_2 exposure in Langmuirs. The scale for exposure is displaced according to the respective alkali precoverage.

As already recognized by Langmuir [21], additional changes in the WF appear when exposing an alkali–covered surface to O_2. When O_2 is admitted before a certain coverage of alkali is reached, the WF generally increases, while for O_2 exposures after the above alkali coverage, an additional minimum appears. This kind of behaviour is illustrated in Fig. 5 for Rb and K adlayers on GaAs(110). The additional minimum in $\Delta\phi$ can be attributed microscopically to the presence of an alkali–bonded oxygen species, namely, $O^=$. Depending on coverage and chemical state of the alkali atoms, it reflects the formation of compounds of different stoichiometry (Na_2O, K_2O, Rb_2O, Rb_3O, Cs_2O, $Cs_{11}O_3$). The existence of $O^=$ at the additional minimum produced by Cs on GaAs(110), is demonstrated in Fig. 6, where, in order to stabilize the intermediate species, the temperature has been lowered to 150 K. The $O^=$ ion is detected from the peak at 2.8 eV below E_F after exposure to 0.4–0.6 L of O_2.

These observations are smoothly explained by assuming the reaction to occur via a local mechanism involving formation of alkali–oxygen compounds, which can transfer oxygen to the substrate. The transfer of oxygen occurs rapidly at 300 K and for coverages $\approx$ 1 ML. It can be substantially slowed down by lowering the temperature or increasing the AM coverage. This microscopic mechanism is basically a solid state reaction, that can not be properly termed catalytic, since the "catalyst" actually forms compounds with one of the reactants and the final product (oxide) is proportional to the "catalyst" concentration [2]. In order to verify this hypothesis a number of AM oxides with well defined compositions has been prepared, and their spectroscopic characterization has been performed with UPS.

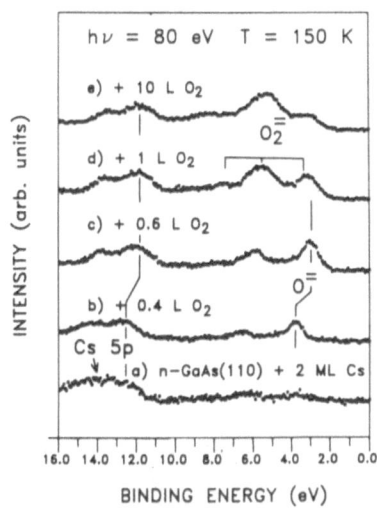

Fig. 6.   Valence band PE spectra for 2 ML of Cs on GaAs(110) exposed to O_2 at 150 K. The rigid shift of the spectra is due to band bending changes. The peaks corresponding to the different alkali-bonded oxygen species are indicated in the figure.

3.3. The role of alkali–oxides in the AM–promoted oxidation

Photoemission from the 0–2p levels produces a multiplet splitting characteristic of the $O^=$, $O_2^=$ and O_2^- ions [22] present in the highly ionic alkali oxides, which can be used to identify the various oxides during oxidation. Fig. 7 shows PE spectra of the valence band taken at 150 K after codeposition for increasing oxygen/alkali ratio. Fig. 7a shows that after a codeposition of 5 ML of Cs and 0.6 L of O_2, a single structure appears at 2.7 eV below E_F. This is the characteristic line of the $O^=$ ion [22]. The composition of this oxide is Cs_2O. GaAs itself shows no sign of oxidation at 150 K. Codeposition of 10 ML of Cs and 5 L of O_2 results in a spectrum dominated by a triplet structure (dash-dotted) assigned to $O_2^=$ ions, presumably present in Cs_2O_2. Smaller contributions of $O^=$ and O_2^- (broken line) are detected in the deconvolution. This latter, related to CsO_2, dominates the spectrum for an oxygen-rich codeposition of 5 ML of Cs and 150 L of O_2 (Fig. 7c).

As previously demonstrated for Rb oxides [23], oxidation of GaAs is possible with only the alkali oxide providing the oxygen, which can be thermally transferred to the substrate. The thickness of the oxide produced in this way is strictly proportional to the

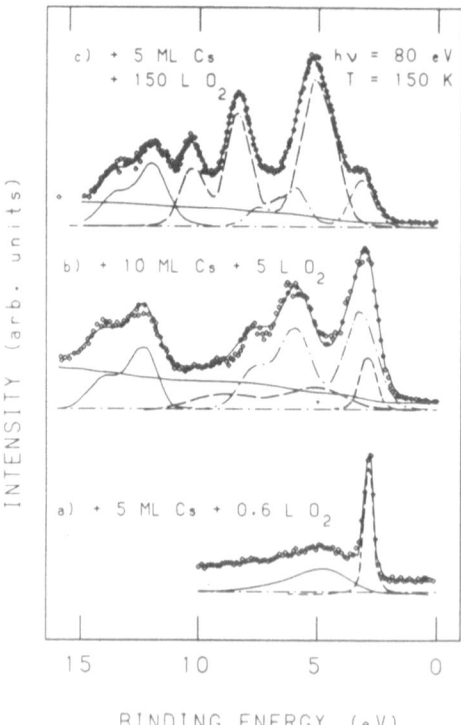

Fig. 7.   Valence band photoemission spectra of a cleaved n–GaAs(110) surface held at
150 K and covered with (a) 5 ML of Cs and 0.6 L of O_2, (b) 10 ML of Cs and
5 L of O_2, and (c) 5 ML of Cs and 150 L of O_2. All spectra have been
normalized to the same intensity.

amount and the stoichiometry of the initial alkali oxide [4]. This process is illustrated in
Fig. 8 for the CsO_2 overlayer in Fig. 7c. At 150 K there is no sign of substrate oxidation.
The valence band shows the multiplet splitting of CsO_2 and a small contribution from
Cs_2O_2 right at the GaAs interface [4]. Accordingly, two 0-1s core levels appear. The first
signs of GaAs oxidation appear at 250 K, as a chemically shifted As-3d core level and a
broadened Ga-3d level. The 0-1s line of CsO_2 has disappeared. The decomposition of the
alkali oxides to form substrate oxides has begun already at 250 K. At 400 K, the
oxidation is massive. The As-3d core level is shifted by 3.6 eV, indicating the formation
of As_2O_3 [18]. Ga oxide (Ga_2O_3, shifted by 0.9 eV) is clearly visible at the higher B.E.
side of the Ga-3d line. Upon annealing to 580 K, this latter increases at the expense of
As_2O_3, while the Cs-5s signal disappears. In the valence band only oxides of GaAs are
seen. Complete removal of Cs is achieved at 800 K. At the same temperature is observed
a complete disappearance of the signal due to As_2O_3. The bulk As_2O_3 is volatile at
800 K, in contrast to the Ga_2O_3, which is still present at the surface at this temperature.
At 820 K a clean GaAs(110) surface is recovered with a slight enrichment of As and a
band bending induced by the defects created during the thermal treatment. It is not
possible to remove totally the alkali without affecting the GaAs oxide.

The present results indicate that the mechanism of the AM-promoted oxidation of
semiconductors is, strictly speaking, non-catalytic. Rather than favouring the direct
dissociation of the O_2 molecules impinging anywhere on the surface [19], the AM adlayer
traps the oxygen into alkali-oxide compounds, that can be thermally decomposed and act
as an efficient source of oxygen to oxidize the substrate.

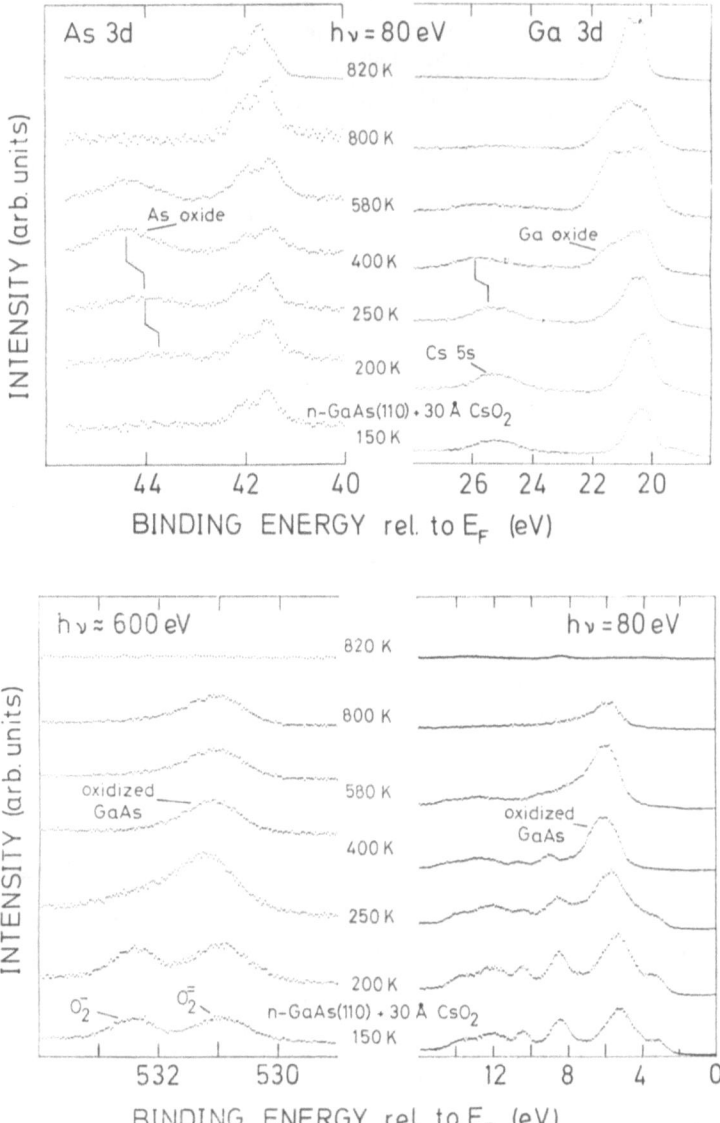

Fig. 8.   Photoemission spectra of the Ga-3d, As-3d, 0-1s core levels, and valence band for 30 Å of CsO$_2$ deposited at 150 K on n-GaAs(110) and subsequently annealed. The sample was maintained at each temperature for two minutes. The photon energy was 80 eV, except for the 0-1s spectra, which have been recorded with 600 eV.

4. RECENT FINDINGS AND PERSPECTIVES

With respect to the basic mechanism, a recent STM study [24] has revealed the local character of the solid state reaction produced during the AM-promoted oxidation of Si. Technological applications in the field of semiconductors have been stimulated by recent SIMS findings of a residual level of K contamination in the SiO_2 layer produced by AM-oxidation lower than the standard in the industry for device-quality SiO_2 [25]. The local character of the reaction, however, can produce a high level of stress at the SiO_2/Si interface that may be a problem for applications [24]. This latter problem is being studied currently. Applications in photochemistry have also been explored. It has been found that K preadsorbed on Si(111) surfaces is very efficient in promoting photodissociation of $Mo(CO)_6$, even at wavelengths longer than 800 nm [26].

The AM-promoted oxidation, on the other hand, has been extended to metal substrates, such as, Pt under UHV conditions [27]. The "catalyst" is an hydrated KOH layer thermally decomposed at 480 K. The phenomenon has also been observed for Ta and W [28]. It is presently under active research due to its eventual use in the production of the oxide layer in Josephson junctions.

In summary, we think that while most of the scientific problems have been studied, characterized, and clarified, much work is going to be devoted to applications in the next future.

REFERENCES

[1] C.R.M. Grovenor "Microelectronic Materials", Adam Hilger (Bristol, 1989).
[2] For a review of previous work see, R. Miranda in "Physics and Chemistry of Alkali Metal Adsorption". Ed. H.P. Bonzel, A.M. Bradshaw and G. Ertl. Elsevier, Amsterdam, 1989, p. 425.
[3] E.G. Michel, M.C. Asensio and R. Miranda in "Metallization and Metal–Semiconductor interfaces". Ed. I.P. Batra. Plenum Press, New York (1989).
[4] J.E. Ortega, Ph.D. Thesis, UAM (1990), unpublished.
[5] J. Derrien and F. Arnaud d'Avitaya, Surf. Sci. 65, 668 (1977).
[6] P.N. First, R.A. Dragoset, J.A. Stroscio, R.J. Celotta and R.M. Feenstra, J. Vac. Sci. Technol. A7, 2868 (1989).
[7] T.M. Wong, D. Heskett, N.J. diNardo and E.W. Plummer, Surf. Sci. 208, L1 (1989).
[8] M. Prietsch, M. Domke, C. Laubschat, T. Mandel, C. Xue and G. Kaindl, Z.Phys. B. 74, 21 (1989).
[9] Bulk work functions for the alkali metals are 2.42 (Na), 2.23 (K), 2.16 (Rb) and 1.8 eV (Cs).
[10] I. Langmuir and K.H. Kingdon, Phys. Rev. 21, 380 (1923).
[11] D. Heskett, T.M. Wong, A.J. Smith, W.R. Graham, N.J. di Nardo and E.W. Plummer, J. Vac. Sci. Technol. B7, 915 (1989).
[12] J. Topping, Proc. Roy. Soc. London, A114, 67 (1927).
[13] T.M. Wong, N.J. di Nardo, D. Heskett and E.W. Plummer, Phys. Rev. B. 41, 12342 (1990).
[14] C.Y. Su, P.W. Chye, P. Pianetta, I. Lindau and W.E. Spicer, Surf. Sci. 86, 894 (1979).
[15] J.E. Ortega, J. Ferrón, R. Miranda, C. Laubschat, M. Domke, M. Prietsch and G. Kaindl, Phys. Rev. B 39, 12751 (1989).
[16] F. Bartels and W. Monch, Surf. Sci. 143, 315 (1984).
[17] G. Remmers, M. Priestch, C. Laubschat, M. Domke, T. Mandel, J.E. Ortega and G. Kaindl, J. Chem. Phys. to be published.
[18] G. Landgren, R. Ludeke, Y. Jugnet, J.F. Morar and F.J. Himpsel, J. Vac. Sci. Technol. 2, 351 (1984).

[19] H.I. Starnberg, P. Soukiassian and Z. Hurych, Phys. Rev. B **39**, 12775 (1989).
[20] H.J. Ernst and M.L. Yu, Phys. Rev. B **41**, 12953 (1990).
[21] D.S. Villars and I. Langmuir, J. Am. Chem. Soc. **53**, 486 (1931).
[22] C.Y. Su, I. Lindau and W.E. Spicer, Cem. Phys. Lett. **87**, 523 (1982).
[23] R. Miranda, M. Prietsch, C. Laubschat, M. Domke, T. Mandel and G. Kaindl, Phys. Rev. B **39**, 10387 (1989).
[24] A.L. Vazquez de Parga, C. Ocal, J.E. Ortega and R. Miranda, Vacuum, in press.
[25] J. Ferrón, E.G. Michel and R. Miranda, unpublished.
[26] W. Ho in Ref. 1, p. 159
[27] G. Pirug, R. Dziembaj and H.P. Bonzel, Surface Sci. **221**, 553 (1989).
[28] L. Galán, private communication.

IRREVERSIBLE ADSORPTION OF METAL ATOMS IN ELECTROCATALYSIS

J. M. Feliu and J. M. Orts

Departamento de Química-Física, Univ. de Alicante
Apartado 99, 03080 Alicante, Spain

ABSTRACT

Bi, As, Sb and Pb adatoms remain irreversibly adsorbed on platinum single crystal electrodes. These adatoms undergo reversible redox reactions in the adsorbed state. The measurement of the charge involved in this redox reaction, jointly with that involved in hydrogen adsorption/desorption on the adatom-free platinum sites are used to in situ characterize these systems and check their stability in different experimental conditions.

INTRODUCTION

Adatom modified single crystal surfaces have been widely studied in the field of heterogeneous catalysis in solid/gas interfaces, due to their fundamental interest, as well as the possibility of improving the general performance of catalysts.

The catalytic effects of submonolayer quantities of foreign adatoms have also been pointed out in a great diversity of electrocatalytic reactions. Classically, the way of obtaining such improvements in the current study profited from the so-called underpotential deposition (UPD)[1]. In this phenomenon, the deposition of a metal M on a different metallic substrate takes place from a solution of its cations at electrode potentials positive to the Nernst potential for the $M/M(ox)_{sol}$ redox couple (that is, to the bulk deposition potential).

$$M(ox)_{soln} \Longleftrightarrow M(red)_{ads} \qquad (1)$$

The difficulties presented by the UPD technique for the achievement of a controlled surface modification by metallic deposition arise from the intrinsically reversible nature of the deposition process of the adsorbed metallic species. So, the coverage of the electrode by the deposited metal is found to depend on the electrode applied potential, if the concentration of M(ox) is sufficiently high, and on potential and time, if it is low. This lack of control in the adatom

Fundamental Aspects of Heterogeneous Catalysis Studied by Particle Beams,
Edited by H.H. Brongersma and R.A. van Santen, Plenum Press, New York, 1991

57

coverage over the whole potential range is a serious drawback for the UPD technique. Another difficulty is the need of significant cation concentrations in the working solution in order to maintain the reaction equilibrium. In the absence of M(ox) cations in solution, the whole M submonolayer may dissolve.

For a proper surface characterization a constant and controlled adatom coverage is needed in the actual experimental conditions. Such a requirement can be fulfilled by systems in which the foreign atoms are irreversibly adsorbed at the working electrode. Among them, some systems are known where, by putting a solution of certain metalic salts in contact with a platinum electrode in open circuit conditions, the spontaneous formation of submonolayers of irreversibly adsorbed metal atoms is observed[2-4]. These adlayers are stable even in electrolytes without the cations which gave rise to the formation of the layer. The most easily controlable systems present a redox surface reaction in the adsorbed state

$$M(ox)_{soln} \quad ----> \quad M(ox')_{ads} \quad <======> \quad M(red)_{ads} \quad (2)$$

where ox and ox' may or may not correspond to the same oxidation state or chemical composition.

In this paper our interest will be focused on the reversible surface reaction, i.e. the last reaction in scheme 2, which is independent of how the irreversible adsorption process takes place. This reaction can take place in conditions where the concentration of M(ox) in solution is esentially zero.

RESULTS

Metallic atoms, such as Bi, Sb, As, and Pb, give this spontaneous irreversible adsorption step on platinum[2-4], from a solution of their salts. The irreversibility of the adatom adsorption step allows us to maintain a constant

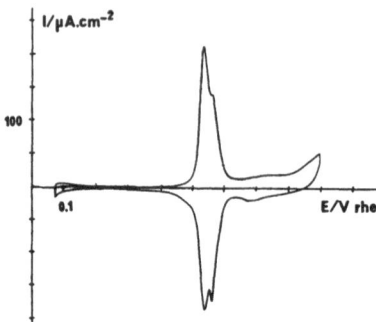

Fig. 1. Voltammetric profile of the surface redox reaction for irreversibly adsorbed As on Pt(111) electrode. θ_{As}= 0.32. Electrolyte: 0.5 M H_2SO_4. Sweep rate = 50 mV/s in all figures. Coverages are defined as the number of adatoms per Pt atom.

adatom coverage in the absence of solution species. The resulting adsorbed species undergoes a redox process when sweeping the electrode potential in cyclic voltammetry experiments. Figure 1 shows the voltammograms for the redox process of a Pt(111)-As electrode, at full coverage (no hydrogen adsorption).

In this redox reaction the adatom modifies its oxidation state from zero (totally discharged atom) to a higher oxidation state during the positive-going potential sweep, both species being adsorbed on the platinum electrode surface. This transformation is reversed in the negative-going sweep. The oxidized states correspond to oxygenated species, as follows from the invariance of the redox potential in the RHE scale when pH is varied.

This valence change does not modify the total adatom coverage, and the voltammetric curve remains constant in succesive cycles. The only effect is the modification of the relative populations of the reduced and oxidized states as a consequence of the charge transfer. These surface reactions have been proven to be structure sensitive processes, as shown in Figure 2 for the case of adsorbed bismuth on several platinum single crystal surfaces.

In Figure 2 we can also see that submonolayer coverages of adatoms allow the bare platinum surface sites to adsorb hydrogen. The amount of electric charge involved in the surface reaction can be measured, and used, jointly with the amount of charge corresponding to hydrogen adsorption, to in situ characterize the surface system. In this way, the stoichiometry of the surface reaction can be determined, as well as the number of platinum substrate atoms blocked for hydrogen adsorption per individual metallic adatom.

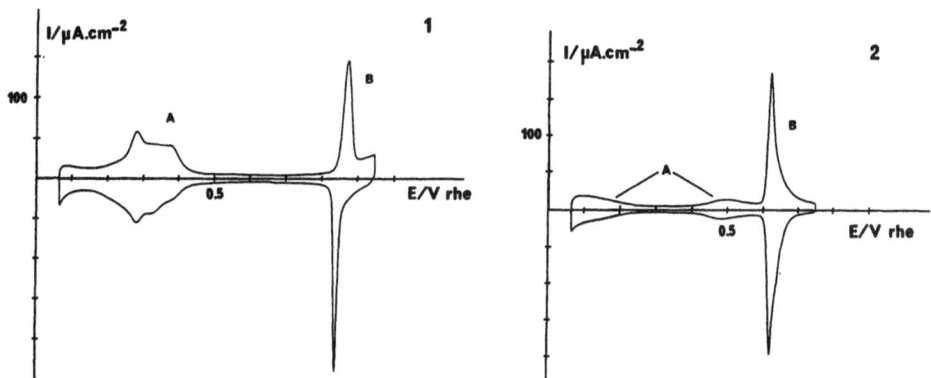

Fig. 2. Voltammetric profiles showing hydrogen adsorption/desorption (A) and adsorbed bismuth surface reaction (B) on 1) Pt(100), $\theta_{Bi}= 0.17$; 2) Pt(111), $\theta_{Bi}=0.26$. Electrolyte:0.5M H_2SO_4.

The relationship between redox charge and hydrogen adsorption charge is usually linear, with the slope being a simple arithmetic fraction. In the case of Bi on Pt(111) (Figure 3),a slope of -2/3 indicates that each bismuth adatom interchanges two electrons and blocks three hydrogen adsorption sites. In the same way, As undergoes a three electron process while blocking three hydrogen adsorption sites. The observation , in both cases, of strongly bonded hydrogen suggests an island distribution of the adatom layer.

We have previously stated that in electrodes obtained by irreversible adsorption adatom coverage remains constant over a wide potential range. Nevertheless, if sufficiently high potential values are reached, significant adatom loss is observed. In this way it is possible to obtain the desired adatom coverage, stable in the lower potential range, which is the interesting one in Electrocatalysis. It is also possible to desorb completely the adatom and study the final state of the platinum surface. In most cases, the voltammetric profile remains unchanged but the adsorption of Sb on Pt(111) at high coverages causes the appearance of (110) defects and a loss of (111) bidimensional long range order due to alloy formation. Besides electrode potential, other variables related with the electrolyte medium,(for example, pH and anionic competitive adsorption) can severely modify the stability of the adlayer.

Irreversible adsorption of lead on Pt(111) electrodes in different media ilustrates the effect of pH on the stability of the adlayer. This particular system

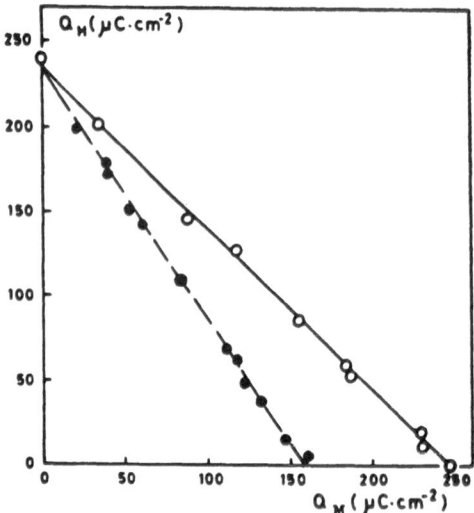

Fig. 3. Relationship between hydrogen adsorption/desorption charge and adatom redox surface reaction charge on Pt(111) electrode in 0.5M sulphuric acid electrolyte.(———————) Pt(111)-As; (— — —) Pt(111)-Bi.

appears to be more stable the higher the pH value, probably due to the greater stability of the Pb(II) adsorbed oxygenated species in alkaline medium. Figure 4 shows the maximum amounts of adsorbed lead on Pt(111) electrodes in 0.1 M HClO4 and 0.1 M NaOH electrolytes, respectively.

The presence of specifically adsorbed anions can also lead to a decrease of the adlayer stability, especially at relatively high potentials. In the case of Pt(111)-As electrodes, it has been verified that the presence of strongly adsorbed Br$^-$ anions in the electrolyte (0.001 M in 0.5 M H$_2$SO$_4$), makes it practically impossible to avoid a continuous adatom loss due to the competition between Br$^-$ and the oxidized form of the adatom for the adsorption sites (Figure 5). No adatom loss was observed in the same potential range in bromide-free sulphuric acid electrolyte.

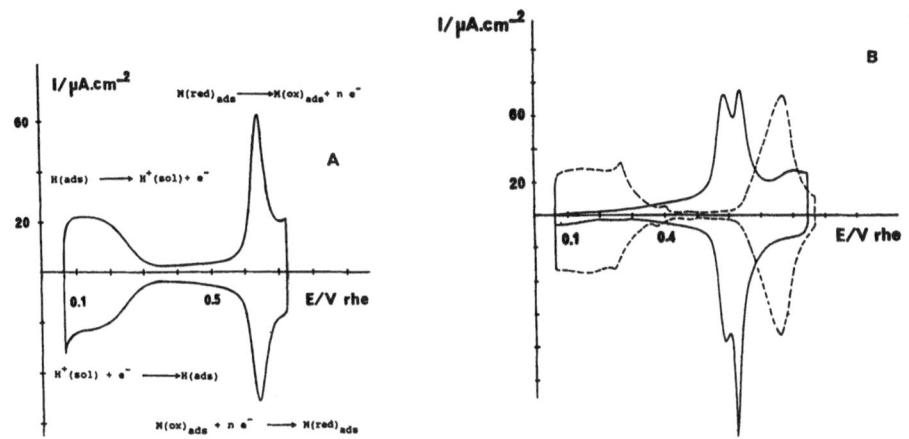

Fig.4. Voltammetric profiles for maximum adsorbed Pb coverages on Pt(111) electrodes: A) 0.1 M HClO$_4$, B) 0.1M NaOH (full line). Included in B is the adsorption/desorption of a hydrogen monolayer on bare Pt(111) in 0.1M NaOH (dotted line).

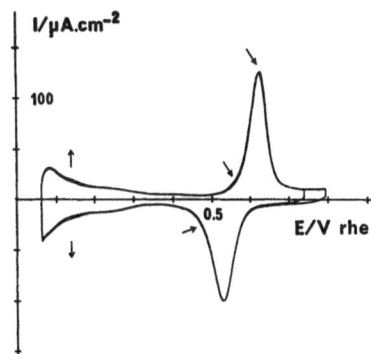

Fig. 5. Continuous adatom loss in Pt(111)-As system due to competitive adsorption of bromide anions (0.001 M in 0.5 M sulphuric acid). Arrows indicate the increase in hydrogen adsorption charge and the decrease in arsenic redox process charge.

CONCLUSIONS

Irreversible adsorption of metallic adatoms presents significant advantages over the classical UPD technique as a means of obtaining adatom-modified electrode surfaces. Besides allowing constant and controlled coverages of the adatom under the experimental conditions, it alsoprovides information about oxidation states, distribution of the adatoms on the substrate and stability , which can be useful in Electrocatalysis.

REFERENCES

1. D. M. Kolb. Physical and Electrochemical Properties of Metal Monolayers on Metallic Substrates. In "Advances in Electrochemistry and Electrochemical Engineering", vol. 11. (H. Gerischer and C. W. Tobias, Eds.). John Wiley, New York, 1978. p. 125-271.

2. J. Clavilier, J. M. Feliu and A. Aldaz, J. Electroanal. Chem. 243 (1988) 419.

3. J. M. Feliu, A. Fernández-Vega, A. Aldaz and J. Clavilier, J. Electroanal. Chem. 256 (1988) 149.

4. J. Clavilier, J. M. Orts, J. M. Feliu and A. Aldaz, J. Electroanal. Chem., in press.

ELECTROCATALYSIS ON WELL-DEFINED PLATINUM SURFACES

A. Fernández-Vega, M.J. Llorca and A. Aldaz

Departamento de Química-Física, Universidad de Alicante, Facultad de Ciencias.Apartado 99 03080 Alicante (Spain)

ABSTRACT

In this work well defined Pt surfaces have been used with an electrocatalytic purpose. Formic acid oxidation has been selected as a model reaction in which surface controlled bimetallic systems obtained by means of the irreversible adsorption technique have been tested. Glucose oxidation has been used to study the role of long-range surface order in this reaction. These results give basic knowledge for the design of selective metal electrocatalysts.

INTRODUCTION

The main effort in Heterogeneous Electrocatalysis as Catalysis in general, has proceeded on the selection and design of the most efficient material (active and selective) for a particular reaction, both in fundamental and applied systems.

The electrooxidation of many organic molecules is very sensitive to the properties of the electrode surface. For example, various modifications in the surface composition of the base catalyst have been used with the aim of increasing its reactivity. On the other hand, it is well known that, certain reactions are "structure sensitive" in that they show different behaviour for different crystalline orientations. In this way an improvement in the electrocatalytic effect can be achieved by selecting the most appropiate crystalline structure of the electrode substrate.

It is known that the adsorption of adatoms can be used in order to enhance the rate of some electrochemical reactions. The irreversible adsorption technique on well defined Pt surfaces allows simultaneous control of the structural and chemical conditions of the electrode surface. The electrocatalytic properties of these bimetallic electrodes has been applied to the case of formic acid oxidation which has been selected for this study as a test reaction.

Also, glucose oxidation has been studied to show the long range order effects involved in this process.

Fundamental Aspects of Heterogeneous Catalysis Studied by Particle Beams,
Edited by H.H. Brongersma and R.A. van Santen, Plenum Press, New York, 1991

63

RESULTS

Formic acid

The electrooxidation of formic acid on Pt is a model for a structure sensitive reaction [1]. A dual path mechanism has been proposed according to the general scheme:

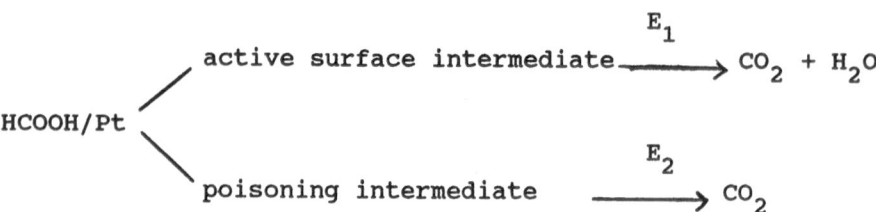

$$\text{HCOOH/Pt} \left\langle \begin{array}{l} \text{active surface intermediate} \xrightarrow{\quad E_1 \quad} CO_2 + H_2O \\[2em] \text{poisoning intermediate} \xrightarrow{\quad E_2 \quad} CO_2 \end{array} \right.$$

with $E_2 > E_1$, both paths being structure sensitive.

Pt(100) surface:

Among the three basal platinum orientations, the Pt(100) electrode appears to be very active both in the oxidation of the formic acid through path (1), and also in the formation of the poisoning intermediate through path (2). This last point is corroborated by the highly blocked oxidation current obtained in the positive voltammetric sweep.

Antimony is adsorbed irreversibly on the surface of platinum. On Pt(100) the adsorbed antimony undergoes a surface redox reaction at approximately 0.66 V. (Fig 1). Formic acid oxidation is clearly enhanced in the presence of antimony (Fig 2) and takes place in that part of the potential range where the adatoms are in their reduced form, so no bifunctional catalysis takes place.

Fig 3 presents the formic acid oxidation peak currents as a function of antimony coverage. No significant current is obtained for adatom coverages below 0.2 . A maximum in the activity is obtained with an antimony coverage of roughly 0.45 and in a narrow range.The maximum current obtained at this coverage is practically the same as that for the intrinsic activity of pure Pt(100) measured by pulsed voltammetry [2].The drop of the oxidation rate above 0.45 suggests that the reaction takes place mainly on Pt(100) sites free from adsorbed antimony.

The reaction path 2 can be studied through the isolation of the poisoning intermediate (P.I.) by measuring the amount of charge involved in the oxidation of the layer of adsorbed molecules after contact of the surface with a formic acid solution at open circuit.

The experiments attempting to isolate this poisoning intermediate were carried out by applying this dissociative adsorption technique. The results as a function of antimony coverage are shown in fig 4. Increasing amounts of PI (identified as CO [3]) with decreasing amounts of antimony adspecies are observed for coverages below 0.35.

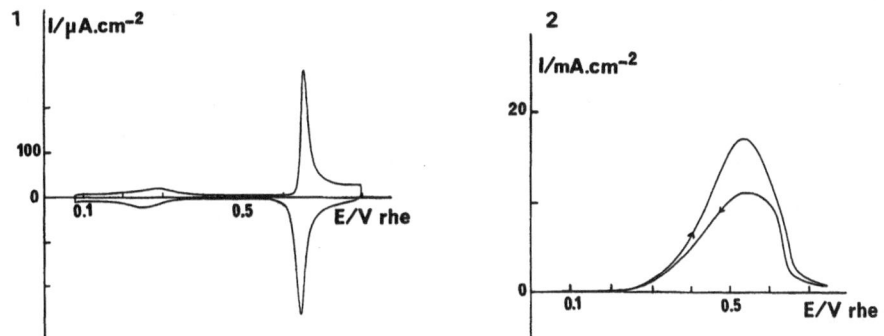

Fig 1. Stabilized voltammogram of a Pt(100) electrode with Θ_{Sb}=0.41. 0.5 M H_2SO_4, 50 mV/s in all figures.

Fig. 2. First voltammogram obtained with the same electrode in 0.25 M HCOOH + 0.5 M H_2SO_4.

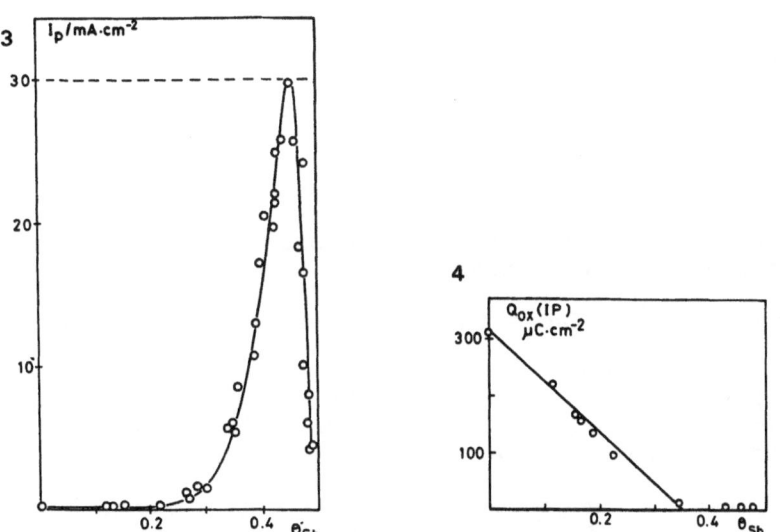

Fig. 3. Plot of the formic acid oxidation current density versus antimony coverage. (- - -) intrinsic activity for a Pt(100) surface obtained by pulsed voltammetry.

Fig. 4. Plot of the poison oxidation charge versus the antimony coverage.

The results suggest that the (100) sites modified by the presence of antimony block the ability for the pathway via the poisoning intermediate. The adatom acts as a third body which phisically impedes pathway 2 at high coverages. So the pathway 1 can proceed at very high rate on the free Pt sites. Similar blocking results have been obtained when Bi is the adsorbed adatom.

Pt(111) surface:

Pt(111) is the electrode surface which undergoes the least effect of self-poisoning but its activity for reaction 1 is the lowest too.

Bismuth atoms, irreversibly adsorbed on Pt(111), produce an enhanced catalytic effect on the direct oxidation reaction of formic acid. As in the preceeding case, the reaction takes place at potentials where the adatoms are in their reduced (metallic) form. In fig 5, values of maximum current density have been plotted as a function of bismuth coverage. A broad peak is observed in the coverage range of 0.2-0.27 . The fast decrease of current above 0.27 indicates that formic acid oxidation takes place on the uncovered Pt sites only. In this case, current density values higher than those obtained in the pulse voltammogram are attained. This result suggest a deep modification of the electronic properties of the surface which is now more active for formic acid oxidation via pathway 1. Similar results are obtained when arsenic is the adsorbed adatom.

For both, arsenic and bismuth adatoms, attempts to isolate the surface poison formed by dissociative adsorption have failed even at low coverages (Fig 6). However, it has been possible to detect the poison after the CO adsorption from a CO solution. So, it seems that the adatom acts on this modified surface selectively, inhibiting the reaction path via the poisoning intermediate.

On these (111) surfaces the presence of significant amounts of strongly adsorbed hydrogen implies that long-range order Pt(111) surface domains are present on the surface. In this case third body effects can not be detected and the absence of adsorbed poison is related to a deep change in the properties of the surface.

These results suggest an important change in the electrocatalytic properties of the adatom modified Pt(111) surfaces which are significantly more selective and active for the reaction pathway via active intermediate.

Glucose

It is well known that the electrochemical oxidation of formic acid takes place, to a considerable extent, on defects such steps rather than on terraces [4]. However, with big molecules like glucose, the situation could be quite different.

The use of stepped surfaces allows control of both theterrace width and the type of step site. The electrochemical

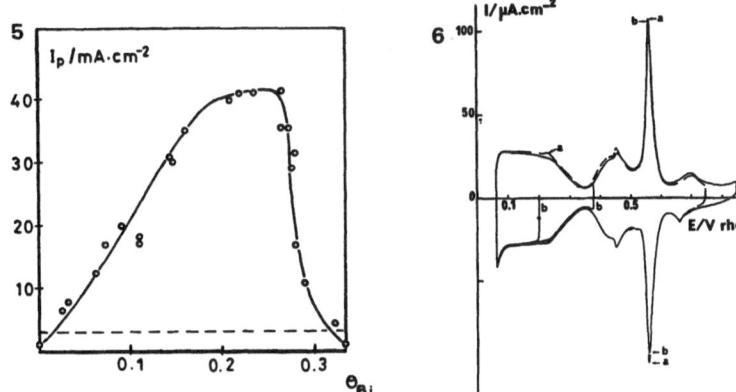

Fig. 5. Formic acid oxidation current density versus bismuth coverage. (- - -) intrinsic activity for a Pt(111) surface obtained by pulsed voltammetry.

Fig. 6. Negative test for the accumulation of the poison on a Pt(111) electrode with a low arsenic coverage $\Theta_{As} = 0.1$.

adsorption of hydrogen is a very useful test reaction in order to characterize the surface structure.

Figure 7 represents the voltammogram of a stepped surface with (100) terraces, Pt(610)= 6(100) x (110). The comparison of this voltammogram with that of Pt(100)[5] allows us to identify the (110) step site contribution to the weakly adsorbed hydrogen states (below 0.2 V), while the region between 0.2 and 0.45 V may be attributed to the strongly adsorbed hydrogen states. The narrow peak at 0.27 V corresponds to the edge sites, characteristic of a monoriented surface of (100) symmetry. The absence of the peak at 0.37 V confirms the lack of two dimensional long range order in this stepped surface.

The voltammogram shown in fig. 8 represents glucose oxidation on Pt(610). The peak centered at 0.39 V is asigned to glucose oxidation. The second oxidation peak at 0.79 V is ascribed to the oxidation of a poisoning intermediate , formed during the preceeding process. The effect of this species, identified as CO [6], is the blocking of active surface sites. The broad peak at 0.32 V of the blank is not visible in this voltammogram. Nevertheless, the reversible peak at 0.27 V, although decreased in magnitude is observed unambiguosly. These results indicate that glucose adsorbs preferentially on the wide terraces,rather than on step sites, suggesting the importance of a bidimensionally well-ordered surface in this process. The dependence of the current density on the terrace density (as obtained with several stepped surfaces) is not a linear relation, as it would be expected, but increases strongly with the (100) terrace contribution.

It may be concluded that certain reactions have geometric considerations in addition to a determined active site. Long range order surface stereochemistry plays a role in this reaction, but such an effect is not observed with small organic molecules.

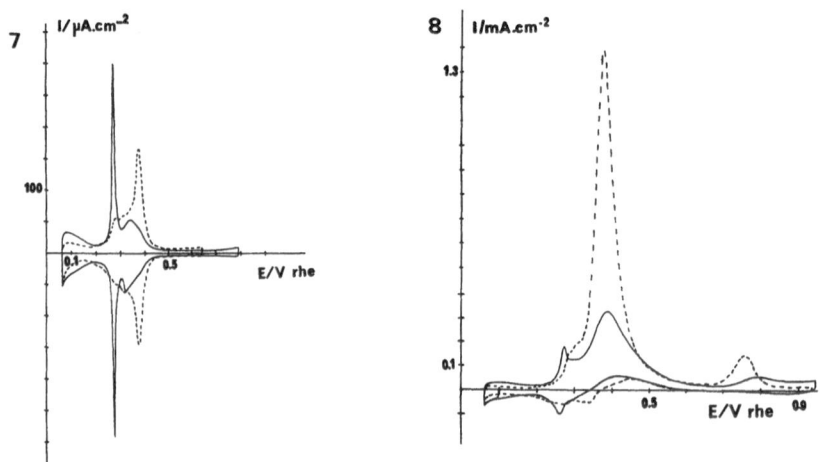

Fig. 7. Voltammogram of a Pt(610) after flame treatment. (- - -) Stabilized voltammogram of Pt(100) in the same conditions.-
Fig. 8. First voltammogram of 0.02 M glucose oxidation on Pt(610). (- - -) Idem on Pt(100).

REFERENCES

1.- J. Clavilier, R.Parsons, R. Durand, C. Lamy and J.M. Leger. J. Electroanal.Chem.,124 (1981)321.
2.- J. Clavilier. J. Electroanal. Chem.,236 (1987) 87.
3.- J.Clavilier and S.G.Sun. J. Electroanal. Chem., 199 (1986) 471
4.- S. Motoo and N.Furuya. Ber Bunsenges Phys.Chem., 91 (1987) 457.
5.- J.Clavilier, D.Armand, S.G.Sun and M.Petit. J.Electroanal. Chem., 205 (1986) 267.
6.- I.T.Bae, X. Xing, C.C. Liu and E. Yeager. J. Electroanal. Chem.,284 (1990) 335.

EFFECT OF SURFACE-CATALYZED NO FORMATION ON THERMAL LOAD IN HYPERSONICS

Luigi Guarino[*] and Claudio Bruno[†]

[*] Aermacchi, UTEV, Via Sanvito 80, 21100 Varese, Italy

[†] University of Rome, Dip. Meccanica, Via Eudossiana 18
00184 Roma, Italy

INTRODUCTION

A craft flying at Mach number M > 5-7 creates a bow shock dissociating air into a weakly ionized plasma composed of O_2, N_2, O, N, NO, NO^+, e^- for low earth orbit re-entry, and many more ionized species for larger M and denser air (see Figure 1).

The reactive plasma past the shock, at $T_\infty \sim$ 2000-7000 K, is convected

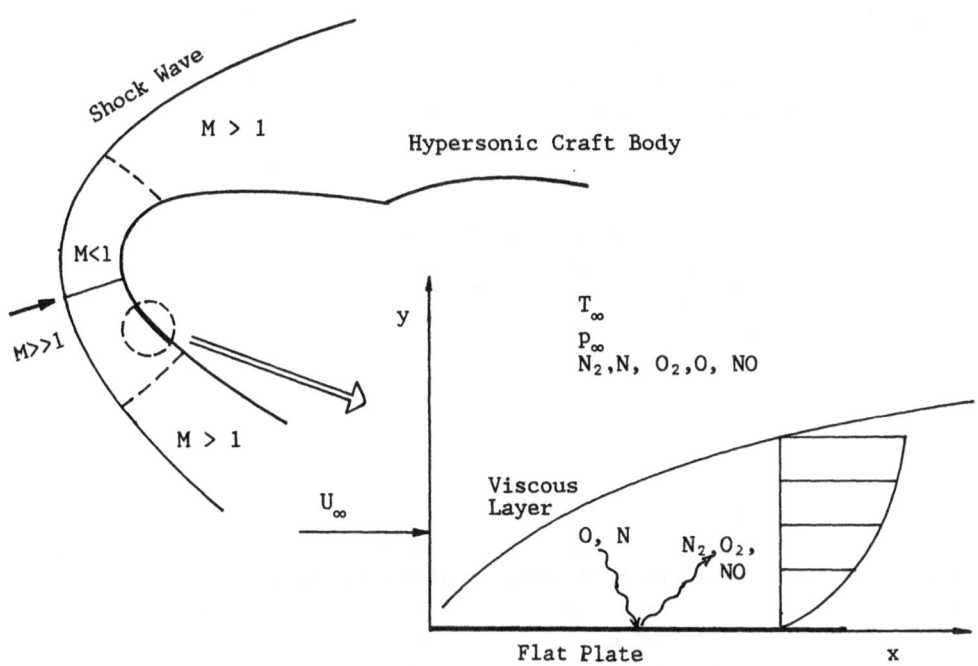

FIG. 1 Physical Problem and its modelization: Flat Plate.

Fundamental Aspects of Heterogeneous Catalysis Studied by Particle Beams,
Edited by H.H. Brongersma and R.A. van Santen, Plenum Press, New York, 1991

and diffuses toward the cooler craft surface (wall temperature T_w = 800 to 2000 K). Energy is transported to the surface by molecular conduction, species diffusion, and also deposited as recombination energy, since the surface catalyzes exothermic O, N recombinations and, possibly, N+O → NO reactions. Gas-phase reactions among all species are also very active.

Surface O, N recombinations are known[1] and their effect on heat flux during re-entry has been investigated[2]; discrepancies between predictions and inflight measurements have never been satisfactorily reconciled[2-3], suggesting that mechanisms other than "direct" O, N recombinations to O_2, N_2 may be present.

The purpose of this paper is to model and examine the effect of "cross" recombination between N and O leading to surface NO formation. Evidence for this exists[4-5], but no model for it has been proposed so far. A second goal is to raise interest in this area among experimenters dealing with heterogeneous kinetics, with the ultimate goal of achieving measuring capability for many active desorbing species associated with this problem.

THERMAL LOAD

O, N recombinations and NO formation are highly exothermic. In the continuum regime, the net energy flux q_w (thermal load) entering the craft surface may be approximated by[6]

$$q_w = K\partial T/\partial n - \sum_i h_i Y_i (\vec{V}_i + \vec{V}) \cdot \vec{n} + \vec{q}_r \cdot \vec{n} - \varepsilon_w \sigma T_w^4 \qquad (1)$$

where K is the thermal conductivity, $\vec{n}$ the normal to the surface, h_i, Y_i, $\vec{V}_i$ are the total enthalpy (including enthalpy of formation), mass fraction and diffusion velocity of the i-th species, and $\vec{V}$ is the bulk velocity; $\vec{q}_r$ is the heat flux from the radiating gas (important only at M > 20) and ε_w the surface emissivity. Surface reactions drive q_w through the $\vec{V}_i$, since the i-th species boundary condition at the surface is[6]:

$$Y_i (\vec{V}_i + \vec{V}) \cdot \vec{n} = w_i \qquad (2)$$

imposing the mass transfer due to diffusion and convection to be equal to the surface production w_i. Current thermal protection systems (=TPS) are coated with SiO_2 and nonporous, so that in (1-2) the $\vec{V} = 0$. In essence, q_w depends on T, Y_i (and $\vec{V}$) of the entire gasdynamic field, that must be found by solving the reactive Navier-Stokes equations.

AIR CHEMISTRY

Solving reactive Navier-Stokes equations requires a gas kinetics and its set of rates. High temperature exothermic reactions may yield diatomic species vibrationally excited. However, high temperature VV, VT transfer for O_2, N_2 are still not established to the extent that appropriate modeling can be cast in conventional Arrhenius kinetics form. A rule of thumb suggests ∿ 3500 K as the maximum gas temperature where these processes may be neglected in hypersonics. This limits current modeling, including the present, to M ∿ 14. In this range, Bortner's kinetic scheme and Arrhenius kinetics[7] (O_2, N_2 dissociations plus the Zel'dovich mechanism for NO) are adequate. Rate data were taken from Park[8].

SURFACE CATALYSIS

The w_i rates in (2) need modeling. Currently this is based on a Eley-Rideal mechanism with adsorption controlling[3]. Accordingly, w_O, w_N are modeled using first-order kinetics: e.g., $w_O = k_{wO} Y_O \rho$, where k_w = catalyticity is equal to $(\gamma/(\gamma-2))(2\pi W/RT_w)^{-\frac{1}{2}}$ and γ, the recombination probability[3] is from experiments[1-2] Simultaneous O,N recombinations, leading possibly to NO, were never measured[9]. The novel, Eley-Rideal-type, surface NO mechanism proposed here is:

$$N* + O \rightarrow NO + * \qquad \text{rate: } k_{w5,O} \equiv Z\, k_{wO} \qquad (3)$$
$$N + O* \rightarrow NO + * \qquad \text{rate: } k_{w5,N} \equiv Z'\, k_{wN} \qquad (4)$$

where Z, Z' need to be measured. Species may desorb excited (O_2 as singlet delta, N_2 vibrationally, as, presumably, NO); to keep the present model simple, however, thermal equilibrium with the surface was assumed.

EXAMPLE: FLAT PLATE IN HYPERSONIC FLOW

To illustrate the effects of (3-4) on q_w, the simple geometry of Fig. 1 was used. Conditions at infinity (x<0; y large) are those for an equilibrium shock layer within the subsonic bubble at M = 12 and $\sim$ 70 Km. As they are unknown, Z, Z' were varied between 0 and 1; since $Y_N \ll Y_O$, only Z affects results, a sample of which follows. Figure 2 compares normalized species profiles for Z = 0 and Z = 1. In both cases wall O, N go nearly to zero, and surface O_2 production is evident (Y_N is instead $\sim 10^{-5}$, and surface N_2 $\sim$ zero). Surface NO at Z = 1 is significant, affecting O_2 as well. Figure 3 shows the percent increase in recombination heat flux, q_d, as a function of Z. At lower T_w the increase due to Z $\neq$ 0 is substantial; at higher T_w, competition for O atoms to produce O_2 and NO affects also gas-phase profiles and reduces q_d (and total q_w). Complete results will be reported soon; the effect of surface structure (i.e., W and SiO_2) is currently being investigated by an analysis of absorption and desorption based on simplified surface potentials[10]. Ab initio calculations will follow in the near future.

CONCLUSIONS

The effect of "cross recombination" leading to surface NO desorption may be significant and affect total thermal load, besides NO production. Measurements of Z are needed to quantify NO rates under conditions typical of hypersonic flight (T_∞ = 2000-3500 K, T_w = 800-1500 K, U_∞ = 300-1000 m/s and SiO_2, borosilicate and Si_3N_4 surfaces).

Detection of desorbing NO, besides O_2, N_2, is challenging; however, heterogeneous production of active species and recombination are topics of great current interest. Among applications, besides that illustrated here, are: surface NO influence on combustion in future hypersonic engines; the interaction between surface-produced species and the O_3 layer; industrial plasma torches; and the production of high energy neutral beams.

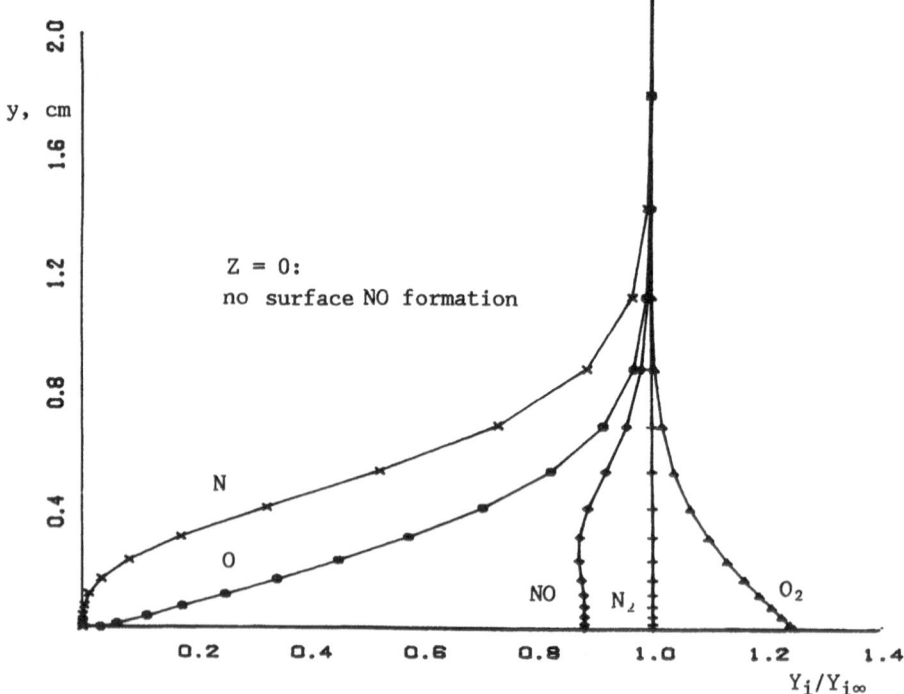

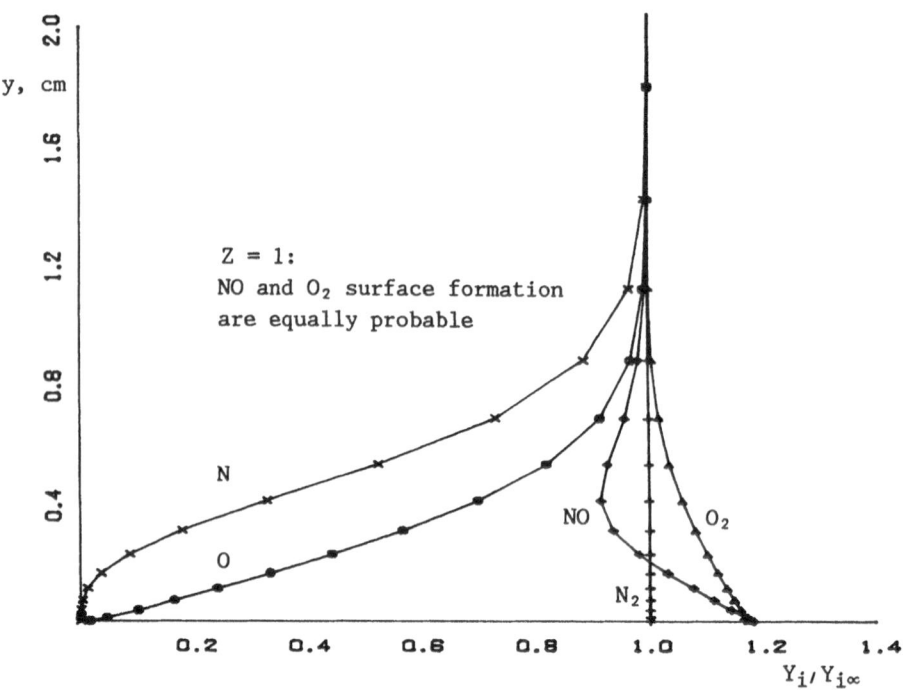

FIG. 2 Mass Fractions Profiles, Normalized with Respect to their
Values at Infinity, at x = 0.48 m from the Plate Leading
Edge. T_∞ = 3000 K, p_∞ = 0.49 atm, U_∞ = 350 m/s, T_w = 1500 K.

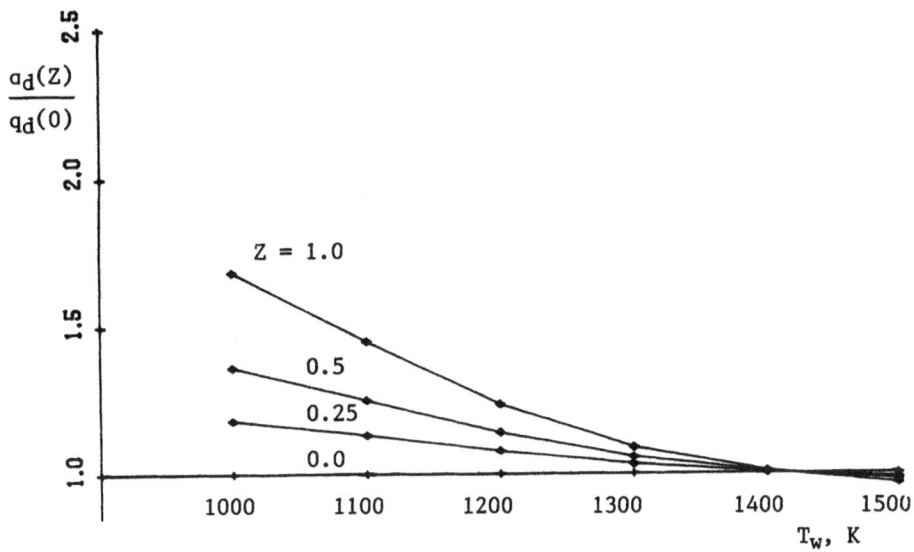

FIG. 3 Ratio Between Surface Recombination Energy Flux $q_d(Z)$
and $q_d(0)$ as a Function of Wall Temperature T_w.
T_∞ = 3000 K, p_∞ = 0.49 atm, U_∞ = 350 m/s.
Energy Flux Integrated Over the Entire Plate Length L=0.5 m.

REFERENCES

1. Scott, C.D., Catalytic Recombination of Nitrogen and Oxygen on High Temperature Reusable Surface Insulation, AIAA Paper 80-1477 (1980).
2. Curry, D.M., Rochelle, W.C., Chao, D.C. and Ting, P.C., Space Shuttle
   Orbiter Nose Cap Thermal Analysis, AIAA Paper 86-0388 (1986).
3. Scott, C.D., "Catalytica Effects", Lecture Notes, 2nd US-Europe Short
   Course in Hypersonics, USAF Academy, Colorado Springs (1989).
4. Ibberson, V.J., Plasma Chemical Reactions, High Temp-High Pressures,
   1:243 (1969).
5. Mutel, B.,Dessaux, O., and Goudmand, P., Energy cost improvement of the
   nitrogen oxides synthesis in a low pressure plasma, Rev. Phys.
   Appl., 19:461 (1984).
6. Bruno, C., Modeling Catalytic Recombination Heating at Hypersonic Speeds,
   AIAA Paper 89-0309 (1989).
7. Blottner, F.G., Viscous Shock Layer at the Stagnation Point with Nonequilibrium Air Chemistry, AIAA J., 8:193 (1969).
8. Park, C., "Nonequilibrium Hypersonic Aerothermodynamics", J.Wiley, New
   York (1990).
9. Scott, C.D., personal communications (1989).
10. Nasuti, F., Aeronautical Engineering Dissertation, Dept. of Mech. and
    Aeron. Eng., University of Rome, in progress (1990).

ELECTROCHEMICAL PROBING OF STEP AND TERRACE SITES ON Pt $\left[n(111)x(1\bar{1}1)\right]$ AND Pt$\left[n(111)x(100)\right]$

A. Rodes[*], K. El Achi, M.A. Zamakhchari and J. Clavilier

Laboratoire d'Electrochimie Interfaciale du CNRS
1, Place Aristide Briand, 92195 Meudon Cedex, France

[*]C.E.C. fellow. On leave of absence from the Department
of Physical Chemistry, University of Alicante, Spain

ABSTRACT

Stepped surfaces with (111) terraces may be divided in two classes: those with (100) steps and those with (111) steps. The hard-sphere model for these surfaces taken as a guide for the analysis of their hydrogen electrosorption voltammograms was proved to work correctly for the counting of step and terrace sites. Moreover it yields information about the atomic configuration of the step sites, i.e. the position of the ledge atomic row relative to the underlying terrace rows.

I. INTRODUCTION

The properties commonly used for characterization of the crystalline surface structure of a metal are screened when the latter is in contact with a liquid phase. This is the case of electrochemical systems for which until recently in situ information of this type was lacking.

The electrochemical way of approaching metal-liquid interfaces seems obvious for in situ surface chemical analysis. Crystalline surface structure information was thought for a long time to be out of the scope of electrochemistry (1). It is only recently that this approach was proved to be feasible (2-5). This requires stable metals in a wide range of physicochemical conditions, clean surfaces and availability of an electrochemical surface sensitive probe with a sufficient level of specificity. These requirements are difficult to concile together with a given electrode material.

Platinum is the most attractive candidate because of its wide spectrum of electrocatalytic properties. The electrochemical study of

Fundamental Aspects of Heterogeneous Catalysis Studied by Particle Beams,
Edited by H.H. Brongersma and R.A. van Santen, Plenum Press, New York, 1991

75

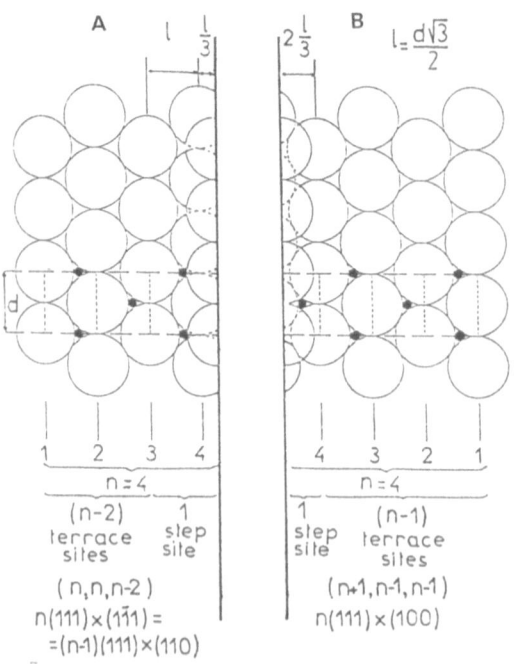

FIGS. 1A,B. Top view of the hard sphere model for (A) n(111)x(1$\bar{1}$1) and (B) n(111)x(100) stepped surfaces with adsorbed hydrogen (●).

platinum stepped surfaces is an example of how electrochemistry can be considered as a quantitative tool for in situ crystalline surface characterization by using hydrogen adsorption-desorption as surface sites probing species.

II. EXPERIMENTAL

Platinum single crystals : spherical single crystals with a diameter of 1.8 mm are molten from a platinum wire. They are oriented, cut and polished on an optical bench. Because the step density is sensitive to small angular deviations from the nominal orientation a technique has been designed avoiding this difficulty. Final angular accuracy is within 3' of the nominal orientation (4). Ordering and cleaning of the surfaces were performed as described in (6,7).

Electrochemical measurements : the reversible hydrogen adsorption-desorption process gives rise to the exchange of one electron per adsorbed hydrogen atom. Hydrogen coverage can be controlled by varying the potential of the electrode relative to the solution. The resulting current is then recorded. This is the basis of voltammetry when the potential is a linear function of time. This technique is sensitive enough for detection of a few per cent of a monolayer. Its application to a set of platinum stepped surfaces with step and terrace sites of given symmetries makes possible to check which one of the features in their voltammograms may be assigned to step and terrace sites respectively.

Pt n(111)x(1$\bar{1}$1) and Pt n(111)x(100). surfaces are investigated by the electrochemical method in sulphuric acid solutions.

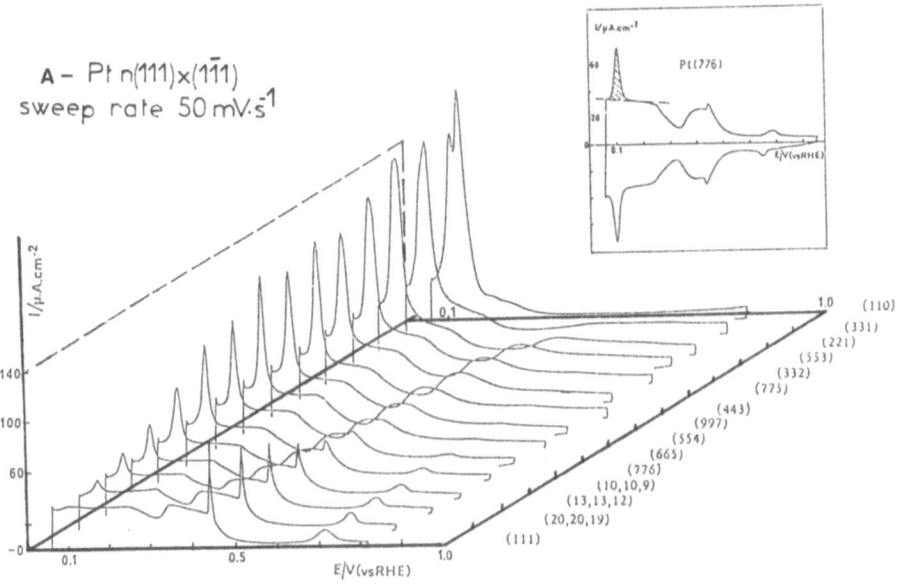

A – Pt n(111)x(1̄1̄1)
sweep rate 50 mV·s⁻¹

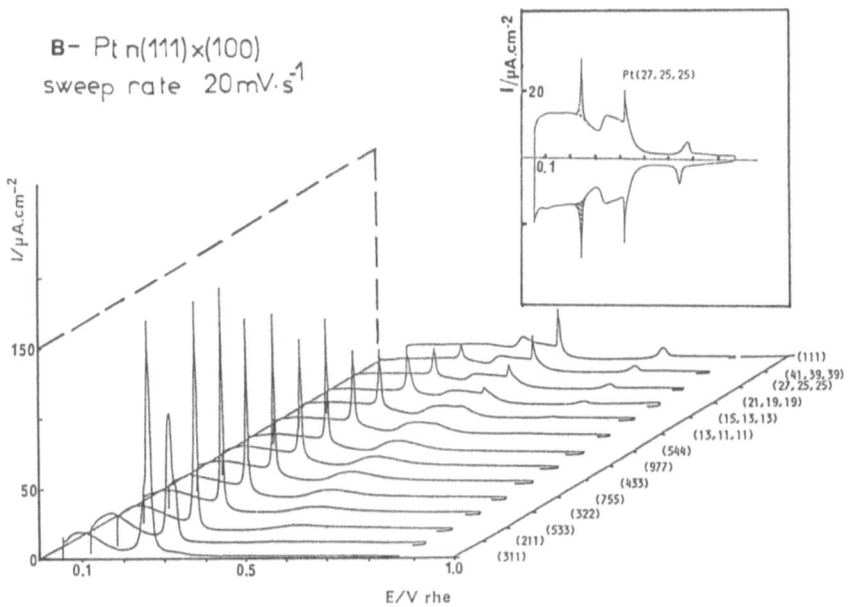

B – Pt n(111)x(100)
sweep rate 20 mV·s⁻¹

FIGS. 2A,B. Reversible electrodesorption of hydrogen from (A) Pt n(111)x(1̄1̄1) and (B) Pt n(111)x(100) in 0.5 M H_2SO_4.

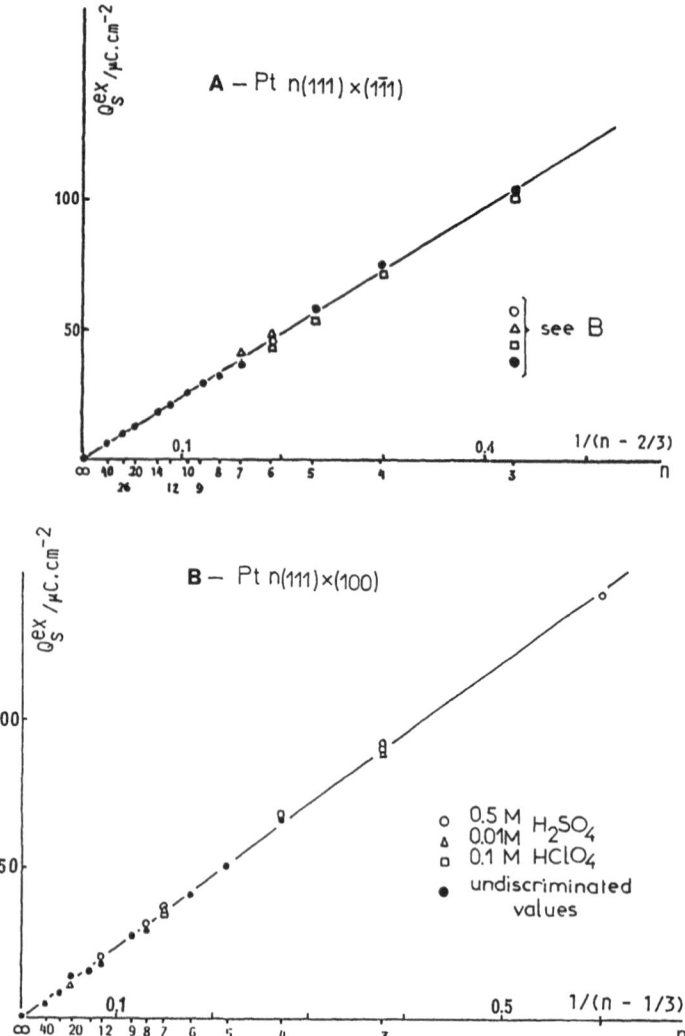

FIG.3A,B. Theoretical (solid line) and experimental amounts of adsorbed hydrogen on step sites, Q_s^{ex}.

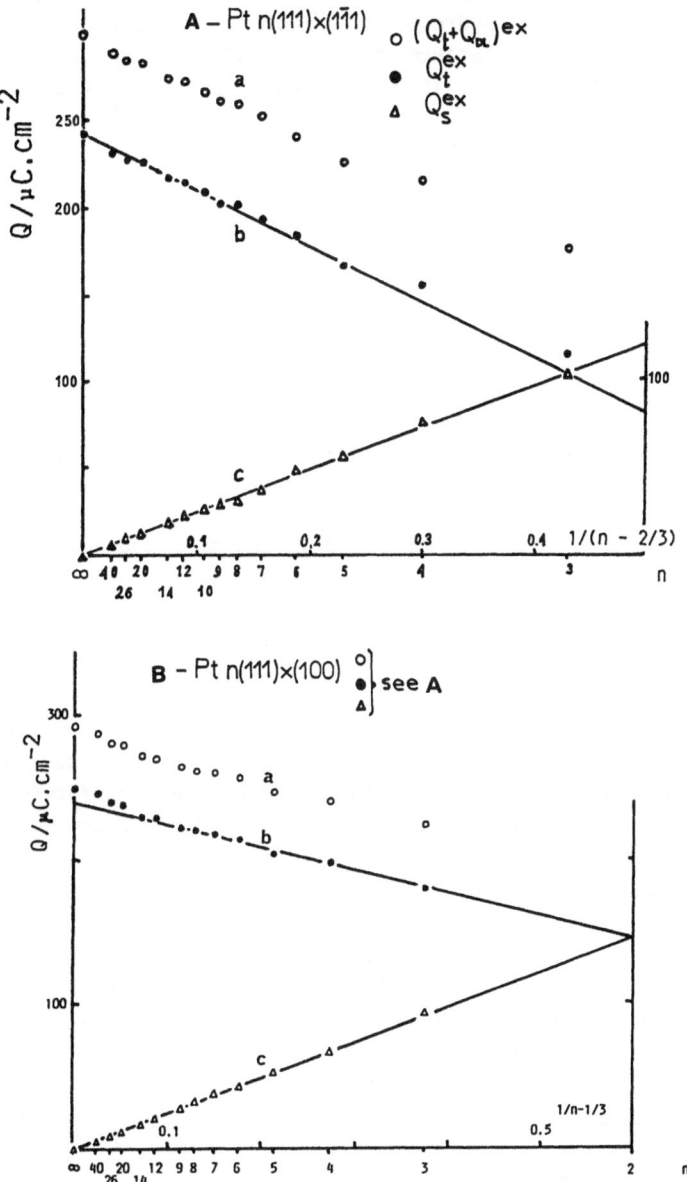

FIGS. 4A,B. Theoretical (solid lines) and experimental amounts of adsorbed hydrogen on terrace (curves a,b) and step (curve c) sites in 0.1 M H_2SO_4.

III. HARD SPHERE MODEL OF THE STEPPED SURFACES

The top views for the two types of stepped surfaces are represented in figs. 1A and 1B for (111) or (110) and (100) monoatomic steps. For both models the following quantities are considered : Q_s is the electric charge equivalent to the amount of hydrogen adsorbed on the step sites at saturation and Q_t that for hydrogen saturation of terrace sites, i.e. one hydrogen atom per surface platinum atom. The following relations hold for Pt n(111)x(1$\bar{1}$1) assuming the (110) step site symmetry (B_5 sites)

$$Q_s = (2e/\sqrt{3}\ d^2)/(n-2/3) \tag{1a}$$

$$Q_t = (2e/\sqrt{3}\ d^2) - (8e/3\sqrt{3}\ d^2)/(n-2/3) \tag{2a}$$

$$Q_t = (2e/\sqrt{3}\ d^2) - (4/3)\ Q_s \tag{3a}$$

while for Pt n(111)x(100)

$$Q_s = (2e/\sqrt{3}\ d^2)/(n-1/3) \tag{1b}$$

$$Q_t = (2e/\sqrt{3}\ d^2) - (4e/3\sqrt{3}\ d^2)/(n-1/3) \tag{2b}$$

$$Q_t = (2e/\sqrt{3}\ d^2) - (2/3)\ Q_s \tag{3b}$$

where e is the charge of the electron, d is the atomic diameter and n is the number of atomic rows belonging to the terrace.

IV. RESULTS

The voltammograms of the two sets of stepped surfaces are reported in fig. 2A and 2B. Features appearing when the step density is increased may be ascribed to step sites while those disappearing may be related to terrace sites. Peak potentials in fig.2A and 2B are specific of the (110) and (100) step sites symmetry, respectively. The hatched areas under the peaks in the insets show how Q_s is determined and separated from the terrace sites contribution, Q_t, which includes the charge for the double layer, Q_{DL}.

Fig.3A and 3B shows the variation of Q_s^{ex} as a function of $1/(n-2/3)$ or $1/(n-1/3)$. The solid lines correspond to eqns. 1a and 1b respectively. Q_s^{ex} is verified to be independent of the concentration and nature of the electrolyte as an intrinsic quantity does. Fig. 4A and 4B show the variation of $(Q_t+Q_{DL})^{ex}$, curve a, Q_t^{ex} (once subtracted the charge of the double layer), curve b, and Q_s^{ex}, curve c. The solid lines correspond to eqns. 2a and 1a (fig.4A) and 2b and 1b (fig.4B).

Fig. 5A and 5B represent the experimental charge for hydrogen adsorbed on terraces as a function of the experimental charge of hydrogen adsorbed on the step sites for all the electrolyte concentrations. In each graph a set of values has been reported for 0.5 M H_2SO_4 without correction of the double layer showing that this correction is nearly a constant

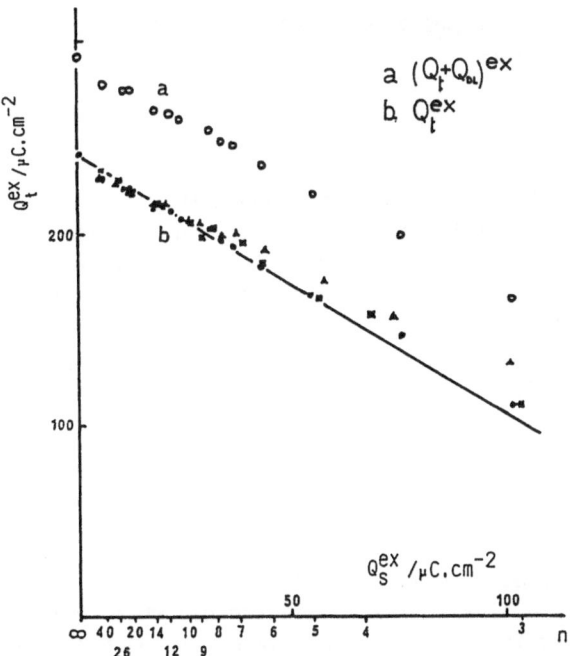

<u>FIG.5A.</u> Double layer uncorrected (curve a) and corrected (curve b) amount of adsorbed hydrogen on terrace sites as a function of Q_S^{ex} for Pt n(111)x(1$\bar{1}$1) at sulphuric acid concentrations ranging from 2.0 M to 0.01 M.

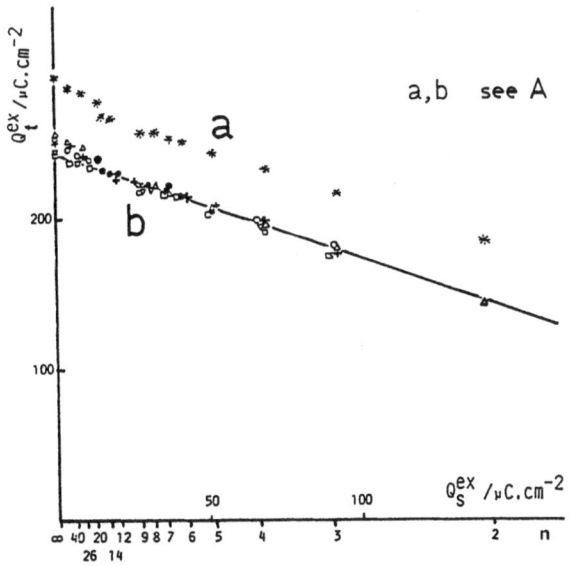

<u>FIG.5B.</u> Same quantities that in Fig.5A for Pt n(111)x(100).

term. Solid lines correspond to eqns. 3a and 3b. The slopes $-4/3$ and $-2/3$ are characteristic of the position of the ledge atoms relatively to those of the underlying terrace.

V. CONCLUSION

These experiments prove that electrochemistry is an efficient tool for a qualitative and quantitative determination of the type of sites present on a surface. The change in the fine structure of that part of the voltammogram relative to the terraces may be used as an indicator for the scaling of the (111) ordered domains (7). The resolution of the method is high enough to discriminate the difference between the positions of the ledge atoms forming (111) or (100) step sites which are only shifted by plus or minus one third of an atomic row.

VI. REFERENCES

1. E.YEAGER. Surface Science 101 (1980) 1
2. B.LOVE, K.SETO and J.LIPKOWSKI. J.Electroanal.Chem. 199 (1986) 219
3. R.ADZIC, A.TRIPKOVIC and V.VESOVIC. J.Electroanal.Chem. 204 (1986) 329
4. J.CLAVILIER,D.ARMAND,S.G.SUN and M.PETIT. J.Electroanal.Chem.205(1986)267
5. S.MOTOO and N.FURUYA. Ber.Bunsenges.Phys.Chem. 91 (1987) 457
6. J.CLAVILIER, K.EL ACHI, M.PETIT, A.RODES and M.A.ZAMAKHCHARI. J.Electroanal.Chem. (in press).
7. J.CLAVILIER, K.EL ACHI and A.RODES. Chem.Phys. 141 (1990) 1
8. A.RODES, K.EL ACHI, M.A.ZAMAKHCHARI and J.CLAVILIER. J.Electroanal.Chem. 284 (1990) 245.

QUANTUM CHEMISTRY OF SURFACE

CHEMICAL REACTIVITY

R.A. van Santen

Schuit Institute of Catalysis
University of Technology
P.O. Box 513
5600 MB EINDHOVEN
The Netherlands

Abstract

The quantum chemist's and surface physicist's view of the surface chemical bond are illustrated by means of chemisorption of CO as an example. Between adsorbate and metal surface bonding as well as antibonding orbital fragments are formed. For molecules the adsorption geometry is a sensitive function of the balance of the repulsive atop directing interaction, resulting from the occupation of antibonding orbital fragments, and the high coordination directing bonding interaction. The bonding contribution to the surface chemical bond energy relates to the surface group orbital local density of states around the Fermi level, as long as the orbital interactions are weak. The concepts of Pauli repulsion and group orbital LDOS can be used to provide a quantum-chemical basis to metal promotion.

Introduction

In recent years theoretical understanding of the surface chemical bond has significantly increased. Whereas a decade ago the quantum chemistry of organic molecule reactivity and also of organometallic chemistry had reached a more advanced level that that of surface reactivity this knowledge gap is rapidly closing. This is not only the result of the increasing interest of quantum chemists for problems on surface reactivity, but also due to the wealth of rich and detailed molecular and electronic information provided for by experimental surface scientific researchers. Two developments are important from a quantum-chemical point of view.The availability of semi-empirical methods resulting in a conceptual framework to analyse quantum-chemical calculations. Secondly the possibility to use first principle theoretical techniques as the local density approximation as well as other highly sophisticated ab-initio methods in the study of adsorption models of direct interest to the surface chemist. The aim of this paper is to present the essential

Fundamental Aspects of Heterogeneous Catalysis Studied by Particle Beams,
Edited by H.H. Brongersma and R.A. van Santen, Plenum Press, New York, 1991

83

features of the electronic basis of surface chemical reactivity. The theoretical-chemical application of surface chemical bonding theory that we will highlight is related to formal chemisorption theory as developed in surface physics, but focusses rather on typical quantum-chemical concepts as the electron occupation of bonding and antibonding orbital fragments than on properties as the local density of states or surface states well known to the surface physicist. We will see that both approaches complement each other. The local density of states concept stemming from surface physics is also indispensable to the surface chemist. The idea of a surface molecule relates to the surface physicists concept of surface states.

We will start with a quantum chemists view of chemisorption of CO to a transition metal surfaces. The surface chemical bond of CO is theoretically as well as experimentally very well understood. Of extreme importance has been Blyholder's (1) view of the surface chemical bond. It is analogous to the Frontier Orbital (2) concept widely used by chemists. Bonding is considered to be the result of the interaction of Highest Occupied Molecular (HOMO) Orbitals in one of the interacting fragments and the Lowest Unoccupied Molecular Orbitals (LUMO) on the other fragment and visa versa. On a surface this results in a sum of two terms. A contribution due to the interaction between occupied adsorbate orbitals and empty metalsurface orbitals (donation) and a second contribution due to the interaction of occupied metal surface orbitals and empty adsorbate orbitals (backdonation). Whereas this provides a satisfactory description for the attractive part of the interaction potential, it does not provide a prescription to estimate the repulsive part of the interaction potential. The coordination of molecules to surfaces cannot be understood without a proper understanding of the latter. More recent developments, mainly based on extension of the Extended-Hückel method but verified with first principle calculations, provide a consistent view of the surface chemical bond incorporating Pauli repulsion, useful to the surface chemical theorist (3). This leads to a significant modification of the original Blyholder approach. In the next section the theoretical basis will be presented and applied to a few other examples in the final section. In the third section the Newns-Anderson (4) approach to chemisorption will be introduced. This represents the surface physicist's approach to the surface chemical bond.

The framework provided for by formal chemisorption theory can be readily shown to be very similar to the theoretical chemical ideas discussed above. It will appear that chemisorption of molecules to transition metals has to be considered intermediate between the so-called weak adsorption and surface molecule strong adsorption limits, As in chemical bonding theory also formal chemisorption theory can be shown to be consistent with the formation of bonding as well as antibonding adsorbate-surface fragment orbitals.The electron occupation of these bonding and antibonding fragment orbitals determines the adsorbate-surface bond strength. The broadening of the adsorbate orbitals will be found to relate to the local density of states of the metal surface orbitals close to the maximum of the adsorbate orbital density. Extending HOMO-LUMO theory to metalsurfaces local density of states around the Fermi level (5), the theoretical results of this section will be illustrated by a discussion of O_2 chemisorption to the silversurface (6) and the CO bond strength on different transition metalsurfaces (7).
In the final section the concepts presented in the earlier two sections are applied to changes in chemical reactivity due to coadsorption of promoters. It will appear that some promoters change the attractive part of the interaction potential, whereas others reduce Pauli repulsion, that contributes to the repulsive part of the adsorbate-surface interaction potential. This will be illustrated by discussing the changes in CO chemisorption to a metal surface or metalparticles due to the presence of coadsorbed cations. It also provides an opportunity to discuss the importance of local electrostatic effects and

polarization to the surface chemical bond. Coadsorption of sulfur primarily effects the attractive part of the interaction energy. Finally the electronic basis of the promoting action of coadsorbed subsurface oxygen or chloride on the ethylene epoxidation rate catalysed by silver will be discussed (8). In this example the coadsorbate changes a repulsive interaction into an attractive one.

Molecular orbital theory of CO chemisorption

The interaction between the molecular orbitals of CO and those of the surface will lead to new molecular orbitals, that can be decomposed into bonding and antibonding orbitals with respect to the originally undisturbed fragment orbitals. In quantum chemistry the bonding or antibonding nature of orbital fragments is deduced from the bond-order overlap population density. Let the free adsorbate orbitals be given by φ_i and the metalsurface orbitals be given by ϕ_k, then the molecular orbitals Ψ_λ of the interacting system can be written as:

$$\Psi_\lambda = \sum_i c_i^\lambda \varphi_i + \sum_k c_k^\lambda \phi_k \tag{1a}$$

$$= \sum_i c_i^\lambda \varphi_i + \sum_{k,j} c_k^\lambda d_j^k \chi_j^m \tag{1b}$$

$$= \sum_i c_i^\lambda \varphi_i + \sum_j c_j^{-\lambda} \chi_j^m \tag{1c}$$

Orbitals χ_j^m are the atomic orbitals localized on the metal atoms. It has been assumed that a metalorbital can be written as a linear combination of metal atomic orbitals. In the case of a transition metal elementary molecular orbital theory would choose the metalvalence d-, s- and p-atomic orbitals to form the basisset from which the metal orbitals are composed. Of interest are the bond-order overlap population densities (BOOPD) between adsorbate molecular orbitals φ_i and surface metal atomic orbitals on the surface metal atoms that coordinate with the adsorbate:

$$\pi_{ij}(E) = Re \sum_\lambda c_i^{\lambda*} c_j^\lambda S_{ij} \delta(E - E_\lambda) \tag{2a}$$

$$\text{with}: \quad S_{ij} = \langle \varphi_i \,|\, \chi_j^m \rangle \tag{2b}$$

S_{ij} is the overlap of an adsorbate orbital with a surface metal atomic orbital. Figures 1 and 2 show π_{ij}'s computed according to the Extended Hückel method for a CO molecule atop adsorbed to a 29 atom cluster of Rh atoms simulating the Rh (111) surface. For details we refer to (7). Figures 1 present the BOOPD $\pi_{5\sigma,d_{z^2}}$, $\pi_{5\sigma,s}$ and $\pi_{5\sigma,p}$. These are the only non-zero BOOPD's of s symmetry. Figures 2 present the corresponding non-zero type BOOPD's, $\pi_{2\pi^*,d_{xz}}$. In CO the 5σ orbital, pointing away from the carbon atom to the surface atom is the CO HOMO. The two degenerate $2\pi^*$ orbitals are the CO LUMO's. In the figures E_F denotes the position of the HOMO of the combined CO, cluster system. One notes that at low energies the values of π_{ij} are always positive, but may become negative if the energy increases. The covalent contribution to the bond energy has the form:

$$E_{cov} = 2 \sum_k N_k \sum_{i<j} Re\, c_i^{k*} c_j^k H_{ij} \tag{3}$$

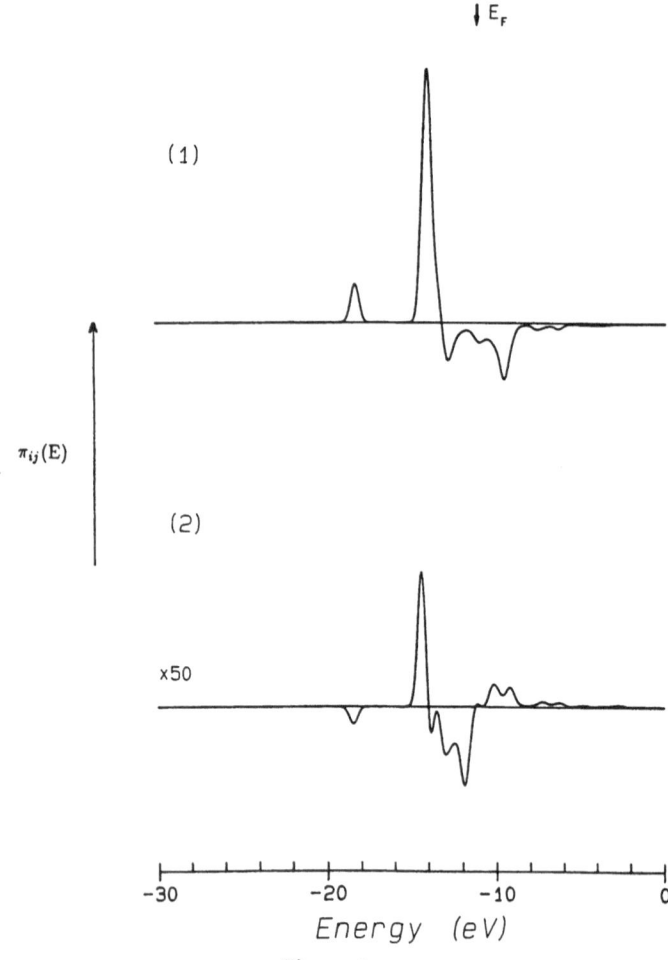

Figure 1a

σ-symmetry bond-order overlap population densities of atop coordinated CO with surface metal orbital on Rh (111), $\pi_{5\sigma,d_z}$.

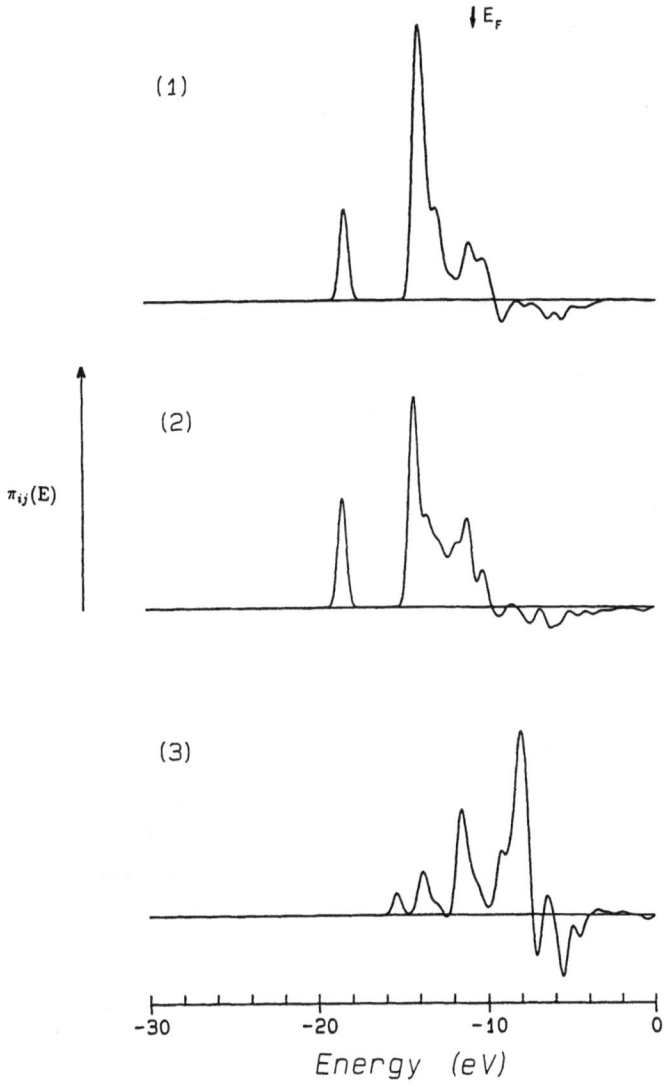

Figure 1b

σ-symmetry bond-order overlap population densities of atop coordinated CO with surface metal orbital on Rh (111), $\pi_{5\sigma,s}$.

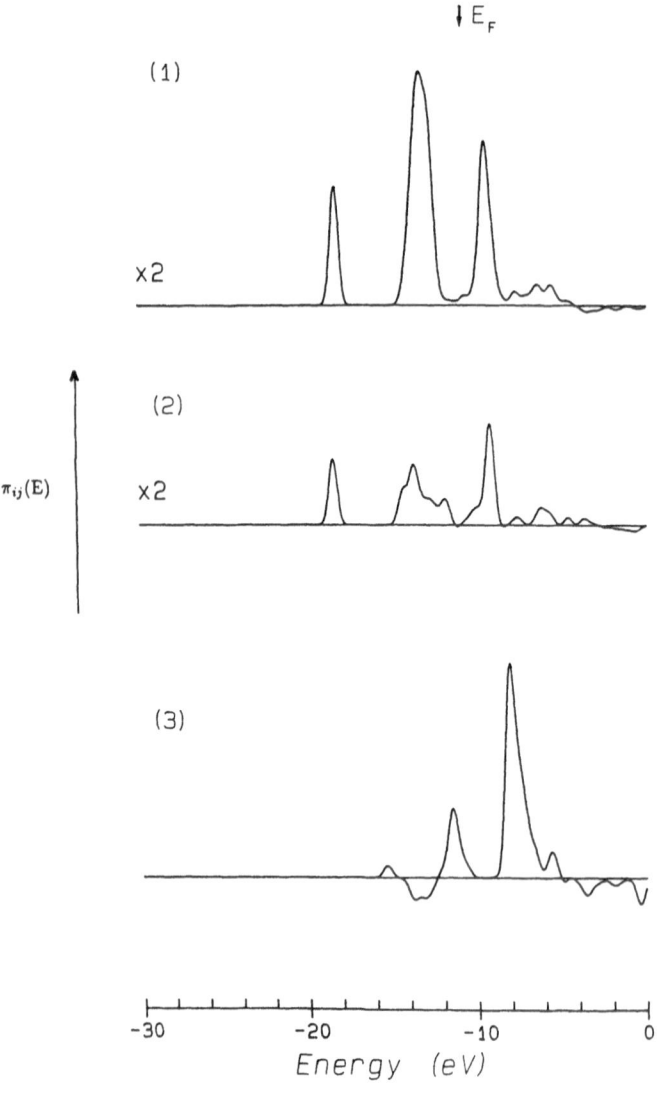

Figure 1c

σ-symmetry bond-order overlap population densities of atop coordinated CO with surface metal orbital on Rh (111), $\pi_{5\sigma,p_z}$.

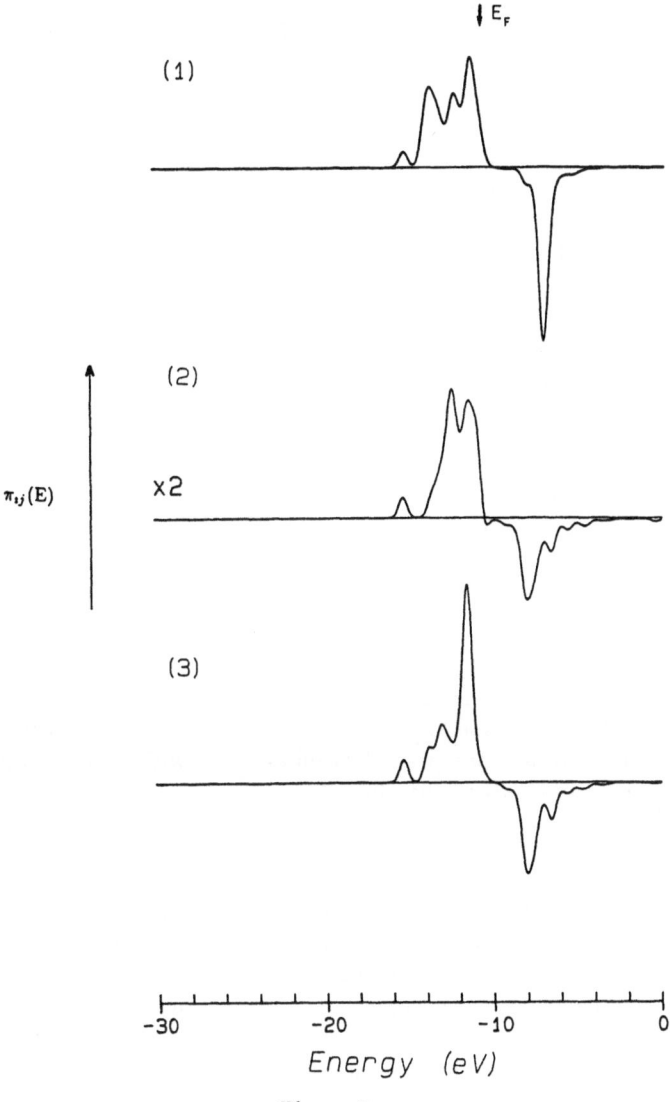

Figure 2

$2\pi^* - d_{xz}$ bond-order overlap population density of atop coordinated CO with d_{xz} metal orbital on Rh (111).

Table 1a. Surface atomic orbital gross populations: 1-fold adsorption of CO on Rh(111).

	orbital	1-fold		
		(111)	(100)	(110)
before adsorption	s	0.293	0.266	0.301
after adsorption		0.518	0.498	0.529
before adsorption	p_x	0.162	0.134	0.102
after adsorption		0.176	0.151	0.127
before adsorption	p_z	0.113	0.111	0.096
after adsorption		0.383	0.387	0.400
before adsorption	$d_{x^2-y^2}$	0.802	0.825	0.854
after adsorption		0.802	0.825	0.862
before adsorption	d_{z^2}	0.922	0.911	0.875
after adsorption		0.667	0.649	0.676
before adsorption	d_{xz}	0.891	0.843	0.923
after adsorption		0.890	0.852	0.909

N_k is the electron occupancy of orbital k and H_{ij} the non-diagonal matrix element of the interaction Hamiltonian. Because of the approximate relationship:

$$H_{ij} \sim S_{ij} \tag{4}$$

a positive value of $\pi_{ij}(E)$ corresponds to a bonding contribution to the bond strength and a negative value of $\pi_{ij}(E)$ an antibonding bond weakening contribution. Note that only for the $5\sigma, d_{z^2}$ interaction antibonding orbital fragments become occupied by electrons. Low lying antibonding orbital fragments are present in $2\pi^*, d_{xz}$ orbital fragments, but they are not occupied. The interaction with s and p valence atomic orbitals is also found to be always bonding.

If the d-valence electron occupation of the transition metal varies the contribution of antibonding $5\sigma, d_{z^2}$ orbital fragments will change. An increase in d-valence electron occupation will weaken the $5\sigma, d_{z^2}$ interaction, a decrease will strengthen it until also bonding orbitals become depopulated. Using metal Rh parameters figure 3 shows that changes of this interaction dominate the computed alteration in bond strength of CO with d-valence electron occupancy. If the Fermi level is chosen such that all antibonding $5\sigma, d_{z^2}$ orbital fragments become occupied, this interaction becomes repulsive. This is for instance the case for the IB metals Cu, Ag and Au. For a spherical symmetric charge distribution this repulsive interaction E_{rep} is proportional to the number of neighbour atoms n (see also 21c):

$$E_{ij}^{rep} \approx n\, S_{ij}^2 \tag{5}$$

Of course depletion of the antibonding orbitalfragments decreases the repulsive interaction and below some orbital occupation it is converted into an attractive interaction. The antibonding orbital fragments of CO coordinated atop deplete more rapidly than the antibonding $5\sigma, d_{z^2}$ orbital fragments formed between CO and metal surface orbitals when chemisorbed in high coordination sites (3). Therefore when the metal d-valence electron occupancy decreases initially the $5\sigma, d_{z^2}$ interaction for CO atop coordinated atop becomes more attractive than that in the higher coordination site.

Table 1b. Surface atomic orbital gross populations: 2-fold adsorption of CO on Rh(111).

		2-fold	
	orbital	σ	π
before adsorption	s	0.327	0.270
after adsorption		0.592	0.317
before adsorption	p_x	0.163	0.167
after adsorption		0.303	0.171
before adsorption	p_y	0.154	0.177
after adsorption		0.171	0.177
before adsorption	p_z	0.115	0.113
after adsorption		0.249	0.173
before adsorption	$d_{x^2-y^2}$	0.939	0.647
after adsorption		0.886	0.643
before adsorption	d_{z^2}	0.940	0.885
after adsorption		0.931	0.870
before adsorption	d_{xy}	0.863	0.720
after adsorption		0.854	0.717
before adsorption	d_{xz}	0.948	0.804
after adsorption		0.787	0.820
before adsorption	d_{yz}	0.900	0.861
after adsorption		0.889	0.857

This changes when the d-valence electron occupation decreases further, when also bonding $5\sigma, d_{z^2}$ orbital fragments become depleted. Then the attractive interaction favouring high coordination-sites starts to dominate. These changes are not easy to verify for d orbitals, because with a change in coordination there is also a large change in the angular dependence of the overlap, so that the interaction with other d atomic orbitals is taken over from that with the d_{z^2} orbitals. This is illustrated in tables 1a and 1b The preference for low coordination when antibonding orbital fragments become initially depleted and the shift to preference of higher coordination with further increase of electron occupancy is due to the fact that in high coordination sites interactions occur with different orbital fragments than in low coordination sites. A σ type orbital will in a high coordination site interact with σ symmetric linear combination of the metal atomic orbitals. This is the group-orbital $\phi_g(n)$, n is the number of surface atoms. When one computes the local density of states ρ_g:

$$\rho_g^{\sigma,m}(E) = \sum_k |\langle \phi_g^{\sigma,m}(n) | \phi_k \rangle|^2 \, \delta(E - E_k) \qquad (6)$$

one finds that its average electron density shifts to lower energy when n increases (3,9), (see also figure 10). As a consequence at the edge of the electron-density contribution antibonding orbital-fragments have usually a higher density coordinated in the atop position than in higher coordination sites. At lower fragment orbital occupation higher coordination sites are usually favoured because the bonding orbital fragments have the larger electron density in the bottom of the electron-density contribution in high condination sites.

For symmetric atomic orbital electron densities:

$$H_{ij}^g \sim \sqrt{n}\, \beta_{ij}' \tag{7}$$

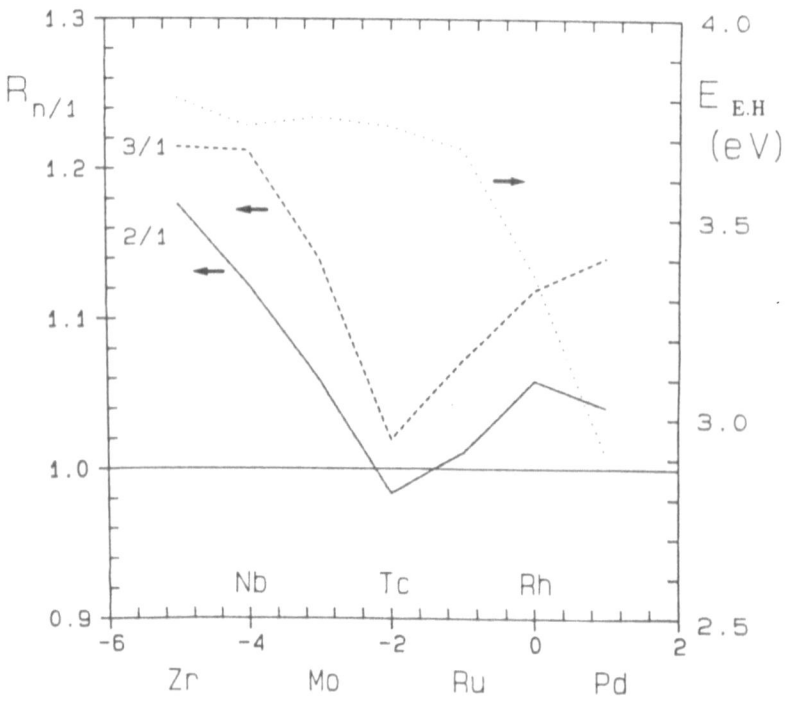

Figure 3

Extended Hückel part of the chemisorptive bond strength ($E_{E.H}$) of atop adsorbed CO on Rh (111) as a function of the occupation of the metal valence electron (dashed line), and the ratio $R_{2/1}$ ($R_{3/2}$) of $E_{E.H}$ of CO adsorbed 2-fold (3-fold) to CO adsorbed on Rh (111) as a function of the occupation of the metal valence electron band (solid line). The elements correspond to the total number of valence electrons according to the periodic system.

H_{ij}^g is the interaction Hamiltonian matrix element between adsorbate orbital i and a group orbital consisting of atomic orbitals of type j. β_{ij}' is the corresponding interaction matrix element for atop adsorption. Substition of (7) into (3) results in an additional contribution to covalent stabilization for high coordination sites. The initially decreasing ratio of threefold versus onefold interaction energy of CO shown in fig. 3 with increasing d-valence electron occupation agrees with this analysis. No minimum in this ratio would have been found, if the interaction with the CO $2\pi^*$ orbitals did not change.

As follows from figure 2 only bonding $2\pi^*$, d_{xz} fragment orbitals are occupied. This usually favours coordination to high coordination sites. In higher coordination sites the $2\pi^*$ will interact with metalsurface group orbitals of π symmetry. Again also an approximate relation as (8) is found. Co the interaction with the $2\pi^*$ orbital will not only be attractive for any d-valence electron occupation, but also favours high coordination sites. The minimum in the ratio of threefold versus atop bond energies derives from the decreasing attractive contribution due to orbitals of σ symmetry. With the particular parameters used at high d-valence electron occupation the interaction with orbitals of π symmetry dominates the differences in adsorption site energies. Interestingly CO prefers atop coordination to Rh and Co but higher coordination to Pd and Ni. Apparently on Pd and Ni the interaction with the $2\pi^*$ orbitals dominates. On Pt CO prefers atop adsorption. Because of the higher workfunction of Pt and the large d-orbital extension the balance of the σ and π type interaction remains in favour of the σ-type interaction.

As a result CO prefers atop coordination to a Pt metal surface. As has been shown elsewhere and will discussed shortly in the next section the BOOPD analysis can also be usefully applied to analyse the results of first principle calculations.

Formal chemisorption theory

Chemical bonding theory has to compute BOOPD's and study the occupation of bonding and antibonding orbital fragments to predict adsorption geometries. BOOPD's are experimentally not directly accessible, but useful information to the spectroscopist is provided by the computation of adsorbate orbital local density of states (LDOS), analogous as defined for the metal grouporbitals in (6):

$$\rho_i(E) = \sum_{\lambda} |\langle \varphi_i | \Psi_\lambda \rangle|^2 \, \delta(E - E_\lambda) \tag{8}$$

Note that the group orbital LDOS defined in (6) is calculated by projecting on undisturbed surface metal orbitals, whereas the adsorbate LDOS defined according to (8) is projected on the molecular orbitals of the interacting system.

A major result of formal chemisorption theory (4) is **a prediction of the dependence of** $\rho_i(E)$ on E:

$$\rho_i(E) = \frac{1}{\pi} \frac{\Gamma_i(E)}{(\alpha_i + \Lambda(E) - E)^2 + \Gamma_i^2(E)} \tag{9a}$$

α_i is the energy of the adsorbate level in the absence of the interaction with the metal surface, $\Lambda_i(E)$ is the level shift function and $\Gamma_i(E)$ the level width function. Explicit expression for $\Lambda_j(E)$ and $\Gamma_i(E)$ are:

$$\Gamma_i(E) = \pi \sum_{k} |\langle \varphi_i | H | \phi_k \rangle|^2 \, \delta(E - E_k) \tag{9b}$$

$$\Lambda_i(E) = \oint dE' \frac{\Gamma_i(E')}{E' - E} \tag{9c}$$

$\oint$ denotes the principal part of the integral.

Expression (9a) is characteristic for the resonant interaction of a discrete level with orbitals that have a contineous energyspectrum (10,11). If $\Gamma_i(\alpha_i) \ll |E_{max} - E_{min}|$, E_{max} and E_{min} being the upper and lower bound of the energyspectrum of orbitals ϕ_k, the interaction between adsorbate and surface metal orbitals can be considered weak. Then $\Lambda_i(E)$ and $\Gamma_i(E)$ can be assumed to be energy independent.

Within this approximation the energy distribution of the adsorbate level is Lorentzian. The interaction of the 4σ orbital is representative for this situation. One observes in figure 4 the very small level shift of the 4σ CO orbital compared to the corresponding energy in the non interacting molecule. The distribution is symmetrical around its maximum as predicted by eq. (8). The orbitals below $\alpha_i + \Lambda_i$ are bonding orbitals, the orbitals above $\alpha_i + \Lambda_i$ are antibonding orbitals. The difference in energy between the maximum of the bonding orbital fragments and antibonding orbital fragments is $\sim \Gamma(\alpha_i + \Lambda_i)$. Since bonding as well as antibonding orbital fragments are occupied the overall 4σ-metalsurface interaction will be repulsive. The electron occupancy of the 4σ orbital does not change.

Table 2.   Gross population Extended-Hückel of CO adsorbed on Rh surfaces.

orbital	adsorbate	atop			bridge	3-fold	step
		(111)	(100)	(110)	(111)	(111)	(111)
1π	ads.	0.998	0.998	0.998	0.996	0.996	0.932
	free	1.0	1.0	1.0	1.0	1.0	1.0
5σ	ads.	0.884	0.882	0.895	0.873	0.873	0.840
	free	1.0	1.0	1.0	1.0	1.0	1.0
$2\pi^*$	ads.	0.136	0.125	0.135	0.228[1]	0.261	0.368
	free	0.0	0.0	0.0	0.0	0.0	0.0

1): average value of $2\pi^*$ orbitals with (occupation of 0.278) and with π_y (0.178) symmetry.

As is shown in table 2 this is different for the CO 5σ and $2\pi^*$ orbitals. As follows from figures 1 and 2 the relative position of these orbitals with respect to the Fermi level are such that upon interaction the antibonding 5σ-surface metal orbital fragments are pushed above the Fermi level, resulting in a depletion of 5σ electron occupancy. The bonding $2\pi^*$-surface metal orbital fragments are pulled below the Fermi level. Figures 5 and 6 show the corresponding local densities of states. Wheres in the atop position the 5σ LDOS remains nearly symmetrical, a broadening towards higher energies is seen in higher coordination sites. The deviation from the Lorentzian line shape is larger for the $2\pi^*$ LDOS. It implies that the weak adsorption assumption according to which the energy dependence of $\Lambda_i(E)$ and $\Gamma_i(E)$ is ignored is not valid any more. Using the definition of the surface metal grouporbital density (5,6), (9a) can be rewritten as:

$$\Gamma_i(E) = \pi \sum_g |\langle \varphi_i | H' | \phi_g \rangle|^2 \, \rho_g^m(E) \tag{10}$$

The energy dependence of $\Gamma(E)$ is related to the metal metal surface group orbital local density of states, that interacts with adsorbate orbitals φ_i. For a transition metal the group orbitals ϕ_g can be considered as a particular linear combination of d, s and p valence atomic orbitals of the same symmetry as orbital φ_i. $\rho_g^m(E)$ has only a finite value as long as $E_{max}^g > E > E_{min}^g$. The boundaries of the energy spectrum of each metal surface group orbital are not necessarily the same. The d-valence electron orbitals have a relatively narrow bandwidth, whereas the s, p valence electron band is very broad. For a metal lattice with one s-atomic orbital per metal atom $\rho_g^m(E)$ has the general form:

$$\rho_g^m(E) = \frac{1}{\sqrt{\bar{n}}\,|\beta|}\, f_g^m(E) \tag{11}$$

$\bar{n}$ is the number of surface metal atom nearest neighbour metalatoms and β the overlap energy integral between two metal nearest neighbour atoms. With (7) and (11), $\Gamma_i(E)$ can written in the normalized form:

$$\Gamma_i(E) = \pi \sum_j \frac{n_{ij}\beta'^2_{ij}}{\sqrt{\bar{n}_j}\,|\beta_j|} \cdot f_{g(j)}^m(E) \tag{12a}$$

$$= \pi \sum_j \mu_{ij}\,\sqrt{n_{ij}}\,\beta'_{ij}\,f_{g(j)}^m(E) \tag{12b}$$

μ_{ij} is a measure of the interaction strength.

$$\mu_{ij} = \frac{\sqrt{n_{ij}}\,\beta'_{ij}}{\sqrt{\bar{n}_j}\,\beta_j} \tag{12c}$$

With (12b), expression (8) is rewritten as:

$$\rho_i(E) = \frac{\sum_j \mu_{ij}\,\sqrt{n_{ij}}\,|\beta'_{ij}|f_{g(j)}^m(E)}{(\alpha_i + \Lambda_i(E) - E)^2 + \left[\sum_j \mu_{ij}\sqrt{n_{ij}}\,\beta'_{ij}\,f_{g(j)}^m(E)\right]^2} \tag{13}$$

We will analyse the general behaviour of (13) as a function of coupling parameter μ_{ij}, restricting ourselves initially to the case of one adsorbate orbital and a metal lattice consisting of one s-atomic orbital per metal atom. The general behaviour of (13) is sketched in figure 7. $\rho_i(E)$ is expected to have in general several maxima. One maximum corresponds to the value of E, that satisfies:

$$\alpha_i + \Lambda_i(E) - E = 0 \tag{14}$$

Its solution is given by $\bar{\alpha}_i$, the upwards or downwards shifted adsorbate level. The direction of the shift depends whether it corresponds to a bonding or antibonding orbital fragment of the interacting system. In figure (7) it has been chosen to be a bonding level. The other maxima derive from $f_{g(j)}^m(E)$. In figure 7 it has been assumed that the valence electron band has one maximum. For a general discussion we refer to (3). When $\mu \ll 1$, ρ_i is Lorentzian around $\bar{\alpha}_i$ with half width $2\Gamma_i(\bar{\alpha}_i)$. With increasing μ the energy distribution becomes asymmetric and a second maximum at $\bar{\alpha}_m$ appears. At a small value of μ, the weak chemisorption limit fragment orbital levels below $\bar{\alpha}_i$ are bonding and above $\bar{\alpha}_i$ are antibonding. We recognize in figures 5 and 6 the increasing asymmetry and the appearance of a separate antibonding and antibonding density when the coordinationnumber with the surface metal atoms increases. As discussed in the previous section the bond strength contribution to the adsorbate-surface chemical bond depends on the distribution of the electrons over bonding and antibonding adsorbate metalsurface fragment orbitals. In formal chemisorption theory the expression for the covalent contribution to the bond strength is given by (4,9):

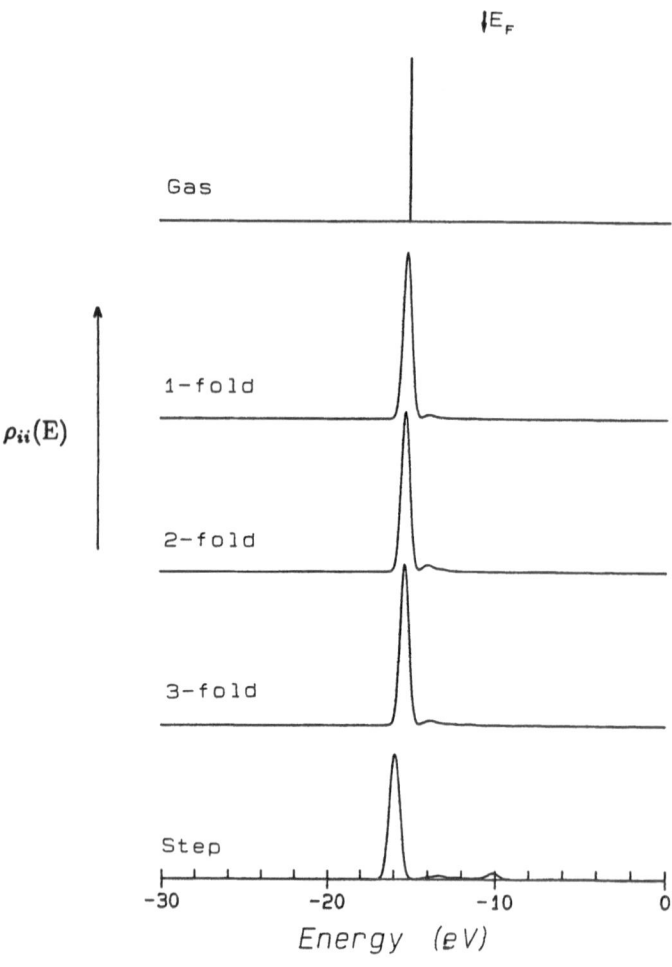

Figure 4

LDOS of the 1π CO molecular orbital in the gas phase, of CO atop adsorbed on Rh (111), of CO 2-fold adsorbed on Rh (111), 3-fold adsorbed on Rh (111) (distance of CO to step 1.551Å). The Fermi level is indicated by E_F.

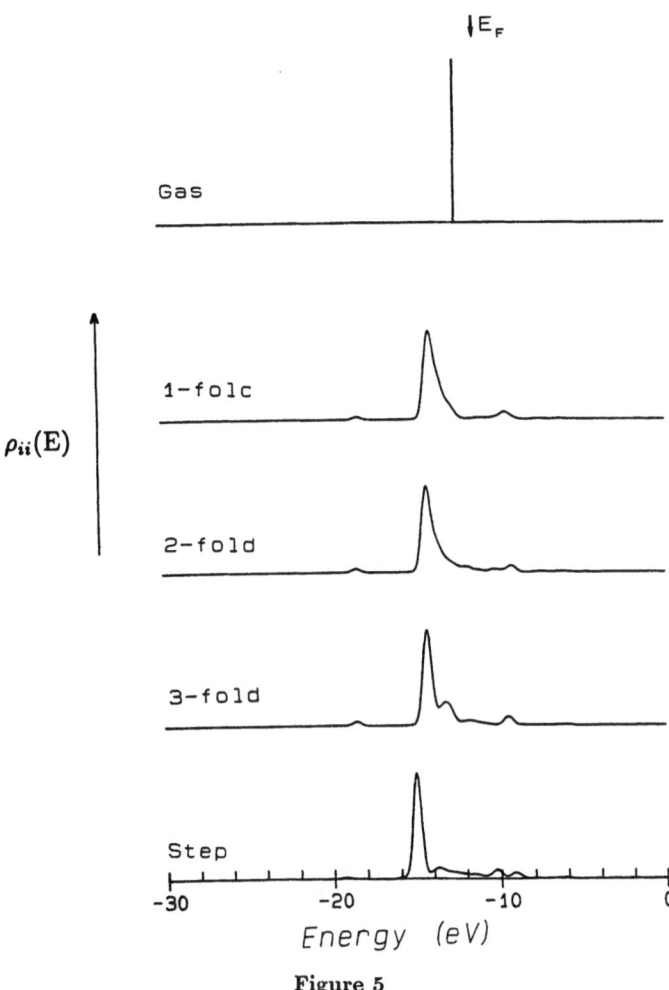

Figure 5

LDOS of the 5σ CO molecular orbital in the gas phase of CO atop adsorbed on Rh (111), of CO 2-fold adsorbed on Rh (111), 3-fold adsorbed on Rh (111) and 3-fold adsorbed on stepped Rh (111) (distance of CO to step 1.551Å). The Fermi level is indicated by E_F.

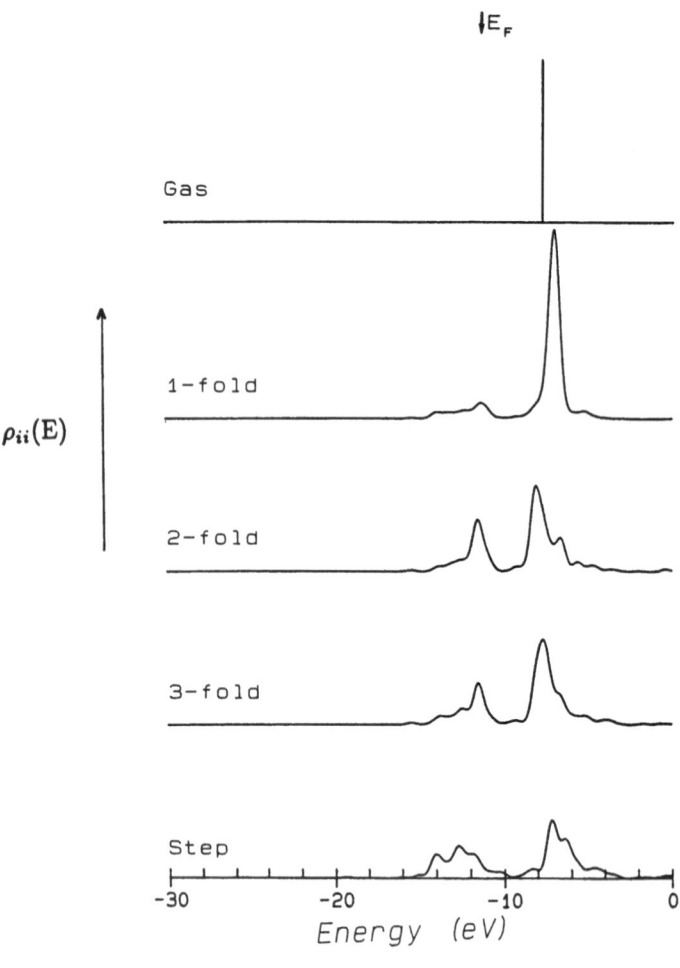

Figure 6

LDOS's if the $2\pi^*$ CO orbitals in the gas phase of CO atop adsorbed on Rh (111), of CO atop adsorbed on Rh (111), of CO 2-fold adsorbed on Rh (111), 3-fold adsorbed on Rh (111) and 3-fold adsorbed on stepped Rh (111) distance of CO to step 1.551Å). The Fermi level is indicated by E_F.

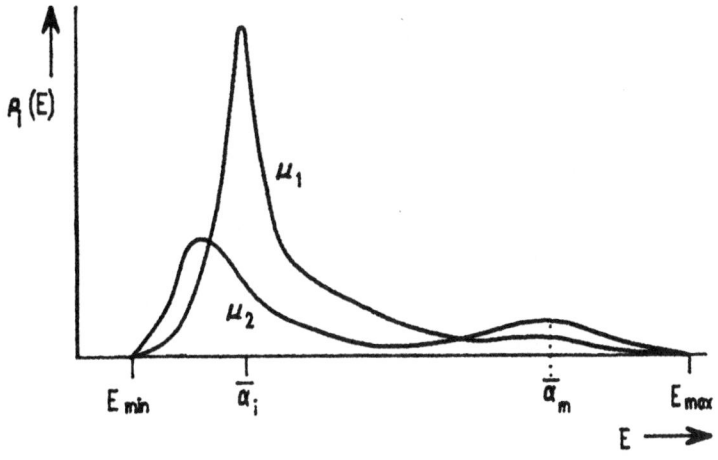

Figure 7

$\rho_i(E)$ as a function of μ (schematic),$\mu_2 > \mu_1$.

$$\Delta E_i = \frac{2}{\pi} \int_{-\infty}^{E_F} dE \eta_i(E) \qquad (15a)$$

$$\operatorname{tg} \eta_i(E) = \frac{\Gamma_i(E)}{\alpha_i + \Lambda_i(E) - E} \qquad (16a)$$

$\eta_i(E)$ can be considered the phaseshift of the surface electrons scattered by the adsorbate. For weak adsorption the behaviour of $\eta_i(E)$ is sketched in figure 8.

One recognizes that at values of $E < \overline{\alpha}_i$ there is bonding contribution to the energy, but at energy values above $\overline{\alpha}_i$ the contribution is antibonding.
In the weak adsorption limit, the attractive covalent contribution to the bond energy is approximately given by:

$$\Delta_{cov}^{w} = \frac{2}{\pi} \int_{-\infty}^{E_F} dE \frac{(\overline{\alpha}_i - E)\Gamma}{(\overline{\alpha}_i - E)^2 + \Gamma^2} \qquad (17a)$$

$$\approx \frac{2\Gamma(E_F)}{\pi} \ln \frac{(\overline{\alpha}_i - \overline{E}_F)^2 + \Gamma^2(E_F)}{(\overline{\alpha}_i - E_{min})^2} \qquad (17b)$$

Remember that $\Gamma(E_{min}) = \Gamma(E_{max}) = 0$. Note that $\ln \left| \frac{(\overline{\alpha}_i - E_F)^2 + \Gamma^2}{(\overline{\alpha}_i - E_{min})^2} \right|$ is minimum when $E_F = \overline{\alpha}_i$. At this value of E_F all bonding orbitals are filled. At a higher value of E_F the bond energy decreases because antibonding levels become occupied. $\Gamma(E_F)$ follows from (12):

$$\Gamma(E) = \pi n \beta'^2 \rho_g^s(E) \qquad (18a)$$

$$= \frac{\pi n \beta'^2}{\sqrt{n} |\beta'|} f_g^m(E) \qquad 18b)$$

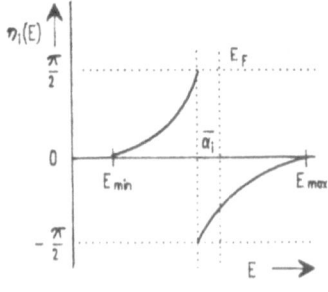

Figure 8

The phaseshift $\eta(E)$ as a function of energy.

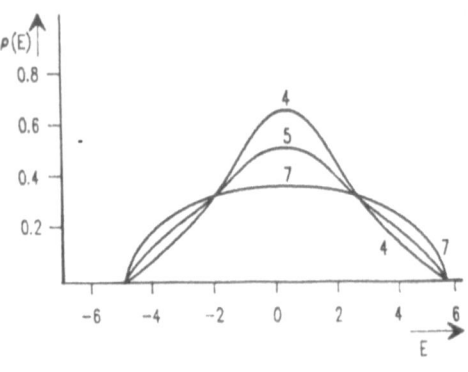

Figure 9

LDOS of a surface atom in the Bethe lattice approximation. Numbers denote the number of surface atom neighbour atoms. Bulk coordination is 8.

As mentioned earlier the linewidth function $\Gamma(E)$ relates with the corresponding metal surface grouporbital local density of states. In the general case the local density of states contains significant structure. Its approximate behaviour is sketched in figure 9 (the density is correct up to its second moment). The width of the surface local density of states is proportional to $\sqrt{n}$, $\overline{n}$ is the number of metalatom neighbours of a surface atom. Because $\rho^s(E)$ is normalized the maximum of $\rho^s(E = E_{max})$ is inverse proportional to $\sqrt{\overline{n}}$:

$$\rho^s(E_{max}) \sim \overline{n}^{-1/2} \tag{19}$$

As long as E_F is not close to the edges of $\rho^s(E)$:

100

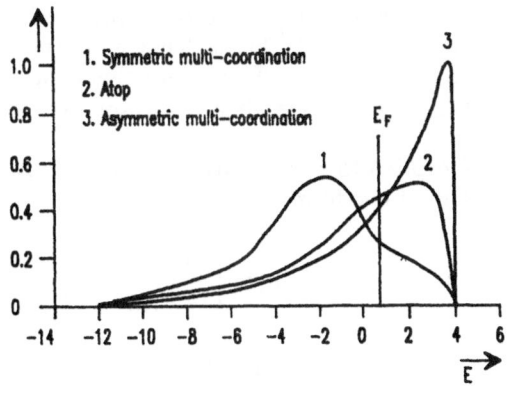

Figure 10

Metal surface group orbital local densities of states (schematic).

$$\Delta_{cov} \approx -\frac{\pi \, n \, \beta'^2}{\sqrt{n} \, |\beta|} \tag{20}$$

Expression (20) is expected to hold when an adsorbate interacts with a valence electron band that is half filled. The attractive contribution to the covalent energy is lowest on surface atoms with a high degree of coordinative saturation. It increases when the degree of coordinative saturation of the surface atomes increases. When the energy dependence of $\rho_g(E)$ is ignored, the bond strength increases with adsorbate coordination-number. As sketched in figure 10 the maximum density of a group-orbital corresponding to a symmetric combination of atomic orbitals is lower than the LDOS of an individual atomic orbital and that of an asymmetric combination of atomic orbitals is higher than the LDOS of an individual atomic orbital (11). This affects the bonding part to the bond energy ΔE^w_{cov}, because according to (17b) it relates to $\Gamma(E_F)$. In the weak adsorption-limit according to (18a) ΔE^w_{cov} depends on the grouporbital local density of states at the Fermi level. Therefore dependent on Fermi level position the relative stability of coordination of adsorption sites changes. In agreement with the discussion in section 2, a high valence electron band occupation favours atop adsorption for σ-symmetry adsorbate orbitals, but a low valence electron band occupation threefold coordination. The group VIII transition metal Pt has a nearly completely filled d-valence electron band. We argued in section 2 that changes of the interaction of the CO 5σ, and the surface metal d orbitals dominate coordination of CO to Pt. The analysis presented here not only agrees with this, it also predicts that CO binds most strongly atop to the (111) surface (7). Only at lower d-valence electron occupation coordination to the more open surfaces becomes favoured.

Elsewhere (6) we demonstrated that the perpendicular threefold coordination of O_2 to the (111) surface of silver optimizes the interaction of the occupied O_2 1π orbital with the asymmetric group orbital local density of states around the Fermi level. It was also shown that bonding of O_2 to the silver surface is only well understood if not only the interaction with the s,p valence electron band density close to the Fermi level, but also formation of bonding and antibonding orbital fragments with d-valence electrons is included. The narrow d-valence electron band and the many d-valence atomic orbitals results in a high LDOS of completely filled orbitals. They interact significantly with the partially occupied π-valence orbitals of O_2. This interaction turns out to be repulsive when O_2 is adsorbed parallel but attractive when O_2 is adsorbed perpendicular to the (111) surface.

According to formal chemisorption theory the attractive contribution to the covalent energy is found to be the sum of two terms:

$$\Delta E_{cov}^{at} = 2 \sum_{i,j} n_{ij} \, \beta_{ij}'^{\,2} \, \overline{\rho_{g(j)}^{s} \,(E_F)} \ln \left\{ \frac{(\overline{\alpha}_i - E_F)^2 + \Gamma_{ij}^2(E_F)}{(\overline{\alpha}_i - E_{min}^j)^2} \right\}$$

$$ - 2 \sum_{j} \frac{n_{ij} \, | \, \langle \varphi_i \, | \, H' \, | \, \varphi_{dj} \, \rangle \, |^2}{E_F - E_{dj}} \tag{21a}$$

The last term has to be included if adsorption to a IB metal is studied. Expression (21a) has to be applied with care to a transition metal with d-valence electron holes. The expression for $\overline{\rho_{g(j)}^s}(E_F)$ to be used is $\rho_{g(j)}^s(E_F)$ averaged over an energy interval in the order of $\Gamma(E_F)$.

Expression (21a) shows explicitly the interesting result that in the limit of weak adsorption, there exist a relation between the group orbital local density of states at the Fermi level and chemisorption energy. In the derivation of (21a) adsorbate and surface molecular orbitals have been assumed to be orthogonal. We have analysed the consequences of non-orthogonality extensively elsewhere (3). The form of expressions (21a) does not essentially change. However an additional repulsive interaction is now introduced. We have shown (3) and already discussed shortly in section 2 that the repulsive interaction equals:

$$E_{rep} = 4 \sum_{i,k}^{occ} \langle \varphi_i \, | \, H' \, | \, \phi_k^m \rangle \, | \, \langle \phi_k^m \, | \, \varphi_i \rangle$$

$$\approx 4 \sum_{k}^{occ} \sum_{i,j} \left\{ n_{ij} \, \beta_{ij}' \, S_{ji} \, q_{g(j)}^k \right\} \tag{21b}$$

Since $\beta_{ij}' \sim S_{ij}$, the repulsive part to the bond energy is proportional to S_{ij}^2.

$$\approx 4 \sum_{i,j} n_{ij} \, | \, S_{ij} \, |^2 \, q_{g(j)} \,(E_F) \tag{21c}$$

with:

$$q_{g(j)} \,(E_F) = \int_{-\infty}^{E_F} dE \, \rho_{g(j)} \,(E) \tag{22}$$

$q_{g(j)}(E_F)$ is the electron occupation of surface group orbital $\phi_{g(j)}$.

Applications to metal promotion

GENERAL BACKGROUND

Coadsorption of adatoms or alloying may change the nature of the adsorbate-metal surface chemical bond for steric as well as electronic reasons. The ensemble effect (12) is used to explain changes in bond energy due to surface dilution or rearrangement effects. As discussed earlier atoms from an adsorbing molecule will in general bind to several surface atoms. If one blocks surface atoms by coadsorption with inert atoms, or by alloying with non-reactive elements, the average ensemble size of the surface atoms decreases and hence the probability for adsorption to high coordination sites also decreases. This has been extensively discussed elsewhere (12). Here we will be concerned with changes in chemical bonding due the consequences of differences in composition on atompositions in the second coordination sphere with respect adsorbate atoms (see fig.11).

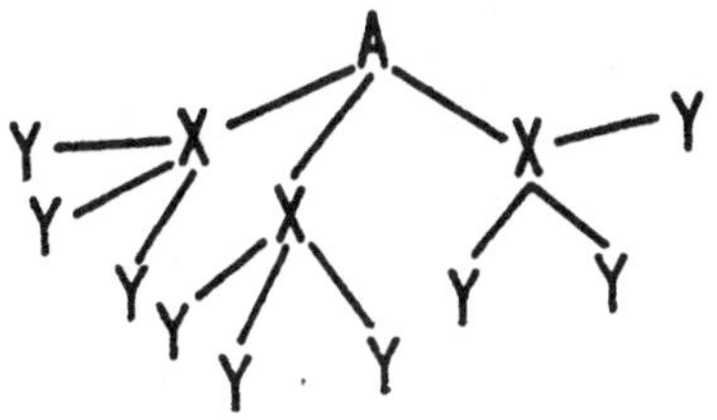

Figure 11

Surface ligand effect. The bond strength between A and atoms on positions X varies due to changes in composition on positions Y.

This may be called surface ligand effect and can be due to changes in covalent bonding or electrostatic effects, resulting from alloying or coadsorption. Changes in surface topology may also be the cause of a surface ligand effect. Several different ways to categorize metal promoters have been proposed. The most natural seems to be to distinguish between electronegative and electropositive coadsorbates or elements (13). Whereas not incorrect, it appears that the overall effect of promoters may differ dependent on the relative position of promoting atom and adsorbate. We will discuss this specifically for,the electropositive promoting elements. Here we will distinguish electronic changes in surface chemical bonding due to the presence of promotors according to whether they affect the repulsive or attractive part of the potential energy curve of the adsorbate-surface chemical bond. In the weak adsorption limit the expression for the attractive and repulsive part of the potential energy curve are given by (21a) and (21c). The attractive part of the potential energy is mainly determined by the grouporbital local energy density of states at the Fermi level or the relative position of the adsorbate orbitals with respect to the surface Fermi level position. The repulsive part of the potential energy curve stems from Pauli-repulsion between doubly occupied orbitals. It depends on the overlap of adsorbate and metal surface wavefunctions as well as the occupation of the surface grouporbital interacting with adsorbate. Due to the presence of a promotor the spatial distribution of the surface-electrons may change, resulting in a change in overlap of adsorbate and surface wavefunctions. Also the orbital electron occupation may change. The latter may result in a change in the electron distribution over bonding and antibonding orbital fragments. We will discuss an example where the promoting atom reduces the occupation of antibonding orbital fragments, so that a repulsive interaction is changed into an attractive one.

Note that the attractive contribution to the bond strength depends on the group orbital local density of states at the Fermi level, but that the repulsive part of the potential energy curve depends on the total electron occupation of the surface grouporbitals. Usually several changes in the bonding parameters occur when a promoting atom or ion coadsorbs. We will discuss a few examples that illustrate the usefulness of an interpretation of promotoreffects in terms of changes of the attractive and repulsive part of the bond strength potential energy curve. Sulphur and alkali coadsorption are examples where the attractive part of the adsorbate-surface metal potential energy curve changes. Metal particles adsorbed to cations, situations that occur on catalyst supports or in zeolites may show reduced Pauli-repulsion. The enhanced reactivity of oxygenations adsorbed to the silversurface in the presence of subsurgace oxygen or chlorine resulting in selective ethylene epoxidation is an example where a repulsive interaction is changed in to an attractive one (8). The electronic basis to the reduction of reactive ensemble atoms size by alloying with an insert element is the Pauli repulsion between adsorbate and insert atom.

Changes of the attractive part of the potential energy curve

Experimental studies on the effect of S, C or P adsorption on CO adsorption show (14) that the rate of CO adsorption is significantly more affected by the presence of these adsorbates than can be explained on the basis of geometric constraints due to a decrease of the effective reactive surface atom ensemble size. Feibelman e.a. (15) computed the variation in electron density on a surface due to the presence of coadsorbed S. As reproduced in figure 12 they find that the total electron density only changes on the transition-metalatom that coordinates with sulfur. The local density of states at the Fermi level is significantly changed also on atoms that are not coordinated to S! This applied to expression (21a) for the bond energy explains the reduced interaction with CO. Electron-energy density fluctuations mainly of the s,p-valence electrons are induced by adsorption of S to a transition metal. They depend on the kind of coadsorbate. This has been studied by Joyner e.a. (16).

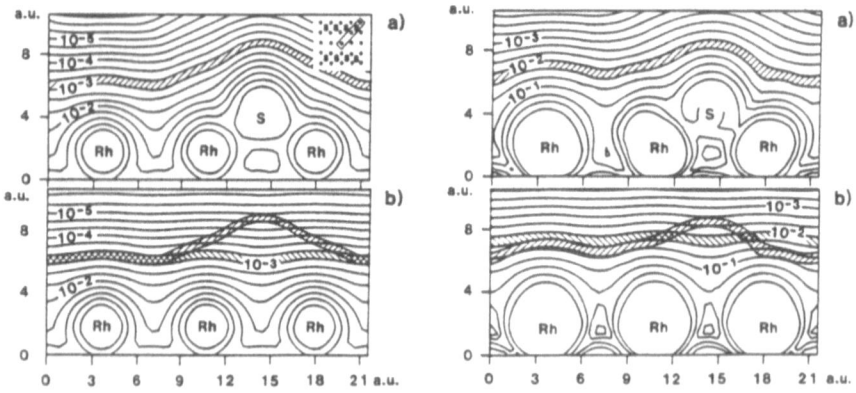

Figure 12

(a) Valence charge densities (in atomic units) for two-layer Rh (001) films (a) with and (b) without a S (3 x 1) adlayer. To facilitate comparison the region between contours of charge density $\sim 10^3$ a.u., about 4 a.u. above the Rh nuclei has been hatched. The hatched region from the S/Rh plot has been transcribed onto that for clean Rh[15].
(b) Fermi level LDOS, in (eV x a^3)$^{-1}$ for two-layer Rh (001) films (a) with and (b) without a S (3x1) adlayer. Regions of equal LDOS have been hatched in the two plots. For comparison the hatched region has been transcribed from the upper to the lower plot[15].

Coadsorption of alkali atoms affects the surface-chemical bond for very different reasons. When an alkali atom adsorbs at low concentration onto a transition metal surface it will develop a small positive charge (N.1e). This is because the difference in alkali-ionization potential and transition metal workfunction is larger than their covalent interaction. The positive adatom charge will be screened by the transition-metal electrons, so that a small negative charge appears on the atoms to which the alkali atom is adsorbed. For a molecule adsorbed close to the alkali atom the result will be a lowering of the adsorbate orbitals with respect to the metal Fermi level. The dipole generated by the adsorbed alkali atom will lower the workfunction. In (21a) this changes the value of $(\bar\alpha_i - E_F)$ to $(\bar\alpha_i' - E_F)$. $\bar\alpha_i' - \bar\alpha_i$ is the potential an electron in absorbate orbital i experiences due to the electrostatic potential generated by the screened alkali atom charge. The results of calculation by Freeman e.a (17). for CO adsorbed to Ni illustrate this nicely (fig. 14).

As follows from equation (21a) two cases can be distinguished.

Case a: $\bar\alpha_i > E_F$. This is the contribution to the bond energy due to "electron backdonation". Metal surface orbitals interact with the unoccupied adsorbate orbitals α_i. Because the attractive potential of alkali lowers the adsorbate orbital levels with respect to E_F:

$$\bar\alpha_i' - E_F < \alpha - E_F$$

and the contribution to the adsorbate bond strength increases.

Case b: $\bar\alpha_j < E_F$. This is the contribution to the bond energy due to "electron donation". Unoccupied metalsurface orbitals interact with occupied adsorbate orbitals α_j . The lowering of the adsorbate orbital levels with respect to E_F gives relation:

$$E_F - \bar\alpha_j < E_F - \bar\alpha_j'$$

The corresponding contribution to the adsorbate bond energy decreases.

Table 3. Bond energy values of CO adsorbed on Pt.

val. bands.	mol. orb.	top	bridge	top	bridge
		no K		K	
s	5σ	−0.439	−0.293	−0.385	−0.184
	$2\pi^*$		−0.256		−0.474
d	5σ	−0.454	−0.263	−0.241	−0.154
	$2\pi^*$	−0.444	−0.492	−0.531	−0.628
ΔE		−1.337	−1.304	−1.157	−1.440

Bond energy contributions and total Bond energy ΔE (eV) of CO adsorbed to (111) face of platinum. Effect of alkali coadsorption.

$$\alpha_{5\sigma}(K) = \alpha_{5\sigma} + \Delta V_{CO}$$
$$\alpha_{2\pi^*}(K) = \alpha_{2\pi^*} + \Delta V_{CO}$$
$$\alpha_{Pt_{s,ds}}(K) = \alpha_{Pt_{s,d}} + \Delta V_{Pt^s}$$
$$\Delta V_{CO} = -0.5 eV$$

In table 3 results of tightbinding calculations of CO adsorbed to the (111) surface of Pt are presented. A comparison of the relative energies of atop and three-fold coordinated CO is given in the presence and absence of coadsorbed potassium atoms. The presence

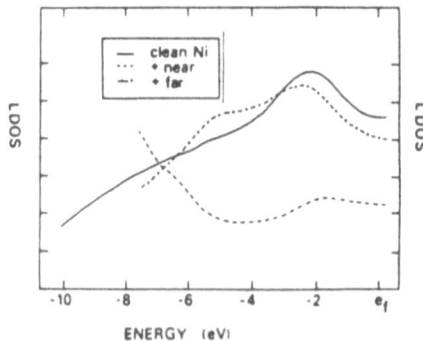

Figure 13

LDOS change on Nickel atom next nearest to adsorbed S atom[1]

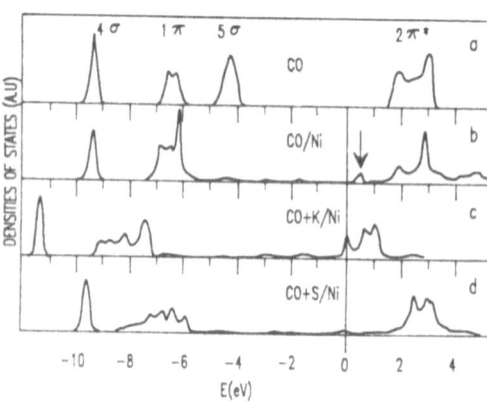

Figure 14

Local density of states of CO adsorbed to Ni[17].
(a) unsupported CO film; (b) CO / Ni
(c) (CO+K) / Ni; (d) (CO+S) / Ni.

of potassium is simulated by a lowering of the CO adsorbate levels with respect to the Fermi level as indicated in table 3. As predicted the interaction with the $2\pi^*$ orbital of CO (backdonation) is found to increase and that with the 5σ orbital of CO (donation) decreases. The overall result is a change in coordination of CO from the atop position to three fold coordination. Especially when the interaction with the s-valence electron band dominates electron backdonation favours high coordination because in order to interact with the metalsurface the CO $2\pi^*$ orbital requires antisymmetric surface metal grouporbitals fragments are occupied. When the interaction with the highly occupied d-valence electron bond is important the donative contribution of σ symmetry favours atop coordination. In the atop configuration the repulsion due to the occupation of antibonding orbital-fragments is minimized (7).

Changes of the repulsive part of the potential energy curve

On a metal surface an alkali-ion will adsorb on the same relative position with respect to the metal surface as the adsorbate. This may be different when one considers small metal particles. In figure 15a we show a configuration where a Mg $^{2+}$ ion is located at the other site of a metal particle than an adsorbed CO molecule. Such a situation may arise in zeolites or metal particles distributed on a MgO carrier. The difference in electron distribution of an Ir_4 particle comparing a situation with no Mg^{2+} ion present and one with a Mg^{2+} ion is shown in figure 15b. The computed result is from a Hartree-Fock-Slater-LCAO calculation (19).

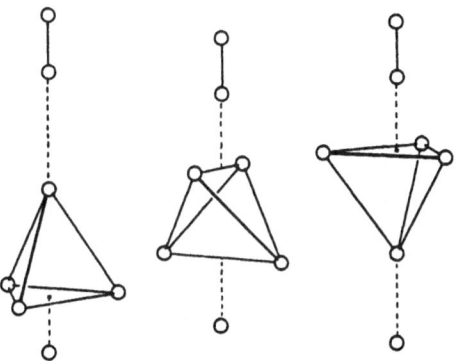

Figure 15a

Relative position of Mg^{2+} and adsorbed CO on a Ir_4 particle.

One observes polarization of the Ir_4 particle. Electron density is pulled towards the Mg^{2+} ion, in order to screen its charge. As a results there is a decrease of electron density on the atom to which adsorbed CO is attached. Table 4 shows a comparison of the Pauli repulsion computed for CO adsorbed atop, twofold or threefold with the Mg^{2+} ion present or absent on the opposite adsorption site of the tetrahedral Ir_4 particle (14). One observes a significant decrease in Pauli-repulsion. The reduction in electron density between CO and neighbouring Ir atoms reduces the Pauli-repulsion between doubly occupied CO-5σ orbitals and occupied Ir_4 orbitals, since according to (21c) Pauli repulsion reduces, when orbital overlap is decreased. The overall result on the CO adsorption strength is no change in adsorbate bond energy, because not only the repulsive part to the bond energy but also the attractive contribution, due to electron backdonation into the CO $2\pi^*$ orbital changes.

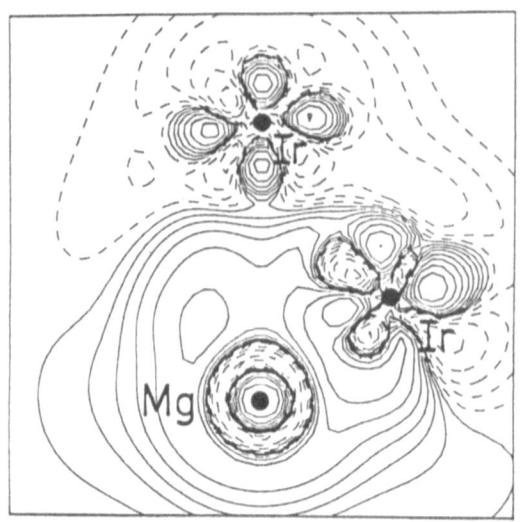

Figure 15b

Change in electron density on Ir_4 cluster close to a Mg^{2+} ion[19].

- - - - - - loss in density

— — —— gain in density.

Table 4a

Adsorption energies and their decomposition (in eV) of CO on a tetrahedral Ir_4 cluster and a Mg^{2+} ion at the opposite side.

geometry	adsorption energy	steric repulsion	interaction energy
1-fold	0.98	0.18	1.16
2-fold	2.35	5.62	7.93
3-fold	2.42	4.78	7.19

Table 4b

Change in the adsorption energies of CO and their decomposition (in eV) upon introduction of the Mg^{2+} ion. Shown are the values with minus values without the cation.

geometry	adsorption energy	steric repulsion	interaction energy
1-fold	0.05	−1.53	−1.48
2-fold	−0.07	−1.86	−1.96
3-fold	0.03	−3.16	−3.11

The CO molecule has a larger distance with respect to the Mg^{2+} ion than the Ir_4 particle. The electrostatic field of the Mg^{2+} ion lowers the Ir_4 electron orbitals more than those of the adsorbate electron orbitals. The result is an increase in the CO $2\pi^*$ orbital energy with respect to the Ir_4 HOMO orbital energy and the electron backdonation contribution decreases. So the loss in repulsion is compensated for by a loss in attraction.

The presence of the cation is only reflected in an increase of the CO stretch frequency due to the decrease in the CO $2\pi^*$ orbital occupancy. Studies of H_2 adsorption show a similar decrease in repulsion interaction as observed for CO, but now overall an increase of the interaction energy is found (20). It may explain the enhanced activity in hydrocarbon conversion activity of metal particles close to positively charged cations in zeolites (21). Significant H_2 bond weakening is found.

The last two examples to discuss derive also from metal catalysis. First we will discuss the electronic basis of the role of subsurface oxygen or chlorine in ethylene epoxidation and we wish to conclude by explaining that the electronic basis to the secondary ensemble effect consists in essence of Pauli-repulsion between adsorbate atoms and inert surface-atoms. Experimental studies of the selectivitiy of ethylene epoxidation catalysed by silver have shown that the epoxidation activity is enhanced by the presence of subsurface oxygen atoms close to the adsorbed oxygen atom to be inserted into the ethylene π bond (22). HFS-LCAO calculations on small silver metal particles demonstrate the reason for this (8). We will present results for 4 silver atom clusters. Whereas details are slightly different, on larger particles the essential physics does not change. A silver atom cluster has been chosen such that atoms are located as on the Ag (110) surface. The different cluster compositions are shown in figures 16. Figure 16a en 16b show an adsorbed oxygen atom in the presence and the absence of subsurface oxygen. Figures 16c and 16d show the configuration studied with ethylene present. The results of the calculation are best analysed by computation of Bond-Order Overlap Population Densities (BOOPD's) (see eq. 2a). Figures 17a and 17b show the BOOPD's of the adatom O-Ag cluster-atoms in the absence and presence of subsurface oxygen. A positive value means a bonding contribution to the bond energy and a negative value an antibonding contribution. E_F denotes the highest occupied molecular orbital. One observes that the adatom O-Ag cluster bond is weakened by the presence of subsurface oxygen. When subsurface oxygen atoms are present antibonding O-Ag orbital fragments become occupied. It results from the reduced difference in energy between the bonding and antibonding O-Ag orbital fragments. The coupling parameter μ (eq. 12c) decreases because Ag has effectively an increased number of neighbours. This reduces the Ag local density of states around the Fermi level of the cluster in fig. 16b. The bond weakening of the oxygen-silver bond due to subsurface oxygen has a very large effect on the interaction with ethylene. The corresponding BOOPD's are shown in figures 18a and 18b. The stronger silver-oxygen bond results in a repulsive interaction with ethylene when no subsurface oxygen is present. This changes in the presence of subsurface oxygen. One observes the depopulation of the antibonding silver-oxygen orbital fragment. The weaker oxygen-silver bond results in a stronger interaction with ethylene and a larger difference in energy between bonding and antibonding oxygen-carbon orbital-fragments.

The concepts discussed can also be used to provide an electronic basis to the concept of ensemble effects in alloy catalysis. Soma-Nota and Sachtler (23) explained the shift of chemisorbed CO from bridging to atop coordination comparing adsorption to Pd with a Pd-Ag alloy to a secoundary ensemble effect. CO interacts weakly with Ag. Alloying with silver does not change the s-p valence electron band very much, but the interaction with d-valence electrons changes significantly. The average energy of the d-valence atomic orbitals on silver is lower than that of the Pd d-valence atomic orbitals. In addition on the Ag atom the d-atomic orbitals are completely occupied. If in a high coordination site a Pd atom becomes substituted by an Ag atom, the interaction with the CO $2\pi^*$ orbitals as well as the 5σ orbitals changes. Backdonation into the CO $2\pi^*$ orbital is reduced because of the larger difference in energy between the CO $2\pi^*$ orbital and part of the d-valence orbitals. Pauli repulsion increases. The CO 5σ orbital now interacts with atoms with doubly occupied d-atomic orbitals giving a repulsive

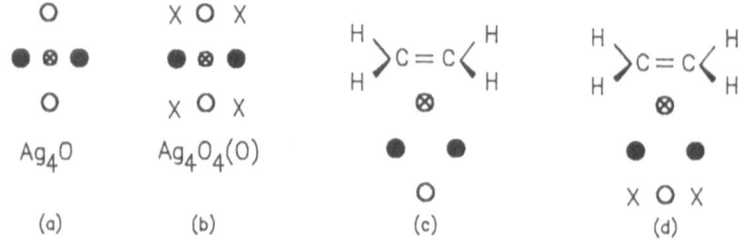

Figure 16

The silver clusters studied; ⊗ chemisorbed oxygen, ● silveratom in outerlayer, O silveratom in inner layer, x subsurface oxygen atom;
(a and b) top view without ethylene; (c and d) side view with ethylene[8].

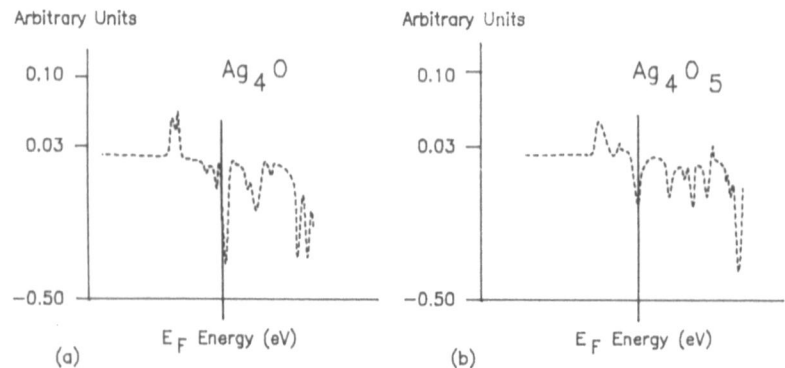

Figure 17

Bond-order overlap population densities of bond Ag-O_{ads};
(a) AgO, (b) Ag_4O_4 (O_{ads})[8].

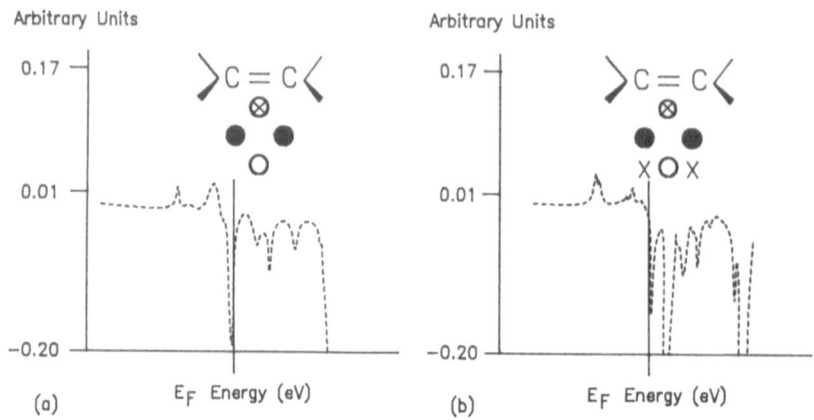

Figure 18

Bond-order overlap densities of ethylene C and O_{ads} orbitals;
(a) without subsurface oxygen, (b) with subsurface oxygen[8].

interaction. Both changes favour preferential adsorption with a site consisting of only Pd-atoms. This will result in a lower coordination number of CO. The importance of the repulsive interaction of the CO 5σ orbitals with doubly occupied orbitals of group IB metals becomes especially apparent from the preferred atop coordination of CO to Cu. This low workfunction metal should have an increased backdonating interaction with the high coordination directing CO $2\pi^*$ orbitals. CO, however, is found to prefer the atopposition. The low average energy of the d-valence electrons is unfavourable to the backdonating interaction and results in a repulsive interaction with the CO 5σ orbital (5).

References

1. G. Blijholder, J. Phys.Chem. 68, 2772 (1064).
2. K. Fukui, Science 218, 747 (1982);
   R.B. Woodward, R. Hoffmann, Ang.Chem.Int.Ed.Engl. 8, 781 (1969);
   Th.A. Albright, J.K. Burdett, M.H. Whangbo, Orbital interactions in chemistry, J. Wiley & Sons (1985).
3. R.A. van Santen, E.J. Baerends, in "Theoretical Models of Chemical Bonding", part. 4, Z.B. Maksic (ed.) Springer Verlag, in press.
4. P.M. Newns, Phys.Rev. 178, 1123 (1969); P.W. Anderson, Phys.Rev. 124, 41 (1964); J.R. Schrieffer, in "Dynamic Aspects of Surf. Physics" (8.0 Goodman, ed). Proc. Int. School of Physics "Enrico Fermi", Course XVII (Ed. Comp. Bologna, 1974);
   T.B. Grimley, CRC Crit.Rev. Solid State Sci. 239 (1976).
5. R.A. van Santen, J. Mol. Struct. 173, 157 (1988).
6. M.C. Zonnevylle, R. Hoffmann, P.J. van der Hoek, R.A. van Santen, Surf.Sci. 223, 233 (1989).
7. A. de Koster, A.P.J. Jansen, R.A. van Santen, J.J.C. Geerlings, Far. Disc. Chem. Soc. 87, 263 (1989);
   A. de Koster, R.A. van Santen, Surf.Sci. 233, 366 (1990).
8. P.J.M. van den Hoek, E.J. Baerends, R.A. van Santen, J. Phys.Chem. 93, 6469 (1989).
9. R.A. van Santen, J. Chem.Soc. Far. Trans. I 83, 1915 (1987).
10. A. Messiah, Quantum Mechanics II, North-Holland Publ.Co, Amsterdam, 1969.
11. R.A. van Santen, Physica 62, 51 (1972).
12. W.M.H. Sachtler, R.A. van Santen, Adv. Catal. 26, 69 (1977).
13. L. Ya Margolis, Adv. Catal 14, 429 (1963).
14. D.W. Goodman, Acc. Chem. Res. 17, 194 (1984).
15. P.J. Feibelman, D.R. Hamann, Phys. Rev. Lett. 52, 61 (1984).
16. J.M. Maclaren, J.B. Pendry, R.W. Joyner, P. Mechan, Surf.Sci. 175, 263 (1986); J.M. Maclaren, J.B. Maclaren, J.B. Pendry, R.W. Joyner, Surf.Sci. 178, 856 (1986).
17. E. Wimmer, C.L. Fu, A.J. Freeman, Phys.Rev.Lett. 55, 2618 (1985).
18. R.A. van Santen, Proc. 8th Int. Congr. on Catal., vol IV, Springer-Verlag, Berlin, 1984, p. 97.
19. A.P.J. Jansen, R.A. van Santen, J. Phys. Chem. in press.
20. E. Sanchez-Marcos, A.P.J. Jansen, R.A. van Santen, Chem. Phys. Lett. 167, 399 (1990).
21. P. Gallezot, Catal. Rev.-Sci Eng. 20, 121 (1979).
22. R.A. van Santen, H.P.C.E. Kuipers, Adv. Catal. 35, 265 (1987).
23. Y. Soma-Nota, W.M.H. Sachtler, J. Catal. 32, 315 (1974); Y. Soma-Nota, W.M.H. Sachtler, J. Catal. 34, 162 (1974).

ION SCATTERING AS A TOOL FOR A BETTER UNDERSTANDING OF CHEMICAL REACTIONS

ON SURFACES

W. Heiland

Universität Osnabrück
D-4500 Osnabrück, FRG

ABSTRACT

For some surface reactions, i. e. O_2 and CO_2 on metals, it has been
shown that negative molecular states play an important role as intermediate states in the chemisorption process. In case of CO_2 the bent CO_2 -
species is the intermediate species which negotiates the catalytic oxidation of CO on metals like Ni. Ion scattering experiments on the other
hand show a host of negative atomic and molecular species especially when
scattering at low perpendicular velocities from surfaces. The finding of
O_2^- and CO_2^- in such experiments is clear and supporting evidence for the
existence of these molecules as intermediates in the corresponding chemisorption and surface reaction experiments. Since in the ion beam experiments the particle velocity is an accessible variable, and the trajectory
lengths on the surface can be estimated from classical trajectory calculations, the ion beam experiments afford a possibility to estimate lifetimes of negative molecular states on surfaces.

INTRODUCTION

Reactions of molecules with surfaces can be divided into different
steps. A usual sequence of steps for a molecule approaching from the gas
phase may be physisorption, followed by dissociative chemisorption which
can be visualized in a one dimensional potential model as proposed by
Lennard-Jones[1]. The dissociation is in this model due to a curve crossing
of the physisorbed molecular state and the chemisorbed atomic (or molecular) state. If the reaction is reversed associative desorption will occur,
i. e. the chemisorbed species have to form a molecular bond again and the
newly formed molecule will desorb thermally activated.

In recent years it became clear that the one-dimensional potential
model is insufficient in many cases to describe the actually observed experimental results. Multidimensional potential energy surfaces (PES) have
been developped which afford a better understanding of the processes discussed, i. e. chemisorption and associative desorption. Fig. 1 shows as an
example the PES for $CO_2 \rightleftharpoons CO + O$ on a metal surface[2]. The incoming CO_2 is
physisorbed in a linear molecular state, than transformed into a $CO_2^{\delta-}$
state which may dissociative in chemisorbed CO and O. This reaction has

been studied in detail by photo-electron spectroscopy[3] as will be discussed below. The CO_2^- has been identified by ion scattering experiments too[4].

The process of negative molecule formation termed "surface harpooning" has been developed theoretically by a number of authors[5-7]. The first direct experimental proof for the effect was publised by Pan Haochang et al[8] for the case of O_2 scattering from Ag(111). The harpooning is a charge transfer process belonging to a class of processes quite familiar in ion-surface scattering, neutralisation by Auger processes and resonant processes[9,10], and, especially in the context here, the capture into and the loss from affinity levels of atoms[11,12]. The affinity level of atoms or molecules is energetically broadened and shifted down by the image potential such that the level can be populated. In case of molecules the population of the affinity level causes at least the change of the vibrational excitation owing to the differences in the molecular potential well[5-7]. In case of CO_2 the molecule will change from the linear configuration into a bent modification[3] according to Walsh's symmetry rules[13]. Beside these effects which mean a transfer of kinetic energy into potential energy or configurational changes, the formation of the negative molecule changes the interaction potential, i. e. coulombic forces are switched on between the image charges in the metal and the charges outside. The coulomb forces affect of course the trajectories of scattered particles[14,15].

In the following the specific example of CO_2 interaction with Ni single crystal surfaces will be discusses in some details based on recent experimental results[3,4].

EXPERIMENTAL RESULTS
PHOTOELECTRON SPECTROSCOPY

From the wealth[3,16-23] of experimental evidence for negatively charged precursors in molecular dissociation on surfaces the photoelectron spectroscopy[3] experiment for the case CO_2 on Ni(110) will be discussed here. If the surface is cooled to 80K CO_2 adsorbs as a linear molecule lying flat on the surface. In the temperature range from about 100K to 230K a new species shows its signature in the angular resolved photoelectron spectra (ARUPS): By comparison with the theoretical results[24] this is identified as a bent $CO_2^{\delta-}$ species (Fig. 2). Further evidence for the identification as a bent species is obtained from high resolution electron energy loss spectroscopy (HREELS) which shows that the adsorbed species is has C_{2v} symmetry. Further qualitative support for the negative charge of the adsorbed species is based on the observed increase of the work function by 1eV. At around 140k the CO_2 feature disappear from the ARUP spectra, i. e. the transformation into $CO_2^{\delta-}$ is complete with respect to the species remaining on the surface. Some physisorbed CO_2 desorbs even at that low temperatures. Further heating converts the CO_2 into adsorbed CO and O. This transition, i. e. the dissociation, is most readily identified by HREELS (Fig. 3), where the losses at 1895 and 2015cm^{-1} at 270K correspond to CO species in two different coordination sites, bridge and terminal sites respectively[25,26]. The loss at 470cm^{-1} is due to chemisorbed oxygen[27]. The losses at 140k are in agreement within the point group C_{2v} identified as the bending mode of $CO_2^{\delta-}$ at 750cm^{-1} and the symmetric stretching mode at 1130cm^{-1}. The loss at 410cm^{-1} is probably due to a metal-molecule vibration mode, which supports the suggestion that $CO_2^{\delta-}$ is bound to the metal with direct oxygen coordination, i. e. both O atoms of the molecule ar bound to the metal, the C sticking out, rather than the C being bound to the metal and the two O-atoms sticking out. Both configurations are possible within the C_{2v} point group symmetry.

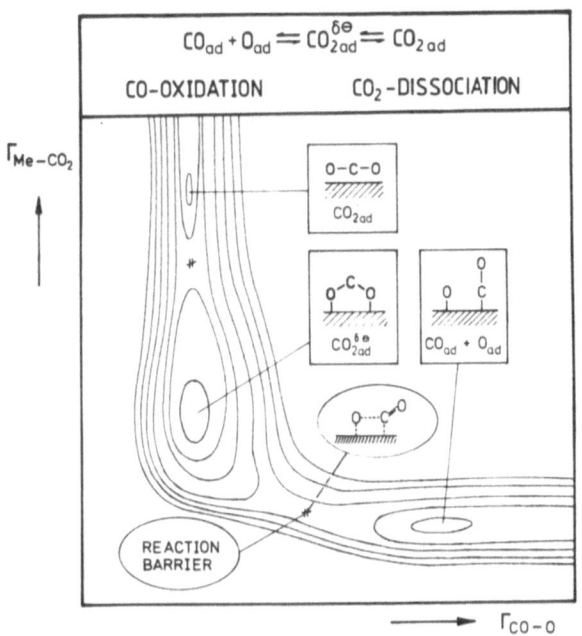

Fig. 1. Two dimensional potential energy diagram
for CO_2 metal interaction (vertical axis)
and CO-O dissociation (horizontal axis)[2,3].

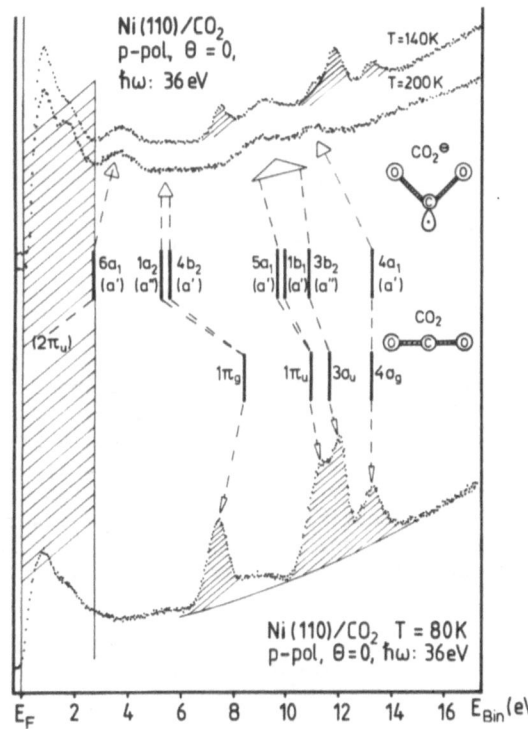

Fig. 2. Assignment of the photoelectron spectra at
80, 140 and 200K to undisturbed molecular
CO_2 and adsorbed $CO_2^{\ominus}$ on the basis of ab
initio calculations[3,24].

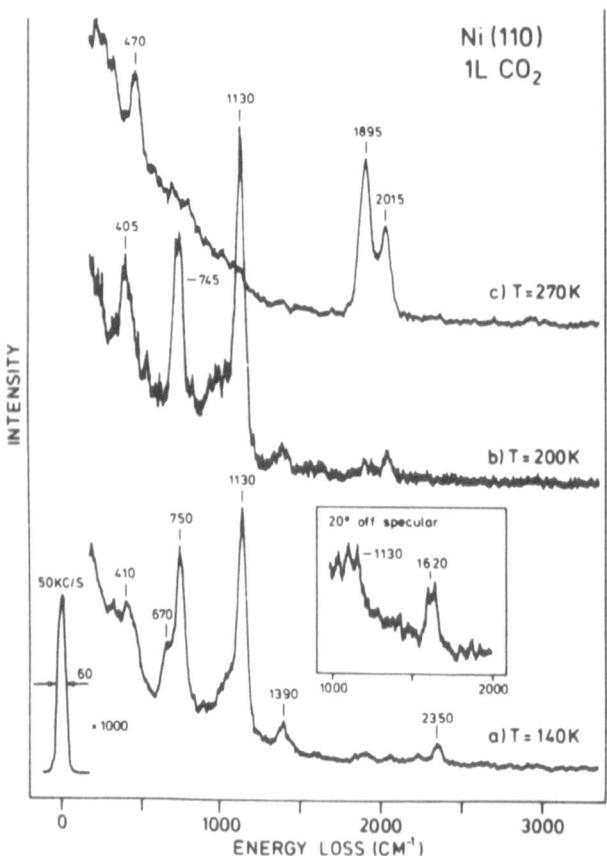

Fig. 3. Vibrational electron energy loss spectra
of a Ni(110) surface exposed to 1 L CO_2
at 140K. (a) Surface temperature 140K;
(b) surface heated to 200K; (c) surface
heated to 270K. All spectra are taken in
specular scattering direction. The insert
shows an off specular spectrum in the energy range 1000 to 2000 cm^{-1} at 140 K,[3].

In the context of this paper an important aspect of ion scattering at small grazing angles is the low perpendicular velocity component which scales as sin ψ if ψ is the grazing angle of impact. The energy scales with sin$^2 \psi$, i. e. at 1 keV total energy and $\psi = 5°$ 'the energy perpendicular to the surface is 7.5eV, at 500eV reduced to 4eV, etc. By virtue of the parallel velocity the ions are fast with respect to vibrational velocities, i. e. a molecule scattering at these energies is essential "frozen" with respect to vibration and rotation. The interaction times with the surface depend on the actual trajectory lengths, which are of the order of 10Å (see e. g. ref. 28). Hence collision times are of the order of 10^{-13} s which is fast compared to electronic exchange times which are of the order of 10^{-15} s (e. g. ref. 9, 10, 29). The elastic potential the particles scatter from are essentially flat[30] which excludes in case of molecules the dissociation by hard, elastic collisions[31]. The scenario is set for the study of the electronic interaction between molecules and surfaces using ion beams in an energy range where "reasonable" experimental condictions prevail, i. e. the ion sources and ion optics for the beam forming are financially affordable.

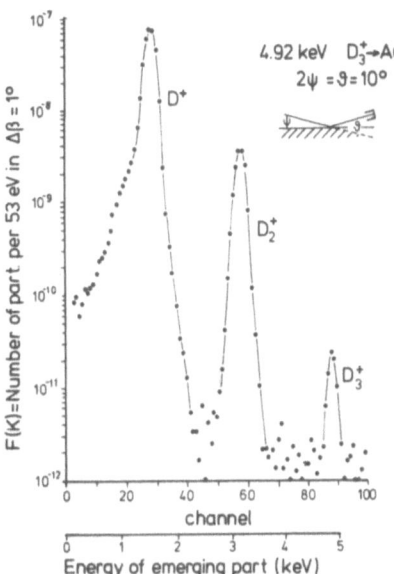

Fig. 4. Energy spectra of scattered ions from D_3^+ impinging on Au at grazing incidence. The grazing angle is 5°, the initial energy 4.92keV, the scattering angle 10°, $\Delta \beta = 1°$ is the angle of acceptance[32].

The first "molecular survival" experiments were reported for D_3^+ scattering[32]. Fig. 4 shows the ion energy spectra of D_3, D_2^+ and D^+ when D_3^+ impinges at 5keV at $\psi = 5°$ on an Au surface. The survival of D_3^+, the "dissociation" into D_2^+ and its survival are remarquable in point of view of the binding energies (a few eV) compared to the total energy of 5keV, and even compared to the perpendicular energy of 38eV in that case.

Further studies using H_2^+ and N_2^+ revealed that the dissociation is due to electron capture rather than due to electron loss[33] which would cause a "Coulomb-explosion". Electron capture into the triplet state of H_2 would cause via transformation from the center of mass system to the laboratory frame to a smooth gaussian – like energy distribution in agreement with the experiment. In case of N_2^+ also "trapping" of N on a Ni-surface was observed.

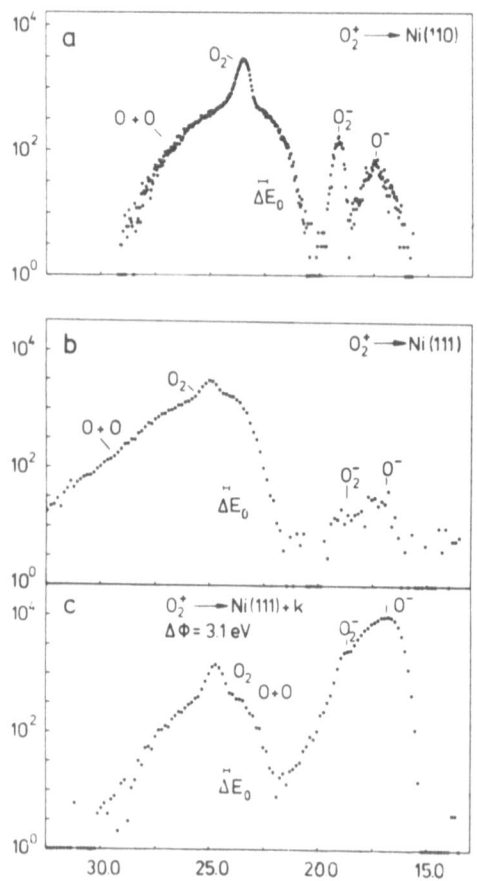

Fig. 5. Time of Flight Spectra of O_2, O_2^- and O^- from O_2^+ incident on Ni(110), Ni(111) and Ni(111)+K. The beam energy is 460eV, the grazing angle of incidence is 5°. The negative ions are found at shorter flight times due to post acceleration[35].

Further insight into the interaction, process was obtained – as predicted – by measuring the neutral flux of scattered particles[34]. As known for atomic ions, the majority of the scattered species, when the incident particle is a molecular particle, are neutrals – atoms and molecules (Fig. 5).

The broad hump of the neutral distribution are dissociated atoms, the narrow peak riding on this distribution are neutralised molecules. The O_2^- and O^- first observed by Pan Hoachang et al[8], are the due to the "surface harpooning".

The charge exchange processes leading to neutralisation, dissociation and negative ions are schematically shown in Fig. 6. Since the ionisation energies of most molecules are of the order of 10eV electron capture into the neutral ground state will be an Auger-neutralisation process (AN)[9,10]. Capture into anti-bonding states ist most likely a resonant process (RN). In case of N_2 and CO these states are "excited" states, i. e. they are not to be reached by capture only. Therefore the dissociation of N_2 and CO is very low compared to H_2, O_2 and CO_2 as shown below[35]. The affinity levels, if existent, will by brought down by the image potential interaction and then filled by a resonant capture (RC). The inverse process is a resonant loss (RL). Transition rates for these processes, albeit for atomic species, have been estimated theoretically[9,10,12,29,36,37]. It is agreed that transition times of the order of 10^{-15} s are a good rule of thumb.

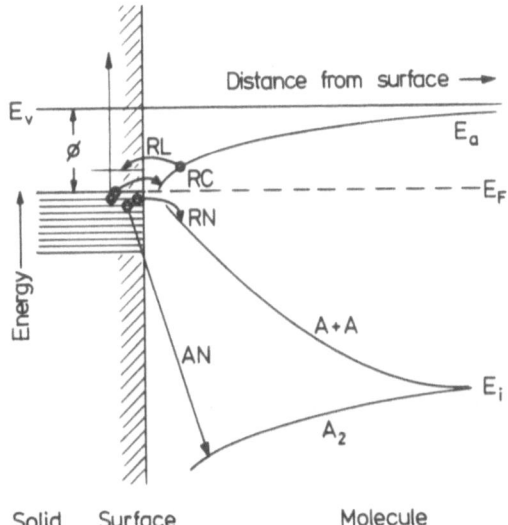

Fig. 6. Schematic diagramm of molecular states
in the vicinity of a metal surface.
RL = resonant loss, RC = resonant capture, RN = resonant neutralisation,
AN = Auger-neutralisation.

A direct experimental access to the time scales was obtained by measuring the autoionization of He** after a two electron capture into He++. It is found that the whole process can occur while the particles approach the surface at 1keV and grazing incidence[38]. It is therefore safe to argue that even slower molecules will approach a surface as negative moieties whenever an affinity level exists. Dissociation due to the forming of the negative state is than very likely. The latter point was proven recently by using O_2 as a primary particle instead of O_2^+ with the result that O^- and O_2^- is observed in both cases with almost identical yields[39].

After the O_2^- experiment on Ag(111) other ion-surface combinations yielded negative molecular ions, e. g. I_2^-, CO_2^- and NO^{-40-44}. In the time of flight spectra (Fig. 7) the CO_2^- peak is clearly identified beside the O^- peak. The O^- is formed after the dissociation by resonant capture of an electron. Further insight into the dissociation process is obtained by changing the work function of the surface by K adsorption[35].

Fig. 8 shows the dependence of the molecular survival defined as $Y = (A_2 + A_2^-)/(A_2 + A_2^- + A/_2 + A^-/_2)$ versus the work function change. The K adsorption, i. e. the lowering of the work function, enhances the dissocia-

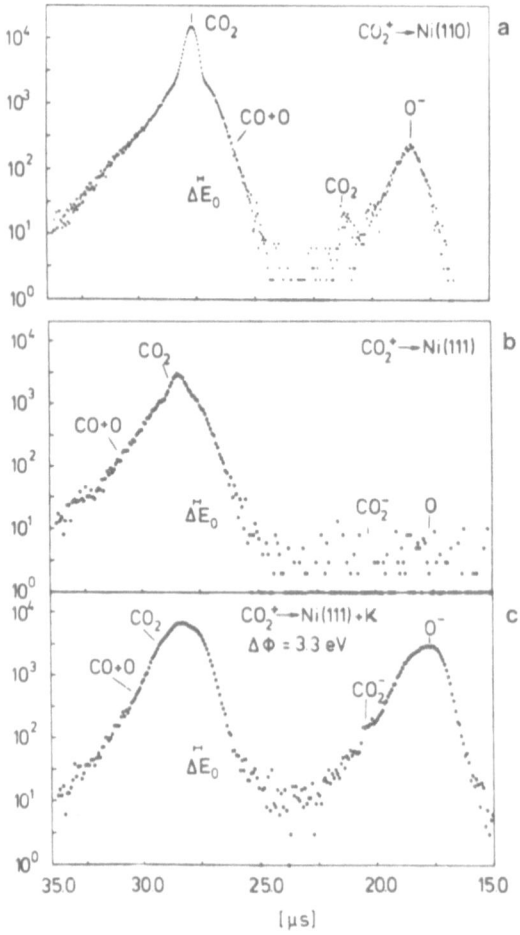

Fig. 7. Time of Flight spectra for CO_2^+ interacting with Ni(110), Ni(111) and Ni(111) + K. The primary energy is 460eV. The angle of incidence 5°. The work function change is measured by a Kelvin-probe[35].

tion. Since the lower work function facilitates the resonant capture into the affinity level, the increase of the dissociation with lowering of the work function clearly shows that dissociation proceeds via the negative molecular state as found in the chemisorption experiments. By comparison, the molecular survival of CO and N_2 is essentially not affected by the K

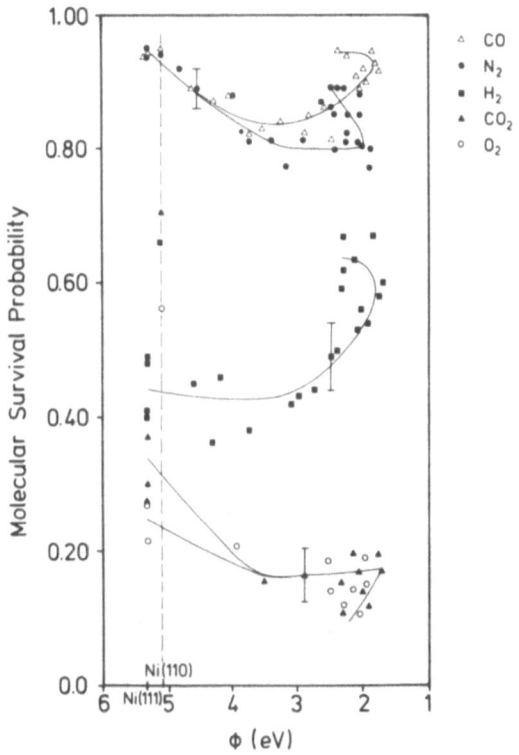

Fig. 8. Molecular survival vs workfunction for CO, N_2, H_2CO_2 and O_2. The work function of Ni(111) is changed by K-adsorption. Ni(110) data are for the clean surface only[35].

adsorption, wheras the survival of H_2 increases with lowering of the workfunction. Since the H_2 dissociation is caused by capturing an electron into the antibonding triplet state, the probability for this process decreases as the electrons of the solid are shifted up due to the K-adsorption. From the velocity and the trajectory length of the scattered CO_2^- we estimate a lifetime $\tau \simeq 10^{-13}$s for the negative species on the surface.

SUMMARY

The experiments with molecular ions scattered at grazing incidence from surfaces allow the conclusion that negative molecular states are indeed important as precursors in surface reactions and dissociative chemisorption. The results clarify part of the role played by adsorbed K in these reactions, i. e. by lowering the work function the probability for "harpooning" increases. Since the scattering experiments have a built in clock, estimates of the lifetimes of negative molecular species at surfaces become possible.

ACKNOWLEDGEMENT

Thanks are due to many students and colleagues at Osnabrück and the F. O. M. Institute, Amsterdam, for many helpful discussions and contributions to the field. Financial support by the Deutsche Forschungsgemeinschaft is greatfully acknowledged.

REFERENCES

1. J. E. Lennard-Jones, Trans. Far. Soc. 28:333 (1932)
2. J. C. Tully, Advanc. Chem. Phys. 42:63 (1980)
3. B. Bartos, H. J. Freund, H. Kuhlenbeck, M. Neumann, H. Lindner and K. Müller, Surf. Sci. 179:59 (1987)
4. S. Schubert, U. Imke, W. Heiland, K. J. Snowdon, P. H. F. Reijnen and A. W. Kleyn, Surf. Sci. 205:L793 (1988)
5. D. Feibelman, Surf. Sci. 160:139 (1985)
6. J. W. Gadzuk and J. Nørskov, J. Chem. Phys. 81:2828 (1984)
7. S. Holloway, J. Vac. Sci. Tech. A5:476 (1987)
8. Pan Hoachang, T. C. M. Horn and A. W. Kleyn, Phys. Rev. Lett. 57:3035 (1986)
9. H. D. Hagstrum, Phys. Rev. 96:336 (1954)
10. H. D. Hagstrum in "Inelastic Ion-Surface Collisions", N. H. Tolk, J. C. Tully, W. Heiland and C. W. White, eds., Academic Press, New York (1977).
11. J. N. M. van Wunnik and J. Los, Phys. Scr.T6:27 (1983)
12. D. M. Newns, K. Makoshi, R. Brako and J. N. M. van Wunnik, Phys. Scr. T6:5 (1983)
13. A. D. Walsh, J. Chem. Soc. 226 (1953)
14. E. Hulpke and K. Mann, Surf. Sci. 133:171 (1983)
    E. Hulpke, Surf. Sci. 52:615 (1975)
15. A.-D. Tenner, K. T. Gillen, J. T. C. M. Horn, J. Los and A. W. Kleyn, Phys. Rev. Lett. 52:2183 (1984)
16. C. T. Campbell. Surf. Sci. 157:43 (1985). Surf. Sci. 173:L641 (1986)
17. D. Andersson, B. Kasemo and L. Wallden, Surf. Sci. 152/152:576 (1985)
18. B. Bartos, H.-J. Freund, H. Kuhlenbeck and M. Neumann, Springer Series in Surf. Sci. 8:164 (1987)
19. G. Illing, D. Heskett, E. W. Plummer, H. J. Freund, J. Somers, Th. Lindner, A. M. Bradshaw, V. Buskotte, M. Neumann, U. Starke, K. Heinz, P. L. Andres, D. Saldin and J. B. Pendry, Surf. Sci. 206:1 (1988)
20. J. Wambach. G. Odörfer, H.-J. Freund, H. Kuhlenbeck and M. Neumann, Surf. Sci. 209:159 (1989)
21. R. G. Copperthwaite, P. R. Davies, M. A. Morris, A. W. Roberts and R. A. Ryder, Catalysis letters 1:11 (1988)
22. A. C. Luntz, J. Grimblot and D. E. Fowler, Phys. Rev. B39:12903 (1989)
23. A. C. Luntz, M. D. Williams and D. S. Bethune, J. Chem. Phys. 89:4381 (1988)
24. H.-J. Freund and P. Messmer, Surf. Sci 172:1 (1986)

25. J. C. Bartolini and B. Tardy, Surf. Sci. 97:377 (1980)
26. B. A. Guerney and W. Ho, J. Vac. Sci. Tech. A3:1541 (1985)
27. S. Lehwald and H. Ibach in "Vibrations at Surfaces", R. Candano, R. Gilles and A. A. Lucas, eds., Plenum Press, New York, p. 137 (1982)
28. H. Derks, A. Närmann and W. Heiland, Nucl. Instr. Meth. B44:125 (1989)
29. A. Närmann, R. Monreal, P. M. Echenique, F. Flores, W. Heiland and S. Schubert, Phys, Rev. Lett. 64:1601 (1990)
30. U. Imke, J. H. Rechtien and P. H. F. Reijnen, Surf. Sci. 221:454 (1989)
31. S. Bitensky and E. S. Parilis, Nucl. Instr. Meth. B2:364 (1984)
32. W. Eckstein, H. Verbeek and S. Datz, Appl. Phys. Lett. 27:527 (1975)
33. W. Heiland, U. Beitat and E. Taglauer, Phys. Rev. B19:1677 (1977)
34. B. Willerding, W. Heiland and K. J. Snowdon, Phys. Rev. Lett. 53:2031 (1984)
35. S. Schubert. U. Imke and W. Heiland, Vacuum 41:252 (1990)
    S. Schubert, Thesis, Osnabrück (1990)
36. P. Nordlander and J. C. Tully, Phys. Rev. Lett. 61:990 (1988)
37. P. Nordlander and J. C. Tully, Surf. Sci. 211/212:207 (1989)
38. P. A. Zeijlmans van Emmichoven, P. A. A. F. Wouters and A. Niehaus, Surf. Sci. 195:115 (1988)
39. A. W. Kleyn, Proc. ICPEAC 16, New York (1989)
    A. W. Kleyn, Vacuum 41:248 (1990)
40. A. Danon and A. Amirav, Phys. Rev. Lett. 61:2961 (1988)
41. S. Schubert, U. Imke and W. Heiland, Surf. Sci. 219:L576 (1989)
42. P. H. F. Reijnen and A. W. Kleyn, Chem. Phys. 139:489 (1989)
43. P. H. F. Reijnen, P. J. van den Hoek, A. W. Kleyn, U. Imke and K. J. Snowdon, Surf. Sci, 221:427 (1989)
44. S. R. Kasi, H. Kang, C. S. Sass and J. W. Rabalais, Surf. Sci. Report 10:1 (1989)

OSCILLATING STRUCTURAL CHANGES IN CATALYTIC REACTIONS

Ronald Imbihl

Fritz-Haber-Institut der Max-Planck-Gesellschaft
Faradayweg 4-6, D-1000 Berlin 33, F. R. G.

ABSTRACT

Under appropriate p_{O_2}, p_{CO}, T-conditions kinetic oscillations may arise during catalytic CO oxidation which have been studied on Pt(100), Pt(110) and Pt(210) single crystal surfaces in the 10^{-5} and 10^{-4} Torr range. The mechanism of the oscillations could be traced back to periodic structural transformations of the surface via the operation of an adsorbate-induced surface phase transition. Due to the strong mass transport of 50 % of the surface atoms that is associated with the 1x1 $\rightleftharpoons$ 1x2 phase transition of Pt(110) one may also observe a facetting of the surface during catalytic CO oxidation. The facetted Pt(110) surface was identified as a non-equilibrium structure of the Turing type, since LEED investigations demonstrated that the facets form a regular pattern on the surface with a lateral periodicity of ~ 100 Å. This interpretation could be confirmed by a Monte Carlo simulation which reproduced the formation of regular facet patterns on Pt(110) during catalytic CO oxidation.

INTRODUCTION

In open systems which are far from thermodynamical equilibrium, new phenomena can be observed which are not permitted in a closed system at thermodynamical equilibrium.[1] Typical examples are kinetic oscillations and/or spatial pattern formation. These have been studied extensively both experimentally and theoretically, in the past decades. The prototype of an oscillating chemical system is the Belousov-Zhabotinskii reaction which takes place in homogeneous fluid phase. In heterogeneous catalysis the reaction $CO + \frac{1}{2} O_2 \rightarrow CO_2$ has attracted most of the interest because of the simplicity of its chemistry. Conditions far from thermodynamical equilibrium are established in catalytic CO oxidation by a constant flow of the reactants of which only a small fraction is usually converted to the product CO_2.

Kinetic oscillations in the catalytic oxidation of CO were first discovered with polycrystalline Pt at high pressure (p ~ 1 atm), where, however, the analysis of the results was complicated by the influence of contaminants, non-isothermal conditions and by the lack of suitable experimental techniques for in-situ investigations. The situation is conceptually much simpler if one uses Pt single-crystal surfaces and studies kinetic oscillations in an ultra-high vacuum (UHV) environment under well-defined conditions. The conditions of these investigations, which were generally performed in the 10^{-5} and 10^{-4} Torr range,

Fundamental Aspects of Heterogeneous Catalysis Studied by Particle Beams,
Edited by H.H. Brongersma and R.A. van Santen, Plenum Press, New York, 1991

are significantly different from those of the high pressure studies, as the low pressure ensures strictly isothermal conditions. In addition cleaning cycles suppress the formation of Pt oxides. The existence of an oxidation-reduction cycle, which has been proposed for kinetic oscillations under high pressure conditions, can be ruled out for the Pt single-crystal studies. [2,3]

The results of the investigations show that the mechanisms of kinetic oscillations on various Pt single-crystal surfaces can all be explained by one common principle. [2,3] The variations in the reaction rate are due to periodic structural changes in the surface, which modify the catalytic activity via a change in the oxygen sticking coefficient s_{O_2}. Reversible structural changes proceed via the mechanism of an adsorbate-induced surface phase transition (PT), namely the CO-induced lifting of the hex reconstruction on Pt(100) and the CO-induced lifting of the 1x2 "missing row" reconstruction on Pt(110). Associated with the surface PT is a mass transport of Pt atoms which can amount up to 50 % of the surface atoms in the case of the 1x1 $\rightleftharpoons$ 1x2 PT of Pt(110). Therefore the structural changes are not restricted to the surface layer, but one may observe a facetting of the surface. Like the PT a facetting is associated with a reversible change in the catalytic properties of the surface.

The facetting of Pt(110) in $CO + O_2$ leads to the formation of a regular spatial pattern on the surface ($\lambda \sim 100 - 200$ Å) which persists only under reaction conditions. The formation of a spatially periodic, but stationary patterns under non-equilibrium conditions is known as the "Turing instability" of which so far only very few experimental examples exist. [1] Besides such stationary structures one also observes another type of spatial pattern formation in which propagating reaction fronts are observed on the oscillating surface.

The existence of chemical waves in which the coupling between CO surface diffusion and the surface reaction causes the formation of reaction fronts is characteristic of reaction-diffusion systems. One may also obtain more complex spatial patterns, e.g. standing waves, spirals, target patterns etc. which were detected on Pt(110)/$CO + O_2$ and correspond to different modes of temporal oscillations in the reaction rate. [10] All of these phenomena arise since the local oscillators on the surface need to be synchronized in order to achieve measurable variations in the reaction rate. The synchronization process under isothermal conditions can either proceed via surface diffusion of a mobile adsorbate or via partial pressure changes in the gas phase, in which case a homogeneously oscillating surface results. [4]

An understanding of the microscopic mechanisms has enabled the development of mathematical models of the different oscillating systems. The numerical simulation reproduced almost all of the experimentally observed features so that a detailed understanding of the observed phenomena is available. An overview of the systems studied so far and of the basic phenomena which have been observed is presented is in this report.

CO OXIDATION ON Pt SINGLE CRYSTAL SURFACES

The phase transition model

It has been well established in numerous studies that the catalytic CO oxidation takes place via the Langmuir-Hinshelwood (LH) mechanism which proceeds along the following steps:

$$
\begin{array}{llllll}
CO & + & * & \rightleftharpoons & CO_{ad} & (1) \\
O_2 & + & 2\,* & \rightarrow & 2\,O_{ad} & (2) \\
O_{ad} & + & CO_{ad} & \rightarrow & CO_2 & + \quad 2\,* \quad (3)
\end{array}
$$

* denotes a free adsorption site

The kinetics of catalytic CO oxidation show a hysteresis in the reaction rate upon variation of p_{CO}. The rate jumps from the active branch associated with an oxygen-covered surface to an inactive branch in which a CO adlayer inhibits oxygen adsorption and hence CO_2 formation. This region of kinetic instability is also in the parameter range where kinetic oscillations may occur. The LH mechanism alone, however, does not produce kinetic oscillations. Although the non-linearities which exist in the kinetic equations derived from the LH mechanism, e.g. the quadratic dependence of the O_2 adsorption rate on the number of free sites can predict multiple steady states and hence the hysteresis found in the instability region, these non-linearities are not strong enough to cause kinetic oscillations. These have to be provided by an additional mechanism for which numerous suggestions have been made.

In order to cause the reaction to jump periodically between the active and the inactive branch of the reaction one needs a mechanism which modifies the catalytic activity of the surface. Since the oxygen sticking coefficient s_{O_2} on Pt surfaces is very sensitive to the surface structure, the catalytic activity can be modulated by a reversible adsorbate-induced surface PT.

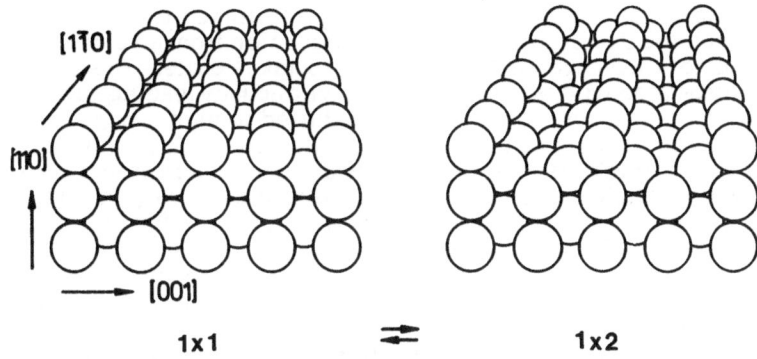

[1̄10]

[1̄1̄0]

[001]

1x1   ⇌   1x2

Fig. 1. Structural model of the reconstructed and the unreconstructed Pt(110) surface.

The clean Pt(110) surface in its stable state exhibits a 1x2 "missing row" reconstruction which can be reversibly lifted upon CO adsorption, as shown by the structural model displayed in fig. 1. Experimentally it is known that s_{O_2} is higher on the 1x1 than on the 1x2 surface and the mechanism of kinetic oscillations can then be sketched as follows: Starting with a CO covered 1x1 surface, s_{O_2} and hence the reaction rate will be high. Due to the strong consumption of adsorbed CO by the surface reaction, θ_{CO} will decrease. As θ_{CO} decreases below the critical value for the PT, the surface will relax into its less active reconstructed configuration. With s_{O_2} being low, the rate of CO adsorption will exceed the reaction rate, and θ_{CO} will rise again. Above the critical value for the PT the active 1x1 surface is established again, and a new cycle may start. The experimental verification of this mechanism has been carried out with in-situ LEED experiments by showing that the oscillations in the reaction rate are accompanied by periodic variations in the intensities of the 1x2 beams as demonstrated in fig. 2.

The same mechanism was verified for Pt(100)/CO + O_2, where the change in s_{O_2} brought about by the 1x1 $\leftrightarrows$ hex PT exceeds two orders of magnitude. Consistent with the proposed PT model no oscillations were detected on Pt(111) as this surface is structurally stable in its 1x1 state.

Complex structural transformations

In the simplified picture of the 1x1 $\leftrightarrows$ 1x2 PT driving kinetic oscillations on Pt(110) the mass transport of Pt atoms which is necessary for the formation of the 1x2 "missing row" structure, has been neglected completely. This is justified as long as thermal reordering removes the surface roughness, but under suitable conditions the following may happen:

If one exposes a Pt(110) surface to a constant flow of CO and O_2 under conditions close to kinetic oscillations, one observes in LEED a continuous structural transformation of the surface as demonstrated by a series of beam profiles shown in fig. 3. The integral order beams first broaden and then split along the [110] direction, and after a period of ~ 20 min. a stationary state is reached. The analysis of the diffraction pattern reveals that facets with (210) orientation have been formed. [5]

The conditions under which the Pt(110) surface facets can be visualized in a plot of the reaction rate r_{CO_2} vs. p_{CO} shown in fig. 4. The kinetics of the catalytic CO oxidation on Pt(110) exhibit a high and a low reaction rate branch characteristic of the LH mechanism. Facetting is restricted to the transition region between the two branches, where the transition from an oxygen-covered 1x2 surface to a CO-covered 1x1 surface also takes place. The facetting of Pt(110) is associated with an increase in catalytic activity leading to the dashed rate curve in fig. 4, whose rate maximum is shifted to higher p_{CO}. Experimentally the facetting process can therefore be followed in-situ simply my measuring the increase of the reaction rate at constant p_{CO} as indicated by the arrow. The increase in catalytic activity is due to the formation of (100) steps in the facetting process, since these exhibit a higher s_{O_2} than the flat (110) surface.

A detailed LEED beam profile analysis of the facetted surface carried out with a high resolution instrument has shown that facets of uniform size and orientation build a regular array characterized by a lateral periodicity of 70 lattice units along the [110] direction. [7] The facetted Pt(110) surface therefore represents an example for spatial pattern formation in a non-equilibrium system similar to other systems in physics.

The elementary step in facetting is clearly the 1x1 $\leftrightarrows$ 1x2 PT, since the mass transport of 50 % of the surface atoms necessarily creates steps. The oscillatory mechanism causes the surface to undergo the PT many times, and in this way steps accumulate, leading to a facetting of the surface. This mechanism has been demonstrated in a Monte Carlo simulation based on the steps of the LH mechanism and on the properties of the 1x1 $\leftrightarrows$ 1x2 PT. [8]

Shown in fig. 5 are profiles of the Pt(110) surface which is subjected to a continuous flow of CO and O_2. A periodic structure with a lateral periodicity of 40 lattice units develops, as more and more steps are created as a result of local CO coverage fluctuations which drive the 1x1 $\leftrightarrows$ 1x2 PT. In order to demonstrate that the facetted surface is in fact a Turing structure, e.g. only stabilized by an ongoing surface reaction, the gas flow of CO and O_2 was stopped after 4000 cycles. As in the experiment the thermal reordering process which is then dominant alone, leads to restoration of the flat surface.

Since the facetting modifies the catalytic activity, periodic changes in the degree of facetting may also cause kinetic oscillations. This has been observed with Pt(110)/CO + O_2, where the steepness of the facets varied periodically with the oscillations in the reaction rate. [6] On Pt(210) which can serve as a model surface for high-index planes of Pt, a two-stage mechanism for kinetic oscillations was verified. In CO + O_2 the Pt(210) surface first facets into a (310) and a (110) orientation with the latter then driving the kinetic oscillations via the well-known 1x1 $\leftrightarrows$ 1x2 PT mechanism. [9]

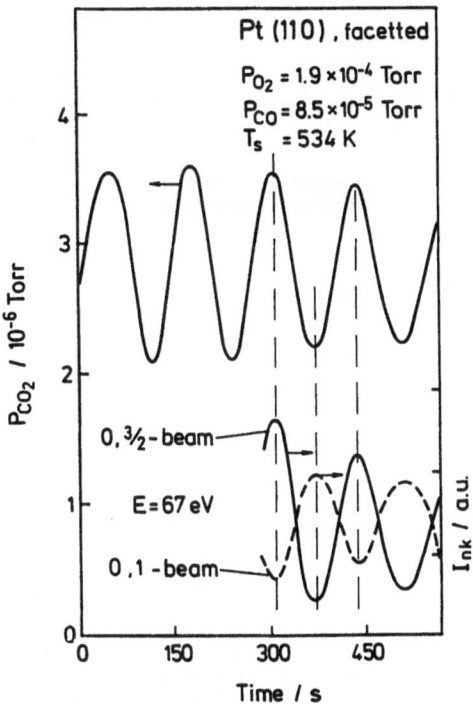

Fig. 2. Periodic structural changes during kinetic oscillations on Pt(110) as monitored through the variation of the 0,3/2 beam of the 1x2 surface. (After ref. 4).

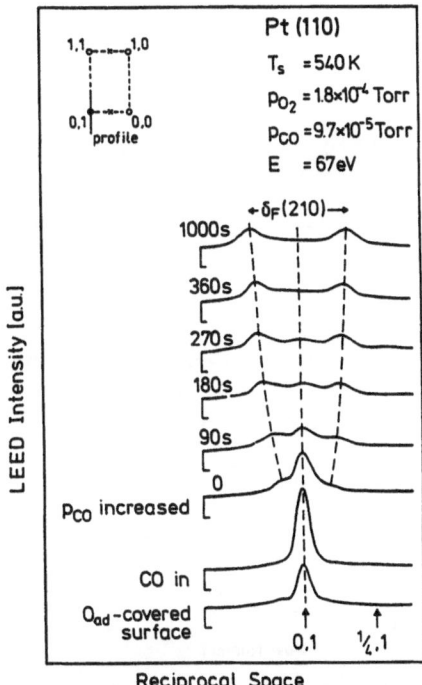

Fig. 3.   Development of facetting during catalytic CO oxidation on Pt(110). LEED beam profiles served to monitor the progress of facetting. (After ref. 5).

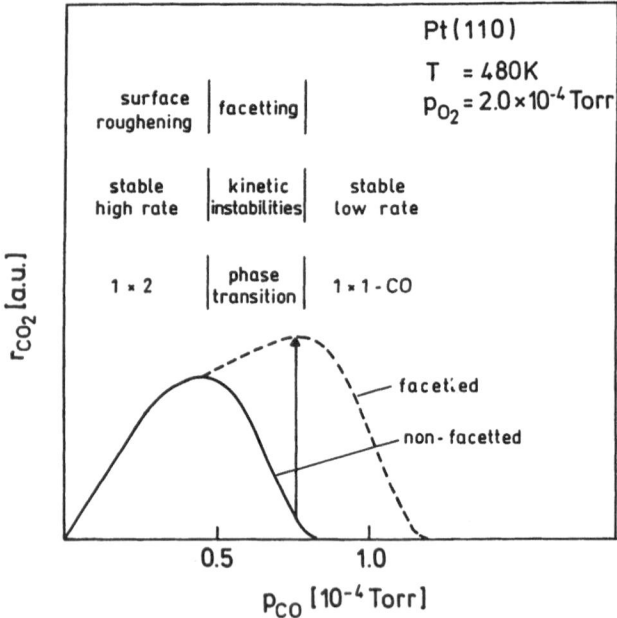

Fig. 4. Relation between the conditions for facetting and the kinetics of the catalytic CO oxidation on Pt(110). The full line indicates the rate curve for the non-facetted surface, while the dashed curve indicates the increase in catalytic activity after strong facetting of the surface. The different regions indicated on top of the rate curve all refer to the non-facetted Pt(110) surface. (After ref. 7).

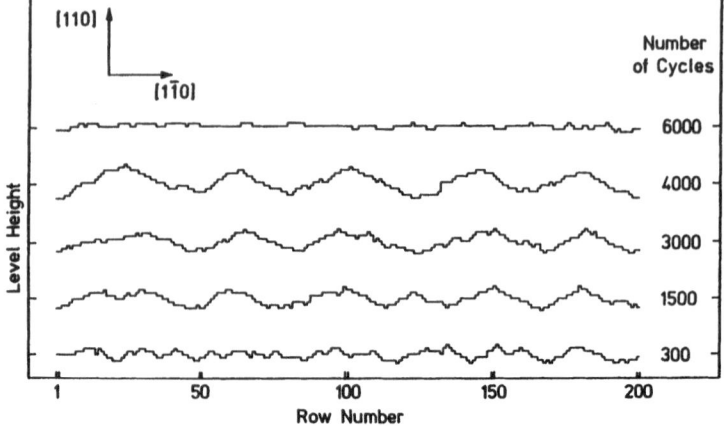

Fig. 5. Monte Carlo simulation showing the development of a regular facet structure during the catalytic CO oxidation on Pt(110) followed by a restoration of the flat surface after stopping the gas flow at 4000 cycles. (After ref. 8).

CONCLUSIONS

The study of kinetic oscillations on Pt single crystal surfaces has demonstrated that it is possible to gain a detailed microscopic picture of the mechanism using the techniques of surface science. The change in catalytic activity proceeds via reversible structural transformations on Pt surfaces. These involve either a single layer as in the case of an adsorbate-induced PT or they may extend to several layers in a facetting process.

REFERENCES

1. G. Nicolis and I. Prigogine, "Self-Organization in Nonequilibrium Systems", Wiley, New York (1977).
2. R. Imbihl, The study of kinetic oscillations in the catalytic CO oxidation on single crystal surfaces, in: "Optimal Structures in Heterogeneous Reaction Systems", Vol. 44, p. 26, P. J. Plath, ed., Springer Series in Synergetics, Berlin (1989).
3. G. Ertl, Oscillatory catalytic reactions at single crystal surfaces, Adv. in Catal., 37:XXX (1990).
4. M. Eiswirth, P. Möller, K. Wetzl, R. Imbihl and G. Ertl, Mechanism of spatial self-organization in isothermal kinetic oscillations during the catalytic CO oxidation on Pt single crystal surfaces, J. Chem. Phys., 90:510 (1989).
5. S. Ladas, R. Imbihl and G. Ertl, Microfacetting of a Pt(110) surface during catalytic CO oxidation, Surf. Sci., 197:153 (1988).
6. S. Ladas, R. Imbihl and G. Ertl, Kinetic oscillations and facetting during the catalytic CO oxidation, Surf. Sci. 198:42 (1988).
7. J. Falta, R. Imbihl and M. Henzler, Formation of periodic facet structures in the catalytic CO-oxidation on Pt(100), Phys. Rev. Lett., 64:1409 (1990).
8. R. Imbihl, D. Kaletta and A. E. Reynolds, Computer simulation of a nonequilibrium surface structure: The facetting of Pt(110) in $CO + O_2$, Phys. Rev. Lett., submitted.
9. M. Sander, Thesis, Technical University, Berlin (1990).
10. S. Jakubith, H. H. Rotermund, W. Engel, A. von Oertzen and G. Ertl, Spatio-temporal concentration patterns in a surface reaction: Propagating and standing waves, rotating spirals, and turbulence, Phys. Rev. Lett., submitted.

MOLECULAR DYNAMICS

FOR REACTIONS OF HETEROGENEOUS CATALYSIS

A.P.J. Jansen

Laboratory for Inorganic Chemistry and Catalysis
Eindhoven University of Technology
P.O. Box 513, 5600 MB Eindhoven, The Netherlands

ABSTRACT. An overview is given of Molecular Dynamics, and numerical integration techniques, system initialization, boundary conditions, force representation, statistics, system size, and simulations duration are discussed. Examples from surface science are used to illustrate the pros and cons of the method. Two new methods are presented with which it is possible to compute reaction rates and reaction mechanisms in spite of activation barriers that are much higher than thermal energies, and results are shown for Xe desorption from Pd(100).

INTRODUCTION

Molecular Dynamics is one of the main methods that have led to computational physics and chemistry. It is an incredible powerful method: it gives very detailed information on dynamics, without being restricted to a particular class of systems. Since the first papers on Molecular Dynamics,[1,2] the number of atoms and the length of Molecular Dynamics simulations have been increased enormously, and it has become possible to investigate phenomena which can be treated by analytical methods only with great difficulty: e.g., non-equilibrium phenomena.[3]

The development of Molecular Dynamics has been very closely related to the development of computer hard- and software, but it is not just doing a lot of calculations. It has given rise to important new theoretical developments, especially in classical mechanics and classical statistical physics.[4] We like to mention here new methods to compute free energies,[5,6] and new mechanics like the one of Nosé.[7,8] Also the whole field of non-linear processes and chaos has evolved alongside Molecular Dynamics work.[9-13]

An important driving force is the ease with which Molecular Dynamics can be used nowadays to study large bio-organic molecules. Many good easy-to-use programs have become available for simulating these molecules.[14-16] The reason why they are relatively easy to handle is that reliable interatomic potentials, often called force fields for these molecules, are available. For systems with metal atoms, and especially transition metal atoms, these interatomic potentials are still missing. The delocalization of the electrons in metals makes multi-body potentials necessary, which are computationally very time-consuming. For simple metals the Car-Parrinello method, a combination of Molecular Dynamics and Hartree-Fock, may be used to circumvent

Fundamental Aspects of Heterogeneous Catalysis Studied by Particle Beams,
Edited by H.H. Brongersma and R.A. van Santen, Plenum Press, New York, 1991

133

this problem.[17-19] The subtle interplay of s, p, and d bands however still prevents the use of this method to transition metals.

The absence of good potentials is one of the reasons that Molecular Dynamics has not been applied to any great extent to processes of heterogeneous catalysis. In addition there is the problem that catalysis is usually interested in reactions that have a high activation barrier. These reactions could not be studied well with Molecular Dynamics until recently.

In this paper we start with a general overview of Molecular Dynamics. For a more extensive survey we refer the interested reader to reference 3. We shall give some examples from studies on processes of heterogeneous catalysis to illustrate what information can be obtained. We shall pay particular attention to the problem of the activation barrier, and present a new method to solve it.

THE METHOD

In Molecular Dynamics we try to solve an equation of motion for atomic or molecular systems numerically. In most cases this equation is Newton's equation

$$m\ddot{x} = F, \tag{1}$$

where m is a mass, x a coordinate, and F a force, but it can also be Hamilton's equations, Lagrange's equation,[20] or the Schrödinger equation.[21] In this paper we shall be concerned mainly with Newton's equation. The numerical recipe to solve equation (1) consists of a method that gives us numerical values for the coordinate at points in time $t_0, t_1, \ldots, t_n, t_{n+1}, \ldots$. In most cases the points in time are equidistant.

Numerical solution of Newton's equation

The simplest algorithm to obtain the solution of equation (1) is given by[22]

$$x_{n+1} = x_n + v_n \Delta t + \frac{1}{2}(F_n/m)(\Delta t)^2 \tag{2a}$$

and

$$v_{n+1} = v_n + (F_n/m)\Delta t, \tag{2b}$$

where Δt is the difference between subsequent points in time for which we calculate the coordinate, v is the velocity, and the index indicates that we are looking at quantities at times t_n and $t_{n+1} = t_n + \Delta t$. We obtain equation (2) by making a Taylor expansion around t_n. It is called the Euler-predictor method of second order. It is second order because it gives exact solutions if x is a function of time of second order. It is not a very good method, because it is not invariant under time reversal, and because its order is too low. The former drawback can lead to an artificial loss or gain of energy of the system. The latter can lead to erratic behaviour near a potential energy minimum.[23] As the second derivative of the potential around minima is always positive, it should be included in equations like (2), which requires at least a third-order algorithm.

We have to realize that of all better algorithms the Runge-Kutta algorithms cannot be used, because they involve more than one evaluation of the force per time step, which costs a lot of computer time. Two good algorithms are the Verlet algorithm[22,24]

$$x_{n+1} = 2x_n - x_{n-1} + (F_n/m)(\Delta t)^2 \tag{3}$$

and the leap-frog algorithm[22,25]

$$x_{n+1} = x_n + v_{n+(1/2)}\Delta t, \tag{4a}$$

with

$$v_{n+(1/2)} = v_{n-(1/2)} + (F_n/m)\,\Delta t. \qquad (4b)$$

Both are third-order algorithms, invariant under time reversal. Actually, if we eliminate the velocity in equation (4), we see that they are equivalent. Although there are higher-order algorithms, they are only seldom necessary.[22,26]

The time step Δt must be small to give accurate trajectories, and it must be long to minimize the CPU-time for the simulation. In general a time step of a few femtoseconds is used. With such a time step even the fastest vibrations are simulated accurately. The total length of a simulation is usually tens or hundreds of picoseconds. This depends on the duration of the process of interest, and on the relaxation times of modes that influence that process.

Initialization of the system

If we use an algorithm of high order we need the coordinates and velocities at various points in time. These are not available at the beginning of the simulation, so we have to start with a lower-order algorithm, and choose some starting coordinates and velocities. A problem is that we may accidently choose these coordinates and velocities inappropriately. It can happen that a small part of the system gets too much energy. We then have to give the system some time to distribute the energy by running the simulation for a time longer than some relaxation time: i.e., we let the system find its equilibrium. The relaxation time can be found be computing the velocity autocorrelation function $\langle v(t)v(0)\rangle$, also from the simulation, and look at the exponential decrease.[27]

The number of atoms and boundary conditions

Although Molecular Dynamics simulations have been done with 161 604 atoms,[4,28] one usually simulates systems with only some thousands or fewer atoms. As there are more atoms in a real system, boundary conditions become important.

The simplest procedure is to take a cluster. This is a good approach for very localized processes. We choose the cluster in such a way that the boundary atoms do not affect the process of interest. In general this means that large clusters are necessary.

If, for example, energetic particles are scattered from a surface, then a cluster may well dissociate. This can be prevented by embedding the cluster: i.e., surrounding it with atoms that are fixed, or atoms that can move, but not far from their equilibrium positions. The moving atoms allow for some energy transfer from and to the cluster. It is also possible to couple the cluster to a set of coordinates, called heat bath, ghost particles,[29] or harmonic chain hierarchy,[30] to get the same effect.

An alternative to the cluster approach is formed by periodic boundary conditions. A system with translation symmetry is divided into large unit cells, and corresponding atoms in different unit cells are assumed to be moving identically.[31] The linear dimensions of the unit cells must be larger than the range of the interactions between corresponding atoms. It is not possible to study long-range phenomena. Surfaces have only translation symmetry in at most two dimensions. This means that the third dimension has to be treated as a cluster.

The forces

The main source of disagreement between the simulation and the experiment is the forces. They are, in general, derived from a potential which has been determined via experiments or via quantum chemical calculations.

Principally we need the potential energy of a system for all possible configurations of the atoms in it. It is clear that this energy is not available experimentally.

Instead we assume a form for the potential, and fit its parameters to experimental results. Well-known forms are atom-atom potentials of the Lennard-Jones 12-6, Morse, Buckingham exp-6, harmonic, and Coulomb form.[32] The drawback of all these forms is that they derive from certain special cases: e.g., around a minimum the potential can be approximated by a harmonic form,[20] and the leading term of the long-range dispersion interaction is proportional to R^{-6}, where R is an interatomic distance.[33] If our system does not resemble one of the special cases, a bad potential will result.

There are also forms that are based on the symmetry of the system: e.g., the Fourier expansion of an atom-surface potential.[34-36] Such forms, however, are often complicated.

More and more *ab initio* quantum chemical calculations are used to obtain potentials, but they are still restricted to small systems.[32,37] The energy differences that determine the trajectories of atoms are several orders of magnitude smaller than the electronic energies. This means that tremendously accurate calculations are needed.

Statistics

A Molecular Dynamics simulation generates trajectories for the atoms, which shows directly how a process takes place. To compare the simulation with an experiment one often has to do some statistics. For example, to obtain the direction in which an atom is scattered from a surface we have to do many simulations and get a distribution for the scattering direction.

Thermodynamic variables are not so straight-forward. If there are no external forces and there is no friction, then the total energy does not change: i.e., the system belongs to a microcanonical ensemble.[38,39] Strictly speaking, in such a system one cannot talk about *the* temperature or *the* pressure. The temperature T, for example, given by[4]

$$\frac{1}{2}NkT = \frac{1}{2}\sum_{i=1}^{N} m_i \dot{x}_i^2,$$ (5)

where N is the number of degrees of freedom, fluctuates. If we use more atoms the fluctuations become smaller: i.e., we approach the thermodynamic limit. Newton's equation does not provide a method with which we can do a simulation at a specified temperature and pressure. Therefore, a number of extensions of it has been proposed. By coupling to additional coordinates both constant temperature and constant pressure simulation are possible.[7,8,40-42] Even the shape of a unit cell, if there are periodic boundary conditions, can be allowed to change. In a variant of Molecular Dynamics, called Stochastic Dynamics, Langevin equations are used, which means that effectively the system is coupled to a heat bath.[29,30] Also scaling of velocities and unit cell dimensions has been proposed.[23]

Statistical physics is formulated in terms of ensemble averages: i.e., averages over many systems. In Molecular Dynamics the same average is used if one does many simulations. In addition, if one is interested in an equilibrium value for a thermodynamic variable, one averages over the time steps of the simulation.

EXAMPLES

In this section we present some examples to show the fertility of Molecular Dynamics. Simulations of secondary-ion mass spectrometry by the group of Garrison form a nice example of standard Molecular Dynamics.[43] The surface is represented by a cluster of five or six layers, each containing 100 to 150 atoms. On top of this cluster admolecules can be put. An atom or ion with an energy of about 1 keV is shot at the surface at the beginning of the simulation. During the simulation the motions of all atoms are monitored. By simply looking where the atoms go it is possible to determine what fragments are formed, and in what direction they are scattered. This "looking at the atoms" can be taken quite literally as the identification of the fragments that are formed is much simplified by using a graphic workstation. The results of many simulations are finally compared to experiments. These simulations have yielded great insight into the SIMS process.

Standard Molecular Dynamics using clusters or periodic boundaries requires often a large number of atoms to get converged results. In many cases we are only interested in a local process, so that most of the computer time is spent on computing uninteresting trajectories. The before-mentioned Generalized Langevin Equation formalism does not have this drawback.[30] Only a small number of atoms—the interesting ones—are simulated via Newton's equation. The influence of the surrounding atoms is taken into account by coupling to a set of more or less arbitrary coordinates the motions of which are described by a generalized Langevin equation: i.e.,

$$m\ddot{x} = -fx - \beta\dot{x} + R(t), \tag{6}$$

where f is a force constant, β is a friction coefficient, and $R(t)$ is a random force. The point is that only a small number of extra coordinates are needed to simulate the environment. The friction terms and the random forces simulate energy flows from and to the atoms of interest. They are related via a fluctuation-dissipation theorem, so that we really have a coupling to a heat bath, and we are simulating a canonical ensemble.

The Generalized Langevin Equation formalism has been applied, most notably by Tully, to a variety of surface processes: N_2 scattering from Ag(111),[44] diffusion on surfaces,[45] the reaction $C+O\rightarrow CO$ on Pt(111),[46] and scattering of Ar clusters from Pt(111).[47] The analysis of the results is the same as for the standard Molecular Dynamics.

Although we mentioned the Schrödinger equation in the initial description of Molecular Dynamics, up till now we only talked about classical mechanics. The numerical solution of the Schrödinger equation is much more difficult. Instead of a limited set of variables one deals with a function.[48] Therefore surface processes have been studied with Quantum Molecular Dynamics, notably by the group of Kosloff, for only one and two atoms. In all studies the surface is taken to be rigid, and only the adatom or admolecule is moving. Quantum effects are important for small molecules. This means that especially hydrogen, atomic or molecular, has been studied,[49] but also a study on N_2 adsorption has been published recently.[50,51] The analysis of the results is different from those of a classical study. We obtain probabilities even from one simulation. These probabilities tell us where the atoms are. Other properties are more difficult to obtain.

INFREQUENT EVENTS

Many chemical reactions have an activation barrier. If this barrier is about the same or lower than the thermal energy standard Molecular Dynamics can be used. If the thermal energy is very small compared to the activation energy, then Molecular Dynamics will then not generate reactive trajectories. The reason is that the system has to attempt many times to overcome the activation barrier before it succeeds, hence the name "infrequent events," whereas Molecular Dynamics looks at a few attempts only.[52,53] This is especially true for catalytic processes. Although a catalyst reduces certain activation barriers, the thermal energies at which catalytic processes generally take place are still much lower than the activation barriers. Consequently, standard Molecular Dynamics cannot be used, and the methods we will describe below are particular advantageous for studying reactions on catalysts.

For reaction rates and other thermodynamic variables there is a variety of methods that can be employed.[52,54] They all alter the equations of motion so that reactive trajectories become more frequent. Instead of Newton's equation, Hamilton's equations are used.

$$\dot{q} = \frac{\partial H}{\partial p} \tag{7}$$

and

$$\dot{p} = -\frac{\partial H}{\partial q}, \tag{8}$$

where H is the Hamiltonian of the system, q is a generalized coordinate, and p is its conjugate momentum.[20] We have developed the compensating Hamiltonian method in which the original Hamiltonian H and a new Hamiltonian $\tilde{H}$, of Hamilton's equations, are related via

$$H = \tilde{H} + U, \tag{9}$$

where U is the compensating Hamiltonian.[55] For a canonical ensemble the statistical average $\langle X \rangle_H$ of an arbitrary quantity X, calculated with the original Hamiltonian, is given by

$$\langle X \rangle_H = \langle e^{-U/kT} X \rangle_{\tilde{H}} / \langle e^{-U/kT} \rangle_{\tilde{H}}, \tag{10}$$

where the statistical averages on the right are calculated with the new Hamiltonian.[39] A similar relation holds for reaction rates. The simulations are done with the new Hamiltonian.

We calculated the desorption rate for a Xe atom on a Pd(100) surface from 50 K to 1000 K. The activation barrier is $E_{act}/k = 3756$ K. As compensating Hamiltonian we used

$$U = \tau \left[\frac{p_z^2}{2m_{Xe}} + V_C(z) \right], \tag{11}$$

where $0 \leq \tau < 1$ (depending on the temperature), z is the normal coordinate of the Xe atom, and V_C is the potential along the normal at a 4-fold site. The activation barrier for desorption is in V_C. Depending on τ, part of the activation barrier is taken out of the new Hamiltonian that is used in the simulation, so that the desorption becomes easier. The change in the potential is shown in figure 1. The reason for the kinetic term will be explained later. We find good agreement with the experimental thermodesorption spectrum when we compare figure 2 with the experimental spectrum for Xe/Pd(111), which has a comparable adsorption energy.[56]

The peak in TDS is about 120 K. It is very illustrative to calculate the number of attempts the Xe atom has to make to desorb at that temperature. The calculated

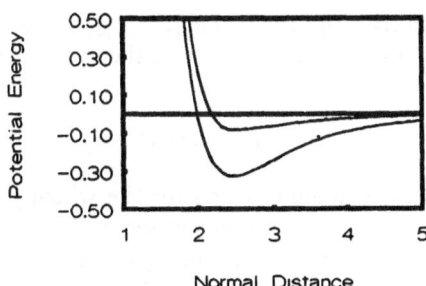

Fig. 1. Potential energy term (in eV) along the normal at a 4-fold site in the new Hamiltonian $\tilde{H}$ as a function of normal distance (in Å) at $T = 1000\,\text{K}$ ($\tau = 0$; lower curve), and at $T = 200\,\text{K}$ ($\tau = 0.75$; upper curve).

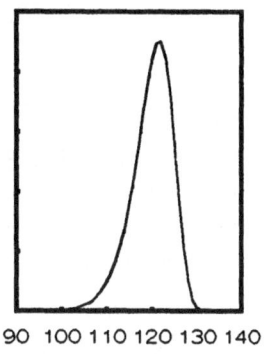

Temperature

Fig. 2. Calculated thermo-desorption spectrum using desorption rates from the compensating Hamiltonian method, and a temperature increase of $3\,\text{K.s}^{-1}$.

desorption rate is $0.849\,\text{s}^{-1}$, and the frequency of the normal vibration is $1.55 \cdot 10^{12}\text{s}^{-1}$. This gives about $1.8 \cdot 10^{12}$ attempts for the desorption. With our computer (a VAX-8530) and our program it would have taken on average about 180 000 years CPU-time to find desorption using standard Molecular Dynamics.

By changing the equations of motion we no longer can get information on the mechanism of a reaction. Keck has suggested to start trajectories at the transition state.[57-59] The problem is then to find the probability at a specific temperature for the exact positions and velocities of all the atoms in the system at the starting point. For some simple systems this problem has been solved.[57-59] We shall show below how the compensating Hamiltonian method can be used to obtain solutions for any system.

The problem is best formulated by viewing the system in its phase space. Assume that we have a function S defined on the phase space of the system so that $S(q_1,\ldots,q_N;p_1,\ldots,p_N) = 0$ defines the activation barrier in phase space.[55-59] The starting point $(q_1',\ldots,q_N';p_1',\ldots,p_N')$ of a trajectory of the system on $S = 0$ has to be chosen with the probability $P(q'_1, \ldots, q'_N; p'_1, \ldots, p'_N)$ proportional to $\rho(q'_1, \ldots, q'_N; p'_1, \ldots, p'_N)$, where ρ is the phase space density of the system. Or

$$P = \frac{\rho}{\int dq\, dp\, \rho\, \delta(S)}, \tag{12}$$

where the integration in the denominator is over the whole phase space, and δ is a delta function. In this way trajectories are generated with the correct probability. We have a temperature dependence via the phase space density. If we assume that we have a canonical ensemble:[38,39] i.e., $\rho \propto \exp(-H/kT)$, then equation (12) becomes

$$P = \frac{e^{-H/kT}}{\int dq\, dp\, e^{-H/kT}\, \delta(S)}. \tag{13}$$

Only for simple systems has it been possible to generate starting points at the activation barrier with probability P.[57-59] We will present here a new method with which it is possible to get P for any system.

We start by looking at $\tilde{P}$: i.e., the same probability as above for the new Hamiltonian $\tilde{H}$. We find

$$\tilde{P} = \frac{e^{-\tilde{H}/kT}}{\int dq\, dp\, e^{-\tilde{H}/kT}\, \delta(S)}. \tag{14}$$

The crucial step is now to define the compensating Hamiltonian U as a function of S. As we have a lot of freedom in choosing S, the only restriction this imposes is that the compensating Hamiltonian U has to be constant on the activation barrier in phase space. We find, substituting equation (9) into equation (13),

$$P = \frac{e^{-\tilde{H}/kT}e^{-U(0)/kT}}{\int dq\, dp\, e^{-\tilde{H}/kT}e^{-U(S)/kT}\, \delta(S)} = \tilde{P}. \tag{15}$$

The probability that a reactive trajectory crosses the activation barrier at a certain point in phase space is the same for the original Hamiltonian H as for the new Hamiltonian $\tilde{H}$. The procedure to generate good trajectories is then as follows. We start with a system that has not crossed the activation barrier, and integrate it forwards in time with the new Hamiltonian $\tilde{H}$ until it crosses the activation barrier. At that moment we switch to the original Hamiltonian H and integrate further forwards, and also backwards. We finally reverse the backward part of the trajectory and put it before the forward part. Equation (15) takes care that the trajectory we obtain is generated with the correct probability. The switch to the original Hamiltonian H has to be made without changing to another point in phase space. This may mean that velocities have to be changed, because they may be related differently to the momenta for the two Hamiltonians.

Neither the choice of the compensating Hamiltonian, nor that of the activation barrier $S = 0$ is very critical. Any compensating Hamiltonian will do if it decreases the activation barrier, although some are preferable because they give Hamilton's equations that are simpler. The activation barrier $S = 0$ only has to fulfil two requirements. First, every reactive trajectory has to cross it. Second, the majority of the trajectories that do cross it should be reactive. The latter requirement means that $S = 0$ does not necessarily have to correspond to configurations of no return in configuration space.

We computed desorptive trajectories for Xe/Pd(100) at a low ($T = 200\,$K) and a high ($T = 1000\,$K) temperature. The function S is equal to the term in square brackets in equation (11). If the Pd atoms are fixed at their equilibrium positions, then for motions of the Xe atom normal to the surface S is constant, because there is no energy transfer between the Xe atom and the surface. The activation barrier $S = 0$ separates then the bound from the non-bound trajectories. This is also the reason for the kinetic energy term in the compensating Hamiltonian U. That desorption is really a fast, but also a rare, event can be seen if we look at the time between the moment the normal distance of the Xe atom to the surface is minimal for the last time and the moment that the activation barrier is crossed (see figure 3). At the high temperature this time is only 100 fs: much shorter than a vibrational period in the normal direction, which is 645 fs. At the low temperature this time increases to 245 fs, but it still is very short, especially compared to the average time it takes a Xe atom to desorb which is 12 500 fs at $T = 1000\,$K and 7.73 μs at $T = 200\,$K.

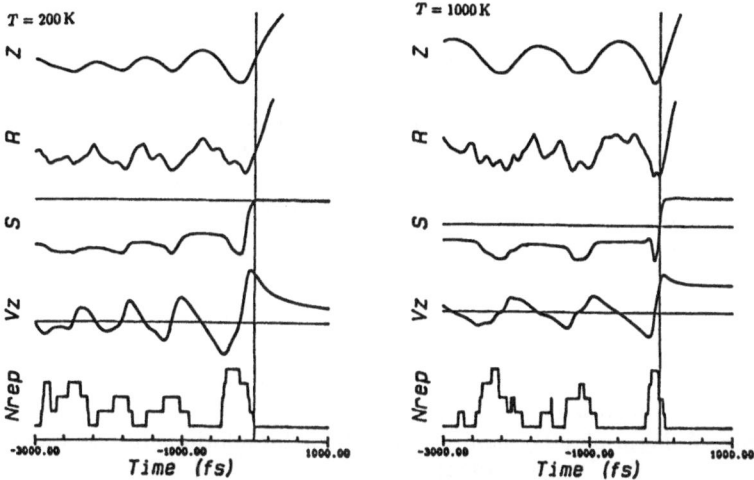

Fig. 3. From top to bottom: normal distance of the Xe atom, distance between Xe and nearest Pd atom, the function S, the velocity of the Xe atom, and the number of Pd atoms that repel the Xe atom. Shown are variations from 3 ps before to 1 ps after desorption. Horizontal lines show where the ordinate is zero.

The trajectories of the Xe atom tell us about the mechanism. At $T = 1000\,$K the Xe atom easily diffuses over the surface, because the barrier for diffusion is only 890 K. Desorption is caused by a collision with at least two, but often more, surface atoms simultaneously. There is no preference for some site from which to desorb. At $T = 200\,$K diffusion is via jumps from one 4-fold site to another. It is an infrequent event at this temperature. Desorption, in most cases, is from a 4-fold site, although we also found trajectories in which the Xe atom first diffuses before it desorbes.

CONCLUSIONS

Two new methods have been presented for Molecular Dynamics simulations when there are activation barriers that are much higher than the thermal energy, as is generally the case in catalysis. The first can be used to compute statistical averages like reaction rates; the second can be used to compute reaction mechanisms.

The first method is a generalization of the work of Grimmelmann et al..[52] The incorporation of the kinetic energy term in equation (11) is however new, and relates the compensating Hamiltonian to the cause of the desorption; i.e., the thermal vibrations of the surface atoms. The second method is a transformation of the idea to start the reactive trajectories at the transition state to phase space. It resolves the problems of what to choose for the velocities at the transition state, and allows the reaction to proceed via configurations different from the transition state. It has been applied to simple systems by Keck,[57-59] except for the use of the compensating Hamiltonian in equation (15), which makes the method generally applicable.

The method will be employed to study coverage dependence of desorption, and the mechanism of the dissociation of CO. Also the influence of defects will be studied.

REFERENCES

1. B.J. Alder and T.E. Wainwright, Phase Transition for a hard Sphere System, *J. Chem. Phys.* 27:1208 (1957).

2. B.J. Alder and T.E. Wainwright, Studies in Molecular Dynamics: II. Behaviour of a Small Number of Elastic Spheres, *J. Chem. Phys.* 33:1439 (1960).

3. G. Ciccotti and W.G. Hoover, eds., *Proceedings of the International School of Physics "Enrico Fermi", Course XCVII*, North-Holland, Amsterdam (1986).

4. W.G. Hoover, *Molecular Dynamics*, Springer, Berlin (1986).

5. D. Frenkel, Free-Energy Computation and First-Order Phase Transitions, in *Proceedings of the International School of Physics "Enrico Fermi", Course XCVII*, G. Ciccotti and W.G. Hoover, ed., North-Holland, Amsterdam (1986).

6. W.F. van Gunsteren and P.K. Weiner, eds., *Computer Simulation of Biomolecular Systems*, ESCOM, Leiden (1989).

7. S. Nosé, A Molecular Dynamics Method for Simulations in the Canonical Ensemble, *Mol. Phys.* 52:255 (1984).

8. S. Nosé, A Unified Formulation of the Constant Temperature Molecular Dynamics Methods, *J. Chem. Phys.* 81:511 (1984).

9. E. Fermi, J. Pasta, and S. Ulam, Los Alamos Report No. LA-1940 (1955).

10. E. Fermi, *Collected Papers of Enrico Fermi*, University of Chicago Press, Chicago (1962).

11. J.L. Tuck and M.T. Menzel, The Superperiod of the Nonlinear Weighted String (Fermi-Pasta-Ulam) Problem, *Adv. Math.* 9:399 (1972).

12. G. Benettin, Ordered and Chaotic Motions in Dynamical Systems with Many Degrees of Freedom, in *Proceedings of the International School of Physics "Enrico Fermi", Course XCVII*, G. Ciccotti and W.G. Hoover, ed., North-Holland, Amsterdam (1986).

13. J.M.T. Thompson and H.B. Stewart, *Nonlinear Dynamics and Chaos*, Wiley, Chichester (1986).

14. U.C. Singh, P.K. Weiner, J.W. Caldwell, and P.A. Kollman, *AMBER (UCSF) version 3.0*, Department of Pharmaceutical Chemistry, University of California, San Francisco (1986).

15. B.R. Brooks, R.E. Bruccoleri, B.D. Olafson, D.J. States, S. Swaminathan, and

M. Karplus, CHARMM: A Program for Macromolecular Energy, Minimization, and Dynamics Calculations, *J. Comput. Chem.* 4:187 (1983).

16. W.F. van Gunsteren and H.J.C. Berendsen, *Groningen Molecular Simulation (GROMOS) Library Manual*, BIOMOS, Groningen (1987).

17. R. Car and M. Parrinello, Unified Approach for Molecular Dynamics and Density-Functional Theory, *Phys. Rev. Lett.* 55:2471 (1985).

18. R. Car and M. Parrinello, The Unified Approach to Density Functional and Molecular Dynamics in Real Space, *Solid State Commun.* 62:403 (1987).

19. R. Car and M. Parrinello, Structural, Dynamical, and Electronic Properties of Amorphous Silicon: An Ab Initio Molecular-Dynamics Study, *Phys. Rev. Lett.* 60:204 (1988).

20. H. Goldstein, *Classical Mechanics*, Addison-Wesley, Reading (1980).

21. A. Messiah, *Quantum Mechanics*, North-Holland, Amsterdam (1969).

22. H.J.C. Berendsen and W.F. van Gunsteren, Practical Algorithms for Dynamics Simulations, in *Proceedings of the International School of Physics "Enrico Fermi"*, Course XCVII, G. Ciccotti and W.G. Hoover, ed., North-Holland, Amsterdam (1986).

23. W.F. van Gunsteren, Classical Molecular Dynamics Simulations: Algorithms and Application, Stochastic Dynamics, and Free Energy, in *Mathematical Frontiers in Computational Chemical Physics*, D.G. Truhlar, ed., Springer, New York (1988).

24. L. Verlet, Computer "Experiments" on Classical Fluids. I. Thermodynamical Properties of Lennard-Jones Molecules, *Phys. Rev.* 159:98 (1967).

25. R.W. Hockney and J.W. Eastwood, *Computer Simulations Using Particles*, McGraw-Hill, London (1981).

26. C.W. Gear, *Numerical Initial Value Problems in Ordinary Differential Equations*, Prentice-Hall, New York (1971).

27. P.A. Madden, Simulation of Properties of Spectroscopic Interest, in *Proceedings of the International School of Physics "Enrico Fermi"*, Course XCVII, G. Ciccotti and W.G. Hoover, ed., North-Holland, Amsterdam (1986).

28. F.F. Abraham, W.E. Rudge, D.J. Auerbach, and S.W. Koch, Molecular-Dynamics Simulations of the Incommensurate Phase of Krypton on Graphite Using More Than 100000 Atoms, *Phys. Rev. Lett.* 52:445 (1984).

29. J.C. Tully, Dynamics of Gas-Surface Interactions: 3D Generalized Langevin Model Applied to FCC and BCC Surfaces, *J. Chem. Phys.* 73:1975 (1980).

30. S.A. Adelman, Chemical Reaction Dynamics in Liquid Solution, *Adv. Chem. Phys.* 53:61 (1983).

31. K. Binder, Introduction: Theory and "Technical" Aspects of Monte Carlo Simulations, in *Monte Carlo Methods in Statistical Physics*, K. Binder, ed., Springer, Berlin (1986).

32. J.N. Murrell, S. Carter, S.C. Farantos, P. Huxley, and A.J.C. Varandas, *Molecular Potential Energy Functions*, Wiley, Chichester (1984).

33. A. van der Avoird, P.E.S. Wormer, F. Mulder, and R. Berns, Ab Initio Studies of the Interactions in Van der Waals Molecules, in *Topics in Current Chemistry*, Vol. 93, F.L. Boschke, ed., Springer, Berlin (1980).

34. W.J. Briels, A.P.J. Jansen, and A. van der Avoird, Dynamics of Molecular Crystals, *Adv. Quant. Chem.* 18:131 (1986).

35. W.A. Steele, The Physical Interaction of Gases with Crystalline Solids: I. Gas-Solid Energies and Properties of Isolated Adsorbed Atoms, *Surf. Sci.* 36:317 (1973).

36. H. Hoinkes, The Physical Interaction Potential of Gas Atoms with Single-Crystal

Surfaces, Determined from Gas-Surface Diffraction Experiments, *Rev. Mod. Phys.* 52:933 (1980).

37. G.C. Schatz, Analytical Representation of Electronic Potential-Energy Surfaces, *Rev. Mod. Phys.* 61:669 (1989).

38. R. Becker, *Theorie der Wärme*, Springer, Berlin (1978).

39. D.A. McQuarrie, *Statistical Mechanics*, Harper, New York (1976).

40. H.C. Andersen, Molecular Dynamics Simulations at Constant Pressure and/or Temperature, *J. Chem. Phys.* 72:2384 (1980).

41. M. Parrinello and A. Rahman, Polymorphic Transitions in Single Crystals: A New Molecular Dynamics Method, *J. Appl. Phys.* 52:7182 (1981).

42. S. Nosé and M.L. Klein, Constant Pressure Molecular Dynamics for Molecular Systems, *Mol. Phys.* 50:1055 (1983).

43. D.W. Brenner and B.J. Garrison, Gas-Surface Reactions: Molecular Dynamics Simulations of Real Systems, *Adv. Chem. Phys.* 76:281 (1989).

44. A.C. Kummel, G.O. Sitz, R.N. Zare, and J.C. Tully, Direct Inelastic Scattering of N_2 from Ag(111). III. Normal incident N_2, *J. Chem. Phys.* 89:6947 (1988).

45. J.C. Tully, G.H. Gilmer, and M. Shugard, Molecular Dynamics of Surfaces Diffusion. I. The Motion of Adatoms and Clusters, *J. Chem. Phys.* 71:1630 (1979).

46. J.C. Tully, Dynamics of Gas-Surface Interactions: Reaction of Atomic Oxygen with Adsorbed Carbon on Platinum, *J. Chem. Phys.* 73:6333 (1980).

47. G.-Q. Xu, R.J. Holland, S.L. Bernasek, and J.C. Tully, Dynamics of Cluster Scattering from Surfaces, *J. Chem. Phys.* 90:3831 (1989).

48. H. Tal-Ezer and R. Kosloff, An Accurate and Efficient Scheme for Propagating the Time-Dependent Schrödinger Equation, *J. Chem. Phys.* 81:3967 (1984).

49. M. Hand and S. Holloway, The Scattering of H_2 and D_2 from Cu(100): Vibrationally Assisted Dissociative Adsorption, *Surf. Sci.* 211/212:940 (1989).

50. M. Asscher, O.M. Becker, G. Haase, and R. Kosloff, A Quantum Mechanical Mechanism for the Dissociative Chemisorption of N_2 on Metal Surfaces, *Surf. Sci.* 206:L880 (1988).

51. G. Haase, M Asscher, and R. Kosloff, The Dissociative Chemisorption Dynamics of N_2 on Catalytic Metal Surfaces: A Quantum Mechanical Tunneling Mechanism, *J. Chem. Phys.* 90:3346 (1989).

52. E.K. Grimmelmann, J.C. Tully, and E. Helfand, Molecular Dynamics of Infrequent Events: Thermal Desorption of Xenon from a Platinum Surface, *J. Chem. Phys.* 74:5300 (1981).

53. J.C. Tully, Dynamics of Gas-Surface Interactions: Thermal Desorption of Ar and Xe from Platinum, *Surf. Sci.* 111:461 (1981).

54. C.H. Bennett, Mass Tensor Molecular Dynamics, *J. Comput. Phys.* 19:267 (1975).

55. A.P.J. Jansen, Compensating Hamiltonian Method for Chemical Reaction Dynamics: Xe Desorption from Pd(100), *J. Chem. Phys.* (submitted).

56. K. Wandelt and J.E. Hulse, Xenon Adsorption on Palladium. I. The Homogeneous (110), (100), and (111) Surfaces., *J. Chem. Phys.* 80:1340 (1984).

57. J.C. Keck, Variational Theory of Chemical Reaction Rates Applied to Three-Body Recombinations, *J. Chem. Phys.* 32:1035 (1960).

58. J.C. Keck, Statistical Investigation of Dissociation Cross-Sections for Diatoms, *Discuss. Faraday Soc.* 33:173 (1962).

59. J.C. Keck, Variational Theory of Reaction Rates, *Adv. Chem. Phys.* 13:85 (1967).

IS THERE A DISTRIBUTION OF TRANSITION STATE ENERGIES IN THE

REACTION COORDINATE OF CO OXIDATION ON PT FOIL?

George W. Coulston[†] and Gary L. Haller

Department of Chemical Engineering, Yale University
New Haven, Connecticut 06520, USA

[†] Current Address:
Physics Department, University of Kaiserslautern
D6750 Kaiserslautern, West Germany

INTRODUCTION

The oxidation of CO on the noble metals is one of the few surface catalyzed reactions studied from a dynamic perspective and we have recently reviewed this literature [1]. By dynamics, we mean the study of reactions at a level that provides information about the potential energy surface (PES) governing the motion of the nuclei throughout the reaction, usually being satisfied to consider a small region of the PES known as the transition state. Carbon monoxide oxidation is particularly well suited for dynamic studies because product CO_2 is thought to desorb immediately after having passed through the transition state and, therefore, the partitioning of reaction energy in the product degrees of freedom should be related more to the structure of the transition state than to the dynamics of desorption of an accommodated CO_2 molecule. Indeed, excess translational, rotational, and vibrational energy has been detected in the CO_2 produced on the noble metals, direct evidence for the rapidity of the desorption process.

Since the amount of energy which must be available for partitioning increases rapidly with the level of vibrational excitation in the molecule, e.g. each quantum of antisymmetric stretch in CO_2 corresponds to approximately 7 Kcal/mol, the population distribution of the vibrational levels becomes a measure of the distribution of transition state energies. For example, if a constant fraction β of the available reaction energy goes into product vibrational modes, then the measured product vibrational temperature would be β times the temperature characterizing the distribution of transition state energies, which need not be the surface temperature. The above example assumes, of course, that the vibrational distribution of the product and the distribution of transition state energies can be described by Boltzmann distributions which, in general, is probably not correct although the implication that the two distributions are connected is more generally correct. It thus becomes useful to characterize the population distribution of the product vibrational levels and, in what follows, we discuss some of our recent work on the dynamics of CO oxidation on Pt foil where we have determined directly this distribution for product CO_2.

Fundamental Aspects of Heterogeneous Catalysis Studied by Particle Beams,
Edited by H.H. Brongersma and R.A. van Santen, Plenum Press, New York, 1991

EXPERIMENTAL

The primary tool in this study involved the collection of infrared chemiluminescence, whereby the populations of excited rovibrational states in nascent CO_2 were determined in a nearly collisionless environment. The experiments were performed in a He cryogenically pumped UHV chamber, equipped with Auger, ion sputtering, and mass spectrometric capabilities. The chamber was optically coupled to an FTIR spectrometer which was operated at $0.012 cm^{-1}$ resolution for the high resolution experiments and at $8 cm^{-1}$ resolution for the low resolution experiments. To obtain good S/N it was necessary to signal average the high resolution spectra for 12 hrs, while the low resolution spectra took less than 5 min to collect. The system was stable over the 12 hr period as evidenced by the low resolution vibrational spectra, which were identical before and after the high resolution experiments. It is important to note that our liquid N_2 cooled InSb detector could only detect emission of photons associated with the loss of a single antisymmetric stretch quantum in CO_2, although, we were able to partially characterize the symmetric stretch and the bending vibrational modes by looking at their combination bands with the antisymmetric stretch. The reactants, CO and O_2, were delivered to the 814K Pt foil surface using a free jet nozzle source with a total reactant flux of $6.7 \times 10^{18} cm^2/s$ and with an associated density above the surface of $1 \times 10^{14} cm^{-3}$. Under these conditions, our most conservative estimates [2] indicated that 96% of the CO_2 in the field of view has avoided all CO_2-CO_2 collisions prior to detection, while nearly 80% have had at most a single CO_2-reactant collision. It is the CO_2-CO_2 collisions that are most relevant to the question of collision induced vibrational relaxation and, in this regard, it is clear that we are operating under near collisionless conditions.

RESULTS AND DISCUSSION

The high resolution spectrum, comprised of thousands of rovibrational transitions, could be analyzed for the relative populations of the 15 lowest lying vibrational levels that contained at least one quantum of antisymmetric stretch vibration. This was accomplished by taking a J by J average of the population of a given vibrational level relative to that of the 00^01 antisymmetric stretch fundamental. Such an averaging procedure was possible because the rotational distribution was found to be independent of the level of vibrational excitation [3]. The result is shown in Figure 1 in the form of a vibrational Boltzmann plot. The relative populations can be seen to fall upon a common line consistent with an apparent vibrational temperature of 1580K$\pm$40K. This result is qualitatively similar to our result on Rh [3], i.e. the relative populations lie on a common line when plotted in Boltzmann form , but they are dramatically different from our results on Pd [3, 4] which showed a clear preference for population of those levels with at least one quantum of both symmetric and antisymmetric stretch, e.g. the apparent symmetric stretch temperature was more than 1000K hotter than either the apparent bending or antisymmetric stretch temperatures for reaction on Pd. Whereas our reported apparent temperature for reaction on Pt is based on a direct measurement of the relative vibrational level populations, Mantell et al. [5] report apparent temperatures for reaction on Pt which were determined by fitting a model to a $0.06 cm^{-1}$ resolution spectrum. In qualitative agreement with our result, they found $T_{anti} = 1500K$, $T_{sym} = 1300K$, $T_{bend} = 1600K$, and $T_{anti} = 1600K$, $T_{sym} = 1700K$, and $T_{bend} = 1750K$ for reaction at 730K and 900K, respectively.

We have been careful to use the terminology *apparent* temperature in our description of the vibrational distributions for CO_2 because use of the term temperature would imply that the straight line drawn through the data of Figure 1 could be extrapolated to all higher lying vibrational levels. In order to ascertain the validity of such an extrapolation we collected a low resolution spectrum of CO_2 produced under the

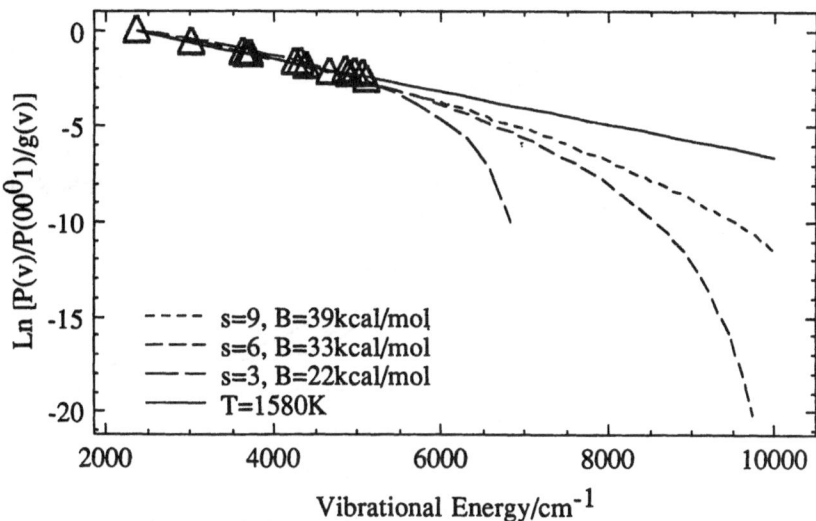

Figure 1. Vibrational Boltzmann plot for CO_2 produced on Pt foil at 814K. The open triangles are measured relative populations, and in order of increasing energy they are the 00^01, 01^11, 02^01, 02^21, 10^01, 03^11, 03^31, 11^11, 00^02, 04^01, 04^21, 04^41, 12^01, 12^21, and 20^01 levels. In the notation $v_1v_2^1v_3$, v_1 is the symmetric stretch quantum number, v_2 the bending quantum number, v_3 the antisymmetric stretch quantum number, and the superscript 1 is the projection of angular momentum on the molecular axis. The lines show the various extrapolations used in the models which are discussed in the text.

same conditions as the high resolution spectrum that was used to construct Figure 1 and attempted to predict the contour of the low resolution vibrational band using simulations based on two different models. In both models, the intensity of a particular rovibrational transition was related to the population of the upper state through [2]

$$i(v_1v_2^1v_3;J) = \alpha\ S(J^{'}:J^{''})\ v_3\ v^4\ D(J)\ N(v_1v_2^1v_3) \qquad (1)$$

where $N(v_1v_2^1v_3)$ is the density of molecules in the initial state which, in our experiments, must have at least one quantum of antisymmetric stretch, v is the photon energy, $S(J':J'')$ is the rotational linestrength, $D(J)$ is the rotational population distribution and α is a factor that accounts for the effect of anharmonicity on the lifetimes of the vibrational levels, the value of which was determined from data compiled in the literature for CO_2 [2, 6]. Both models used the measured relative populations of all those levels plotted in Figure 1 and assumed that the rotational distribution measured for the 00^01 vibrational level was the same for all vibrational levels. The validity of this latter assumption was easily verified for all of the vibrational levels plotted in Figure 1. The two models differed only in the assumptions used to determine the populations of the higher lying vibrational levels, i.e. $N(v_1v_2^1v_3)$ in Equation 1.

The first model assumed that the relative populations of all higher lying vibrational levels could, indeed, be described by the straight line that passed through the levels plotted in Figure 1, i.e. that our *apparent* temperature of 1580K was a real temperature. The simulated spectrum based on this model, shown in Figure 2 along with the measured low resolution spectrum, is a poor reproduction of the measured spectrum, especially when one considers that, above 2350 cm^{-1}, the simulated band shape is

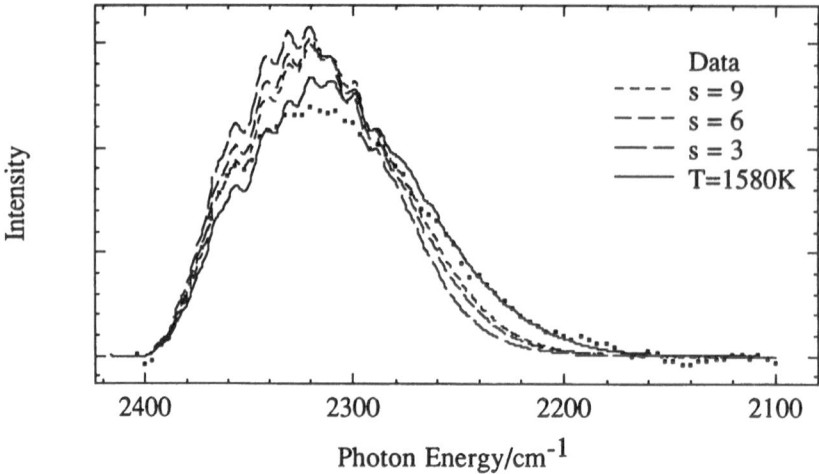

Figure 2. Low resolution spectrum (8cm^{-1}) of CO_2 produced on Pt at 814K along with simulations based on the models described in the text.

based entirely on measured relative populations and precisely known spectroscopic constants [6]. The deviation in this region is due to the fact that we scaled the model to have the same integrated intensity as the measured spectrum and it thus appears that the model puts too much intensity in the spectrum at the longer wavelengths, i.e. where the higher lying vibrational levels contribute to the spectrum. Mantell et al. [5] achieved better agreement between an equilibrium model and their measured low resolution spectra for CO_2 produced on Pt at 730K and 900K than we find here. However, their model differed from ours in that the relative populations of all vibrational levels were determined using Boltzmann distribution functions, whereas our model used the measured relative populations for the first 15 levels. We also compared the results of our equilibrium model to the low resolution spectrum of CO_2 produced in a CH_4-air diffusion flame and obtained reasonable fits assuming a temperature of 2050K [2]. This temperature was slightly lower than the adiabatic flame temperature (2200K) which represents an upper bound to the actual temperature, the difference being attributed to H_2O in the fuel and to heat transfer to the walls of the burner. Thus, based on our results, we conclude that the higher lying vibrational levels are not equilibrated with those plotted in Figure 1 and, furthermore, that the correct extrapolation must be concave down.

We developed a second model which met the requirement that the extrapolation be concave down. This second model, similar to the so-called prior distributions applied to the study of gas phase reactions [7], differed from the equilibrium model in that the system was assumed to have a well defined energy and the entropy of the system was not maximized. We envisioned the following process: the adsorbed CO and oxygen entered into a transition state complex involving some number of surface oscillators s and this complex contained an amount of energy B greater than the energy of ground state CO_2 in the gas phase; the CO_2 exited the transition state and entered directly into the gas phase with the energy B being partitioned into those degrees of freedom that comprised the transition state. The probability of partitioning the energy in a particular way was taken to be proportional to the total density of states [2]

$$P(B,G,E_{rot}, E_{surf}) = \frac{A\sqrt{(B - G - E_{rot} - E_{surf})}\,(E_{surf})^{s-1}}{(s-1)!} \qquad (2)$$

which is equal to the product of the surface densites of states with the vibrational, rotational, and translational densities of states in CO_2. In Equation 2, the constant A is necessary for normalization. Since the spectrometer is sensitive to the density of CO_2 in the field of view and since we are interested in the total number of ways of partioning energy G into the CO_2 vibrational modes, we divide Equation 2 by the speed of the molecules and integrate over E_{surf} with the constraint that the total system energy is equal to B. The result is given in Equation 3.

$$N(B,G,E_{rot}) = A\int_0^{(B-G-E_{rot})} (E_{surf})^{s-1} dE_{surf} = \frac{A(B-G-E_{rot})^s}{s!} \qquad (3)$$

To use Equation 3 in a systematic way we considered three cases: s=3, 6 and 9 loosely corresponding to 1, 2, and 3 Pt atoms in the transition state. The value of B for a given s was chosen such that Equation 3 passed through the measured relative populations plotted in Figure 1. As the number of surface oscillators is decreased (forcing B to decrease in Equation 3) the curves in Figure 1 fall off at lower vibrational energy since the energy partitioned into CO_2 cannot exceed B. One should not be mislead by the trend shown in Figure 1, i.e. in the limit of large s the prior distribution does not correspond to a Boltzmann disbribution since, unlike the case for equilibrium, prior distributions do not maximize the entropy of the system. The simulations using Equation 3 for determining the intensities of the transitions from the higher lying vibrational levels are shown in Figure 2. Once again, the fits are quite poor suggesting that the vibrational energy in product CO_2 is not distributed based on densities of states alone.

Since the prior distributions are defined assuming that there is a well defined transition state energy, the failure of the prior distributions to adequately simulate the data is consistent with there being a distribution of transition state energies. However, this failure is not sufficient to prove there is such a distribution since the development of the reaction dynamics over the full PES needn't result in a partitioning of energy based on densities of states as assumed in the development of the prior distributions. Of course, since these experiments were performed on polycrystalline Pt—probably possesing a distribution of catalyst sites—one might have expected the results to implicate the presence of a distribution of transition state energies. A distribution of transition state energies could also arise from the fact that the critical configuration of the transition state comprises an extended region of classical phase space, such that the transition state becomes ill-defined. Assuming then that a distribution of transition state energies does exist our result that the vibrational populations cannot be described by a Boltzmann distribution indicates that the transition state energies are not characterized by a Boltzmann distribution, that the PES is not such that a constant fraction β of the energy goes into product vibrational modes, or both.

REFERENCES

1. Coulston, G.W. & Haller, G.L., Dynamics of Heterogeneously Catalyzed Reactions, in: "Catalysis: Science and Technology," M. Boudart andJ. Anderson, ed., Springer, (1990).
2. Coulston, G.W., PhD Dissertation, Yale University (1990)
3. Coulston, G.W. & Haller, G.L., in preparation.
4. Coulston, G.W. & Haller, G.L., J. Chem. Phys. 92:5752 (1990).

5. Mantell, D.A., Kunimori, K., Ryali, S.B., Haller, G.L. & Fenn, J.B., <u>Surf. Sci.</u> 172:281 (1986).
6. Rothman, L., <u>Appl. Opt.</u> 25:1795 (1986).
7. Levine, R.D. & Bernstein, R.B., "Molecular Reaction Dynamics and Chemical Reactivity," Oxford University Press, New York (1987).

NUCLEAR MAGNETIC RESONANCE SPECTROSCOPY IN STUDIES OF

CATALYSTS

Harry Pfeifer

University of Leipzig
Linnéstr. 5
Leipzig 7010

INTRODUCTION

During the last decades nuclear magnetic resonance (nmr) spectroscopy has found widespread application in the field of heterogeneous catalysis. Its success is a consequence of the world-wide introduction of zeolites as catalysts and selective adsorbents since their well-defined structure as porous crystallites corresponding to specific surface areas of up to 1000 m^2/g allows the application of nmr spectroscopy in spite of its relatively poor sensitivity compared with infrared spectroscopy or particle beam methods. In the following, at first basic principles of nmr spectroscopy including typical examples for their applicability in studies of solid catalysts and molecules adsorbed thereon are given. Then in the second paragraph the unique possibilities of the ^{1}H mas nmr method are demonstrated to characterize quantitatively Broensted acidity of catalysts and finally in the third paragraph principles of the pulsed field gradient nmr method and its application to study molecular diffusion in porous crystallites are presented including the so-called nmr tracer desorption technique which is a unique method to measure transport resistances at the outer surfaces of porous crystallites.

BASIC PRINCIPLES OF NMR/ TYPICAL EXAMPLES OF APPLICATION

Elementary Theory of NMR

In the following we shall assume for simplicity that the spin quantum number I of the nuclei under study is 1/2 , exceptions are mentioned in the text. If we denote by μ_x , μ_y and μ_z the expectation values of the x, y and z components of the magnetic moment of a nucleus respectively, their values are given in a static magnetic field of intensity B_0 under the conditions of thermal equilibrium by the following equations:

$$\mu_x = \mu_y = 0 \tag{1}$$

$$\mu_z = (\gamma \hbar /2) * \{\exp(r) - \exp(-r)\} / \{\exp(r) + \exp(-r)\}$$

$$r = \gamma \hbar B_0/2kT \tag{2}$$

where γ denotes the magnetogyric ratio of the nucleus (for ^{1}H the value is $\gamma = \gamma_p = 26752.2128(81)*10^4$ s^{-1} T^{-1}), $\hbar$ Planck's constant divided by 2π , k Boltzmann's constant and T the absolute temperature.
For $B_0 = 0$ it follows $\mu_x = \mu_y = \mu_z = 0$ as it should be. For

$$\gamma \hbar B_0 \ll 2kT \tag{3}$$

i.e. for protons and $B_0 = 5$ T a temperature much greater than $5*10^{-3}$ K which is generally denoted as the "high temperature approximation" equ.(2) simplifies to

$$\mu_z = \gamma^2 \hbar^2 B_0/4kT = \mu_{z0} \tag{4}$$

In the following discussions we shall always make use of this approximation which is very well satisfied in all experiments of practical interest.
If the static magnetic field of intensity B_0 is switched on at time t = 0 the thermal equilibrium value of μ_z is achieved only after a certain time has elapsed, and in most cases the dependence can be described by

$$\mu_z(t) = \mu_{z0} * \{ 1 - \exp(-t/T_1) \} \tag{5}$$

where T_1 is the so-called longitudinal or thermal nuclear magnetic relaxation time which depends on the sample under study, on B_0 and on the temperature T. At a magnetic field of intensity $B_0 = 5$ T and a temperature T = 300 K the value of T_1 is 3.6 s for bulk water[1] and 200 ms for water adsorbed on a zeolite NaX of high purity[2].
If in addition to the static magnetic field of intensity B_0 along the z-direction an alternating magnetic field of intensity $2B_1*\cos\gamma B_0 t$ is applied along the x-direction during the time interval from t = $- t_w$ to t = 0 and if it is assumed that for t < $-t_w$ thermal equilibrium exists and that the width of the pulse is sufficiently short

$$t_w \ll T_1 , T_2 \tag{6}$$

where T_2 denotes the so-called transverse relaxation time (see below) it follows for the expectation values of the components of the magnetic moment of a nucleus at time t = 0

$$\mu_z(0) = \mu_{z0}* \cos(\gamma B_1 t_w)$$

$$\mu_x(0) = \mu_{z0}* \sin(\gamma B_1 t_w) * \sin(\gamma B_0 t_w) \tag{7}$$

$$\mu_y(0) = \mu_{z0}* \sin(\gamma B_1 t_w) * \cos(\gamma B_0 t_w)$$

The third factor on the right hand side describes a fast rotation of the magnetic moment about the z-direction, the so-called Larmor precession with frequency $\gamma B_0/2\pi$. For protons in a static magnetic field of intensity $B_0 = 7.05$ T the value of this Larmor frequency is 300 MHz whence nmr belongs to the field of rf (radio frequency) spectroscopy. The second factor on the right hand side of equs.(7) describes a relatively slow deviation of the magnetic moment from the z-direction (nutation): For protons and a typical value of $5*10^{-3}$ T for B_1 the corresponding frequency is ca. 250 kHz which is much lower than the above mentioned 300 MHz. If the width t_w is chosen such that

$$\gamma B_1 t_w = \pi/2 \tag{8}$$

the alternating magnetic field is called a $\pi/2$ pulse since according to equs.(7) it reorients the magnetic moment from the z-direction into a plane perpendicular to it:

$$\mu_z(0) = 0$$
$$\mu_\perp(0) = \{ \mu_x^2(0) + \mu_y^2(0) \}^{1/2} = \mu_{z0} \tag{9}$$

Typical values for t_w are of the order of some microseconds ($t_w = 1.2$ μs for protons and $B_1 = 5*10^{-3}$ T). For

$$\gamma B_1 t_w = \pi \tag{10}$$

which is the condition of a π pulse, the magnetic moment is reoriented into the -z-direction:

$$\mu_z(0) = -\mu_{z0}$$
$$\mu_\perp(0) = 0 \tag{11}$$

The motion of the magnetic moment after the application of a $\pi/2$ pulse, i.e. for t > 0 is given by

$$\mu_z(t) = \mu_{z0} * \{1 - \exp(-t/T_1)\} \tag{12}$$

$$\mu_\perp(t) = \mu_{z0} * \exp(-t/T_2) * \cos(\gamma B_0 t) \tag{13}$$

The time constant T_2 is called transverse nuclear magnetic relaxation time which depends like T_1 on the sample, on the strength B_0 of the static magnetic field and on the temperature T. In general, its value is equal to or less than that of T_1. Equation (13) is essential for the basic nmr experiment where after the application of a $\pi/2$ pulse the voltage induced in a small rf coil surrounding the sample under study and with its axis perpendicular to the static magnetic field is recorded as a function of time. This is the so-called free induction decay (FID). Its Fourier transform is the usual nmr spectrum. In the simple case that for all nuclei of the sample equ.(13) holds the FID is an exponentially decaying sinusoidal function and the nmr spectrum a single Lorentzian-shaped line at the position of the Larmor frequency $\gamma B_0/2\pi$ and with a FWHM (full width half maximum) linewidth

$$\Delta f = (\pi T_2)^{-1} \tag{14}$$

The minimum number of nuclei N_{min} detectable by an nmr

experiment depends on their spin quantum number I, on their magnetogyric ratio γ, on the values of both nuclear magnetic relaxation times T_1 and T_2, on the strength of the static magnetic field B_0, on the temperature T and on the value of the measuring time T_m which is of the order of T_1 multiplied by the number of accumulations (repetitions of the FID experiment):

$$N_{min} \propto r_{si}^{-1} * (r_f)^{1/2} * (T/B_0)^{3/2} * (T_1/T_2 T_m)^{1/2} \qquad (15)$$

where r_f is the relative resonance frequency

$$r_f = \gamma/\gamma_p \qquad (16)$$

and r_{si} the relative signal intensity

$$r_{si} = r_f^3 * 4I(I+1)/3 \qquad (17)$$

both with respect to the proton ($\gamma = \gamma_p$ and $I = 1/2$). Some values for these both quantities are collected in Table 1:

Table 1. Values for the relative resonance frequency (equ.(16)) the relative signal intensity (equ.(17)) and the natural abundance (n.a.) of various nuclei.

nucleus	r_f	r_{si}	n.a.
^{1}H	1	1	99.98%
^{13}C	0.251	0.0159	1.11%
^{15}N	0.101	0.00104	0.37%
^{27}Al	0.261	0.21	100.00%
^{29}Si	0.199	0.00784	4.7%
^{129}Xe	0.277	0.0221	26.4%

Data for other nuclei may be found e.g. in reference[3]. In order to make use of equ.(15) some empirical results for the absolute values of N_{min} shall be given assuming a signal-to-noise ratio of 10, a measuring time $T_m = 600$ s , a magnetic field intensity $B_0 = 7.05$ T and a temperature T = 300 K:
(1) ^{1}H nmr of surface OH groups. In this case typical values of T_1 and of T_2 (mas enhanced value, see below) are ca. 1 s and ca. 1 ms respectively, and

$$N_{min} \approx 10^{18}$$

For 0.2 g of a porous catalyst (zeolite) with typically $4*10^{20}$ cavities per gram, this value corresponds to ca. 0.01 OH groups per cavity, and for 0.2 g of a catalyst with a specific surface area of S m^2/g to ca. 5/S OH groups per nm^2.

(ii) ^{13}C nmr of adsorbed carbon monoxide enriched to 100% in ^{13}C. In this case typical values of T_1 and of T_2 are ca. 0.1 s and 20 ms, respectively and

$$N_{min} \approx 4 * 10^{18}$$

corresponding to ca. 0.05 CO/cavity or ca. 20/S carbon monoxide molecules per nm^2 of the catalyst.

The Influence of Local Magnetic Fields in NMR

Up till now it was assumed that all nuclei are under the influence of the same magnetic field B_0 so that the free induction decay (FID) is given by (see equ.(13))

$$FID(t) = \exp(-t/T_2) \cos(\gamma B_0 t) \qquad (18)$$

In general, however this assumption is not valid due to the so- called chemical shielding, the magnetic dipolar inter-action and/or the indirect spin coupling.

Isotropic Chemical Shielding: Through the electron cloud surroundig the nucleus under study, the external magnetic field B_0 is shielded to a certain degree, and in the case of cubic symmetry of the electron clouds and/or fast reo-rientation of the molecules containing the nuclei under study it is possible to write

$$B_{loc} = B_0 (1 - \sigma_K) \qquad (19)$$

where σ_K is the so-called isotropic chemical shielding which is a characteristic parameter (fingerprint) for the corresponding electron cloud. Hence, instead of equ.(18) the FID is given by

$$FID(t) = \Sigma p_K \exp(-t/T_{2K}) \cos\{ \gamma B_0(1 - \sigma_K)t\} \qquad (20)$$

with p_K as the relative concentration of the nuclei with a chemical shielding σ_K. Very often the value of T_2 does not depend on K so that the nmr spectrum consists of single lines at positions

$$f_K = B_0 (1 - \sigma_K)/2\pi \qquad (21)$$

with relative intensities p_K which can be determined easily from the amplitudes of the lines since they all have the same lineshape and linewidth.
Instead of listing the values for the line positions ac-cor-ding to equ.(21) which depend on the value of B_0 it is more reasonable to use the so-called chemical shift defined as

$$\delta_K = \sigma_{ref} - \sigma_K \qquad (22)$$

where the reference is tetramethylsilane (TMS) for ^1H, ^{13}C, and ^{29}Si nmr. Some typical values for the chemical shift are given in Tables 2 and 3.

Table 2. Typical values for the chemical shift of ^{1}H nmr (δ_H). More data can be found in reference[4].

compound		chemical shift		
methane	(gas)	0.175	ppm	
water	(gas)	0.3	ppm	
cyclohexane	(liquid)	1.4	ppm	
water	(liquid)	4.8	ppm	
benzene	(liquid)	7.4	ppm	
toluene	(liquid)	2.3	ppm	(methyl)
		7.2	ppm	(ring)
hydroxonium ion	(water)	19	ppm	

Table 3. Typical values for the chemical shift of ^{13}C nmr (δ_C). More data can be found in reference[5].

compound		chemical shift	
methanol	(gas)	− 11.5	ppm
methane	(solution)	− 2.3	ppm
cyclohexane	(liquid)	26.6	ppm
methanol	(liquid)	49.9	ppm
benzene	(liquid)	128.5	ppm
carbon monoxide	(gas)	182	ppm

As a first example for the use of the chemical shift in surface science, the study[6] of adsorption complexes formed between but-1-ene and exchangeable cations in zeolites X shall be mentioned.

The different carbon atoms in this molecule are denoted as follows:

$$H_2C = CH - CH_2 - CH_3$$
$$(1) \quad (2) \quad (3) \quad (4)$$

and in Table 4 experimental values for the difference of the chemical shielding in the neat liquid and for the adsorption complex which is formed in the cavities of the zeolites are given. For comparison also values are included for the complex which is formed in an aqueous solution of silver nitrate.

Table 4. Experimental values for the difference ($\sigma_C^{liq} - \sigma_C^{ads}$) in ppm. The mean error is $\pm$ 0.5 ppm.

	C1	C2	C3	C4
neat liquid	0	0	0	0
T1X	−1.1	−1.6	0	0
AgX	11.5	−1.5	−0.6	−0.6
solved in AgNO$_3$ + H$_2$O	12.2	0.9	0.5	0.2

From these results it must be concluded that in AgX zeolites but-1-ene complexes are formed of the same structure as in solution while in TlX zeolites the but-1-ene is differently bonded to the thallium ion. Results similar to the latter have been found also for alkali ion exchanged zeolites X (cf. reference[6] and the reviews[7,8]).

For the chemical shift of the ^{15}N nmr of acetonitrile adsorbed on LiX, NaX, KX, RbX, and CsX zeolites a nearly linear dependence on the ionization and not on the electrostatic potential of the alkali ions has been found[9] which must be taken as a hint that the bond between the acetonitrile molecule and the alkali ion acting as adsorption site is at least partially of covalent type.

Using the chemical shift as a fingerprint, the appearance of organic compounds can be observed during a catalytic reaction by a cyclic heating of the sample to higher and higher temperatures, while the measurements are performed at room temperature between the periods of heating. This possibility was demonstrated for the first time by Michel[6] in a study of the isomerization of but-1-ene in zeolites NaCaY. Further examples can be found in the reviews[7,8].

Due to its large electron cloud, the chemical shift of the ^{129}Xe nmr signal of adsorbed xenon is a sensitive measure for the microscopic environment. So it could be shown[10] that the difference of the values for the chemical shift of xenon in the gas phase and adsorbed in porous catalysts (e.g. zeolites)

$$\delta_{Xe} = (\sigma_{Xe}^{gas} - \sigma_{Xe}^{ads}) \qquad (23)$$

extrapolated to zero concentration is related to the mean free path length $<l>$ of the adsorbed xenon atoms by the empirical equation

$$\delta_{Xe} = \{ 4.1*10^{-3}(1 + <l>/2.05) \}^{-1} \qquad (24)$$

Here $<l>$ is in Angström and it has to be assumed the absence of two- or threevalenced cations in the pores.

Chemical Shielding Anisotropy (CSA): If the symmetry of the electron cloud is lower than cubic and if the molecules are fixed in space, the local field acting upon the resonating nucleus depends on the orientation of the electron cloud with respect to the external magnetic field, and equ.(19) must be replaced by

$$B_{loc} = B_0 (1 - \sigma_{Kzz}) \qquad (25)$$

where σ_{Kzz} denotes the z-component of the chemical shielding tensor of line K. If the principal values of this tensor are denoted as σ_{K11}, σ_{K22}, and σ_{K33}, the isotropic value is given by

$$\sigma_K = (\sigma_{K11} + \sigma_{K22} + \sigma_{K33})/3 \qquad (26)$$

The resulting lineshape for a powder (statistical but fixed orientation of the molecules) is a characteristic pattern, from which the principal values of the chemical shift tensor can be determined[11]. In Tables 5 and 6 principal values of chemical shift anisotropy tensors for some typical compounds are listed.

Table 5. Principal values of the ^{1}H chemical shift anisotropy tensor in ppm for some compounds. More data
can be found in reference[11].

compound	T/K	δ_{11}	δ_{22}	δ_{33}	$\bar{\delta}$
oxalic acid	300	19.2	17.1	1.5	12.6
gypsum dihydrate	300	17.7	10.9	5.9	11.5
hydrogen sulfide	173	12.7	12.7	1.6	9
ice	77	15	15	-13.5	5.5

Table 6. Principal values of the ^{13}C chemical shift anisotropy tensor in ppm for some organic compounds.
More data can be found in reference [12].

compound	T/K	δ_{11}	δ_{22}	δ_{33}	δ
carbon monoxide	46	293	293	-42	181
toluene					
methyl	87	33	22	5	20
ring	87	222	135	-3	118
methanol	87	74	74	11	53
oxalic acid	300	251	123	112	162

As an example an experimental determination of the SiOH
angle for the surface OH groups of silica gel from the ^{1}H
nmr powder pattern shall be mentioned[13]. The result is a
value of
(140 $\pm$ 5)$^\circ$ for this angle and the observation that even
at the lowest temperatures of measurement (4 K) a fast
rotation of the proton takes place around the Si-O axis[14].
Magnetic Dipolar Interaction: In addition to the chemically
shielded external magnetic field each nucleus experiences
magnetic fields stemming from those neighbours which have a
non-vanishing magnetic moment. While in liquids due to the
fast isotropic reorientation this magnetic dipolar interaction averages to zero, in powdered solids characteristic
lineshapes result. In the case of a homonuclear two-spin
system (like isolated water molecules since the magnetic
moment of the oxygen nucleus can be neglected) a so-called
"Pake" doublet results with a splitting given by

$$\Delta f = 3\ \mu_0\ (4\pi)^{-2}\ \gamma^2\ \hbar/r^3 \qquad (27)$$

where $\mu_0 = 4\pi * 10^{-7}$ Vs/Am and r is the distance between
the both nuclei with the magnetogyric ratio γ . For isolated water molecules with a proton-proton distance r = 0.16
$* 10^{-9}$ m a value of ca. 44 kHz results for the splitting
Δ f. For more-spin systems the lineshape becomes more and
more involved and in contrast to the two-spin-system there
is no real chance to measure nuclear distances apart from

an average over the various distances which can be derived from the second moment of the line[15].

Indirect Spin Coupling (J-Coupling): In addition to the direct dipolar coupling there exists an indirect coupling between two nuclear spins through polarization of a common electron cloud. Although the magnetic field caused by this indirect coupling is much smaller than the magnetic dipole field it plays an important role in liquid state nmr since even a fast isotropic reorientation of the nuclei containing molecules does not eliminate its influence. For a resonating nucleus indirectly coupled to another homo- or heteronuclear spin I = 1/2, the resonance line is split into a symmetric doublet with

$$\Delta f = J \tag{28}$$

Orders of magnitude for the values of the spin coupling constant J are given in Table 7, exact data are collected in references[4,5].

Table 7. Orders of magnitude for the spin coupling constant J (cf. equ.(28))

bond	J / Hz
H - H	0 ... 30
H - C	0 ...250
C - C	0 ...200
C - - C	0 ... 20
C - - - C	0 ... 5

Line Narrowing Techniques in NMR

Magic Angle Spinning of the sample i.e. a mechanical rotation of the sample around an axis inclined by the so-called magic angle 54.74° with respect to the direction of the external magnetic field B_0 eliminates the influence of the magnetic dipolar interaction and of the chemical shift anisotropy upon the nmr line so that "highly resolved" nmr spectra result which allow the identification of the various species via their (isotropic) chemical shift. For a complete elimination however, the mas rate must be much larger than the linewidth of the static line otherwise a residual linewidth remains and spinning sidebands appear in the spectra[16]. This technique has found widespread application in surface science. Important examples are ^{29}Si mas nmr studies on zeolites and related catalysts[8] where silicon nuclei with nAl,(4-n)Si atoms in their second coordination sphere give rise to 5 different lines according to n = 0,1,2,3,and 4, or ^{13}C mas studies on surface compounds of e.g. chemically modified silica[17] where the different carbon atoms can be ascertained by their (isotropic) chemical shift, or ^{27}Al mas nmr spectroscopy which is a sensitive tool for determining the coordination of aluminium[8].

Last but not least the unique possibilities of the ^{1}H mas nmr spectroscopy to characterize Broensted acidity of catalysts should be mentioned (see below).

Multiple Pulse Sequences were developed in order to reduce or even eliminate line broadening or line splitting due to magnetic dipolar interaction while leaving the chemical shift and also its anisotropy unaffected. Since these methods have found not much application in surface science it seems sufficient here to refer only to literature[11].

Hahn's Spin Echo Method. If a $\pi/2$ pulse is applied at time 0 and a π pulse at time T then at time 2T an nmr signal appears which is called Hahn's echo. The amplitude of this signal plotted as a function of T is equivalent to the FID except that in this decay and hence in the spectrum which follows after a Fourier transformation, the chemical shift including its anisotropy, the heteronuclear magnetic dipolar interaction and the heteronuclear spin coupling is eliminated. This method has found an interesting application in the study of ethylene (C_2H_4) adsorbed on metal surfaces[18]:

Ethylene enriched to 90% with ^{13}C was adsorbed on a catalyst of platinum supported on alumina. From the envelope of Hahn's spin echo decay a carbon-carbon bond length of (0.149 ± 0.002) nm could be determined which is close to the single-bond length (0.154 nm) but quite different from the value of 0.134 nm for a double-bond. Through application of a third pulse tuned to the ^{1}H nmr frequency it was possible to show that the ethylene is adsorbed at 300 K as a $\equiv$C-CH$_3$ species which agrees with the result of low-energy electron diffraction for the adorption on Pt single crystal (111) faces[19]. If the sample is annealed in an oven for 3 hours at progressively higher temperatures the nmr signal caused by the $\equiv$C-CH$_3$ species (Pake doublet) decreases in intensity while a bell-shaped signal which is due to the scission products increases in intensity. From the fraction of C-C bonds broken plotted as a function of the annealing temperature the activation energy for this process (150 kJ/mol) could be derived. The fact that the splitting of the Pake doublet does not change with progressive annealing shows that the C-C bond length is not changing up to the point of C-C bond scission. This result is in contrast to the behaviour of acetylene (C_2H_2) adsorbed on the same catalyst where a change of the C-C bond length has been found before the C-C bonds break.

^{1}H MAS NMR SPECTROSCOPY OF CATALYSTS

The study of acidic surface sites capable of donating protons (Broensted acid sites) to adsorbed molecules is one of the most important areas in heterogeneous catalysis. The catalytic activity of a catalyst with regard to a proton catalyzed reaction will be determined by at least three independent parameters for each sort of hydroxyl groups (proton donors):

    (i) The strength of acidity as defined by the rate con stant of the proton transfer to the adsorbed molecule
    (ii) the concentration and
    (iii) the accessibility of the acidic sites.

Strength of Broensted Acidity

Since the rate constant of the proton transfer depends both on the properties of the acidic site ZOH and of the particular molecule M we decompose the protonation reaction into two processes:

$$ZOH \quad = \quad ZO^- + H^+ \qquad (29)$$

$$H^+ + M = MH^+ \qquad (30)$$

which leads to the definition of the strengths of gas phase acidity and basicity as the standard Gibbs free energy change of processes (29) and (30), respectively[20].
In order to guarantee that a higher strength of acidity corresponds to a higher protonation ability, we define the strength of acidity S_a as the reciprocal value of the standard Gibbs free energy change of reaction (29):

$$S_a \quad = \quad 1/ \Delta G^0_{DP} \qquad (31)$$

To compare the strength of acidity with the deprotonation energy of the ZOH group ΔE_{DP}, a quantity which follows from quantum chemical calculations, one must take into consideration that the standard Gibbs free energy change is the sum of the deprotonation energy, of the zero-point energy change and of the thermal Gibbs free energy change which results from the conversion of the three vibrational degrees of freedom of the proton as part of the ZOH group into its three translational degrees of freedom after leaving this group. Assuming that this latter contribution is negligible and that the zero-point energy change is a constant[21], the deprotonation energy ΔE_{DP}, i.e. the energy difference of ZO^- and ZOH can be used approximately instead of ΔG^0_{DP} in equ.(31).
With regard to the chemical shift δ_H of the ZOH group the following qualitative argument may be taken as a hint, that there is a direct relation between δ_H and its strength of acity S_a: A higher value of the chemical shift corresponds per definition (cf. equ.(22)) to a reduced shielding of the external magnetic field and hence to a larger value for the net atomic charge of the hydrogen. On the other hand an enhancement of the net atomic charge will lead to a reduction of the deprotonation energy. In agreement with this suggestion experimental values for δ_H and for the absolute gas phase acidities measured for hydroxyl groups of various organic compounds in the gaseous state show in fact a good correlation[21]. Another proof of our suggestion that δ_H is a suitable and sensitive measure for the strength of acidity is given by the fact[22] that Sanderson's intermediate electronegativity computed for zeolites of varying silicon to aluminium ratio shows the same functional dependence as δ_H. Finally there has been found also a relatively good correlation between values calculated by non-empirical quantum chemical methods for the shielding constant and the deprotonation energy of various OH groups[23,22].
Summarized, it is demonstrated clearly by references[21,22,23] that δ_H is a suitable and sensitive spectroscopic quantity to measure the strength of acidity as defined by equ.(31). In principle however, one should take into ac-

count that in each of these examples isolated hydroxyls of similar type are compared so that it is not clear whether there exists a correlation between δ_H and ΔG^0_{DP} which is of general validity.

Through the use of methane as an inner standard the accuracy of the absolute values of δ_H could be enhanced considerably so that the present error does not exceed $\pm$ 0.05 ppm. The residual line width of the ^1H mas nmr spectra for acidic OH groups which are of bridging type (Si OH Al) is of the order of 0.3 ppm[24]. Taking into consideration the empirical formula[21]

$$f/cm^{-1} \quad = 3906 \quad - 74.5 \; \delta_H/ppm \qquad (32)$$

which connects δ_H with the stretching vibration frequency f of the OH groups, the residual ^1H mas nmr line width of 0.3 ppm corresponds to a residual linewidth of ca. 20 cm^{-1} for the IR band of the bridging OH groups.

In the ^1H mas nmr spectra of dehydrated zeolites and related catalysts containing only oxygen, silicon and aluminium in the framework, five lines can be separated which have been denoted as a,b,c,d, and e:

Line a at 1.8 to 2.3 ppm is caused by non-acidic (silanol) OH groups. In the case of carefully dehydrated silica this line appears at somewhat lower values (1.65 ppm for aerosil 200/Degussa). A distinction between single and geminal OH groups however is not possible, the difference in δ_H is less than 0.1 ppm corresponding to a difference in the stretching vibration of less than 7 cm^{-1}.

Line b at 3.8 to 4.4 ppm is ascribed to acidic OH groups which are known to be of bridging type (Si OH Al).

Line c at 4.8 to 5.6 ppm is also ascribed to acidic OH groups of the bridging type but influenced by an additional electrostatic interaction of the hydroxyl proton presumably with neighbouring oxygen atoms (corresponding to the shift of the so-called LF-band in infrared spectroscopy).

Line d at 6.5 to 7.0 ppm is due to residual ammonium ions.

Line e at 2.5 to 3.6 ppm represents hydroxyl groups associated with extra-framework aluminium species. Due to the limited space available for these OH groups their δ_H value will be determined to a certain degree by an additional electrostatic interaction of the hydroxyl proton with other oxygen atoms. In accordance with this suggestion, for isolated AlOH groups the chemical shift is much lower and in the interval from -0.5 to 1.0 ppm.

For POH groups the value of δ_H is between 1.5 and 4 ppm depending among others on interactions mentioned for line c.

Concentration of Broensted Acidic Sites

With respect to a measurement of the concentration of hydroxyl groups nuclear magnetic resonance spectroscopy has an extremely important advantage compared with infrared spectroscopy since the area of an ^1H mas nmr signal is directly proportional to the concentration of the hydrogen nuclei contributing to this signal irrespective of their bonding state so that any component with a known concentration of hydrogen atoms can be used as a reference (mostly

water). As has been shown for various specimens of catalysts of type SAPO-5 even the relative intensity of an OH
stretching vibration band cannot be taken as a measure for
the concentration of the respective hydroxyl group in
contrast to the nmr signal[22].

Accessibility of Broensted Acidic Sites

The accessibility of hydroxyl groups can be easily determined through a study of the ^{1}H mas nmr spectra after loading
the catalyst with a suitable molecule which however must be
fully deuterated in order to avoid an unwanted additional
^{1}H nmr signal. The signals due to accessible hydroxyl
groups are in general shifted and very often also strongly
broadened. Various examples are presented in reference[22].

PULSED FIELD GRADIENT NMR STUDIES OF ADSORBED MOLECULES

Pulsed Field Gradient NMR Techniques

During a free induction decay two succeeding magnetic field
gradient pulses of different sign shall be applied so that
the intensity of the external magnetic field is given by

$B_0 + g*z$ for $\quad 0 \leq t \leq \delta$ (1.field gradient pulse)

B_0 for $\quad \delta < t < \delta + \Delta$
$B_0 - g*z$ for $\delta + \Delta \leq t \leq 2\delta + \Delta$ (2.field gradient pulse)

where $\pm g$ denotes the intensity and δ the duration of the
magnetic field gradient pulse. Through this repeated but
reversed space dependence of the magnetic field intensity
the amplitude of the free induction decay immediately after
the second field gradient pulse i.e. at $\quad t = \Delta + 2*\delta$
depends on the diffusional motion of the resonating nuclei
during the time interval Δ between the two field gradient pulses which is generally chosen much larger than the
width of the field gradient pulses:

$$\Delta \gg \delta \tag{33}$$

Assuming a powder sample of porous crystallites loaded with
molecules containing the resonating nuclei so that the
relative concentration of these nuclei in the intracrystalline space (p_{intra}) is much greater than in the intercrystalline space (p_{inter}):

$$p_{intra} \gg p_{inter} \tag{34}$$

and that the intercrystalline diffusivity (D_{inter}) is much
greater than the intracrystalline diffusivity (D_{intra}):

$$D_{intra} \ll D_{inter} \tag{35}$$

then it can be shown[25] that the following equation holds:

$$\ln(s_1/s_0) = \{-KD\Delta - Kp_{inter}D_{inter}\Delta /(1+Kp_{inter}D_{inter}\ \tau)\}$$

(36)

where

$$K = (\gamma g \delta)^2$$

(37)

s_1 is the amplitude of the free induction decay immediately after the second field gradient pulse, s_0 the same quantity but for $K = 0$, and τ denotes the mean residence time of a molecule in the pore system of a crystallite.
For relatively small values of K:

$$Kp_{inter}D_{inter}\ \tau \ \ll 1$$

(38)

equ.(36) simplifies to

$$\ln(s_1/s_0) = - (D + p_{inter}D_{inter})K \Delta$$

(39)

which yields the socalled long-range or effective selfdiffusion coefficient $D + p_{inter}D_{inter}$ if one plots the measurable quantity $\ln(s_1/s_0)$ as a function of K. This is the usual pulsed field gradient nmr (pfg nmr) method. In the other limiting case, i.e. for

$$Kp_{inter}D_{inter}\ \tau \ \gg 1$$

(40)

it follows from equ.(36)

$$\ln(s_1/s_0) = - KD \Delta - \Delta /\tau$$

(41)

so that a plot of $\ln(s_1/s_0)$ as a function of K yields both the intracrystalline selfdiffusion constant D and the mean residence time τ. This method has been coined nmr tracer desorption technique.

Intracrystalline Selfdiffusion Coefficients

The minimum value of a selfdiffusion coefficient which can be measured depends on the intensity g of the magnetic field gradients, the magnetogyric ratio γ of the nuclei and their transverse relaxation time T_2. For g_{max} of the order of 20 T/m it follows

$$D \geq 3 * 10^{-14} * (\gamma_p/\gamma)^2 / T_2$$

(42)

with the transverse relaxation time T_2 in seconds.
For hydrocarbons (^1H nmr, i.e. $\gamma = \gamma_p$) adsorbed in zeolites of type NaX the transverse relaxation time is of the order of 10^{-2} s so that the minimum value of D which can be measured is ca. $3 * 10^{-12}$ m^2/s. The minimum measurable r.m.s. free path length is given by

$$(<(z - z')^2>)^{1/2} \geq 0.3 * (\gamma_p/\gamma) * 10^{-6}$$

(43)

corresponding to 0.3 /um for ^1H nmr measurements.
It was one of the most important results of the application of nmr techniques to the study of porous catalysts to show that selfdiffusion coefficients measured by classical

uptake experiments and published in literature and even in textbooks were wrong up to five orders of magnitude[26]. A critical rewiew of the present state can be found in reference[27], and it is now generally accepted that the most reliable results for intracrystalline diffusivities are measured by the nmr technique.

Transport Resistances at the External Surfaces of Porous Crystallites

For spherical crystallites of radius R without an additional transport resistance at the external surface the mean residence time τ of a molecule within the pore system of the crystallite is determined only by the intracrystalline selfdiffusion coefficient D and the value of R, and it can be shown that τ is given by

$$\tau \quad = \quad \tau^D \tag{44}$$

where

$$\tau^D \quad = \quad R^2/(15D) \tag{45}$$

For as-synthesized and shallow-bed treated zeolites this in fact has been observed. Under the conditions of a deep-bed procedure or a hydrothermal treatment however, τ is larger than τ^D and the ratio of both values can be used as a quantitative measure for the strength of the surface barrier[28]. By the same method it is also possible to determine whether and to what extent catalysts are coked at the external surface or in the intracrystalline pore system.

In principle, a transport resistance can be caused either by a complete blocking of some of the windows connecting the internal pore system with the intercrystalline space and leaving the other unaffected or by a reduction of the average diameter of all windows. In the latter case the appearance of a transport resistance should depend on the diameter of the diffusing particles, an effect which has been observed in a comparison of methane and xenon diffusion measurements[30].

REFERENCES

1. N.J.Poulis, W.P.A.Hass in: Landolt-Börnstein II/9, Springer Verlag, Berlin 1962.
2. H.Pfeifer: Phys.Reports 26C(1976)293.
3. Bruker Almanac 1990. Bruker GmbH., D-7512 Rheinstetten 4.
4. W.Brügel: Handbook of NMR Spectral Parameters. Heyden, London 1979, 3 Volumes.
5. W.Bremser, L.Ernst, B.Franke, R.Gerhards, A.Hardt: Carbon-13 NMR Spectral Data. Verlag Chemie, Weinheim 1981.
6. D.Michel, W.Meiler, H.Pfeifer: J. Mol. Cat. 1(1975/76) 85.
7. H.Pfeifer, W.Meiler, D.Deininger: Annual Reports on NMR Spectroscopy 15(1983)291.

8. G.Engelhardt, D.Michel: High Resolution Solid-State NMR of Silicates and Zeolites. Wiley, New York 1987.
9. I.Jünger, W.Meiler, H.Pfeifer: Zeolites 2(1982)309.
10. J.Fraissard: Z. phys. Chemie, Leipzig 269(1988)657.
11. M.Mehring: High Resolution NMR in Solids. Springer Verlag, Berlin 1983.
12. T.M.Duncan: J. Phys. Chem. Ref. Data 16(1987)125.
13. H.Ernst: Z. phys. Chemie, Leipzig 268(1987)405.
14. T.Bernstein, H.Ernst, D.Freude, I.Jünger, J.Sauer, B.Staudte: Z. phys. Chemie, Leipzig 262(1981)1123.
15. H.Pfeifer: NMR - Basic Principles and Progress 7(1972)53.
16. C.A.Fyfe: Solid State NMR for Chemists. C.F.C.Press, Guelph, 1983.
17. G.R.Hays, A.D.H.Clague, R.Huis: Appl. of Surf. Sci. 10(1982)247.
18. P.-K.Wang, J.-P.Ansermet, S.L.Rudaz, Z.Wang, S.Shore, C.P.Slichter, J.H.Sinfeld: Science 234(1986)35.
19. L.L.Kesmodel, L.H.Dubois, G.A.Somorjai: J. Chem. Phys. 70(1978)2180.
20. J.E.Bartmess, J.A.Scott, R.T.McIver: J. Am. Chem. Soc. 1011(1979)6046.
21. H.Pfeifer: Colloids and Surfaces 36(1989)169.
22. H.Pfeifer, D.Freude, J.Kärger: Proc. ZEOCAT '90. Elsevier, Amsterdam 1990 (in press).
23. J.Sauer: J. Mol. Cat. 54(1989)312.
24. E.Brunner, D.Freude, B.C.Gerstein, H.Pfeifer: J. Magn. Res. (in press).
25. J.Kärger, H.Pfeifer, W.Heink: Adv. Magn. Res. 12(1988)1.
26. J.Kärger, H.Pfeifer: Zeolites 7(1987)90.
27. J.Kärger, D.M.Ruthven: Zeolites 9(1989)267.
28. J.Kärger,H.Pfeifer, R.Richter, H.Fürtig, W.Roscher, R.Seidel: Am. Inst. Chem. Eng. Journal 34(1988)1185.
29. J.Kärger, H.Pfeifer, D.Freude, J.Caro, M.Bülow, G.Öhlmann: Proc. 7th International Zeolite Conference, Tokyo 1986. Elsevier, Amsterdam 1986, p.633.
30. J.Kärger, H.Pfeifer, F.Stallmach, H.Spindler: Zeolites 10(1990)288.

CHARACTERIZATION OF HETEROGENEOUS CATALYSTS

BY VIBRATIONAL SPECTROSCOPIES

Helmut Knözinger

Institut für Physikalische Chemie
Universität München
Sophienstrasse 11, 8000 München 2, FRG

INTRODUCTION

Vibrational spectroscopies are certainly among the most promising and most widely used methods for catalyst characterization. This is due to the fact that very detailed information on molecular structure and symmetry can be obtained from vibrational spectra. Also most importantly, several vibrational spectroscopies can be applied under <u>in situ</u> conditions and they can very successfully be used for studies of ill-defined high-surface area porous materials.

Principally photons, electrons and neutrons can be used as probes. A wide variety of special techniques has been developed for photon vibrational spectroscopies, which have to be selected according to the surface and optical properties of the system to be studied. Because of its relative simplicity and wide applicability, infrared transmission-absorption spectroscopy [1-5] and - more recently- diffuse reflectance spectroscopy [6-8] are most frequently used today. Laser Raman spectroscopy [9-11] has also been developed as a powerful tool for surface studies. Internal reflection and optoacoustic spectroscopies [13,14] have found less application. Double resonance excitation of fluorescence [15] may possibly find advantageous application for the vibrational characterization of catalyst surfaces, though no example has yet been reported. This technique would provide the sharp lines characteristic of infrared and Raman spectra but could have an increased sensitivity characteristic of fluorescence spectroscopy. Second harmonic generation (SHG) [16] should also find future application at least for studies of model systems. Two electron vibrational spectroscopies [17] have been developed. Electron energy loss spectroscopy (EELS) can hardly be applied to rough surfaces of polydisperse materials. Inelastic electron tunneling spectroscopy (IETS) on the other hand, is an interesting sophisticated technique which provides high resolution spectra within the entire vibrational frequency range. The experiment, however, requires very costly apparatus and is far less widely applicable than infrared transmission or Raman spectroscopy.

Fundamental Aspects of Heterogeneous Catalysis Studied by Particle Beams,
Edited by H.H. Brongersma and R.A. van Santen, Plenum Press, New York, 1991

167

Inelastic neutron scattering (INS) is an attractive spectroscopy for vibrational analysis of surface species, namely hydrogen -containing surface groups. This technique will be discussed in detail in the contribution by Jobin.

In the present article, the physical basis of those vibrational spectroscopies which are applicable to·the study of real catalyst surfaces will be presented. These are transmission-absorption, diffuse reflectance, emission and Raman spectroscopy. In the second part of the article examples will be given for the application of these techniques for the vibrational characterization of typical catalytic materials.

BASIC PRINCIPLES OF VIBRATIONAL SPECTROSCOPIES

Infrared Transmission-Absorption Spectroscopy

Transmission infrared spectroscopy and its application to surface chemistry has been reviewed in the past. [1-5] The principle is well known from conventional infrared spectroscopy of solids in the transmission mode (thin pressed self-supporting wafers are usually being used for surface studies). The applicability of the transmission technique is determined by the properties of the solid powder to be studied. Thus, samples which exhibit only weak bulk absorption, and the average particle size $\bar{d}$ of which is smaller than the wavelength of the infrared radiation in the region of interest will be optimally suited for the transmission mode. The particle size condition ($\lambda \gg \bar{d}$) which determines the wavelength range of suitably low scattering losses, is usually met in the mid and far infrared region in practice, whereas scattering losses become strongly involved in the near infrared region. On the other hand, most samples show strong bulk absorption in the low wavenumber region (roughly <1000 cm^{-1}). As a result, the accessible wavenumber range for transmission infrared spectroscopy will generally be limited for surface studies to the mid-infrared region ($4000 > \tilde{\nu} > 1000$ cm^{-1}). However, when less than optimal spectroscopic conditions can be accepted spectra may be obtained also in the near infrared. [18] A reduction of scattering losses could be achieved by the immersion technique [19] in which the solid is immersed in a solvent having approximately the same refractive index (e.g. SiO_2 immersed in CCl_4).

Infrared transmission spectroscopy is a bulk rather than a surface specific technique. It is therefore necessary to prove for any detected species that it is a surface group. This can be realized in many cases by following changes in band position on exposure of the solid adsorbent to a suitable adsorptive or by isotopic exchange experiments.

The sensitivity of the technique is dependent on the extinction coefficient of the surface groups which may vary from 5×10^{-18} cm^2 molecule^{-1} for the carbonyl stretching mode in CO ligands to between 10^{-20} and 10^{-19} cm^2 molecule^{-1} for CH stretching modes in saturated hydrocarbon chains. The magnitude of the extinction coefficients renders a high surface-to-volume ratio desirable, the more so as the possible increase in sample thickness is limited by the concomitant increasing energy losses by absorption and scattering.

Assuming typical values of 100 m^{2}g^{-1} for the adsorbent surface area, 20 mg cm^{-2} for the weight of the irradiated geometric area, 10^{-19} cm^2 molecule^{-1} for the extinction coefficient and 5 % for a desirable absorption in order to obtain good quality spectra with standard infrared spectrometers, one estimates a lower limit of 0.02 for the surface coverage θ. This shows that transmission spectra can be obtained even at coverages below one tenth of monolayer. With the application of data acquisition techniques, the sensitivity of the technique can be increased further. Quantitative measurements of surface group densities should be possible, provided the Lambert-Beer law is applicable. It must be kept in mind, however, that this law is valid for optically homogeneous materials and deviations may occur for disperse substances.

A number of different transmission cells have been described in the literature. Typical designs of cells for in situ work at low and high temperatures are discussed in the reviews already mentioned [1,2], by Knözinger[20] and by Basu et al[21]. Gallei and Schadow[22] described a cell which permits work at pressures up to 3 MPa and in ultra-high vacuum at temperaures up to 870 K.

Diffuse Reflectance Spectroscopy (DRS, DRIFTS)

The transmission technique fails when strongly scattering ($\lambda \leq \bar{d}$) materials are to be studied. Diffuse reflectance can be used in such cases provided the solid material does not absorb too strongly in the frequency range to be studied.

Diffuse reflectance techniques have been reviewed by Kortüm[6] and by Wendlandt and Hecht[23] and their application in the infrared regime has been discussed by Griffiths and coworkers[7,8]. These techniques can be used with materials in the form of powders or pellets which are usually terminated by two parallel surfaces. The front surface is illuminated by diffuse light or a parallel beam under a fixed angle of incidence [6]. The reflectance R of such samples is defined as the ratio of backward and forward fluxes at the illuminated front surface. Assuming isotropic scattering, R can be expressed in terms of phenomenologically defined absorption and scattering coefficients K and S, respectively, according to the Schuster-Kubelka-Munk theory [6,23]:

$$R = (1+R_g[b \coth Y-a])(a+b \coth Y - R_g)^{-1} \qquad (1)$$

where $Y = Sbd$, R_g is the reflectance of the background, d = the thickness of the sample layer, a = 1+K/S, and b = $\sqrt{(a^2-1)}$. For infinitely thick samples (several millimeters in the infrared fundamental region [6]), this equation reduces to the simple Kubelka-Munk function:

$$F(R_\infty) = (1-R_\infty)^2[2R_\infty]^{-1} = K/S \qquad (2)$$

with R_∞ being the reflectance at infinite sample thickness. This expression directly determines the ratio K/S, which is equivalent to the true absorption spectrum provided S is independent of the wavelength. This condition may not always be met in the infrared range when the wavelength is greater than the mean particle size and the scattering coefficient becomes wavelength dependent ($S \sim \lambda^{-4}$, Rayleigh scattering)

The diffuse scattered radiation can be collected by a so-called integrating sphere [6,23]. Diffuse reflectance attachments to modern FT-IR spectrometers (DRIFTS), however, use ellipsoidal mirrors. These mirrors should be mounted off-axis so as to avoid contributions from regular reflection.[7,8] An interesting design using a light-pipe for sample illumination has recently been described by Korte et al. [24].

Infrared Enmission Spectroscopy (IRES)

When the solid material is opaque in the frequency range of interest, both transmission and diffuse reflectance fail to be applicable. In this case infrared emission (IRES) may be used advantageously. When electromagnetic radiation interacts with matter, the energy balance is given by

$$a + r + t = 1,$$
(3)

where a, r, and t are the relative contributions to the total energy of absorbed, reflected and transmitted radiation. For strongly absorbing materials, emission and absorption at equal frequency v and temperature T are identical according to Kirchhoff's law:

$$e(v,T) = a(v,T),$$
(4)

and hence, equ. (3) becomes

$$e + r + t = 1$$
(5)

With strongly absorbing materials and a strongly reflecting background (low absorbance and consequently negligible background emission), the transmission is low and the detected emission intensity is entirely due to the sample emission. This wen rationed against blackbody emission I_{BB} gives the infrared emission spectrum:

$$e(v,T) = I(v,T)/I_{BB}(v,T).$$
(6)

IRES certainly is a potentially very useful technique for catalyst characterization the more so as it provides the possiblity for in situ studies on working catalysts. Also the low frequency regime ($\tilde{v} < 1200$ cm^{-1}) which is opaque for many oxides due to bulk absorption, becomes accessible by IRES and allows vibrational spectra of surface groups and chemisorbed species to be studied. Although manufacturers of FTIR spectrometers provide emission attachments today, in-depth studies of catalysts are still scarce. A few interesting examples, e.g. emission infrared spectra of adsorbed molecular metal carbonyl species, have been reported by Mink and Keresztury.[25,26]

Raman Spectroscopy

The application of Raman spectroscopy to surface studies has been reviewed in the past.[9-12] As compared to other vibrational spectroscopic techniques. Raman spectroscopy can widely be used for investigations of surface species on oxides, on supported and bulk metals (including the phenomenon of the so-called "surface enhanced Raman scattering"[27]), on supported oxides and at the water-solid interface. The information depth of Raman spectroscopy depends strongly on the optical properties of the sample and should typically be of the order of magnitude of 2 nm. Raman spectroscopy provides several unique possibilities:

(i)   the interference of gas spectra is negligibly weak. In situ Raman spectra of working catalysts at elevated pressures and temperatures can therefore be obtained:

(ii)   simple glass or quartz cells can be used;

(iii)   the Raman scattering probabilities of the solid adsorbent or support are frequently very low, especially of most typical support oxides such as silica, alumina etc., so that Raman spectra of surface species can be recorded in the low wavenumber region down to approximately 50 cm^{-1}.

The basic experimental requirements for conventional laser Raman spectroscopy including data manipulation and signal averaging are described in review articles. [9-12] The use of optical multichannel detectors [28] leads to significant decrease of total data accumulation times with excellent signal/noise ratios. Suitable in situ Raman cells for surface studies have been described by Brown et al.[29], Cheng et al.[30], and by Vielhaber and Knözinger[31]. A system which permits in situ high temperature (T ≤ 770 K) and low temperature (77 K) spectroscopy has recently been designed by Zeilinger et al.[32]

Some major problems which may be encountered in laser Raman spectroscopy are:

(i)   sample sensitivity to heating effects of the laser beam including desorption of surface species. This phenomenon may become very severe when coloured samples are studied;

(ii)   low sensitivity of the technique;

(iii)   background fluorescence, a problem which is sometimes so severe on oxide surfaces that weak Raman signals remain undetectable.

The heating effects of the laser beam can be reduced by simply applying low laser power levels (≤100 mW). Alternatively, cylindrical lens focus techniques have been applied in order to spread the laser beam into a line image on the sample. As discussed by Freeman et al.[28] this approach may additionally provide increased signal intensity. Finally, sample rotation will decrease the energy flux onto the sample [30] and sample cooling in an inert gas having high thermal conductivity (e.g. helium) will reduce sample heating in the laser spot.[32] The temperature in the laser irradiated area can be determined from the Raman line intensities in the Stokes and Anti-Stokes regimes.

The inherently low Raman scattering cross sections are the reason for the low sensitivity of the technique even if the spectrometer configuration and detection system are optimally chosen. The Raman intensity I_R is proportional to the intensity I_o and to the fourth power of the frequency v_o of the exciting laser radiation:

$$I_R - I_o \ (v_o - \Delta v_R)^4. \tag{7}$$

Hence, increasing I_o and v_o would yield greater Raman intensities. However, I_o must be limited so as to avoid excessive heating of the specimen as discussed above. The choice of high excitation frequencies v_o may be helpful, can, however, also lead to severe fluorescence and saturation of the detector system.

The fluorescence background must partly be attributed to

contaminations by hydrocarbons, which can often be eliminated by heating the samples in an oxygen atmosphere at sufficiently high temperatures, provided the sample would tolerate such severe treatments. Saperstein and Rein[33] reported that the fluorescence background from a 4A zeolite could effectively be reduced by washing in 0.2 N NaOH and calcining at 673 K. Considerable contributions to the fluorescence background from hydroxylated oxide surfaces are due to laser induced electronic excitations of surface hydroxide ions. [34] The fluorescence can be reduced to an extremely low level by thermal dehydroxylation [34], or by ion exchanging the OH$^-$ ions by e.g. molybdate ions. [35] Thus fluorescence seems to be an intrinsic property of oxide surfaces which bear basic hydroxide ions and it can only be overcome by experimental techniques. On the other hand, information on the nature and properties of surface OH$^-$ groups may perhaps be extracted from the fluorescence background.

In some cases a simple change of the exciting frequency may reduce the fluorescence background in the spectral range of interest.[36] Alternatively, time resolved Raman techniques may be applied. These methods, however, are experimentally complicated and depend on the fluorescence decay times of the samples; they can therefore hardly be developed as routine techniques for surface studies.

Frequency modulation (FM) Raman spectroscopy is a relatively simple technique, which produces the final spectrum in form of the first derivative of the Raman intensity. The background is therefore effectively eliminated and the S/N ratio is improved as compared to that of a conventional Raman spectrum recorded under comparable conditions. The fluorescence intensity I_1 is dependent on the excitation frequency v_1 and on the fluorescence frequency v and is given by a convolution of the excitation spectrum $E(v_1)$ (which is independent of v) and the luminescence emission spectrum $L(v)$ (which is independent of v_1):

$$I_1(v_1,v) = L(v) \cdot E(v_1). \tag{8}$$

In an FM experiment the excitation frequency v_1 is modulated so that

$$dI_1/dv_1 = L(v) \cdot dE/dv_1. \tag{9}$$

If the total detected intensity I also contains contributions from Raman scattering I_R, then

$$dI/dv_1 = dI_R/dv_1 + dI_1/dv_1 \tag{10}$$

and if the sample is excited in a relative minimum of the excitation spectrum it follows that

$$dE/dv_1 = 0 \quad \text{and} \quad dI_1/dv_1 = 0.$$

As a consequence equ. (10) reduces to

$$dI_1/dv_1 = dI_R/dv_1, \tag{11}$$

i.e. a first-derivative Raman spectrum results with efficient background reduction. An experimental set-up for FM Raman spectroscopy has been described by Brückner et al. [39].

The most efficient reduction of fluorescence background is

achieved by FT Raman spectroscopy.[40,41] Here, the excitation light source is typically a Nd: YAG laser with an excitation wavelength of 1.064 μm. The essential feature of this technique is that the low excitation frequency in the near infrared cannot excite most samples electronically. FT Raman attachments for FT infrared spectrometers have become commercially available only recently. Thus, to the author's knowledge only one paper dealing with studies of pyridine adsorption has been pulished.[42] However, FT Raman spectroscopy will undoubtedly become a very important addition to surface vibrational spectroscopies in the very near future.

The Raman microscope[43] can be a useful variant when lateral resolution, e.g. in catalysts having inhomogeneous distributions of several Raman active components, is required for the materials characterization. Raman microscopy has been successfully applied in the very recent past to follow the spreading of MoO_3 on the surface of alumina.[44]

SELECTED APPLICATIONS OF VIBRATIONAL SPECTROSCOPIES FOR CATALYSTS CHARACTERIZATION

Surface Hydroxyl Groups on Binary Oxide Surfaces

Surface hydroxyl groups are intrinsic surface oscillators and O-H stretching modes appear as sharp bands in vibrational spectra of binary oxides if the OH groups are isolated,[45] i.e. the surface density is low and the average distance between OH groups is too large for intermolecular interactions to occur. Fig. 1 shows a transmission IR spectrum of SiO_2 after thermal

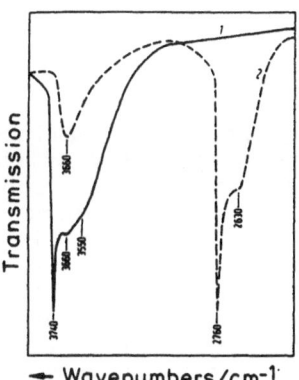

← Wavenumbers/cm⁻¹

Fig. 1. Infrared transmission - absorption spectra of hydroxyl (1) and deuteroxyl (2) groups on a SiO_2 surface heat-treated at 473 K (from ref.[46]).

treatment in vacuum at 473 K. The sharp band at 3740 cm⁻¹ is assigned to isolated silanol groups and the shoulder at 3550 cm⁻¹ is due to so-called vicinal OH groups which undergo H-bond interactions with neighbouring silanol groups. The two groups undergo rapid exchange with D_2O (see Fig. 1, bands at 2760 and 2630 cm⁻¹, respectively). This clearly indicates that the corresponding OH species are surface species (see Scheme 1). In con-

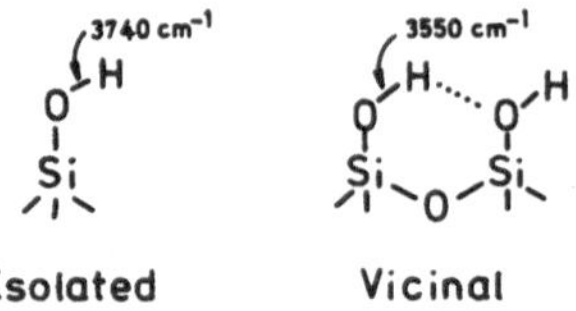

3740 cm^{-1}

O^{H}
|
Si

3550 cm^{-1}

O···O^{H}
|
Si—O—Si

Isolated   Vicinal

Scheme 1

trast, the shoulder at 3660 cm^{-1} appears as an isolated band in the O-H stretching region after exchange. This is clear evidence for the location of the corresponding groups in the bulk of the silica particles. These assignments have been supported by diffuse reflectance spectra in the near infrared region. The first overtone of the isolated silanol groups appears at 7325 cm^{-1} and the [ν(OH) + δ(OH)] - combination band at 4550 cm^{-1}.[47]

Hydroxyl groups on binary oxides can be coordinated to varying numbers of metal centers depending on the structure of the particular oxide and on the type of crystallographic plane under consideration. This leads to a variety of different OH configurations as indicated in Scheme 2:

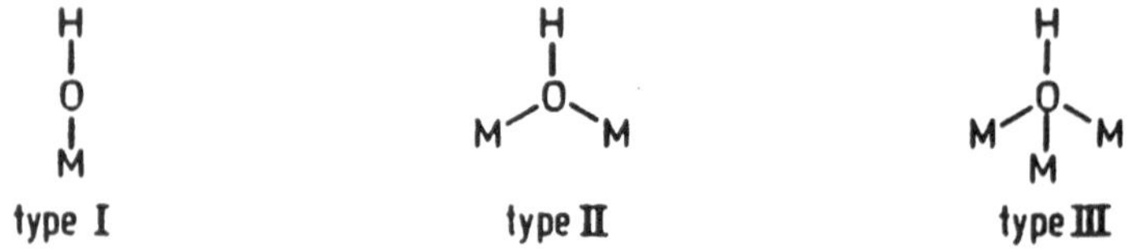

H
|
O
|
M

type I

H
|
O
M   M

type II

H
|
O
M   M
M

type III

Scheme 2

The O-H stretching frequency is determined by its coordination [45,48,49] and by the electronegativity of the metal center.[50] The frequency decreases with increasing coordination number for given metal centers and with increasing electronegativity for fixed coordination numbers.[45] An additional determining factor is the oxygen coordination of the metal center. A particularly interesting and important example is γ-Al$_2$O$_3$ which has a defect spinel structure with Al^{3+} cations being located in tetrahedral and octahedral sites. Spectrum (1) in Fig. 2 shows the five O-H stretching bands of γ-Al$_2$O$_3$ after thermal treatment in vacuum at 773 K which are characteristic for the five possible OH configurations on the low index planes of the alumina structure.[49] The bands are assigned in descending frequency to OH groups being coordinated to a single octahedral (3785 cm^{-1}) or tetrahedral (3775 cm^{-1}) Al^{3+}, to two octahedral Al^{3+} (3725 cm^{-1}), to one octahedral and one tetrahedral Al^{3+} (3715 cm^{-1}), and to three octahedral Al^{3+} (3695 cm^{-1}).[49]

The chemical properties of these surface OH groups (acidic or basic properties, H-bond donor or acceptor strength) is also determined by the bonding of the group to metal ions in the first coordination sphere. The absolute O-H stretching frequency, however, in comparison of different oxides cannot be taken as a direct measure of their acidities. The frequency shift of

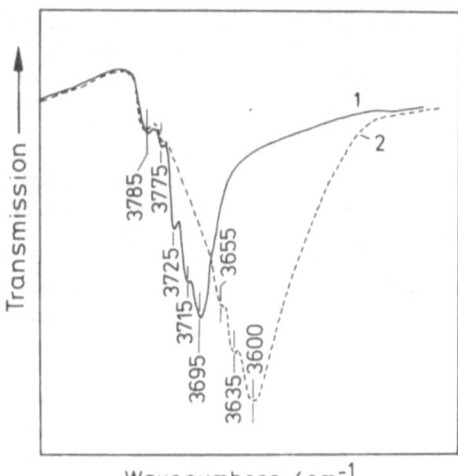

Fig. 2. Hydroxyl infrared spectra of γ-Al₂O₃ after dehydroxylation at 773 K (1) and after subsequent adsorption
at 80 K of 40 torr CO (2); spectra taken at 80 K,
(from ref.[52]).

the O-H stretching mode in interaction with probe molecules, on
the other hand, is a good measure of proton or H-bond donor
strength.

Use of Probe Molecules on Oxide Surfaces

A variety of basic molecules have been used as probe molecules for the characterization of adsorption (acid) sites on
oxide surfaces.[45,51] These molecules can be used to probe the
properties of OH groups (proton-containing sites) as well as
aprotonic sites in the form of coordinatively unsaturated cations (Lewis acid sites), the latter being formed during surface
dehydroxylation via condensation of surface OH groups. The
interaction between probe and surface site can be monitored by
changes in the vibrational spectra of the probe molecule and, if
OH groups are involved, by the effect on their stretching frequencies.

Ammonia and pyridine are favourite probe molecules. The
addition of protons from sufficiently strongly acidic sites can
easily be detected by the characteristic vibrational spectra of
the resulting ammonium and pyridinium ions.[51] It turns out that
such Bronsted sites hardly exist on pure binary oxides, while
they are characteristic for ternary mixed oxides such as e.g.
SiO₂-Al₂O₃. Hydroxyl groups bridging a Si and an Al atom develop
Bronsted acidity and are able to protonate ammonia and pyridine.

Other probe molecules which have been used for the characterization of OH properties are benzene and nitriles.[45,51] Major
spectral changes with these molecules occur in the O-H stretching region while the effects on intramolecular normal modes
are relatively small. These probes are attractive because of
their comparably low basicity relative to ammonia and pyridine.[45] They are therefore more specific and permit to detect more

detail as regards hydroxyl groups of slightly different properties. A particularly useful probe molecule for the characterization of OH group properties is carbon monoxide when adsorbed at low temperature (80 K). An example is shown in spectrum 2 (Fig. 2) of the changes in the O-H stretching bands on the γ-Al$_2$O$_3$ surface when CO is adsorbed at 80 K.[52,53] The spectrum still shows five resolved bands, the three low frequency bands, however, being shifted to lower frequencies relative to the position of the unperturbed original bands (spectrum 1, Fig. 2). The frequency shifts of the individual OH groups increase with decreasing frequency of the unperturbed groups as indicated in Table 1. This trend coincides with that predicted by the effect

Table 1. Hydroxyl Infrared Bands (cm^{-1}) of Alumina and their Shifts Induced by Adsorbed Carbon Monoxide.

| Unperturbed OH's | | | OH's interacting | |
Type I	Type II	Type III	with CO	$\Delta\nu$(OH)
3785			3785	-
3775			3775	-
	3725		3655	70
	3715		3635	80
		3695	3600	95

of the hydroxyl coordinations on their acidity (vide supra). These shifts are due to the formation of H-bonding interactions of the type OH...CO [52,53] and are thus a measure of the H-bond donor strength and of the acidity the individual hydroxyl group.

On SiO$_2$ a frequency shift of the surface O-H stretching mode of 93 cm^{-1} was reported [54] on adsorption of CO at 83 K. A much larger shift of 312 cm^{-1} was observed by Kustov et al.[55] on H-ZSM-5, thus indicating the strong protonic acidity of the bridging Al-(OH)-Si groups of this zeolite material. The carbonyl stretching also responds to the H-bond interaction and shifts to higher frequencies relative to the gas phase frequency of 2143 cm^{-1}. On SiO$_2$ it is observed at 2158 cm^{-1} ($\Delta\nu$=+15 cm^{-1}) while on H-ZSM-5[55] due to the stronger H-bond the band occurs at 2175 cm^{-1} ($\Delta\nu$=+32 cm^{-1}).

Carbon monoxide can also be used as a probe for coordinatively unsaturated (cus) cationic sites (aprotonic Lewis acid sites). An example is shown in Fig. 3 for the adsorption of CO at 80 K on a partially dehydroxylated TiO$_2$ (anatase) surface. In the presence of 10 torr CO two bands at 2154 and 2176 cm^{-1}, are observed. The low frequency band is due to H-bonded CO molecules (see above). This species is weakly adsorbed as indicated by its easy desorption on evacuation at 80 K, while the high frequency band remains almost unaffected. The higher positive frequency shift relative to the gas phase frequency thus indicates the greater adsorption bond strength. The band at 2176 cm^{-1} is assi-

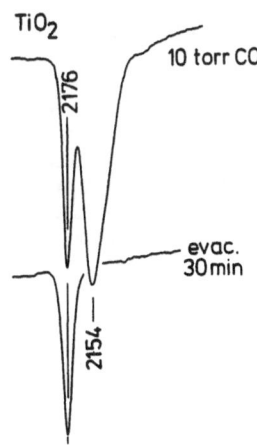

Fig. 3. Carbonyl infrared spectra of CO adsorbed on TiO_2
(anatase) at 80 K in the presence of 10torr CO and
after evacuation.

gned to the coordinatively adsorbed CO molecule on a cus Ti^{x+}
site.[53] The positive frequency shift ($\Delta v > 0$) is due to an increase in bond order of the CO ligand when it coordinates via the
antibonding 5σ orbital at the carbon atom onto a cation center
having negligible d-electron density so that π-back donation
does not play any role. Static and dynamic interactions between
neighbouring adsorbed molecules at high coverage can additionally affect the carbonyl stretching frequency.[56,57] These effects
become negligible when the carbonyl stretching frequency of the
adsorbed CO is measured in the low coverage limit. It has been
shown [53,58] that an empirical correlation holds in this situation
between the CO frequency and the electric field strength at the
coordination site. The field strength F_m is estimated by

$$F_m = s \cdot R_m^{-2}, \tag{12}$$

where s is the strength of the electrostatic bond s = Z/N, Z
being the formal charge at the cus cation, and N its coordination number. R_m is given by the sum of the cation radius in the
given coordination and the van der Waals-radius of the carbonyl
carbon atom. For favourable cases this empirical correlation
permits to determine the nature of the cation site, its coordination number and oxidation state from the carbonyl stretching
frequency of the coordinated CO ligand. The CO frequency is thus
undoubtedly also related to the acid strength of these aprotonic
acid sites. It must be admitted, however, that the simple electrostatic bonding model underlying the empirical correlation
only holds strictly for d^0-ions.

Structure of Mixed Metal Oxides

Bismuth molybdates are active catalysts for selective oxidation.
They occur in various stroichiometries with different structures
which can be distinguished by their Raman spectra.[59,60] It has
also been demonstrated by Raman spectroscopy that lattice oxygen
undergoes ^{18}O-exchange during catalytic oxidation of propene
with $^{18}O_2$.[59,60]

Structures and Acidity of Supported Oxides

The importance of supported oxides as catalysts and catalyst precursors has been discussed.[61] Alumina-supported molybdena is a characteristic example. These materials are typically prepared by impregnation of the alumina support by an aqueous heptamolydate solution, from which the polyanion is adsorbed as shown by Raman spectroscopy.[62,63] The same polyanionic surface species is formed when a physical mixture of MoO_3 and γ-Al_2O_3 is thermally treated in a moist O_2 atmosphere at 720 K.[64,65]

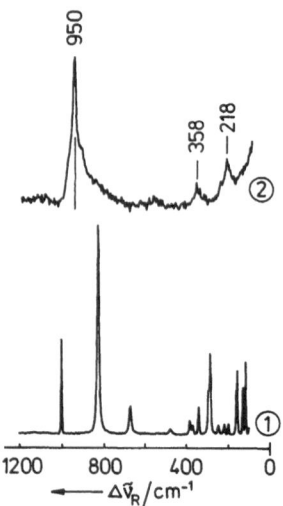

Fig. 4. Raman spectra (1) of a physical mixture of MoO_3 and Al_2O_3 prior to thermal treatment and (2) after calcination at 720 K (30h) in the presence of 24torr H_2O vapour (from ref.[65]).

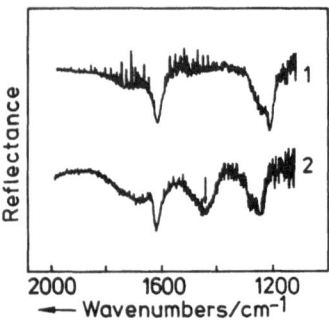

Fig. 5. DRIFT spectra of ammonia adsorbed (1) on TiO_2 (anatase) and (2) on 9 wt% WO_3 supported on TiO_2.

The Raman spectrum (1) of Fig. 4 is that of MoO_3 in the physical mixture prior to thermal treatment while spectrum (2) is obtained after 30h calcination at 720 K and is characteristic of the surface polymolybdate. The formation of this species from polycrystalline MoO_3 in the physical mixture with alumina requires a spreading of the MoO_3 over the alumina surface. Raman microscopy has demontrated [44] that this spreading occurs over macroscopic distances of several hundred micrometers.

The surface polyanion must be chemically anchored to the alumina support surface. Cornac et al. [66] have demonstrated by transmission-absorption spectroscopy that these polyanions do replace the basic surface hydroxyl groups of the alumina surface. The surface molydate species can be described as

$$[Al^+]_x Mo_7 O_{24}{}^{6-} (6-x) H^+$$

suggesting that the heptaanion is anchored to the surface via x Al-O-Mo bonds. The (6-x) protons are required for charge compensation and presumably are to be considered as acidic protons which are responsible for the greater Bronsted acidity of these materials as compared to that of the pure alumina support. Riseman et al. [67] have clearly demonstrated this phenomenon by photoacoustic spectroscopy using pyridine as a probe.

Analogously, the formation of a polytungstate on the surface of a titania support (catalyst for selective catalytic reduction of NO_x by NH_3) leads to increased Bronsted acidity.[68] This can be seen in Fig. 5, which shows DRIFT spectra of NH_3 adsorbed on pure TiO_2 (anatase) and on 9wt% WO_3 on TiO_2. The NH_3 deformtion bands observed on pure TiO_2 are assigned to NH_3 adsorbed coordinatively on Lewis acid sites. In contrast, the titania-supported tungstate shows an additional intense band at 1418 cm^{-1} which is clearly due to the $NH_4{}^+$ ion,[51] this being unequivocal evidence for the induction of Bronsted acidity by the surface tungstate.

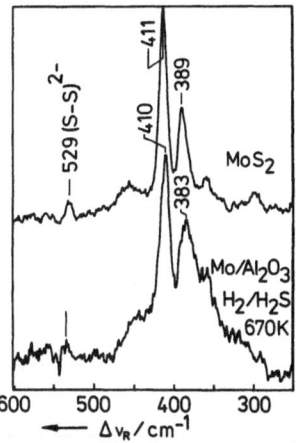

Fig. 6. Raman spectra of polycrystalline MoS_2 and on a sulfided Mo/Al_2O_3 catalyst (from ref.[62]).

Structure and Adsorption Sites of Supported Sulfides

The sulfidation of supported molybdates by H_2/H_2S at temperatures between 470 and 700 K leads to the formation of supported molybdenum sulfides.[62] Besides several other techniques, Raman spectroscopy has provided evidence for the MoS_2-like structure of alumina-supported molybdenum sulfides. The Raman spectra of Fig. 6 demonstrate the close resemblence of the vibrational modes of a supported molybdenum sulfide (bands at 383 and 410 cm^{-1}) and those of polycrystalline MoS_2 (bands at 389 and 411 cm^{-1}). In addition, the band observed at 529 cm^{-1} suggests the presence of disulfide species $(S-S)^{2-}$ being located presumably along the edges of MoS_2-slabs.[62,69] These groups might act as sites for homolytic splitting of dihydrogen.[62,70]

Adsorption sites on the supported sulfide phases can be probed by CO adsorption. Fig. 7 shows transmission-absorption infrared spectra in the carbonyl stretching region of CO adsorbed at 80 K on a sulfided Mo/Al_2O_3 catalyst.[71] The dominant band in the spectra appears at 2108 cm^{-1}. This band and a second less intense band near 2190 cm^{-1} saturate at low CO pressures, while the central band near 2150 cm^{-1} is pressure dependent and can easily be removed on evacuation at 80 K. The latter band is due to H-bonded CO molecules (vide supra) and is not of interest in the present discussion. The high frequency component at 2190 cm^{-1} is to be assigned to $Mo^{4+} \leftarrow$ CO species.[72] The same band position is observed on H_2-reduced Mo/Al_2O_3 catalysts for which XPS shows the presence Mo^{4+} sites. It is interesting to note that the frequency of the stretching mode of this CO ligand being coordinated to a $Mo^{4+}(d^2)$ center is still greater than the gas phase frequency at 2143 cm^{-1}, although some π-back donation may occur. The strongest carbonyl band at 2108 cm^{-1} in the spectra of Fig. 7 is in contrast clearly shifted to a position below the gas phase frequency ($\Delta\nu=-35$ cm^{-1}). This suggests that the corresponding coordination site must be a low-valent Mo site providing significant d-electron density which leads to an

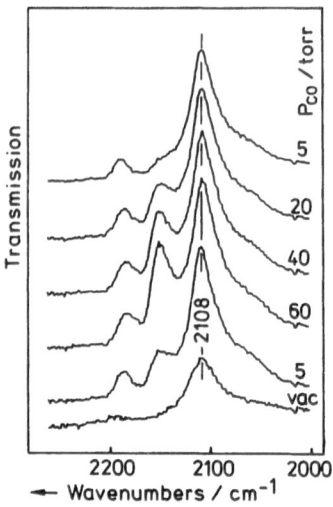

Fig. 7. Carbonyl infrared spectra of CO adsorbed on sulfided Mo/Al_2O_3 at 80 K.

appreciable π-back donation in the coordination bond. It has been suggested [71,72] that the coordination site is to be identified as a $Mo^{2+}(d^4)$ center which is presumably located along the edges of MoS_2-slabs.

Real hydrotreating catalysts contain Co or Ni as promoters which are thought to decorate the edges of MoS_2-slabs.[62,73] (see also ref. [61]). Electronic ineractions between neighbouring Mo and (e.g.) Ni atoms can well be envisaged.[73] Some evidence for such interactions has recently been obtained using CO as a probe.[71] The C-O stretching frequency was found to be 2106 cm^{-1} on an unpromoted sulfided Mo/SiO_2 catalyst. The frequency shifted to higher values, namely 2122 cm^{-1}, on a sulfided $NiMo/SiO_2$. This shift to higher frequency might be interpreted as being due to reduced d-electron density at the Mo center suggesting charge transfer from the Mo site to the Ni promoter atom. In support of this interpretation, a shift by 4 cm^{-1} is observed for the CO ligand coordinated to a Ni site in $NiMo/SiO_2$ as compared to the Mo-free sulfided Ni/SiO_2.

Carbonyl Species on Supported Rhodium Catalysts

Rhodium is an important noble metal in catalysis. It finds e.g. application in CO + H_2 reactions and is a major component in car exhaust catalysts. When used in small concentrations (< 1 wt% Rh) on oxide supports (such as Al_2O_3 or SiO_2), it is present as small particles in high dispersion. Linear and bridged species are formed when CO is adsorbed on metallic Rh^o sites as evidenced by the characteristic carbonyl stretching vibrations at 2068 and 1908 cm^{-1}, respectively.[74] These bands can be seen in Fig. 8 for Rh/SiO_2. The spectrum of Fig. 8 shows, however, an additional pair of bands, namely at 2038 and 2092 cm^{-1}, which has been attributed to an oxidized Rh^+ dicarbonyl species (gem dicarbonyl).[75-77] The appearance of this species points to an interesting phenomenon related to the dynamic character of small metal particles in contact with a CO atmosphere, namely the disintegration of small zero-valent particles by CO and the

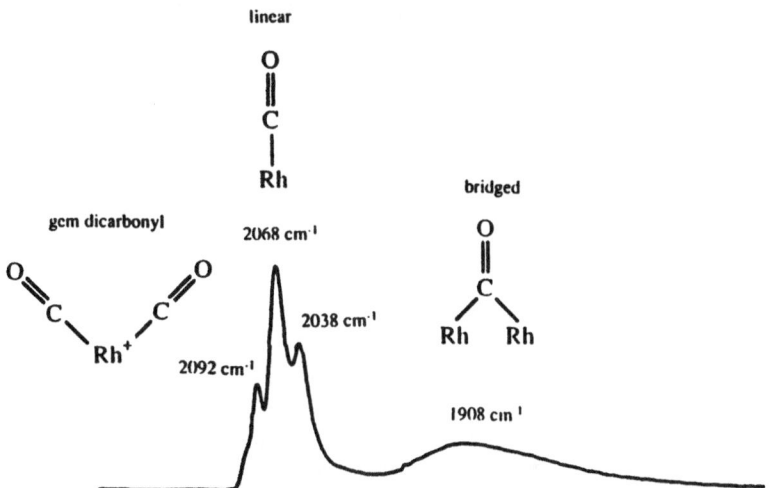

Fig.8. Transmission-absorption infrared spectrum of carbonyl species a Rh/SiO_2 catalyst.

oxidation of presumably mononuclear Rh°-C species on the support surface. The disintegration can be explained by the fact that the average Rh°-CO bond strength is greater than the average Rh-Rh bond strength. This process finds analogies in carbonyl cluster chemistry. The observation that the tendency toward disintegration is much more pronounced for small particles as compared to larger ones, suggests that Rh surface atoms in low coordination position are more susceptible to CO-induced dissociation. Basu et al. [76] provided evidence from transmission-absorption IR spectroscopy in the hydroxyl region of the support oxides (SiO_2 and Al_2O_3) that surface hydroxyl groups are responsible for the oxidation of the Rh°-CO species. They postulated the following overall mechanism, where $[Rh^o]_x$ stands for a metal particle:

$$[Rh^o]_x + CO(g) \rightarrow [Rh^o]_{x-1} + [Rh^o\text{-}CO]_{ads}$$

$$[Rh^o\text{-}CO]_{ads} + [OH^-]_{surf} \rightarrow [O^{2-}\text{-}Rh^+\text{-}CO]_{surf} + \tfrac{1}{2}H_2(g)$$

$$[O^{2-}\text{-}Rh^+\text{-}CO]_{surf} + CO(g) \rightarrow [O^{2-}\text{-}Rh^+(CO)_2]_{surf}.$$

Scheme 3

Van't Blik et al. [79] have shown by EXAFS that the dicarbonyl species is anchored to three surface oxygen ligands.

Rhodium catalysts for (CO + H_2) reactions are generally promoted by transition metal oxides, such as e.g. MnO_x.[80,81] A low-frequency band near 1700 cm^{-1} was observed when CO was adsorbed on such materials.[80,81] Fig. 9 shows transmission-absorption IR spectra in the carbonyl stretching region which were obtained for ^{12}CO and ^{13}CO adsorption on a MnO_x-promoted Rh/SiO_2 catalyst.[77] Besides the characteristic bands at 2073 and 1873 cm^{-1} of terminal and bridging ^{12}CO species, a broad band centered at 1716 cm^{-1} is observed. This band shifts to 1675 cm^{-1} when isotopically labelled ^{13}CO is adsorbed. The shift ($\Delta\nu = -41$ cm^{-1}) corresponds to the theoretical isotope shift and this supports the

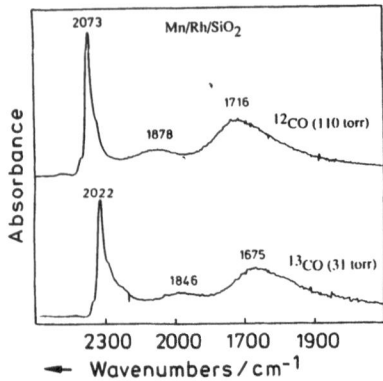

Fig. 9. Transmission-absorption infrared spectra of ^{12}CO and ^{13}CO adsorbed on a MnO_x-promoted Rh/SiO_2 catalyst (from ref.[77]).

assignment of these bands to C-O stretching vibrations. The low-frequency band has been attributed to a C-and O-bonded CO-ligand (Rh-Co→Mn^{x+}) in analogy to molecular Σ-adducts.[82,83] It has been proposed [80,82,83] that this surface adduct may play a role as an intermediate in catalytic CO + H$_2$ reactions. However, no direct spectroscopic proof for the kinetic relevance of this species is yet available. It must be stressed in this connection, that vibrational spectroscopies may well be applied for the detection of intermediates of surface catalyzed reactions. The assignment of observed bands to reactive intermediates must, however, be proven by their response to changes of reaction parameters so as to support their kinetic relevance. Only few examples of this kind have been reported. Thus, the intermediate character of surface π-allyl species in alkene isomerization on ZnO-surfaces was demonstrated by Kokes and Dent.[84]

The formation of the above mentioned C-and O-bonded surface adduct can only occur on areas where Rh-metal and promoter oxide are in intimate contact. This may e.g. be along the perimeter of islands of the promoter oxide on the surface of the metal, a situation which is to be compared to metal particle encapsulation under so-called strong-metal-support-interaction (SMSI) conditions.[85] The encapsulation of small Rh particles by promoter oxide when supported on silica, has in fact been demonstrated by CO adsorption and IR spectroscopy by Kraus et al.[86] For example, carbonyl bands of CO adsorbed on Rh sites could not be detected at temperatures < 270 K on a Rh/Al$_2$O$_3$ catalyst which was promoted by NbO$_x$. At increasing temperatures (T=470 K) in the presence of CO, the characteristic bands of terminal Rh^o-CO and of the gem dicarbonyl species developed. This phenomenon was interpreted as being due to complete encapsulation of Rh particles after H$_2$-reduction by a NbO$_x$ layer, which was stable in CO atmosphere at low temperature. At higher temperature in CO a restructuring of this promoter layer with the possible formation of small oxide islands and partial exposure of the Rh was proposed. This is again an example for the dynamic character of catalyst surfaces and for the fact that their structure may deviate significantly under catalytic conditions from that determined under a non-catalytic situation.

Metal Clusters in Zeolites

Metal clusters can be encaged in zeolite matrices.[87] They can be characterized by CO as a probe and IR spectroscopy. A particularly interesting case is that of Pd carbonyl clusters because carbonyl clusters of this noble metal are not known in organometallic chemistry. Fig. 10 shows a spectrum of Pd in zeolite NaY. The large number of sharp C-O stretching bands in these spectra is suggestive of a well-defined molecular Pd carbonyl species. It has been proposed that a Pd$_{13}$(CO)$_x$ cluster was formed[88] in which the 13 metal atoms are arranged in an icosahedral or cubooctahedral framework. This Pd$_{13}$ cluster is supposed to be entrapped in the large cavity of the NaY structure. It has a diameter of 0.82 nm and can therefore not escape through the windows which have a diameter of 0.74 nm. On the other hand, the diameter of 1.3 nm of the large cavity still permits the interaction with CO ligands.

These CO ligands prevent the cluster from interacting with the cage walls. There is spectroscopic evidence in the carbonyl spectra that during desorption of CO a positive charge builds up on the metal core presumably due to increasing cluster-wall

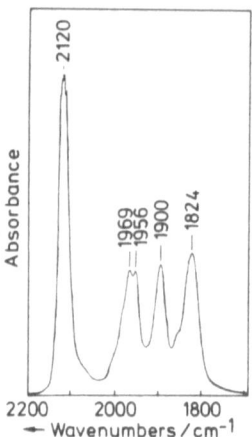

Fig. 10. Carbonyl infrared spectrum of $Pd_{13}(CO)_x$ clusters formed in NaY zeolite after mild reduction at 473 K (from ref.[88]).

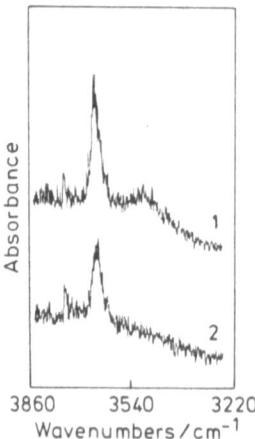

Fig. 11. Infrared transmission-absorption spectra in the hydroxyl stretching region of a NaY zeolite contai ning $Pd_{13}(CO)_x$ clusters: (1) in the presence of CO; and (2) after partial decarbonylation (from ref.[89]).

interactions as the cluster is less sterically shielded by CO ligands.[88,89] As a matter of fact, the hydroxyl stretching band of the acidic Al-(OH)-Si groups is reduced in intensity when the cluster is decarbonylated (see Fig.11), suggesting that perhaps acidic protons interact with the cluster. The following process has been postulated:[89]

$$Pd_{13}(CO)_a + H^+ \rightarrow [H-Pd_{13}(CO)_x]^+ + (a-x)CO, \qquad (13)$$

in which the original cluster is transformed into a positively charged hydride cluster. Alternatively, a simple polarization of the cluster as described by van Santen[90] may also account for the positive charge on the cluster.

When the Pd containing NaY zeolite is thermally treated in flowing H_2 at 770 K, larger extrazeolitic particles are formed which give rise to carbonyl infrared spectra as shown in Fig. 12. The two bands at 2099 and 1971 cm^{-1} are characteristic of terminal and bridging CO species adsorbed on Pd metal and the spectra are almost identical to those observed for silica-supported Pd particles.[74] Hence, depending on reduction (and calcination [88,89]) conditions, molecular clusters (Pd_{13}) or metal particles with broad particle size distributions can be produced on NaY zeolites, and are distinguished by their characteristic infrared carbonyl spectra.

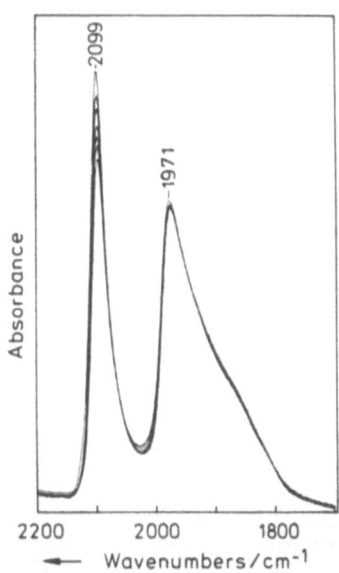

Fig. 12. Carbonyl inrared spectra of CO adsorbed on Pd metal on NaY zeolite produced by reduction at 773 K (from ref.[89]).

Bimetallic CuRu - Catalysts

Copper and ruthenium do not form binary bulk alloys. The two metals are entirely immiscible. Nevertheless, it has been shown[91] that in highly dispersed supported materials Cu "adsorbs" on Ru particles and encapsulates them. This can be shown by infrared spectra of adsorbed CO.[92] Fig. 13 shows carbonyl infrared spectra of CO adsorbed on $CuRu/SiO_2$ containing 5 wt % of each, Cu and Ru which corresponds to a Cu/Ru atomic ratio of 1.6. The band at 2155 cm^{-1} in spectrum (a) is due to H-bonded CO species on the silica support (vide supra). The bands at 2123 cm^{-1} in the presence of CO at 80 K and at 2127 cm^{-1} (spectrum (b)) at 295 K are in the range in which terminal CO species do

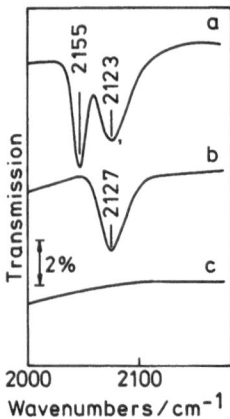

Fig. 13. Infrared transmission-absorption spectra of CO adsorbed on RuCu/SiO$_2$ after reduction at 623 K: (a) in 53torr CO at 80 K; (b) in 53torr CO at 295 K; and (c) after evacuation at 295 K for 5 min (from ref.[92]).

absorb when adsorbed on zero-valent Cu^o metal. The species is relatively weakly bonded as indicated by the fact that it immediately desorbs on evacuation at 295 K (spectrum (c) of Fig. 13). No bands in the frequency regime characteristic Ru-CO species (namely at $\nu < 2100$ cm^{-1}) could be observed. This is a clear indication for the ecapsulation of Ru^o metal particles by a skin of Cu^o metal.

CONCLUDING REMARKS

Only very few and admittedly subjectively selected examples could be discussed in this review. Nevertheless, they should have demonstrated the potential of vibrational spectroscopies for the characterization of real catalyst materials. The structures of supported surface phases can be determined as well as structures of bulk catalyst materials, and surface sites can be analyzed directly or by the use of probe molecules. Few examples do exist to date of an unequivocal detection of catalytic intermediates. This is a challenging problem since intermediates are typically short-lived and do exist in only low concentration. They might, however, be trapped at low temperatures. The major advantage of vibrational spectroscopies as compared to other surface science techniques including those using particle beams is their applicability under in situ conditions in the presence of a gas phase even at elevated pressures. This is due to the low interaction cross sections of photons with matter. Vibrational spectroscopies will undoubtedly continue to find broad application for catalyst characterization in the future. No catalysis laboratory concentrating on characterization can ever go without vibrational spectroscopies.

REFERENCES

1. L.H. Little, "Infrared Spectroscopy of Adsorbed Species", Academic Press, New York, 1966.

2. M. L. Hair, "Infrared Spectroscopy in Surface Chemistry", Dekker, New York, 1967.
3. A. V. Kiselev and V. I. Lygin, "Infrared Spectra of Surface Compounds", Wiley, New York, Toronto, 1975.
4. W. N. Delgass, G. L. Haller, R. Kellerman and J. H. Lunsford, "Spectroscopy in Heterogeneous Catalysis", Academic Press, New York, San Francisco, London, 1979.
5. A. T. Bell and M. L. Hair, "Vibrational Spectroscopies for Adsorbed Species", ACS Symposium Series 137, Amer. Chem. Soc., Washington, D. C., 1980.
6. G. Kortüm, "Reflexionsspektroskopie", Springer, Berlin, 1969.
7. M. P. Fuller and P. R. Griffiths, <u>Anal. Chem.</u> 50: 1906 (1978).
8. P. J. Brimmer and P. R. Griffiths, <u>Appl. Spectrosc.</u> 42: 242 (1988).
9. T. Egerton and A. H. Hardin, <u>Catal. Rev.-Sci. Eng.</u> 11:1 (1975).
10. R. P. Cooney, G. Curthoys and T. T. Nguyen, <u>Advan. Catal.</u> 24:293 (1975).
11. J. M. Stencel, "Raman Spectroscopy for Catalysis", Van Nostrand Reinhold, New York, 1990.
12. G. L. Haller, R. W. Rice and Z. C. Wan, <u>Catal. Rev.-Sci. Eng.</u> 13:259 (1976).
13. Y. H. Pao, "Optoacoustic Spectroscopy and Detection", Academic Press, New York, 1977.
14. A. Rosencwaig, "Photoacoustics and Photoacoustic Spectroscopy", Wiley, Chichester, 1980.
15. J. C. Wright, <u>Appl. Spectrosc.</u> 34:151 (1980).
16. Y. R. Shen, in: "Chemistry and Structure at Interfaces", R. B. Hall and A. B. Ellis, eds., VCH Publishers, Inc., Deerfield Beach, 1986, p. 151.
17. G. Ertl and J. Küppers, "Low Energy Electrons and Surface Chemistry", Verlag Chemie, Weinheim, 1974.
18. A. V. Kiselev and V. I. Lygin, <u>Russ Chem. Rev.</u> 31:175 (1962).
19. C. H. Rochester, <u>Advan. Colloid Interf. Sci.</u> 12:43 (1980); and <u>Progr. Colloid Polym. Sci.</u> 67:7 (1980).
20. H. Knözinger, <u>Acta Cient. Venez.</u> 24, Supl. 2:76 (1973).
21. P. Basu, T. H. Ballinger and J. T. Yates, Jr., <u>Rev. Sci. Instrum.</u> 59:1321 (1988).
22. E. Gallei and E. Schadow, <u>Rev. Sci. Instrum.</u> 45:1504 (1974).
23. W. W. Wendlandt and H. G. Hecht, "Reflectance Spectroscopy", Plenum, New York, 1968.
24. E. H. Korte and A. Otto, <u>Appl. Spectrosc.</u> 42:38 (1988).
25. J. Mink and G. Keresztury, <u>Croat. Chem. Acta</u> 61:731 (1988).
26. J. Mink and G. Kersztury, <u>Proc. SPIE - Inc. Soc. Opt. Eng.</u> 1145 (1989), (CA 112:87360v(1990)).
27. R. Z. Garrell, <u>Anal. Chem.</u> 61:401A (1989).
28. J. J. Freeman, J. Heaviside, P. J. Hendra, J. Prior and E. S. Reid, <u>Appl. Spectrosc.</u> 35:196 (1981).
29. F. R. Brown, L. E. Makovsky and K. H. Rhee, <u>Appl. Spectrosc.</u> 31:563 (1977).
30. C. P. Cheng, J. D. Ludowise and G. L. Schrader, <u>Appl. Spectrosc.</u> 34:146 (1980).
31. B. Vielhaber and H. Knözinger, <u>Appl. Catal.</u> 26:375 (1986).
32. H. Zeilinger, G. Mestl and H. Knözinger, to be published.
33. D. D. Saperstein and A. J. Rein, <u>J. Phys. Chem.</u> 81:2134 (1977).
34. H. Jeziorowski and H. Knözinger, <u>Chem. Phys. Letters</u> 51:519 (1977).

35. H. Jeziorowski and H. Knözinger, J. Phys. Chem. 83:1166 (1979).
36. H. Jeziorowski and H. Knözinger, Chem. Phys. Letters 42:162 (1976).
37. M. Cardona, "Modulation Spectroscopy", in: "Solid State Physics", Suppl. 2, F. Seitz, D. Turnbull and E. Ehrenreich, eds., Academic Press, New York, 1969.
38. F. L. Galeener, Chem. Phys. Letters 48:7 (1977)
39. S. Brückner, H. Jeziorowski and H. Knözinger, Chem. Phys. Letters 105:218 (1984).
40. P. J. Hendra and H. Mould, Intern. Laboratory, 1988.
41. B. C. Chase, Anal. Chem. 59:881 A (1987).
42. P. J. Hendra, C. Passingham, G. M. Warnes, R. Burch and D. J. Rawlence, Chem. Phys. Letters 164:178 (1989).

43. J. Corset, P. Dhamelincourt and J. Barbillat, Chem. Britain 25:612 (1989).
44. J. Leyrer, D. Mey and H. Knözinger, J. Catal. 124:349 (1990).
45. H. - P. Boehm and H. Knözinger, in: "Catalysis - Science and Technology", J. R. Anderson and M. Boudart, eds., Springer, Berlin, Heidelberg, New York, Vol. 4, p. 39, 1983.
46. R. S. McDonald, J. Phys. Chem. 62:1168 (1958).
47. K. Klier, J. Chem. Phys. 58:737 (1973).
48. A. A. Tsyganenko and V. N. Filimonov, J. Mol. Structure 19:579 (1973).
49. H. Knözinger and P. Ratnasamy, Catal. Rev.-Sci. Eng. 17:31 (1978).
50. E. P. Smirnov and A. A. Tsyganenko, React. Kinet. Catal. Letters 7:425 (1977).
51. H. Knözinger, Advan. Catal. 25:184 (1976).
52. M. I. Zaki and H. Knözinger, Mater. Chem. Phys. 17:201 (1987).
53. H. Knözinger, in: "Acid-Base Catalysis", K. Tanabe, H. Hattori, T. Yamaguchi and T. Tanaka, eds., Kodansha, Tokyo, 1989, p. 147.
54. T. P. Beebe, P. Gelin and J. T. Yates, Jr., Surface Sci. 148:526 (1984).
55. L. M. Kustov, V. B. Kazansky, S. Beran, L. Kubelková and P. Jiru, J. Phys. Chem. 91:5247 (1987).
56. D. Scarano, A. Zecchina and A. Reller, Surface Sci. 198:11 (1988).
57. A. A. Tsyganenko, L. A. Denisenko, S. M. Zverev and V. N. Filimonov, J. Catal. 94:10 (1985).
58. M. I. Zaki and H. Knözinger, J. Catal. 119:311 (1989).
59. E. V. Hoefs, J. R. Monnier and G. W. Keulks, J. Catal. 57:331 (1979).
60. J. F. Brazdil, M. Mehicic, L. C. Glaeser, M. A. S. Hazle and R. K. Grasselli, ACS Symp. Ser. 288:26 (1985).
61. H. Knözinger, "Heterogeneous Catalysts and Catalytic Processes", this volume.
62. H. Knözinger, Proc. 9th Intern. Congr. Catal. Calgary, 1988, Vol. 5, M. J. Phillips and M. Ternan, eds., The Chemical Institute of Canada, Ottawa, 1989, p. 20.
63. H. Jeziorowski and H. Knözinger, J. Phys. Chem. 83:1166 (1979).
64. J. Leyrer, R. Margraf, E. Taglauer and H. Knözinger, Surface Sci. 201:603 (1988).
65. J. Leyrer, M. I. Zaki and H. Knözinger, J. Phys. Chem. 90: 4775 (1986).

66. M. Cornac, A. Janin and J. C. Lavalley, *Infrared Phys.* 24:143 (1984).

67. S. M. Riseman, S. Bandyopadhyay, F. E. Massoth and E. M. Eyring, *Appl. Catal.* 16:29 (1985).

68. F. Hilbrig, Dissertation, Universität München, 1989.

69. J. Polz, H. Zeilinger, B. Müller and H. Knözinger, *J. Catal.* 120:22 (1989).

70. R. Prins, V. H. J. de Beer and G. A. Somorjai, *Catal. Rev. - Sci. Eng.* 31:1 (1989).

71. J. Cyrys, J. A. R. van Veen and H. Knözinger, unpublished results.

72. E. Delgado, G. A. Fuentes, C. Hermann, G. Kunzmann and H. Knözinger, *Bull. Soc. Chem. Bel.* 93:735 (1984).

73. H. Topsoe, B. S. Clausen, N.-Y. Topsoe and E. Pedersen, *I & EC Fundamentals* 25:25 (1986).

74. N. Sheppard and T. T. Nguyen, *Advan. Infrared Raman Spectrosc.* 5:67 (1978).

75. F. Solymosi and M. Pásztor, *J. Phys. Chem.* 89:4783 (1985); and 90:5312 (1986).

76. P. Basu, D. Panayotov and J. T. Yates, Jr., *J. Phys. Chem.* 91:3133 (1987).

77. S. A. Stevenson, A. Lisitsyn and H. Knözinger, *J. Phys. Chem.* 94:1576 (1990).

78. B. C. Gates, L. Guczi and H. Knözinger, eds., "Metal Clusters in Catalysis", Elsevier, Amsterdam, 1986.

79. H. F. J. van't Blik, J. B. A. D. van Zon, T. Huizinga, J. C. Vis, D. C. Koningsberger and R. Prins, *J. Amer. Chem. Soc.* 107:3139 (1985).

80. W. M. H. Sachtler, Proc. 8th Intern. Congr. Catal., Berlin, 1984, Vol.1, Dechema, Frankfurt and Verlag Chemie, Weinheim, 1984, p. 151.

81. M. Ichikawa and T. Fukushima, *J. Phys. Chem.* 89:1564 (1985).

82. C. P. Horwitz and D. F. Shriver, *Adv. Organometal. Chem.* 23:219 (1984).

83. H. Knözinger, in: "Homogenous and Heterogeneous Catalysis", Yu. Yermakov and V. Likholobov, eds., VNU Press, Utrecht, 1986, p. 789.

84. R. J. Kokes and A. L. Dent, *Advan. Catal.* 22:1 (1972).

85. S. A. Stevenson, J. A. Dumesic, R. T. K. Baker and E. Rukkenstein, "Metal-Support Interactions in Catalysis, Sintering and Redispersion", Van Nostrand Reinhold, New York, 1987.

86. L. Kraus, M. I. Zaki, H. Knözinger and B. Tesche, *J. Mol. Catal.* 55:55 (1989).

87. P. A. Jacobs, in ref.[78], p. 357.

88. L. L. Sheu, H. Knözinger and W. M. H. Sachtler, *Catal. Letters* 2:129 (1989).

89. L. L. Sheu, H. Knözinger and W. M. H. Sachtler, *J. Amer. Chem. Soc.* 111:8125 (1989).

90. R. A. van Santen, "Quantum Chemistry of Surface Chemical Reactivity", this volume.

91. J. H. Sinfelt, "Bimetallic Catalysts: Discoveries, Concepts and Applications", Wiley, New York, 1983.

92. R. Liu, B. Tesche and H. Knözinger, *J. Catal.*, in print.

Applications of X-ray Absorption Spectroscopy in Catalysis

Richard W. Joyner

Leverhulme Centre for Innovative Catalysis
Department of Chemistry, University of Liverpool
PO Box 147, Liverpool, L69 3BX, UK

1. INTRODUCTION

The X-ray absorption edge jump, which arises when the energy of a photon beam becomes just sufficient to excite core electrons into the continuum, is well known and understood. In 1920, Fricke and Hertz separately, reported the first observations of absorption fine structure at energies above the absorption edge and subsequently Ray, (1929) and Kievet and Lindsay (1930) ·noted that this fine structure could be detected several hundred eV above the edge. It was recognised that these variations in the absorption cross section had their origin in the structure of the material under study and early theories were advanced by Kronig (1931, 1932). These was, however a long standing debate as to whether this extended X-ray absorption fine structure, or EXAFS, was a signature of long range order, as is X-ray diffraction, or was due to short range order around the absorbing atom. It was gradually accepted that short range order alone was responsible, for example as a result of the work of Petersen (1936) and Kostarev (1949). The crucial breakthrough, however, was due to Sayers, Stern and Lytle (1971), who showed that the Fourier transform of the EXAFS oscillations with respect to photoelectron wave vector showed peaks at distances corresponding to the interatomic distances of shells of nearest neighbour atoms. This was undoubtedly the starting point of EXAFS as a modern technique for structure determination. The other main factor of significance has been the increasing availability of synchrotron radiation, which has vastly broadened the range and applicability of the method.

This review is devoted to the application of X-ray absorption spectroscopy in catalysis. In the next section we develop the theory of EXAFS in sufficient detail for general understanding. In section three the basic EXAFS experiment and a number of important modifications are described while section 4 gives a brief discussion of different approaches to data analysis. The final, largest section discusses applications in catalysis. For greater detail the reader is referred to two books, (Teo and Joy, 1981; Koningsberger and Prins, 1988) and a number of reviews which are indicated at appropriate points below.

Fundamental Aspects of Heterogeneous Catalysis Studied by Particle Beams,
Edited by H.H. Brongersma and R.A. van Santen, Plenum Press, New York, 1991

2. THEORY OF X-RAY ABSORPTION

2.1 Near Edge Structure

For theoretical reasons which will emerge shortly, it is convenient
to divide the X-ray absorption spectrum into two regions, as shown in
Fig. 1. That portion which includes any pre-edge features, the edge
itself and extends to 40 - 50eV above the edge is referred to as the near
edge region. Studies of this part are referred to as near edge X-ray
absorption fine structure, (NEXAFS), or X-ray absorption near edge
structure, (XANES). The interpretation of data from this spectral region
is usually only qualitative. The position of the absorption edge itself
provides information about the oxidation state of the element under
study, interpreted in a similar way to the binding energy shifts observed
in X-ray photoelectron spectroscopy. As is the case with XPS,
interpretation of edge shifts may be complicated by the presence of final
state effects, (Wertheim, 1987). The fine structure at the absorption
edge can be considered simply as a form of electronic absorption
spectrum. As such and in favourable cases, information may be obtained
about the oxidation state and coordination environment of the absorbing
atom. Fig. 2 shows an example of the change in the XANES of the chromium
K edge in a Pt / Cr / H-ZSM5 catalyst as a result of reduction in
hydrogen: note that the sharp pre-edge feature is destroyed. Parallel
XPS studies of this catalyst indicate that this is due to the reduction
of Cr(VI) to Cr(III), (Johnston, Joyner and Shpiro, to be published).
For reasons which are unclear, useful XANES information appears to be
restricted largely to the first transition series.

A specialised example of the usefulness of NEXAFS is the study of
adsorbate orientation on single crystal or other well orientated
substrates. Using the polarised nature of synchrotron radiation, it is
possible to identify the symmetry of certain near edge adsorbate features
or resonances, and thence to determine the orientation and possibly the
adsorption site of the adsorbed species (Johnson et al, 1983). Although
requiring the use of single crystals, this approach can be used in situ
at pressures up to 0.1 torr and therefore has applications in fundamental
studies of catalysis, (Zaera et al, 1988).

The impact of NEXAFS on catalysis has been reviewed by Bart (1986).
The majority of X-ray absorption studies in catalysis, however, have
concentrated much more on EXAFS, which is the main subject of this
review.

2.2 The Theory of EXAFS

Before considering the physical origin of EXAFS, it is convenient to
introduce some definitions. The absorption of the X-ray flux incident on
the sample is given by the standard equation:

$$\mu = [Ln(I_0/I)]/t \qquad\qquad (1)$$

where I_0 and I are respectively the incident and the transmitted flux and
t is the sample thickness

In EXAFS the structural information is contained in the oscillations
superimposed on the background absorption or atomic EXAFS, which varies
slowly as a function of the X-ray wavelength. In defining the amplitude
of the EXAFS oscillations it is first necessary to subtract the

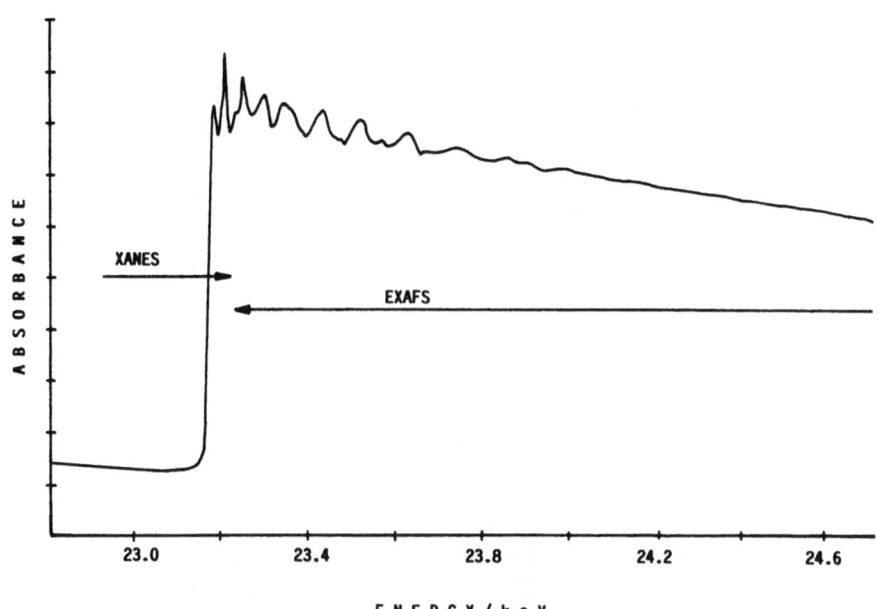

Fig. 1. The X-ray absorption spectrum of a rhodium foil, indicating the near-edge region, (XANES), and the extended fine structure or EXAFS region.

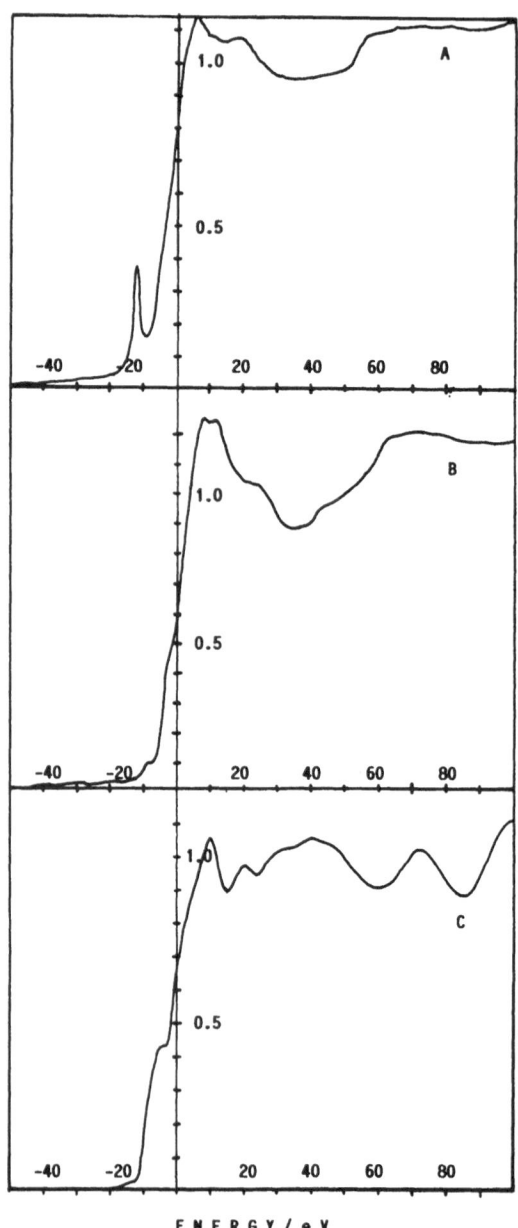

E N E R G Y / e V

Fig. 2.   Near edge spectra, or XANES of the chromium edge.   Curve A is
from a 0.5% Pt, 2.25% Cr, H-ZSM-5 catalyst as received; curve B
is from the same catalyst after in situ reduction in hydrogen,
(1 bar), at 823K; curve C is from a chromium foil. (Johnston,
Joyner and Shpiro, unpublished results).

background due to absorption by other elements in the sample and due to other electronic levels of the same atom. The magnitude of the EXAFS is then:

$$\chi(\mathbf{k}) = (\mu - \mu_o)/\mu_o \qquad (2)$$

where μ_o is the atomic absorption of a free atom and μ the absorption in the sample of interest. $\mathbf{k}$ is the photoelectron wave vector which provides a convenient measure of its energy. $\mathbf{k}$ is defined in atomic units as

$$\mathbf{k} = (2E)^{1/2} \qquad (3)$$

where E is the photoelectron energy measured in Hartrees, (1H = 27.2eV) and the atomic unit of $\mathbf{k}$ is the Bohr radius, (1 Bohr radius = 0.0529nm or 0.529A). Commonly $\mathbf{k}$ is expressed in A^{-1} and energy in eV, where the conversion factor is

$$\mathbf{k}/A^{-1} = 0.512(E/eV)^{1/2} \qquad (4)$$

Values of χ rarely exceed 0.03, or 3% of the atomic absorption, and the oscillations decay into the noise at between 400 and 2000eV above the absorption edge, (ie in the range 10 - ca 20 A^{-1}).

To understand the origin of EXAFS it is necessary to consider the physics which gives rise to a photoelectron. According to Fermi's golden rule, the probability of an X-ray photon being absorbed by a core electron depends on both the initial and final state of the photoelectron. The former is simply the localised core level corresponding to the absorption edge, while the latter can be considered as the interference of the outgoing photon spherical wave with an incoming wave due to the reflection, or backscattering of the photoelectron by all neighbouring atoms. It is this interference, or ability of the emitted photoelectron to interact with its reflection, which is the origin of the EXAFS oscillations and of the structural content of the spectroscopy. If the interference is constructive the extent of the X-ray absorption rises, conversely if the interference is destructive, the absorption cross section falls. Since the photoelectron wavelength varies with energy according to the de Broglie relationship, constructive and destructive interference are alternately observed as the photoelectron energy is varied. EXAFS is typically performed with photoelectron energies between 50 and 1000eV, where the wavelength falls in the range 1.7 - 0.4A^{-1}. The maximum value of the wavelength is less than most interatomic distances in molecules and solids and therefore well suited to structure determination.

To illustrate the factors which govern the amplitude of EXAFS oscillations it is convenient at this point to introduce the form of the EXAFS equation introduced by Sayers, Stern and Lytle:

$$\chi(\mathbf{k}) = \sum_{i=1}^{j} \frac{N_j}{\mathbf{k}.R_j{}^2} \sin(2\mathbf{k}.R_j + 2\delta + \psi_j).|f_j(\pi)| \\ .\exp(-U_j{}^2k^2).\exp(-2V_i R_j/\mathbf{k}) \qquad (5)$$

The observed oscillations in X-ray absorption are treated as a summation over j shells of scatterers, each at a distance R_j from the absorbing atom. For each shell, the EXAFS may be considered as a modulated sine wave, when the abscissa is the photoelectron wave vector. The period of oscillation is determined by the interatomic distance R_j, modified by two sets of phase shifts. These are the phase shift experienced by the photoelectron on emission from the central atom and on reabsorption, 2δ, and the phase shift experienced on backscattering from the neighbouring atom, ψ_j.

The amplitude of the EXAFS oscillation due to a single shell is determined by a number of factors, the most significant of which is the coordination number in the shell N_j. The identity of the backscattering atom is defined by the magnitude of its backscattering factor, $|f_j(\pi)|$, itself a function of photoelectron energy. The amplitude is modulated by two additional terms. V_i is the so called 'imaginary part' of the self energy and takes account of inelastic scattering of the photoelectron; the exponential term involving V_i is more dominant at low photoelectron energies. Thermal effects are accounted for by the inclusion of a Debye - Waller term:

$$U_j^2 = 2\sigma_j^2 \qquad (6)$$

where σ is the root mean square (rms), displacement of the absorbing atom from the backscatterer atom. [NB This is not quite the same as the Debye - Waller factor encountered in X-ray diffraction, which is the rms displacement about a fixed reference point.] The Debye - Waller term becomes increasingly important as the photoelectron energy increases.

Although equation 5 is reasonably complicated, it includes two simplifications of the EXAFS process which need to be considered. It assumes that the photoelectron can be treated as a plane wave, which is an increasingly severe approximation at photon energies below ca 100eV. More complete treatments, in which the curvature of the propagating wave is considered, have been developed (Ashley and Doniach, 1975; Lee and Pendry, 1975), and spherical wave theory can now be included in routine EXAFS theoretical calculations. A more significant drawback is that the theory which gives rise to Eqn. 5 assumes only single scattering of the photoelectron. Multiple scattering is ignored. Although there are situations where multiple scattering becomes significant in the EXAFS region of the spectrum, they are relatively few. At lower energies, in the XANES region, however, multiple scattering is often the dominant process, and represents a serious problem, because theory does not yet exist to allow routine and reliable treatment of this low energy multiple scattering. This is the reason for the division of the spectrum into EXAFS and XANES regions and explains why XANES spectra are normally subjected only to qualitative analysis.

It is useful at this point to consider the outcome if Eqn. 5 is subjected to Fourier transformation with respect to the photoelectron wave vector. The result is a type of radial distribution function, with peaks corresponding to R_j, of magnitudes related to N_j. Fig. 3 shows the Fourier transform of the EXAFS spectrum of metallic rhodium, where peaks corresponding to five shells of nearest neighbours can be observed. The consequence of the phase shift terms is to move all of the peak maxima to lower values than the true interatomic distances. Nonetheless the Fourier transform is a valuable means of inspecting the result of the EXAFS experiment and forms a crucial part of many data analysis procedures.

We are now in a position to discuss the information which an EXAFS measurement can in principle provide. The main feature are:

Local Structure   Because there is no requirement for long range order, the method is applicable to amorphous and poorly ordered samples, which cannot readily be examined by X-ray diffraction; the study of ordered sample is of course also possible;

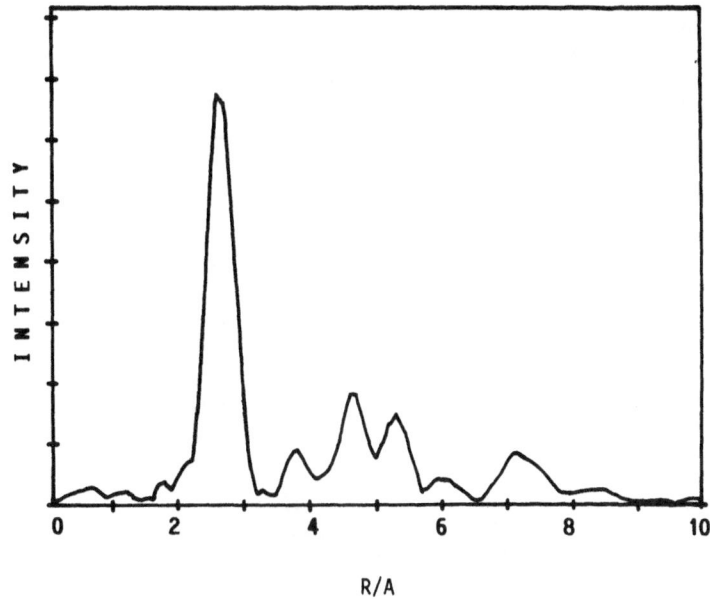

Fig. 3. Fourier transform of the EXAFS of rhodium foil, showing the presence of six shells of neighbours.

Element Specificity   Since the X-ray absorption edge is specific to a particular element, the method provides information on the structural environment only of the chosen element;

Identity of Near Neighbours   Because the backscattering factor is also element specific, it is possible to gain some information on the identity of nearest neighbours if this is not already known. Backscattering factors do not vary rapidly, so that only neighbours from different rows of the periodic table can easily be separated;

Interatomic Distances   Interatomic distances of near neighbours can be determined; the accuracy obtained is never better than 0.01A, (1A = 0.01nm) and only neighbours closer than 5 - 7A to the scattering atom can be studied routinely. The range of neighbours which can be studied depends on the degree of order present in the sample;

Coordination Numbers   Coordination numbers (CN's) are obtained accurate to $\pm$ 10% in the best cases; there is always correlation between the value obtained for the coordination number and the Debye - Waller factor and this reduces the accuracy with which the CN may be determined;

<u>Debye - Waller Factors</u>   There is rarely interest in the value of this
parameter, so the accuracy to which it can be determined is not
clear. If the neighbours in the sample are not arranged in well
defined shells, the result is **static disorder** and an apparent
increase in the Debye - Waller factor. As a result EXAFS cannot
separate shells of the same scatterer where the interatomic
distances are within 0.1A. In principle static disorder can be
identified by examination of the variation of EXAFS with
temperature, but such studies are rarely undertaken.

It is clear that the accuracy with which structures can be
determined by EXAFS is much less than that routinely achieved by X-ray
diffraction. As a result the method is normally only applied to samples
which are too poorly ordered to be studied in the conventional way.
These include the heterogeneous catalysts which are the subject of
present interest, as well as glasses, liquids and a wide range of
biological samples, in particular metalloenzymes and proteins.

3. EXPERIMENTAL ASPECTS

In its basic form, the EXAFS experiment is very simple, consisting
of the measurement of an absorption spectrum. The K edge absorption edge
is normally studied, except for the third row transition series, where
the K edge energy is too high and line broadening would degrade the
spectrum. The L_{III} edge is then preferred. The photon energy required
thus ranges from 285eV for carbon, (photon wavelength ca 44A) up to
11,564 eV for platinum L_{III} and 23,220 eV for rhodium, (photon
wavelength ca 0.5A). This wide range imposes a requirement for an
intense white source across the full range of the spectrum, from the
vacuum ultra-violet to the hard X-ray region. Although Bremsstrahlung
from conventional X-ray generators can be used, it is clear that
synchrotron radiation is much more satisfactory, (see eg Winick and
Doniach, 1980). Modern dedicated synchrotron radiation sources provide
intense radiation over the full range of interest and have superseded
laboratory X-ray absorption facilities almost completely.

EXAFS consists of a family of slightly different experiments, which
we now discuss. Transmission EXAFS is the simplest and is considered in
some detail; fluorescence detection, dispersive EXAFS, and surface EXAFS
or SEXAFS are then briefly described.

<u>3.1 Transmission EXAFS</u>

The experimental arrangement used for conventional or **transmission**
EXAFS is shown in Fig. 4. Monochromatic radiation is obtained by the use
of a pair of single crystals, for example Si(111), Si(220) or Ge(111),
chosen to have a suitable d spacing for the energy range of interest.
Originally two parallel, channel cut crystals cut from a single block
were used. In the last 10 years order sorting monochromators have become
common, in which two separate crystals of the same type are held not
quite parallel to each other. This arrangement has the advantage that
only the fundamental X-ray frequency which satisfies the Bragg relation
is transmitted. Higher harmonics are undesirable as they distort the
absolute value of the absorption cross section and lead to inaccurate
coordination numbers, and they are rejected, having narrower rocking
curves than the fundamental. The spectrum is scanned by changing the
angle of the monochromator with respect to the X-ray beam. This angle
determines the wavelength of the transmitted X-rays and is normally
controlled to within 1 mdeg. Other essential features of the apparatus

are the slits, often fabricated of lead, and the ionisation chambers, which measure the X-ray flux before and after the sample. The vertical distance between the monochromator entrance slits usually determines the resolution of the spectrum obtained and ca 1 eV is readily achievable. The whole operates under computer control and is contained in a lead lined 'hutch' to provide X-ray protection. Optional features include focusing mirrors and monochromators, which increase the photon flux at the sample.

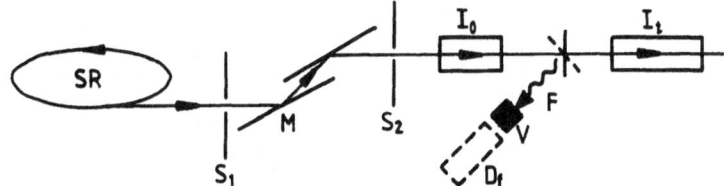

Fig. 4. The experimental arrangement for X-ray absorption spectroscopy in transmission mode. SR, synchrotron radiation source; S1, S2, vertical slits; M, two crystal order sorting monochromator; Io, It, reference and sample ionisation chambers; F, sample. For fluorescence detection the sample is rotated 45 deg. and the fluorescent photons pass through Soller slits, V, into a counter, Df.

The sample may be in any phase, but solids are most common. It should be homogeneous, free from pinholes and have a thickness t, chosen to satisfy the criterion:

$$\mu.t = 1 \qquad\qquad\qquad (7)$$

where μ is the absorption of the sample, given by

$$\mu = \rho. \quad g_i(\mu/\rho)_i \qquad\qquad\qquad (8)$$

ρ is the density of the sample, g_i is the mass fraction of the sample contributed by element i and μ/ρ is the mass absorption coefficient of element i at the X-ray wavelength of interest. Values of μ/ρ may be obtained from tables, (Lonsdale, 1962). For typical catalysts the thickness varies from fractions of a mm up to several mm, depending on the elemental composition and the absorption edge of interest. It is common to use powders or pressed discs of the type routinely employed in infra-red spectroscopy.

The data collection time is chosen to give a desired level of signal to noise ratio. It may vary from a few minutes for preliminary measurements on a concentrated sample up to many hours for detailed study of a dilute material.

The value of X-ray absorption studies in catalysis is greatly enhanced by the ability to perform measurements **in situ**. This is possible because X-rays are strongly penetrating and can pass through suitable windows and gas atmospheres without severe attenuation. Typical

window materials are beryllium, boron nitride and various polymer films, for example mylar or the polyimide Kapton. Windows with good mechanical and chemical integrity can readily be fabricated which have high X-ray transmission (> 90%) at energies above
ca 4 keV, allowing study of all of the transition elements. Windows for softer edges can also be prepared, with increasing difficulty as the X-ray energy is decreased.

Greaves et al (1988), have considered the absorption of the X-ray beam by high pressures. In 20 bar nitrogen it should be readily possible to study all of the elements above vanadium provided the path length of the X-ray through the gas is no longer than 5 cm.

A number of in situ cells of quite different design have been constructed or proposed. The earliest, due to Lytle et al (1979), is available commercially and other designs have been described by Clausen et al (1981a), Joyner and Meehan (1983), and Koningsberger and Cook (1983). Recently Greaves et al. (1988), have described a design for use at pressures up to 20 bar and suitable for operation in a wide range of modes including transmission, fluorescence, dispersive and reflection EXAFS.

3.2. Fluorescence Detection

Transmission EXAFS is effective where it is intended to study high Z element in a low Z matrix, as is the case for many supported metals. Where a high Z support such as ceria or zirconia is present, the absorption of the element of interest will be small in relation to that of the support and the EXAFS will be superimposed on a large background. In such cases it is advantageous to use another means of detection to suppress the background and improve signal to noise ratio. The core hole left by the photoelectron decays rapidly, by the emission of a fluorescent photon or of an Auger electron. The rate of production of either is also a measure of the rate of production of core holes and thus of the X-ray absorption. The advantages of measuring the fluorescent photon yield were first recognised by Jaklevic et al. (1977). By choice of a suitable energy dispersive detector, significant suppression of the background can be achieved, and excellent spectra recorded, particularly with highly sensitive, modern, multi element solid state detectors.

3.3. Dispersive or Time-Resolved EXAFS

In the techniques described above the spectrum is accumulated point by point, taking a single X-ray wavelength at a time, and data collection is therefore slow. In the dispersive version of EXAFS the whole spectral region is measured simultaneously, and so can be recorded much more rapidly. White radiation from the synchrotron source is allowed to fall on a bent crystal, with a radius of curvature greater than that of the Rowlands circle. The result is to disperse the radiation into a 'rainbow' of wavelengths, with different wavelengths originating from different parts of the crystal. The X-ray optics are arranged so that the whole wavelength range of interest is focused through the sample and thence diverges to a position sensitive detector, for example a photodiode array. Using this approach complete spectra can be recorded in < 1s, so that time resolved or kinetic studies become possible.

3.4. Surface EXAFS or SEXAFS

None of the variations of EXAFS described so far is surface sensitive. Their value in heterogeneous catalysis derives from the element specific nature of the technique, since samples are chosen where the element of interest is wholly at the surface, such as vanadia in a vanadia/titania monolayer catalyst, or the noble metal in a highly dispersed supported catalyst. Surface sensitivity can be achieved by choosing a measurement proportional to the X-ray absorbance in the surface layer only. This can be the yield of a low energy Auger electron, where the escape depth is < 10A, or the partial or total yield of secondary electrons. Because electrons interact strongly with matter these measurements require ultra high vacua, and as a result, SEXAFS is a branch of surface science and measurements are normally carried out on single crystal samples. The results are therefore not of direct relevance to catalysis and the reader is referred to the review by Stohr (1988), which promises, 'Everything you always wanted to know about SEXAFS, but were afraid to ask.'

4. DATA ANALYSIS

Spectral analysis is the heart of the successful application of EXAFS. EXAFS is essentially a modelling technique and an obvious, but nonetheless crucial requirement for success is thus that a full range of models is explored and that all possibilities are considered.

Accurate phase shifts, are required before data analysis can commence. In the United Kingdom it is normal practise to calculate these ab initio, using some suitable quantum mechanical package and refining the results by reference to the spectra of compounds of known structure. Elsewhere in Europe and the USA the phase shifts and backscattering amplitudes are usually obtained directly from the spectra of reference compounds, using Fourier transform techniques. These different methods are now examined in more detail. It is assumed that background subtraction has been performed, so that the spectrum is a plot of $\chi(\mathbf{k})$ versus $\mathbf{k}$ or photoelectron energy. As noted above, the magnitude of the EXAFS oscillations decays as the photoelectron energy increases. It is common to multiply $\chi(\mathbf{k})$ by $\mathbf{k}^n$, where $0 \leq n \leq 3$. This accentuates oscillations at higher $\mathbf{k}$, and allows more accurate determination of interatomic distances.

4.1. Ab Initio Methods

Ab initio approaches to the calculation of the EXAFS spectrum treat the photoelectron as a spherical wave, as described eg by Lee and Pendry, (1975). The phase shifts at each energy therefore have to be decomposed into separate angular momentum terms and it is typically necessary to use values of l from 0 to 15, (although higher terms may be required for very heavy atoms). As noted above, these are usually calculated by a quantum mechanical computer package. Phase shifts for the central atom are computed by considering the element of atomic number Z + 1, and performing the calculation with a single electron hole in the K shell. The phase shifts for the central atom and each of the backscattering elements are then incorporated into an optimisation package designed to vary chosen experimental parameters to give the best theory - experiment agreement, (Gurman et al, 1984). With modern computers, least squares minimisation involving a large number of parameters can be carried out interactively at the computer terminal.

The next stage of the fitting procedure is validate the phase shifts by calculating the EXAFS of suitable reference material of known structure, for example a metal foil or a metal oxide. It is usually found that ab initio phase shifts give only adequate agreement with such standard spectra so it is common practise to apply a polynomial correction to the phase shifts as a function of **k** and thus to improve agreement. Fig. 5 shows a theory - experiment comparison for a rhodium foil, where modified ab initio phase shifts have been used in the calculation. Good agreement is obtained with the known crystallographic parameters of rhodium and including five shells of nearest neighbours.

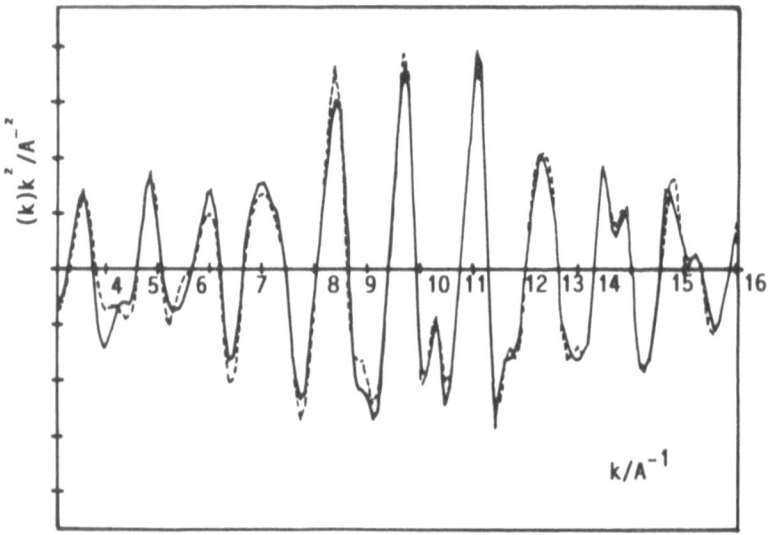

Fig. 5. Measured EXAFS (full line), and calculated EXAFS (dashed line), for a rhodium foil. Six shells of neighbours with the correct crystallographic parameters are included in the calculation, (Johnston, Joyner and Joyner, unpublished results).

Once suitably corrected phase shifts have been obtained, theory - experiment comparisons can be carried out for samples of unknown structure and parameters of interest obtained..

This approach is more tedious than the Fourier transform methods to be discussed below but it has one significant advantage, since it allows a simple answer to be given to the question, 'How many shells of neighbours should be included in any EXAFS fit?' It also permits objective assessment of experimental errors. In any fitting exercise between theory and experiment it is necessary to have a mathematical measure of 'goodness of fit' or fitting index, FI, which the fitting programme seeks to optimise. Typically FI is just the sum over the spectrum of the square of the residuals between theory and experiment. For a direct theory - experiment comparison, a simple statistical test can be used to determine the magnitude of the decrease in the fitting index which is required for a new shell of neighbours to be accepted as statistically significant, (Joyner et al, 1987). Similarly the change in the fitting index observed as a parameter is varied about its optimum valve, can be used to assess the experimental error in that parameter. The fitting index can also be used to study the extent to which errors in different variables interact or are correlated. From Eqn. 6 it is not surprising that there is usually a strong positive correlation between those variables which determine the EXAFS amplitude, in particular coordination numbers and Debye - Waller factors. In contrast, interatomic distances and coordination numbers are not often correlated. Fourier transformation destroys the statistical independence of individual data points, so application of statistical analysis to Fourier transformed data becomes much more difficult. As a result, objective techniques for error assessment have not been developed for use with Fourier transform methods.

4.2. Fourier Transformation Methods

This approach to data analysis depends on the use of Fourier transform methods to isolate data from individual shells, through a combination of forward and reverse Fourier transformation. As shown in Fig. 3, the appropriate Fourier transform of the experimental data resembles a radial distribution function. The contribution of an individual shell to the EXAFS spectrum can be isolated by the following process. A mathematical 'window' function is used to isolate a chosen region of real space spectrum,
(eg 2.0A < R < 2.5A), and the spectrum in this region is then subjected to reverse Fourier transformation. Subject to certain qualifications the result is a single shell EXAFS spectrum, which is much easier to analyse than the complete spectrum. By suitable selection of windows, each shell of neighbours can be analysed individually and the whole spectrum understood.

The EXAFS spectrum for the single shell which is generated in this way differs somewhat from the 'true' spectrum. Firstly, the noise, which represents high frequency components in the Fourier series, is smoothed out. This is an attractive feature of the Fourier transform method provided it does not mask spectral distortion. The finite nature of the data set also results in some modification of the single shell spectrum, although this can be minimised by careful choice of the mathematical form of the window function.

When the single shell spectrum has been isolated, the experimental parameters of interest can be extracted using methods which rely on the plane wave, single scattering theory defined by Eqn. 6. Phase shifts and backscattering amplitudes may be taken from tables (eg Teo and Lee, 1979), or obtained from model compounds. The latter approach is preferred provided models can be chosen which are sufficiently close in structure and chemistry to the sample under study. Determination of experimental parameters requires a knowledge of both the amplitude and the phase of the EXAFS, and these are obtained from the real and imaginary parts of the Fourier transform, (Sayers and Bunker, 1988). The interatomic distance is deduced from the phase difference between sample and reference, which is given approximately by:

$$\sigma_S - \sigma_R = 2\mathbf{k}(R_S - R_R) \qquad (9)$$

Values for the coordination number and the Debye - Waller factor are obtained from a consideration of the ratio of the amplitude terms for sample and reference.

The Fourier transformation approach to data analysis is convenient and accurate and, as noted above, is the most widely used. Although there are concerns about the extent to which phase shifts are independent of chemistry, and therefore transferable, it is clear that both the ab initio and the Fourier transform approach to data analysis rely on model compounds for phase shift information and validation.

5. CATALYTIC APPLICATIONS

X-ray absorption spectroscopy in general and EXAFS in particular has been applied to a very wide range of catalysts, and cannot be covered comprehensively in a brief review. In what follows some interesting and representative examples are presented. A number of reviews have been published, the most detailed being that of Bart and Vlaic, (1987). Others worth consulting include Evans (1989), Lytle (1988), and Koningsberger (1988). Although EXAFS often provides unique information about the catalyst of interest, it should be recognised that analysis by a single technique is rarely adequate and can sometimes be misleading. In most of the studies cited supporting evidence has been obtained from methods such as photoelectron spectroscopy, infra-red spectroscopy, Mossbauer spectroscopy or electron microscopy. Catalysis is concerned with rates of chemical reactions, so reactivity and selectivity data are also central elements of the whole picture.

5.1. Hydrodesulphurisation Catalysts

Hydrodesulphurisation is the important industrial process by which sulphur is removed from appropriate crude oil fractions. In the presence of hydrogen and a catalyst the sulphur in the organic molecules is converted into hydrogen sulphide. The preferred catalyst contains molybdenum and cobalt, supported on alumina and the field has recently been reviewed by Knozinger (1988). Much interest has been attached to the role of cobalt, which was originally thought of as a promoter for the molybdenum.

In an impressive series of studies using a range of experimental methods, Topsoe et al have shown that hydrodesulphurisation activity is associated with a structure containing cobalt, molybdenum and sulphur, known to as the COMOS phase (Topsoe, 1983). As indicated by Bart and Vlaic (1987), EXAFS information was significant in identifying the nature of the molybdenum environment in the COMOS phase, which is closely similar to the layer structure of MoS_2, but main contribution of EXAFS was to identify the location of the cobalt atoms. A number of possibilities were considered, as indicated in Fig. 6 and EXAFS measurements were made on both the cobalt and molybdenum K edges (Clausen et al., 1981a,b; 1986). The only shell of neighbours associated with cobalt is due to sulphur, with a coordination number less than six. The molybdenum EXAFS was not consistent with substitution of cobalt in bulk MoS_2. In addition, XANES studies of unsupported Co - Mo catalysts indicated changes on passivation in air. All of the results therefore favour location of cobalt at the surface of the MoS_2 slabs and specifically at the edge sites. The weight of the EXAFS and other evidence is such that the COMOS picture is very generally accepted. It has become so influential that it has recently been proposed that cobalt should be considered as the active catalytic species and that the molybdenum sulphide merely serves as the support.

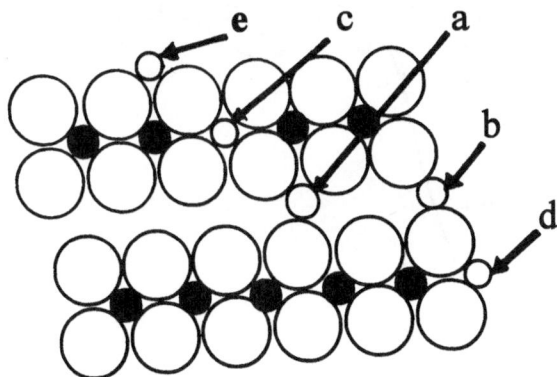

Fig. 6 The layer structure of the COMOS phase; large circles, sulphur;
small filled circles, molybdenum; small open circles, cobalt.
Possible locations of cobalt atoms;- a, intercalated; b, edge
intercalated; c, substituted for Mo; d, at the slab edge; e, on
the basal plane.

5.2. Copper / Zinc Oxide Catalysts for Methanol Synthesis

The commercial catalyst employed worldwide for methanol synthesis
comprises copper, zinc oxide and alumina, the synthesis gas feedstock
being a mixture of carbon monoxide, carbon dioxide and hydrogen. There
is ongoing interest and controversy surrounding the mechanism by which
the catalyst operates and recent proposals have variously claimed that
the active species is the oxidised copper surface, (Chinchen et al.,
1986), is anion vacancies at the zinc oxide surface, (Frost, 1988), or
involves synergy between copper and zinc oxide, (Burch, 1988, 1990). It
is generally agreed that the role of the alumina is to stabilise the
catalyst against sintering and that it does not participate directly in
the catalysis. Earlier it was proposed by Herman et al. (1979), that the
active species are Cu(I) ions located on the zinc oxide, rather than the
metallic copper particles which X-ray diffraction and XPS studies had
indicated were present in the catalyst after reaction. Clausen et al
(1985), investigated this possibility by in situ EXAFS, but found no
evidence for Cu(I) species within the detectability limit of the
technique. The XANES structure is a particularly useful diagnostic here,
since non zero-valent copper species show a sharp peak or 'white line' at
the absorption edge. Fig. 7 compares the XANES for metallic copper and a
copper / zinc oxide catalyst after in situ reduction, (hydrogen /
nitrogen, 1 bar, 573K), (this author, unpublished results). As can be
seen, the two curves are almost identical apart from the greater noise
level for the catalyst, supporting the conclusions of Clausen et al.

5.3. Monolayer Catalysts

When one oxide is covered by a monolayer of another, fascinating
catalytic properties can sometimes result. Catalysts prepared by
supporting monolayers of 'vanadium pentoxide' on the anatase form of
titanium dioxide have excellent performance in the selective oxidation of
butadiene or but-1-ene to maleic acid and o-xylene to phthalic anhydride,
(Higgins and Hayden, 1977, Bond, 1982). There is a true synergy between
the catalyst components because each is effective on its own, (titania is
a good **total** oxidation catalyst), and good catalysts cannot be prepared
on the rutile modification of TiO_2. The optimum vanadia content for the
catalyst is in the range 1 - 5 monolayer.

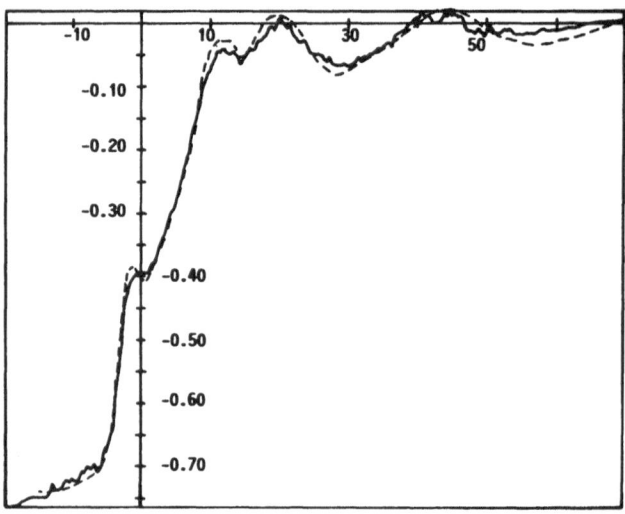

Fig. 7. XANES from a copper foil (solid curve), and a Cu/ZnO catalyst reduced in situ in hydrogen / nitrogen (dashed line).

There has been considerable effort to understand the solid state chemistry of this catalyst and it was suggested, on the basis of the structural similarity between the (010) plane of V_2O_5 and the exposed surface of anatase, that epitaxy between the monolayer and the substrate is the cause of the observed synergy, (Vejux and Courtine, 1978). Kozlowski et al, (1983a,b,c) prepared highly disperse anatase, containing 60A diameter particles, as a means of maximising the sensitivity of an EXAFS study. Examination of the vanadium K edge indicated that no epitaxial relationship existed between the surface monolayer and the substrate. The structure of the surface vanadium was identified as involving two terminal vanadium - oxygen bonds of order ca 1.5, with interatomic distance 1.65 $\pm$ 0.05A, and two V - O - Ti bridging single bonds, of length 1.90 $\pm$ 0.05A. XANES observations supported these conclusions. The terminal V = O species are argued to be crucial to the catalysis. Kozlowski et al. calculate the valence of surface vanadium atoms to be 4.5 $\pm$ 0.5, encouragingly close to the value for vanadium pentoxide. On heating to 873K, bonding of the vanadium entity to the surface is lost and the vanadium recrystallises as V_2O_5.

Kozlowski et al have also analysed the EXAFS of vanadia / alumina monolayer catalysts, which do not exhibit synergy. Vanadium was again shown to be bonded to two types of oxygen, with interatomic distances 1.67 and 1.82A, but with greater static disorder than observed for vanadia / titania. It was concluded that in this case the vanadium - oxygen bond order was only 1.43, and that this, as well as the disorder, may be responsible for the poor catalytic performance. These authors have also related the structural and redox properties of vanadia monolayers on both titania and alumina, (Haber et al., 1986)

Vanadia / titania catalysts have been studied by a wide range of other techniques, including infra - red and Raman spectroscopy, (eg Cristiani et al.,, 1989, Eckert and Wachs, 1989a) and solid state nmr, (Eckert and Wachs, 1989b). Although it is clear that the situation is somewhat more complicated than suggested by Kozlowski et al., the role of V = O species is still thought to be central and there appear to be no suggestions that monolayer / support epitaxy is important.

5.4. Supported Monometallic Catalysts

Supported metal catalysts are particularly amenable to study by EXAFS and have probably been examined in more detail than any other type of catalytic material. The simplest question which can be asked relates to the structure of the metal particles which are formed and how this relates to that of the bulk metal. This is of considerable interest, since a number of theoretical calculations suggest that small particles, containing < ca 100 atoms, will not be stable in the face centred cubic or other close packed structure, (Burton, 1974, Gordon et al., 1979; Khanna et al., 1983). Instead, it is argued that these small particles prefer to minimise their surface energy by forming icosahedra, which have 5 fold symmetry and are therefore not space filling. For small icosahedra all of the exposed surfaces have the the close packed structure of the fcc (111) plane. In contrast, close packed clusters with similar numbers of atoms expose an equal number of (111) and the more open (100) faces; surface energy considerations are thought to be dominant for small particles. Although possible in principle, it is not easy to differentiate between close packed and icosahedral structures by EXAFS, and I have argued that this is best done by consideration of non-nearest neighbour distances, (Joyner, 1980a). Close packed structures will exhibit the characteristic $2^{1/2}a$ and $3^{1/2}a$ distances, (where a is the close packing or nearest neighbour distance). These distances do not occur in icosahedral structures, where the expected distance is ca 1.62a. Although a small number of claims to have identified icosahedral structures have been made, none is considered reliable, (see discussion in Joyner, Shpiro, et al., to appear in J. Catal.). The reasons for this difference between theory and observation are not clear, although it may be significant that calculations do not include the energy of the metal - support interaction.

There is general agreement that some decrease in nearest neighbour interatomic distance occurs in small metallic clusters, compared to the bulk. This was first shown by Via et al (1979), for a platinum / alumina catalyst although a study of small nickel and copper particles prepared by evaporation onto carbon showed more dramatic results, with contractions of up to 10% in each case, (Apai et al., 1979). Contractions for platinum have been confirmed by a number of workers, (Moraweck and Renouprez, 1981; Gallezot et al., 1979) and also reported for iron and chromium, (Monatano et al., 1985), gold, (Balerna et al., 1985) and rhodium, (Johnston et al., 1989). Most changes are less than 0.05A.

There have only been limited attempts to study morphological changes brought about by chemisorption or surface reaction. Several studies have noted that platinum particles 'relax' to their bulk interatomic spacing when chemisorbed hydrogen in present, (eg Moraweck and Renouprez, 1981). Oxygen interaction causes much more dramatic changes, with a significant decrease in Pt - Pt coordination number and the appearance of a Pt - O distance of ca 2.0A, (Fukushima et al, 1980). Similarly it was reported that no platinum - platinum bonding could be detected in the as received 6.3% Pt/ Al_2O_3 reference catalyst designated EUROPT1, (Joyner, 1980b). This result was initially questioned since metallic platinum reflections could be observed by X-ray diffraction. Subsequently it was shown that the EXAFS observation gave the more correct picture and that there is very little metallic platinum in the fresh catalyst. The X-ray diffraction is sensitive only to the presence of order in the sample and does not detect the poorly ordered platinum oxide, which is the majority species. By contrast EXAFS yields a true average.

The most interesting morphological changes have been observed with rhodium, where van't Blik et al. (1985) reported that rhodium particles less than 10A in diameter are completely disrupted by exposure to carbon monoxide. Rhodium - rhodium bonding is completely destroyed and the spectrum observed is characteristic of a rhodium (I) gem dicarbonyl moiety. Recently Johnston et al. (1990), have argued that this species is stabilised by chlorine from the rhodium (III) chloride catalyst precursor, in analogy with the crystalline compound dirhodium tetracarbonyl dichloride, where two rhodium atoms are bonded by a bridge of two chlorine atoms. Unpublished studies by the Author's group suggest that rhodium catalysts prepared from halogen-free precursors do not undergo disruption in the presence of carbon monoxide. Johnston et al. have examined rhodium catalysts in the presence of synthesis gas, (CO/H_2), after exposure to carbon monoxide. Metallic particles of a similar size to those present originally are regenerated under catalytic conditions, (1 bar pressure, T = 473K). There is thus no evidence for the involvement of Rh(I) species in the reactions catalysed. Intriguingly, no clear evidence of chemisorbed carbon monoxide was detected, except for an increase of the Rh - Rh nearest neighbour distance of ca 0.06A.

5.5. Supported Bimetallic Catalysts

Supported catalysts containing two metallic elements are of academic interest and industrial importance. The aim in introducing two metals is to generate synergy, so that the catalytic performance is better than the sum of the individual components, either in terms of activity, selectivity or lifetime. Over many years the studies of bimetallic catalysts by Sinfelt and coworkers, at the Exxon Corporate Laboratory, have been preeminent. A wide variety of methods have been used and we review here only their EXAFS work. Much of this is well described in Sinfelt's book, (1983) and a review, (Sinfelt et al., 1984). Studies are still continuing, and an improved methods of data analysis has recently been described, (Via, 1990).

The copper - ruthenium system may be taken as an archetype. Ruthenium is an excellent catalyst for hydrogenolysis of hydrocarbons, while copper is ineffective; both metals are active hydrogenation catalysts. It may therefore not seem surprising that addition of a small quantity of copper to a ruthenium catalyst suppresses hydrogenolysis activity by several orders of magnitude, while hydrogenation activity is unaffected. The result is, however unexpected, since copper and ruthenium are not miscible in the bulk. Sinfelt et al., (1980) have studied catalysts containing 1% Ru and 0.63% Cu on silica, measuring both the ruthenium and copper K edge EXAFS. The nearest neighbour environments of the elements were deduced to be: ruthenium, total coordination 11, of which 92% is Ru and 8% Cu,; copper, total coordination 9, of which 50% is Ru and 50% Cu. The relatively high values of the total coordination numbers, (note that the bulk value is 12) indicate the existence of large particles, greater than 30A in diameter. The lower total coordination number for copper suggests that it may be located preferentially at the particle surface, which would be consistent with the large extent of ruthenium - ruthenium bonding. This was confirmed by exposing reduced catalysts to oxygen and re-examining the EXAFS. In the bimetallic catalyst, the coordination sphere of the copper was much more disrupted by oxidation than that of ruthenium, although similar degrees of oxidation of each metal were observed in copper only and ruthenium only catalysts.

A clear picture thus emerges of the bimetallic particle: the surface is primarily copper and the bulk almost exclusively ruthenium. This is

208

consistent with the known surface energies of the metals, which strongly favour copper segregation and it is also provides a reasonable explanation of the catalytic changes. A number of other copper containing bimetallic catalysts have been studied by the Sinfelt group, with platinum, iridium, rhodium, rhenium and silver, (Sinfelt et al, 1981; Meitzner et al., 1983, 1985a,b, 1987). In each case substantial segregation of copper to the particle surface was observed, even though some of the metals, eg platinum and silver, are fully miscible with copper.

The platinum / rhenium / alumina catalyst which is widely used in the reforming of gasoline to improve the octane has also been studied by EXAFS. The addition of rhenium was introduced by the Chevron Co., and results in an improvement in the catalyst lifetime. Pt / Re catalysts are not easy to study by EXAFS, since the L_{III} edges of the two elements are separated by only ca 400 eV, so the platinum spectrum is truncated and the rhenium spectrum overlapped by that of platinum. Also, the backscattering amplitudes of the elements are very similar, so separating the contributions of different neighbours to the spectra is very difficult. XANES studies (Bazin et al., 1986), have indicated that rhenium is not reduced to the metallic state in hydrogen, and it has been proposed to form an oxidised link between the platinum particle and the support, thus perhaps preventing sintering.

5.6. Metal in Zeolite Catalysts

There has been considerable interest in the properties of transition metal ions exchanged into zeolites and also in small metal particles located within the zeolite cavities. Morrison et al. (1980), have studied the environment of Co^{2+} exchanged into the sodium A and Y zeolite structures. The changes in coordination environment which occur as water of hydration is removed from the zeolite has previously been deduced from uv - visible spectroscopy, (Kellerman and Klier, 1975) and the earlier conclusions were largely confirmed by EXAFS. In the fully hydrated Cu-Y material the environment is indistinguishable from the cobalt hexaaquo ions in solution; the Co - O distance was $2.06 \pm 0.01A$. Interestingly, in fully hydrated Co-A Morrison et al suggest that the metal ion is bound to only five water ligands. Five coordinate cobalt in zeolite A has been invoked previously, on the basis of gravimetric and uv - visible spectroscopy measurements, but only as a stage in the dehydration sequence, (Akbar and Joyner, 1978).

EXAFS measurements on the fully dehydrated Co-A material show that the cobalt ion is coordinated to three oxygens of the zeolite 6 - ring, in agreement with X - ray diffraction measurements, (Riley and Seff, 1974). It was noted that the EXAFS measurement establishes the parameters of the first coordination shell of the cobalt more accurately than XRD, which determines only the average Co^{2+} position. The same point has recently been made by Doryhee, Greaves, Steel, Townsend, Carr, Thomas and Catlow, (Faraday Disc. Chem. Soc., in press), in a combined EXAFS and X-ray diffraction study of nickel exchanged Y zeolite. In the dehydrated material EXAFS gave a value of 2.02A for the Ni - O bond length, with the ion located in the S1 site, while the XRD value was 2.25A. It is known that the transition metal ion distorts the immediate zeolite framework, and Doryhee et al. conclude that EXAFS give the true value for the Ni - O distance, while XRD averages over the distorted and undistorted environments. It is encouraging to note that there are cases where EXAFS yields valuable data for crystalline materials.

Lastly we mention some recent work on platinum - chromium particles in the zeolite ZSM5, (Joyner et al., 1990). These catalysts are potentially interesting for gasoline reforming, and the chromium additive has a marked influence on the hydrogenolysis behaviour of the catalyst. Fig. 8 shows the changes in ethane and propane hydrogenolysis activity caused by the addition of 0.75% chromium to a 0.5% Pt/ H-ZSM-5 catalyst which has been reduced at 823K. XPS measurements show that this treatment generates some metallic chromium from the Cr(III) precursor, but only in the presence of platinum. EXAFS studies show that only pure platinum particles are observed after reduction at only 620K, but that a platinum - chromium alloy is formed after high temperature reduction. The Pt EXAFS is shown in Fig. 9, and the parameters used in the fit are consistent with small metallic particles, of 15 - 30 atoms, containing ca 25% chromium.

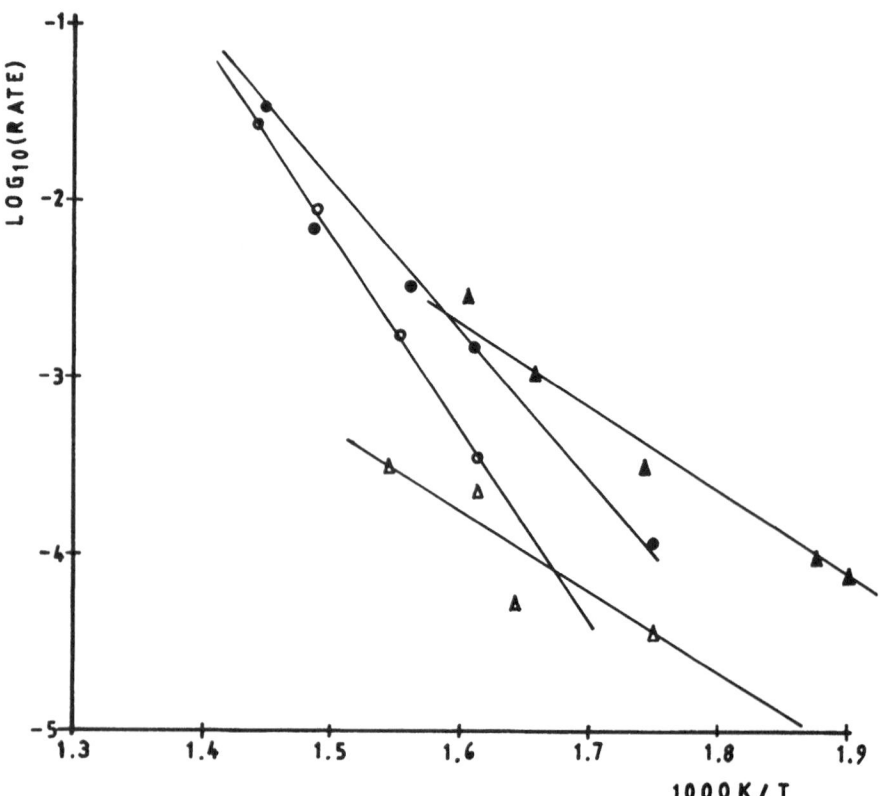

Fig. 8. Arrhenius plot for hydrogenolysis at atmospheric pressure, hydrogen / hydrocarbon ratio 10 / 1 and conversions < 10%. Filled symbols refer to the Pt / H-ZSM-5 catalyst and open symbols to the catalyst with 0.75% chromium. Circles relate to ethane hydrogenolysis and triangles to propane.

The reduction of hydrogenolysis activity as a result of adding chromium is unexpected. Although no data are available for the pure metal, extrapolation of Sinfelt's data for other first row transition metals (1976), suggests that chromium should be a very efficient hydrogenolysis catalyst. We therefore argue that it is not appropriate to explain these observations by invoking the ensemble effect, where the chromium 'dilutes' the active platinum surface. Instead, it seems more likely that electronic interactions in the alloy particle cause partial poisoning of the catalyst by the strongly chemisorbed hydrocarbon fragments.

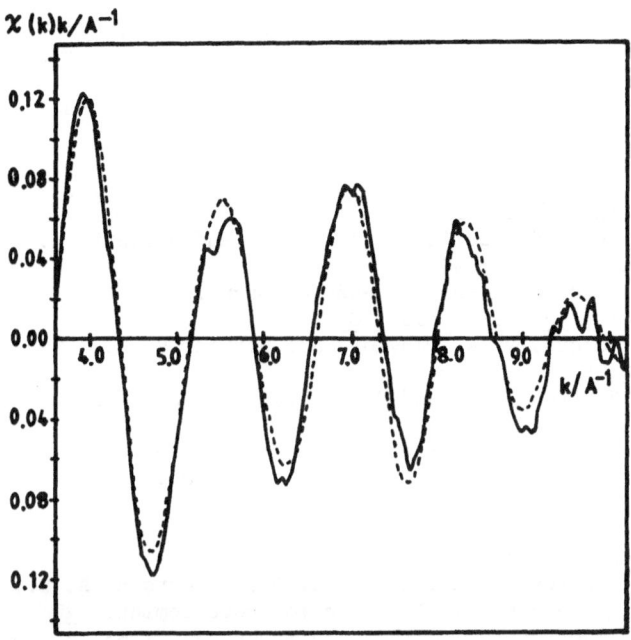

Fig. 9. Platinum L_{III} EXAFS for the Pt / Cr / H-ZSM-5 catalyst after reduction at 823K in hydrogen; solid line, experiment; dashed line, theory, including a Pt - Pt distance of 2.71A and a Pt - Cr distance of 2.65A.

6. REFERENCES

Akbar, S., and Joyner, R. W., 1978, JCS Chem. Commun., pp 548.
Ashley, C. A., and Doniach S., 1975, Phys. Rev., B11; 1279.
Apai, G., Hamilton, J. F., Stohr, J., and Thompson, A., 1979, Phys. Rev. Lett., 43; 165.
Balerna, A., Bernieri, E., and Mobilio, S., 1985, Phys. Rev. B, 31. 5058.
Bart, J. C. J., 1986, Adv. in Catal., 34; 203.
Bart, J. C. J., and Vlaic. G., 1987, Adv. in Catal., 35; 1.
Bazin, D., Dexpert, H., Lagarde, P., and Bournonville, J. P., 1986, J. Physique, 47; C8-293.

Bond, G. C., and Turnham, B. D., (1978), J. Catal., _45_; 128.

Burch, R., Chappell, R. J., and Goulnski, S. E., (1988), Catal. Lett., _1_; 439

Burch, R., Goulnski, S. E., and Spencer, M. S., (1990), JCS Faraday Trans., _86_; 2683.

Burton, J. J., 1974, Catal. Rev., _9_; 209.

Chinchen, G. C., Waugh, K. C., and Whan, D. A., (1986), Appl. Catal., _25_; 101.

Clausen, B. S., Lengeler, B., Candia, R., Als-Nielsen, J., and Topsoe, H., 1981a, Bull. Soc. Chim. Belg., _90_; 1249.

Clausen, B. S., Lengeler, B., Candia, R., Als-Nielsen, J., and Topsoe, H., 1981b; J. Phys. Chem., _85_; 3868.

Clausen, B. S., Lengeler, B., and Rasmussen, B. S., 1985, J. Phys. Chem., _89_; 2319.

Clausen, B. S., Lengeler, B., and Topsoe, H., (1986), Polyhedron, _5_, 199.

Cristiani, C., Forzatti, P., and Busca., G., (1989), J. Catal., _116_; 586.

Eckert, H., and Wachs, I. E., (1989a), J. Phys. Chem., _93_; 6796.

Eckert, H., and Wachs, I. E., (1989b) Mater. Res. Soc. Symp. Proc., _111_; 459.

Evans, J., 1989, in "Catalysis", Vol 8, Ed. G. C. Bond and G. Webb., Royal Soc. Chem., London.

Fricke, H., 1920, Phys. Rev., _16_; 202.

Fukushima, T., Katzer, J. R., Sayers, D. E., and Cook, J., 1980, Proc. 7th Intern. Congr. Catal., Tokyo.

Gallezot, P., Weber, R., Della Betta, R. A., and Boudart, M., 1981, Z. Naturforsch., _34a_; 40.

Gordon, M. B., Cyrot-Lackman, F., and Desjonqueres, M.C., 1979, Surface Sci., _68_; 359; ibid., _80_; 159.

Greaves, G. N., Moore, P. R., Joyner, and Morris, M. A., 1988, Vacuum, _38_; 929.

Gurman, S. J., Binstead, N., and Ross, I., 1984, J. Phys. C. Solid State Phys., _17_; 143.

Herman, R. G., Klier, K., Simmons G. W., Finn, B. P., Bulko, J. B., and Kobylinski, T. P., (1979), J. Catal., _56_; 407.

Hertz, G., 1920, Zeit. f. Physik, _3_; 19.

Higgins, R., and Hayden, P., (1977) in "Catalysis," Ed. C. Kemball, Chem. Soc., London, V1, pp 168.

Jaklevic, J. M., Kirby, J. A., Klein, M. P., Robertson, A. S., Brown, G. S., and Eisenberger, P., 1977, Solid State Commun., _23_; 679.

Johnson, A. L., Muetterties, E. L., and Stohr, J., 1983, J. Amer. Chem. Soc., _105_; 7183.

Johnston, P., Joyner, R. W., and Pudney, P. D. A., 1989, J. Phys.; Condensed Matter, _1_; SB171.

Joyner, R. W., 1980a, in "Characterisation of Catalysts," Eds. J. M. Thomas and R. M. Lambert, J. Wiley, Chichester.

Joyner, R. W., 1980b, JCS Faraday Trans. I, _76_; 357.

Joyner, R. W., Martin, K. J., and Meehan, P., 1987, J. Phys. C. Solid State Phys., _20_; 4005.

Joyner, R. W., and Meehan, P., 1983, Vacuum, _33_, 691.

Joyner, R. W., Minachev, K. M., Pudney, P. D. A., Shpiro, E. S., and Tuleouva, G., 1990, Catal. Lett., _5_, 257.

Kellerman, R., and Klier, K., 1975, in "Surf. Defect. Prop. Solids," Ed. M. W. Roberts and J. M. Thomas, Chem. Soc., London.

Khanna, S. N., Bucher, J. P., Buttet, J., and Cyrot-Lackman, F., 1983, Surface Sci., _127_; 165.

Kievet, B., and Lindsay, G. A., 1930, Phys. Rev., _36_; 648.

Knozinger, H., (1988), in "Proc. 9th Intern. Congr. Catal.," Ed. M. J. Phillips and M. Ternan, Chem. Inst. Canada, Ottawa, Vol. 5, pp20.

Koningsberger, D. C., 1988, in "X-ray Absorption, Principles, Applications and Techniques," Eds. Koningsberger, D. C., and Prins, R., J. Wiley, Chichester.

Koningsberger, D. C., and Cook, J. W., 1983, in "EXAFS and Near Edge Structure," A. Bianconi, L. Incoccia and S. Stipcich, Eds, Springer-Verlag, Berlin.

Koningsberger, D. C., and Prins, R., 1988, "X-ray Absorption, Principles, Applications and Techniques," J. Wiley, Chichester.

Kostarev, A. I., 1949, Zh. Eksper. Theor. Fiz., $\underline{19}$; 413.

Kozlowski, R., Pettifer, R. F., and Thomas, J. M., (1983a), JCS Chem. Commun., pp438.

Kozlowski, R., Pettifer, R. F., and Thomas, J. M., (1983b), J. Phys. Chem., $\underline{87}$; 5172.

Kozlowski, R., Pettifer, R. F., and Thomas, J. M., (1983b), J. Phys. Chem., $\underline{87}$; 5176.

Kronig, R. de L., 1931, Zeit. f. Physik, $\underline{70}$; 20: 1932, ibid. $\underline{75}$; 468.

Lee, P. A., and Pendry, J. B., 1975, Phys. Rev., $B\underline{11}$; 2795.

Lonsdale, K., 1962, "Internatnional Tables for X-ray Crystallography," Kynoch Press, Birmingham.

Lytle, F. W., Wei, P. S. P., Greegor, R. B., Via, G. H., and Sinfelt, J. H., 1979, J. Chem. Phys., $\underline{70}$, 4849.

Lytle, F. W., 1988, in "Proc. 9th Intern. Congr. Catal.," Ed. M. J. Phillips and M. Ternan, Chem. Inst. Canada, Ottawa, Vol. 5, pp?.

Meitzner, G., Via, G. H., Lytle, F. W., and Sinfelt, J. H., 1983, J. Chem. Phys., $\underline{78}$; 882.

Meitzner, G., Via, G. H., Lytle, F. W., and Sinfelt, J. H., 1985a, J. Chem. Phys., $\underline{83}$; 353.

Meitzner, G., Via, G. H., Lytle, F. W., and Sinfelt, J. H., 1985b, J. Chem. Phys., $\underline{83}$; 4793.

Montano, P. A., Purdum, H., Shenoy, G. K., Morrison, T. I., and Schulze, W., 1985, Surface Sci., $\underline{156}$; 228.

Moraweck, B., and Renouprez, A. J., 1981, Surface Sci., $\underline{108}$; 35.

Morrison, T. I., Iton, L. E., Shenoy, G. K., Stucky, G. D., and Suib., S. L., 1980, J. Chem. Phys., $\underline{72}$; 6276.

Petersen, H., 1936, Zeit. f. Physik, $\underline{98}$; 569.

Ray, B. B., 1929, Zeit. f. Physik, $\underline{55}$; 119.

Riley, P. E., and Seff, K., 1974, Inorg. Chem., $\underline{13}$; 1355.

Sayers, D. E., and Bunker, B. A., in "X-ray Absorption, Principles, Applications and Techniques," Eds. Koningsberger, D. C., and Prins, R., J. Wiley, Chichester.

Sinfelt, J. H., 1976, Catal. Rev. Sci. Eng., $\underline{3}$; 175.

Sayers, D. E., Stern, E. A., and Lytle, F. W., 1971, Phys. Rev. Lett., $\underline{27}$; 1204.

Sinfelt, J. H., Via, G. H., and Lytle, F. W., 1980, J. Chem. Phys., $\underline{75}$; 4832.

Sinfelt, J. H., Via, G. H., Lytle, F. W., and Greegor, R. B., 1981, J. Chem. Phys., $\underline{75}$; 5527.

Sinfelt, J. H., 1983 "Bimetallic Catalysts," J Wiley, New York.

Sinfelt, J. H., Via, G. H., and Lytle, F. W., 1984, Catal. Rev. Sci. Eng., $\underline{26}$; 81.

Stohr, J., 1988, in "X-ray Absorption, Principles, Applications and Techniques," Eds. Koningsberger, D. C., and Prins, R., J. Wiley, Chichester.

Teo, B. K., and Joy, D. C., 1981, in "EXFAS Spectroscopy," Plenum, New York.

Teo, B. K., and Lee, P. A., 1979, J. Amer. Chem. Soc., $\underline{101}$, 2815.

Topsoe. H., 1983, in "Surface Properties and Catalysis by Non Metals," Ed. J. P. Bonelle, B. Delmon and E. G. Derouane, Reidel, Dordrecht.

Vejux, A., and Courtine, P., J., (1978), J. Solid State Chem., $\underline{23}$; 93.

Via, G. H., Sinfelt, J. H., and Lytle, F. W., 1979, J. Chem. Phys., $\underline{71}$; 690.

Via, G. H., Drake, K. F., Meitzner, G., Sinfelt, J. H., and Lytle, F. W. 1990, Catal. Lett., $\underline{5}$; 25.

Wertheim, G. K., 1987, in "Solid State Chemistry, Techniques," A. K. Cheetham and P. Day, Ed., Oxford Science Publications, Oxford.

Winick, H., and Doniach, S., Eds., 1980, "Synchrotron Radiation Research," Plenum, New York.

Zaera, F., Fischer, D. A., Shen, S., and Gland, J. L., (1988), Surface Sci., $\underline{194}$; 205.

A MODEL FOR CHARACTERISING THE GROWTH OF RUTHENIUM ON AMORPHOUS ALUMINA BY THE USE OF THE AUGER PARAMETER

Nina Aas, Robin H. West and John C. Vickerman

Centre for Surface and Materials Analysis (CSMA)
Department of Chemistry, UMIST, P.O.Box 88, Manchester
M60 1QD, UK

INTRODUCTION

With XPS enjoying a high popularity in surface science, mainly due to the ease with which quantitative information is obtained, recent developments of the technique have centred around gaining a better understanding of the inherent characteristics of the photoemission process. In its application to catalyst characterisation this has fuelled a discussion on the influence of particle size on binding energy shifts[1-3]. In this context, it has been realised that the X-ray induced Auger peaks in an XPS spectrum in the form of the Auger parameter can be used for a more detailed analysis. The Auger parameter α' for any element is the relative difference between the binding energy of a photoelectron line $E(PE)_{BE}$ and kinetic energy of a related Auger electron emission $E(XAES)_{KE}$ from the same element on the same energy scale.

We report here how the Auger parameter has been applied to the study of the initial ruthenium deposition on an insulating alumina film to produce a model supported ruthenium catalyst. Furthermore, it is shown how current theories allow the calculation of the screening ability of the oxide with respect to the small particles of ruthenium and how this screening influences the ruthenium Auger parameter.

EXPERIMENTAL

The experiments were performed using a VG ESCA 3 Mk II electron spectrometer. The alumina substrate had been prepared by anodic oxidation of aluminium foil (supplied by Cegedur Pechiney, 99.99% purity, 0.1 mm thickness) in 3 wt% tartaric acid which had been adjusted to pH 5.5 with ammonia, using platinum as a cathode. The oxidation was carried out at 1 mA cm^{-2} and 10 V. Under these conditions, the oxide formed is very pure, amorphous and non-porous and the film thickness is around 12 to 15 nm[4-6].

The ruthenium was deposited at room temperature on the alumina substrate by argon ion beam induced sputtering from a ruthenium metal disc (disc of compressed powder, 13 mm diameter, supplied by Cerac)[5,6]. The experimental arrangement was set up in the preparation chamber of the UHV instrument. This allowed in situ studies of the prepared model catalysts without exposure to air. For the analysis, Mg K_α X-radiation was used. The pass

Fundamental Aspects of Heterogeneous Catalysis Studied by Particle Beams,
Edited by H.H. Brongersma and R.A. van Santen, Plenum Press, New York, 1991

energy was 50 eV, the angle of electron escape was 60° with respect to the surface normal.

The photoelectron peaks of interest were Ru $3d_{5/2}$ and O 1s together with the X-ray induced Auger electrons Ru $M_5N_{4,5}N_{4,5}$ and O KLL. For the determination of the peak positions energy referencing to C 1s was not carried out as the Auger parameter by its very nature is independent of charging.

Both the ruthenium disc and the alumina film were cleaned by argon ion etching prior to any experiment. Furthermore, the oxide was heated in vacuo to reduce the number of surface hydroxyl groups.

RESULTS

In order to determine the growth mechanism increasing amounts of ruthenium were deposited on the alumina and the plot developed by Biberian[7] produced. This is shown in fig. 1 where it is seen that the deposited metal overlayer grows by first completing a monolayer which is indicated by the first break. The second break would suggest that the growth is of the layer-by-layer type. As this is less pronounced than the first break it cannot be excluded that islands might form on top of the first monolayer.

As the objective of this work was to characterise both the physical as well as the electronic state of the ruthenium particles on the alumina the XPS results were further used to determine the ruthenium Auger parameter $\alpha'(\text{Ru})$. Fig. 2 shows that the variation in $\alpha'(\text{Ru})$ follows a systematic path. The first depositions produce an increase in $\alpha'(\text{Ru})$ which is followed by a constant level before another increase occurs. This last approach of the second constant level (not shown here) coincides with $\alpha'(\text{Ru})$ of bulk metal which had been determined in a separate experiment. The growth takes place in two stages according to this plot the first of which coincides with the completion of one monolayer as seen in fig. 1.

In a previous publication[5] it was argued that the experimental results shown in fig. 2 could be interpreted qualitatively on the basis of the change in extra-atomic relaxation energy. This idea has here been used to develop a model which is introduced in the following section.

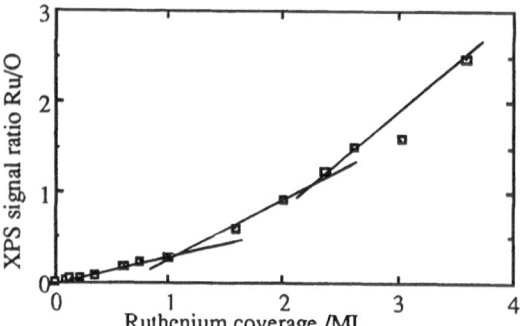

Fig. 1. The variation in XPS signal ratio Ru/O with increasing ruthenium coverage.

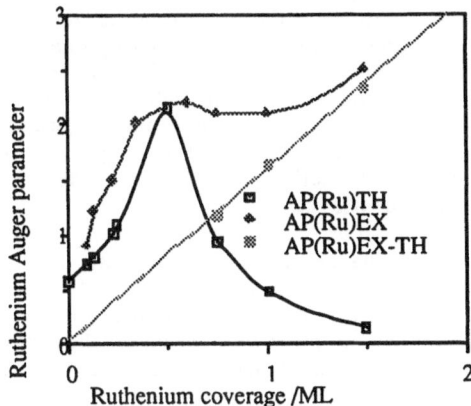

Fig. 2. The variation in the ruthenium Auger parameter with increasing metal coverage. TH are the data calculated using the proposed model; EX are the experimental data and EX-TH constitutes the difference between the two sets of data.

THEORY AND DEVELOPMENT OF THE AUGER PARAMETER

The concept of the Auger parameter was introduced by Wagner in 1972[9] and further developed in 1975[10,11]. It is defined as the difference in KE of the most intense Auger line and the main photoelectron line which makes it independent of charging. Today, it is mainly used in the form of the modified Auger parameter α'[12] which includes the X-ray excitation energy and hence ensures that α' is always positive.

$$\alpha' = E(XAES)_{KE} + E(PE)_{BE} \quad (eV) \tag{1}$$

Since 1972 it has been established that the Auger parameter yields both chemical and structural information in the form of the two-dimensional chemical state plot[13,14] and that the shift in the Auger parameter can be related directly to the difference in extra-atomic relaxation ΔR^{ea}, compared with the free atom state.

$$\Delta\alpha' = 2\Delta R^{ea} \quad (eV) \tag{2}$$

Several sources of extra-atomic relaxation energy are identified, including the electronic polarisability of neighbouring atoms and ions, and the influence of the conduction band electrons.

A simple electrostatic model has been proposed[16,17] for the determination of the electronic polarisation energy for a non-localised screening electron density.

$$\Delta R^{ea} = 14.4 \, [0.5 \, \Delta \, \{n\alpha^*/r^4 \, (1 + D\alpha^*/r^3)\}] \quad (eV) \tag{3}$$

where
α^* = Polarisability of neighbouring atoms ($Å^3$)
n = Number of nearest neighbours
r = Distance to nearest neighbour ($Å$)
D = Geometric Factor (1.15 tetrahedral, 2.37 octahedral[17])

This model is applied here for the first time to the initial deposition and layer formation of ruthenium metal on an amorphous alumina film. It considers the interaction of the oxygen layer with the ruthenium overlayer as isolated ruthenium atoms are deposited on the substrate. At this level of metal loading the probability of ruthenium-ruthenium interactions is minimal. Consequently, only the interaction with oxygen neighbours causes changes in the ruthenium extra-atomic relaxation energy and hence Auger parameter changes of the ruthenium overlayer.

It is proposed that the changes in $\Delta R^{ea}(Ru)$ with loading on the alumina substrate, from equation 3, are determined directly by r(Ru-O). As metal atoms arrive at the alumina surface the mean interatomic distance r(Ru-O) will be greater than the miniumum value r_m(Ru-O) because the arrival points will be randomly positioned on the alumina lattice. Consequently, the initial value of r(Ru-O) is calculated to be $\sqrt{2} \, r_m$. As the metal loading increases, the mean value of r_m(Ru-O) decreases to r_m at a coverage of 0.5 ML. From this level, ruthenium atoms will land on ruthenium atoms with more probability and the mean Ru-O distance will increase linearly with deposition of the metal. These effects on r(Ru-O) are described by the following equations:

$$r(Ru\text{-}O) = 2 \, (1\text{-}\sqrt{2}) \, (r_m/ML)L + \sqrt{2} \, r_m \qquad \text{for } L = 0 \text{ - } 0.5 \text{ ML}$$

$$r(Ru\text{-}O) = (r_m/ML)L + r_m \qquad \text{for } L = 0.5 \text{ - } \infty \text{ ML}$$

where   L is the metal loading in monolayers
ML is 1 equivalent monolayer (3.3×10^{15} atoms cm^{-2} Ru)
r_m is 2.74 Å for ruthenium on alumina

The application of these radii to equation 3, with the following values in addition: $\alpha(O^{2-}) = 2.43$ Å^3, n = 4 and D = 0.5 (2.37) = 1.18 (no oxygen layer above ruthenium), leads to changes in the extra-atomic relaxation energy and in the Auger parameter of the ruthenium as the ruthenium layer builds up which follow theoretically the curve marked in fig. 2. The straight line results from the subtraction of the theoretical from the experimental curve and indicates the increased screening ability of the developing conduction band associated with the ruthenium-ruthenium interaction.

DISCUSSION

The growth mode plot as shown in fig. 1 has been extensively used by many workers. It is well understood and in conjunction with other techniques it provides a good way to assess the physical distribution of a deposit. For this work no other surface science technique was available on the UHV instrument and further information about the ruthenium overlayer was therefore gained from heating experiments and from the application of the Auger parameter. The results concerning the thermal stability of the ruthenium deposit up to 450 °C are reported in ref. 6. These show a clustering of the metal with increasing temperature from which was concluded that the initial deposition led to a more spread out overlayer. This is a clear confirmation of the determined layer-by-layer growth mechanism.

We find further proof of the growth mode by considering the change in the Auger parameter. This is known to describe the electronic state of the element studied through variations in extra-atomic relaxation. In this case, two contributions towards this screening are important: the influence of the oxide substrate and the influence of increasing numbers of ruthenium atoms. Assuming that the ruthenium is deposited by completion of a monolayer and knowing that the majority of sputtered particles are single atoms the model introduced in the foregoing section shows that the main source of screening is derived from the oxygen ion of the alumina. The influence of the aluminium ion is essentially non-existent as its polarisability is very low (ref. 18 quotes. 0.06 Å^3 for Al^{3+}).

As the ruthenium atoms will not be deposited epitaxially on the alumina, a range of three, four and five fold coordination sites will be covered by the ruthenium atoms. Consequently, a four fold coordination (n = 4) approaches the mean value and is used in this model.

A limitation of the model is the inability to accurately determine the distance between the ruthenium and the oxygen atoms. Different r values would invariably lead to different values for the changes in the Auger parameter. However, the overall trend as shown in fig. 2 would not change.

CONCLUSIONS

It has here been shown that the Auger parameter can be used to monitor the growth of ruthenium on an alumina support. The theoretical approach based on Moretti's electrostatic model shows that it is the screening ability of the oxygen ion that contributes towards the increase in the ruthenium Auger parameter in the submonolayer region.

REFERENCES

1. M.G. Mason, *Phys. Rev. B* **27**, 748 (1983)
2. T.T.P. Cheung, *Surf. Sci.* **140**, 151 (1984)
3. T. Huizinga, H.F.J. Van'T Blik, J.C. Vis and R. Prins, *Surf. Sci.* **135**, 580 (1983)
4. D. L. Cocke, E.D. Johnson and R.P. Merrill, *Catal. Rev.-Sci. Eng.* **26**, 163 (1984)
5. N. Aas, B.H. Sakakini, R.H. West and J.C. Vickerman, *Surf. Interface Anal.* **16**, 359 (1990)
6. N. Aas, Ph.D. Thesis, UMIST (1990)
7. J.P. Biberian and G.A. Somorjai, *Appl. Surf. Sci.* **2**, 352 (1979)
8. C.D. Wagner, W.M. Riggs, L.E. Davis, J.F. Moulder and G.E. Muilenberg, "Handbook of X-ray Photoelectron Spectroscopy," Perkin-Elmer Corporation, Physical Electronics Division, Eden Prairie, Minnesota 1979
9. C.D. Wagner, *Anal. Chem.* **44**, 967 (1972)
10. C.D. Wagner, *Anal. Chem.* **47**, 1201 (1975)
11. C.D. Wagner, *Faraday Discuss. Chem. Soc.* **60**, 291 (1975)
12. S.W. Gaarenstroom and N. Winograd, *J. Chem. Phys.* **67**, 3500 (1977)
13. R.H. West and J.E. Castle, *Surf. Interface Anal.* **4**, 68 (1982)
14. C.D. Wagner and A. Joshi, *J. Electron Spectrosc. Relat. Phenom.* **47**, 283 (1988)
15. B.W. Veal and A.P. Paulikas, *Phys. Rev. B* **31**, 5399 (1985)
16. G. Moretti, *Surf. Interface Anal.* **16**, 159 (1990)
17. D.K.G. de Boer, C. Haas and G.A. Sawatzky, *Phys. Rev. B* **29**, 4401 (1984)
18. J.R. Tessman, A.H. Kahn and W. Shockley, *Phys. Rev.* **92**, 890 (1953)

IN-SITU FTIR SPECTROSCOPY OF Cu-Cr CATALYSTS IN CO OXIDATION

J.W. Bijsterbosch, J.C. Muijsers, A.D. van Langeveld, F. Kapteijn, J.A. Moulijn

Department of Chemical Engineering, University of Amsterdam,
Nieuwe Achtergracht 166, 1018 WV Amsterdam, The Netherlands

1. INTRODUCTION

Air pollution is one of the major environmental problems nowadays. The sources of air pollution are automobiles, industrial plants, and power plants. In many urban areas, the large number of automobiles is the major source of emissions of hydrocarbons (HC), carbon monoxide (CO), and nitrogen oxides (NO_x). These substances directly contribute to photochemical smog, acid rain and other forms of air pollution. Concern for the danger to public health and environment has led to emission regulation in western countries [1].

The most used method to reduce automobile emissions is cleaning the exhaust gases by means of a three-way catalyst (TWC). Three reactions have to be catalyzed at the same time:

HC oxidation: $\quad HC + O_2 \rightarrow CO_2 + H_2O$

CO oxidation: $\quad CO + O_2 \rightarrow CO_2$

NO reduction: $\quad NO + CO \rightarrow CO_2 + N_2$

The TWC is based on the noble metals platinum, rhodium, and palladium. Because these metals are scarce and therefore expensive, research for more available and cheaper metals has been started. Copper-chromium-oxides (Cu-Cr) appeared to be very good catalysts for CO oxidation and NO reduction [2,3,4]. At our department carbon, alumina and monolithic supported copper-chromium-oxides have been investigated. The activity of these catalysts for CO oxidation and NO reduction is comparable with the activity of the noble metal based catalysts [4].

Little is known about the active species of the Cu-Cr catalyst under reaction conditions and about the mechanism of the reactions. A better understanding of the behaviour of the catalyst is necessary for improving catalyst preparation. Therefore we started an *in-situ* Fourier Transform Infra Red (FTIR) spectroscopy study concerning the interaction of CO and NO with copper containing catalysts. The aim of this part of the study was to obtain information about the adsorption sites and adsorption intermediates in the CO oxidation reaction. The preliminary results and the *in-situ* transmission cell we have developed and built are discussed in this paper.

2. EXPERIMENTAL

Sample preparation

Three 10 wt% (based on metals) catalysts have been studied in their oxidized state: Cu/Al_2O_3, Cr/Al_2O_3 and $Cu-Cr(1:1)/Al_2O_3$. The Al_2O_3 support was a high purity γ-Al_2O_3; Ketjen 000-1.5E (BET[N_2] surface area 200 m2/g; pore volume 0.5 cm3/g; particle size 100-150 μm). The catalysts were prepared by pore volume impregnation of γ-Al_2O_3 with solutions of copper and/or chromium nitrates (Merck p.a.) in demineralized H_2O followed by drying and calcining. The Cu-Cr catalyst was prepared by co-impregnation. Drying was performed in air,

Fundamental Aspects of Heterogeneous Catalysis Studied by Particle Beams,
Edited by H.H. Brongersma and R.A. van Santen, Plenum Press, New York, 1991

221

by gradually increasing the temperature from 298 to 353 K over a period of 5 hours, followed by an isothermal period of one night at 353 K. The samples were finally dried at 383 K in a flow of air for one hour. Dried samples were calcined in air at a heating rate of 5 K/min to 773 K and were calcined at this temperature for one hour.

After calcination the catalyst was ground in a mortar and 10 mg of the powder was pressed to a thin self supporting disc of 10 mm diameter at 1.2 ton (on the ram) for 5 minutes. The disc was placed into a special stainless steel holder, which was used directly in the infrared cell.

Before each infrared experiment, the samples were subjected to an *in-situ* pretreatment by heating in 20 mbar O_2 up to 673 K at a heating rate of 3 K/min. At 673 K the cell was evacuated for 15 minutes and subsequently cooled in 20 mbar O_2 to room temperature. After evacuation at room temperature a clean, oxidized sample was obtained.

<u>Equipment</u>

The infrared transmission cell in which the catalyst sample is studied is extremely important. A suitable infrared cell must satisfy the following requirements: (1) vacuum tight to pressures below 10^{-6} mbar, (2) easily removable from the infrared spectrometer, (3) heating and cooling facilities, (4) equipped with a gas in- and outlet for controlling the reaction atmosphere, (5) quick sample and window exchange [5]. The *in-situ* transmission infrared cell, which satisfies these requirements is shown in figure 1. A detailed description will be published elsewhere [6].

The cell is equipped with a quartz body and CaF_2 windows. A small external heating element is used to heat the sample. At each end of the quartz tube the cell has two stainless steel holders, in which the windows are mounted. Due to the difference in thermal expansion coefficients between the quartz body and the CaF_2 windows, a method has been devised to make a vacuum tight and flexible seal, even when the sample is heated. Two differentially pumped viton rings are used on both ends as a seal between the body, the stainless steel holder, and the window. The space between the two rings is evacuated with a rotary pump (10^{-2}-10^{-3} mbar). In this way the configuration is less sensitive for a possible leakage. The temperature, measured with a thermocouple at the sample surface, is used for temperature control of the cell. The cell is connected to a 50 l/s turbo molecular pump and a gas inlet system, which alllow either static or flow experiments. With this configuration pressures well below 10^{-7} mbar can be achieved.

Spectra were recorded with a Nicolet MX10 Fourier transform spectrometer. 100 Or 320 scans were added to obtain a spectrum with a resolution of 2 cm^{-1} over the total spectral range (3800-1100 cm^{-1}). Below 1100 cm^{-1} no information was obtained due to strong absorption of the alumina support.

3. RESULTS AND DISCUSSION

Al_2O_3

On γ-alumina usually 5 bands, characteristic for the OH stretching modes, are detected in the wavenumber range 3800-3700 cm^{-1}. Strong absorption is observed below 1100 cm^{-1} due to Al-O and Al-OH stretching modes [7]. CO adsorption on γ-Al_2O_3 is studied by adding different pressures of CO at room temperature, increasing the temperature, and observing the spectrum. Besides the rotational fine structure of gaseous CO around 2143 cm^{-1}, no clear band is observed due to CO adsorption between 2300 and 2000 cm^{-1}.

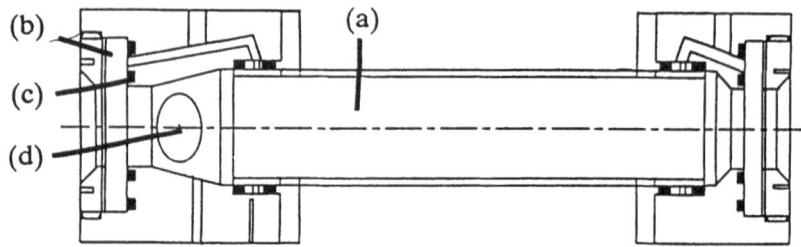

Figure 1 Top view of the cell with (a) quartz body, (b) CaF$_2$ window, (c) viton ring, (d) pump connection

At 2347 cm^{-1} a small band is seen, due to formed, physically adsorbed CO_2, of which the intensity increases with increasing temperature. Bands at lower wavenumbers (~1639, 1567, 1471 and 1250 cm^{-1}) are assigned to chemisorbed CO_2. They become more intense by interaction of CO with OH or other oxidizing groups of the support. The bands around 1639, 1467, and 1250 cm^{-1} represent bicarbonate species, while the exact nature of the band at 1567 cm^{-1} is not clear from literature [8,9,10].Hierl et al. [10] suggested that it could be a monodentate carbonate as well as a carboxylate species, in accordance with London and Bell [11]. At higher CO pressures and at higher temperatures the band at 1250 cm^{-1} shifts to 1233 cm^{-1} and is increasing just as the band at 1639 cm^{-1}, while the other two bands do not change.

Cu/Al_2O_3

When CO is adsorbed on copper containing catalysts changes are observed in three spectral regions: in the hydroxyl stretching (3800-3000 cm^{-1}), carbonate stretching (1600-800 cm^{-1}) and carbonylic (2300-2000 cm^{-1}) region [12]. A typical spectrum is shown in figure 2. The most intense band is rather broad and ranges from 2155 to 2080 cm^{-1} with a maximum at 2123 cm^{-1}. This band is assigned to Cu^+..CO [10]. It increases slightly in time indicating that the catalyst is slowly reduced by CO at room temperature. The weak band at 2179 cm^{-1} is assigned to gaseous CO, which can be removed easily by evacuation, while the main absorption band still has a considerably intensity. Below 1700 cm^{-1} broad bands (1651, 1595, 1451, 1388, and 1229 cm^{-1}) are observed, which represent carbonate-like surface species [10].

Upon heating the Cu^+..CO band becomes more intense up to 200 °C, indicating a surface reduction. Between 200 and 300 °C the transmission of the sample diminishes considerably indicating a further reduction of the catalyst. At 400 °C the band has disappeared. The bands below 1700 cm^{-1} hardly increase by heating. Changes in the OH stretching region are difficult to interpret due to the low signal to noise ratio in that region.

Adsorbing a second amount of CO on the slightly reduced catalyst a more intense band appears at 2123 cm^{-1}. No reduction is observed at room temperature. The band decreases at higher temperature and has disappeared at 350 °C. The bands in the hydroxyl region decrease slightly, while the bands in the carbonate region increase slightly up to 200 °C.

Adsorbing CO and O_2 (2:1) simultaneously, a band due to Cu^+...CO is observed, as shown in figure 3. It is maximal between 50 and 100 °C and disappears at 300 °C, while physically adsorbed CO_2 (2347 cm^{-1}) appears. The carbonate band at 1659 cm^{-1} disappears at 200 °C. Adsorbing a second amount of CO and O_2 on the slightly reduced catalyst, the band at 2123 cm is more intense. At 300 °C the band has disappeared and the catalyst has become oxidized again.

In summary, CO adsorption is a reversible process. At room temperature CO adsorbs on Cu^+ and the formed interaction is stable up to 350-400 °C.

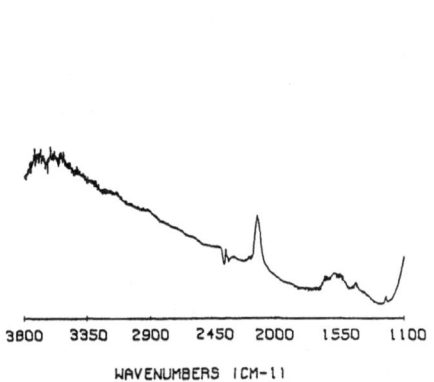

Figure 2 Typical spectrum (3800-1100 cm^{-1} of CO adsorption on Cu/Al_2O_3

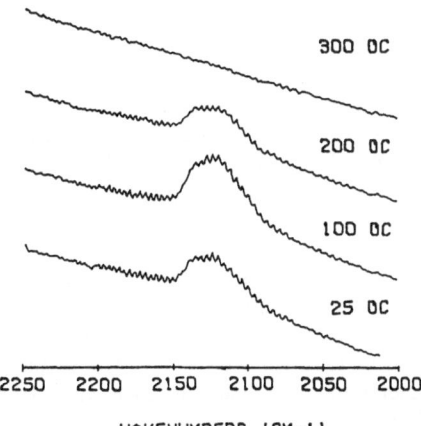

Figure 3 Spectra (2250-2200 cm^{-1}) of Cu/ Al_2O_3 after adsorption of CO and O_2 (2:1)at increasing temperature

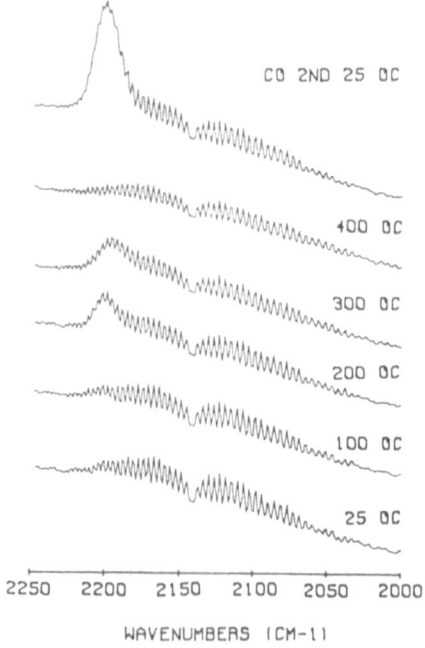

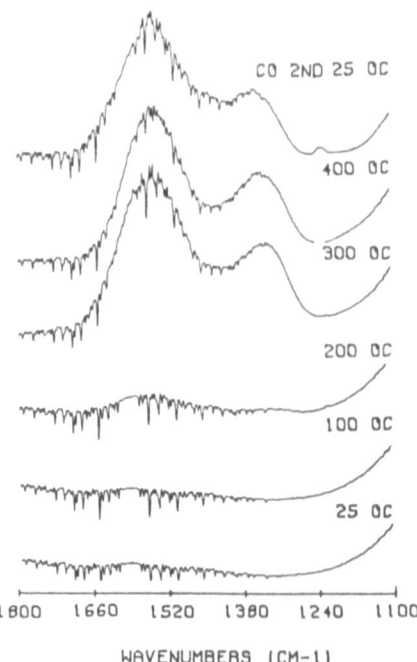

*Figure 4 Spectra (2250-2000 cm^{-1}) of
Cr/Al$_2$O$_3$ after adsorption of
CO at increasing temperature*

*Figure 5 Spectra (1800-1100 cm^{-1}) of
Cr/Al$_2$O$_3$ after adsorption of
CO at increasing temperature*

CO adsorption on Cu^{2+} is not observed, since Cu^{2+}..CO is only stable at 77 K [12]. In the presence of O_2, the Cu^+..CO species is less stable than in the absence of O_2 due to CO oxidation: between 100 and 300 °C the bands, which represent Cu^+..CO and a bicarbonate (a possible intermediate) decrease, whereas the band, which represent CO_2, increases. It appears that Cu^+ is generated by CO, while O_2 is oxidizing this site again.

Cr/Al$_2$O$_3$

When CO is adsorbed on Cr/Al$_2$O$_3$ at room temperature, only the gas phase spectrum of CO is seen. At 200 °C a band around 2200 cm^{-1} is observed and this band has disappeared at 400 °C, as shown in figure 4. It is assigned to CO adsorption on a higher valence state Cr, because CO adsorption on a H$_2$ reduced sample (Cr^{3+}) is not observed. Obviously, the reduction is an activated process. At 300 °C two bidentate carbonate bands (~1550 and 1338 cm^{-1}) become very intense [13], as shown in figure 5. At 400 °C the carbonates are still present even after evacuation at room temperature.

Adsorption of a second amount of CO directly shows an intense band at 2200 cm^{-1} of which the intensity decreases when the temperature is raised. By evacuating at 400 °C the carbonylic band has disappeared and the carbonate bands are diminished. A small carbonate band at 1230 cm^{-1} appears at the second CO adsorption and disappears at 400 °C. In both experiments the band at 2348 cm^{-1} (physically adsorbed CO$_2$) increases with increasing temperature.

Summarized, CO adsorption does not occur on oxidized Cr/Al$_2$O$_3$ at room temperature. Adsorption on a high valence state Cr-ion is observed at 200 °C. CO adsorption does occur at room temperature on a slightly reduced catalyst. The increase of the carbonate species around 300 °C is remarkable.

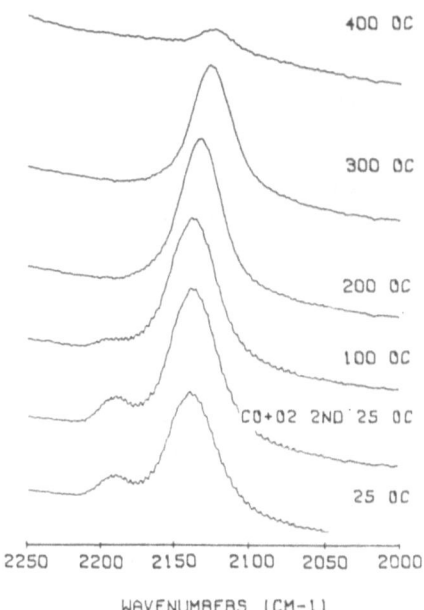

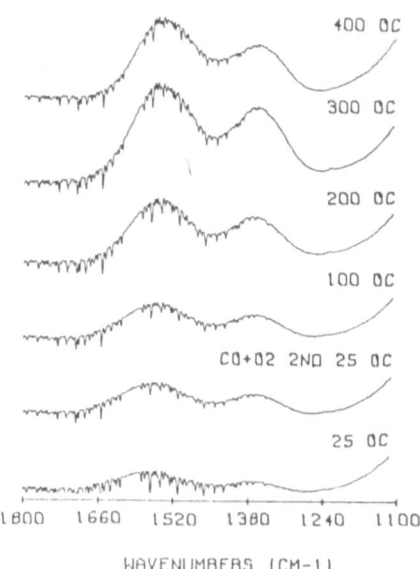

Figure 6 Spectra (2250-2000 cm⁻¹) of Cu-Cr/ Al₂O₃ after adsorption of CO and O₂ (2:1) at increasing temperature

Figure 7 Spectra (1800-1100 cm⁻¹) of Cu-Cr/ Al₂O₃ after adsorption of CO and O₂ (2:1) at increasing temperature

Cu-Cr(1:1)/Al₂O₃

Adsorption of CO at room temperature shows two bands: an intense one at 2142 and a weak one at 2199 cm⁻¹. In time both bands increase and shift (2142-2134 and 2199-2192 cm⁻¹): the catalyst is reduced slightly. The two carbonate bands present (~1560 and 1350 cm⁻¹) also increase. Upon evacuating the band at 2192 cm⁻¹ disappears immediately, while the band at 2134 cm⁻¹ and the carbonates are still present, even after one night evacuating.

Adsorbing a second amount of CO shows the same trends as described before: the process of CO adsorption is reversible. When heating, the band at 2192 cm⁻¹ has disappeared at 200 °C. The Cu^+...CO band decreases and shifts to 2121 cm⁻¹ and is still present at 400 °C. So this Cu^+..CO bond is stronger than the corresponding bond on a Cu/Al₂O₃ catalyst. The carbonate bands are maximal around 200 °C. A shift is observed during the heat treatment: from 1560 and 1350 cm⁻¹ at room temperature to 1544 and 1360 at 400 °C. A sharp band at 1230 cm⁻¹ is present up to 250 °C.

Adsorbing CO and O₂ simultaneously (figure 6 and 7), the same trend as in CO adsorption is seen: a band at 2142 cm⁻¹, which increases in time, and a weak band at 2192 cm⁻¹, which disappears by evacuating. The bands are more intense than the bands on Cu/Al₂O₃ in a similar experiment. CO+O₂ adsorption on a slightly reduced catalyst shows more intense bands at the same wavenumbers. The band at 2192 has disappeared at 150 °C, while the main band diminishes slowly and shifts to 2123 cm⁻¹. At the same time the carbonates (1545 and 1364 cm⁻¹) and CO₂ (2348 cm⁻¹) are increasing. The adsorption of CO in presence of O₂ is comparable with the adsorption of CO in absence of O₂. The Cu^+..CO band is shifted to a higher wavenumber compared with the adsorption on Cu/Al₂O₃. The band at 2192-2199 cm⁻¹ can be assigned to CO adsorption on a chromium species. This reduction is not activated, like the reduction of CO on Cr/Al₂O₃. The frequency shifts are probably due to electronic interactions between Cu and Cr in CuCr₂O₄, which is present on the catalyst besides CuO. The shift of the Cu^+..CO band to lower wavenumbers during the experiment can be caused by a change in adsorption sites. During an experiment CO prefers to adsorb on Cu^+ present at the surface of CuO instead of CuCr₂O₄. It is possible that CuCr₂O₄ is a low temperature active specie creating a relative large number of Cu^+ sites. At higher temperatures Cu^+ on CuO become more important.

4. CONCLUSIONS

As shown, the infrared transmission cell is excellently suited for this study. Ultra high vacuum conditions can be achieved quickly and the leaking rate is very low. On the copper containing catalysts adsorption of CO on Cu^{2+} is not observed, although Cu^{2+} is present: $Cu^{2+}..CO$ is not stable at room temperature [12]. CO reduces Cu^{2+} to Cu^{+} at room temperature. O_2 is able to reoxidize Cu^{+} to Cu^{2+}. Cu^{0} is not formed under this conditions. Cu^{1+} sites seem important in this reaction. Copper-chromium-oxides, creating Cu^{+} sites, are probably responsible for the higher activity of the Cu-Cr catalyst compared to the Cu catalyst.

ACKNOWLEDGEMENT

The financial support of PSA-France is gratefully acknowledged.

REFERENCES

1   K.C. Taylor, in 'Catalysis and Automotive Pollution Control' (eds. A. Crucq and A. Frennet), Elsevier Amsterdam (1987) 97
2   M. Shelef, K. Otto, H. Ghandi, J. Catal. 12 (1968) 361
3   F. Severino, J. Laine, Ind. Eng. Chem. Prod. Res. Dev. 22 (1983) 396
4   S.Stegenga, A.J.C. Mierop, C. de Vries, F. Kapteijn, J.A. Moulijn, Proc. 19th Biennal Conference on Carbon, Penn State University U.S.A. (1989) 396
5   J.B. Peri in 'Catalysis, Science and Technology' part 5 (eds. J.R. Anderson and M. Boudart) Springer-Verlag Berlin (1984) 171
6   A.D. van Langeveld, J.W. Bijsterbosch, F. Kapteijn, J.A. Moulijn, to be published
7   H. Knözinger, P. Ratnasamy, Catal. Rev.-Sci. Eng. 17 (1978) 31
8   N.D. Parkyns, J. Chem. Soc. (A) (1967) 1910
9   L.H. Little, C.H. Amberg, Can. J. Chem. 40 (1962) 1997
10 R. Hierl, H. Knözinger, H.-P. Urbach, J. Catal. 69 (1981) 475
11 J.W. London, A.T. Bell, J. Catal. 31 (1973) 32
12 G. Ghiotti, F. Bocuzzi, A. Chiorino, in 'Adsorption and Catalysis on Oxide Surfaces' (eds. M. Che and G.C. Bond) Elsevier Amsterdam (1985) 235
13 A. Zecchina, S. Coluccia, E. Guglielminotti, G. Ghiotti, J. Phys. Chem. 75 (1971) 279

DEPTH PROFILING OF Rh/CeO$_2$ CATALYSTS : AN ALTERNATIVE METHOD FOR DISPERSION ANALYSIS

A. R. González-Elipe, J. P. Holgado, R. Alvarez, J. P. Espinós, A. Fernández and G. Munuera

Instituto de Ciencias de Materiales de Sevilla (CSIC-Univ. de Sevilla) and Dpto de Química Inorgánica. P.O. Box 1119. 41071 Sevilla. Spain

ABSTRACT

Depth profiles of Rh in two Rh/CeO$_2$ catalysts prepared from RhCl$_3$ and Rh(NO$_3$)$_3$ have been obtained by means of Ar$^+$ sputtering and XPS analysis. A mathematical model describing the sputtering behaviour of powder materials has been used to simulate the depth profiles of the samples. In this way, semiquantitative information about the dispersion of the rhodium has been obtained for the precursor, the calcined and reduced states of the two samples, showing that a progressive agglomeration of the rhodium occurs during the calcining and reducing treatments. The results show that this agglomeration depends on the presence of chlorine in the precursor.

INTRODUCTION

The use of ion sputtering is a common tool in surface analysis which, in combination with surface sensitive techniques such as AES, XPS, ISS or SIMS, provides information on the "indepth" distribution of elements in solid samples [1]. This technique has been extensively used for compact materials where some inhomogeneity in composition occurs at the topmost layers. However, the use of the sputtering techniques to study powder specimen or materials with a globular structure (i.e. cermets or composites) has

Fundamental Aspects of Heterogeneous Catalysis Studied by Particle Beams,
Edited by H.H. Brongersma and R.A. van Santen, Plenum Press, New York, 1991

227

been more limited . In this context we have recently worked out a model [2] to simulate the depth profiles obtained for catalysts where a few percents by weight of an active phase in the form of small particles is dispersed on the surface of bigger particles of a support.

In the present work we have used this method to characterize two Rh/CeO_2 catalysts prepared from two different precursor salts. The analysis has provided information on the distribution of rhodium in these two samples in their precursor, calcined and reduced states. Specially in the two former situations, where the most conventional techniques for the characterization of metallic catalysts cannot be used, this appears to be a valuable method to obtain a semiquantitative evaluation of the dispersion degree of rhodium.

EXPERIMENTAL

Two Rh/CeO_2 samples have been prepared by incipient wetness impregnation of CeO_2 (Rhône-Poulenc, $S_{BET}=109$ $m^2.g^{-1}$) with $RhCl_3$ and $Rh(NO_3)_3$ aqueous solutions to get a metal loading of 3% by weight. Once prepared, these precursors were dried in air at 383K for 24h, calcined at 673K in 100 $ml.min^{-1}$ oxygen flow for 4h and stored in a desiccator until their use. The two samples were studied in their precursor form (labeled $Rh/CeO_2(N)$ and $Rh/CeO_2(Cl)$), and in the calcined ($Rh/CeO_2(N)-O$ and $Rh/CeO_2(Cl)-O$) and reduced ($Rh/CeO_2(N)-H$ and $Rh/CeO_2(Cl)-H$) states. The reduction treatment, performed "in situ" in the XPS spectrometer consisted of heating at 773K, first in vacuum for 1h to remove most of the water and other impurities and then under flowing H_2 (10 $ml.min^{-1}$) for 1h.

XPS spectra were recorded in a LHS-10 spectrometer (Leybold) working in the pass energy constant mode at 50 eV and using the Mg K_α line as excitation source. The spectra were recorded and stored in a HP 1000E computer on line with the spectrometer where they could be submitted to background subtraction using a straight baseline and area calculations. The samples, in the form of pellets, were placed on a Mo holder that could be heated resistively while controlling the temperature by a thermocouple spot-welded at its rear. All the treatments, including Ar^+ sputtering, were carried out in the pretreatment chamber of the spectrometer. Ar^+ bombardment was performed with an ion penning source working at 3.5 Kv.Calibration with a Ta_2O_5 standard gave a sputtering rate of ca. 12 Å min^{-1} for this source.

MODEL FOR THE SPUTTERING BEHAVIOUR OF CATALYSTS

A detailed description of the model for the sputtering behaviour of catalyst samples have been presented in a previous paper [2]. Fig.1 schematically shows the basic considerations of the model. According to this figure, for the electron energy analyzer placed perpendicular to the sample, a catalyst sample can be approximated by a globular structure formed by cubic particles of the active phase embedded in a compact matrix of the support. By assuming a layer by layer sputtering behaviour, and under the approximation that effects such as preferential sputtering, ion implantation, roughening, etc, do not occur [3], the following two equations can be derived to account for the depth profiles obtained from the XPS intensities of the active phase of catalyst samples with a single type of particles of size d:

$$I_s \propto A\ d^2\ [1-\exp((a-d)/\lambda_m)] \qquad [1]$$

$$I_b \propto D\ d^2\ \{a + \lambda_m\ [\exp(-d/\lambda_m)-\exp((a-d)/\lambda_m)]+ \\ +\lambda_s\ [1-\exp(-d/\lambda_m)]\} \qquad [2]$$

where λ_m is the mean free path of the electrons through these particles, λ_s the mean free path of these electrons through the support matrix and A and D two adjustable parameters which depend on the actual texture of the catalyst support. These equations stand for the contribution as a function of the sputtered thickness (a) of the particles on the surface (I_s) and in the bulk (I_b) of the idealized structure in Fig.1. The total XPS intensity is $I=I_s + I_b$ for $a \geq d$, which defines the range of application of the equations. Then the steady state intensity is reached for a=d, where I_s vanishes (the particles at the surface would be sputtered away) and $I=I_b$ reduces to a constant value given by

$$I_b \propto D\ d^2\ \{d + \lambda_m\ [\exp(-d/\lambda_m)- 1]+ \lambda_s\ [1-\exp(-d/\lambda_m)]\}$$

The relative values of A and D define the weights of the I_s and I_b contributions to I. In our previous paper [2] we have developed two formulas for these two parameters by assuming that the particles of the support were small spheres without porosity. Using these formulas the values calculated for A and D are 43.65 and 0.017 for the case of the Rh/CeO_2 samples. Deviations from these values should be expected if the support has a different structure. Using these calculated values Fig. 2 shows the plot of I as a function of the sputtered thickness "a" (i.e. depth profiles) calculated for a series of Rh/CeO_2 catalysts assuming different size of the

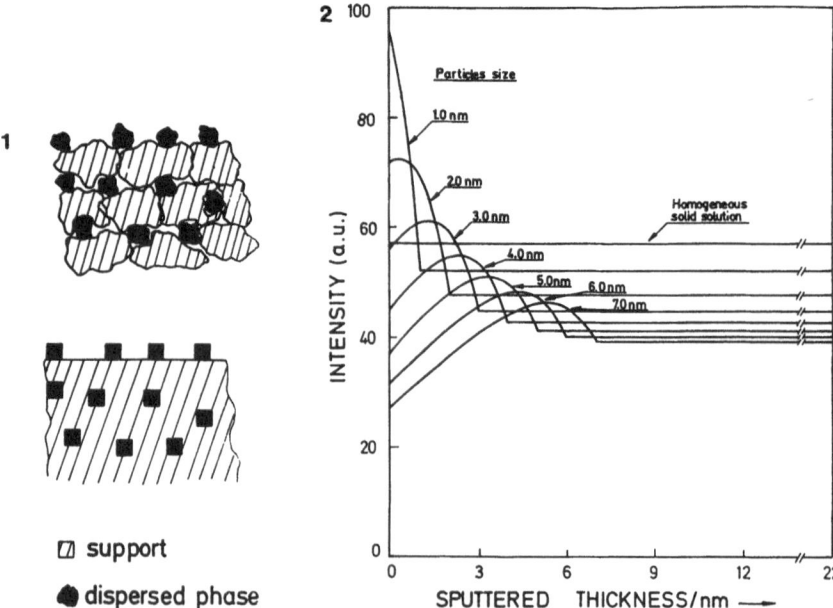

Fig. 1. Top: Representation of a catalyst sample
formed by big particles of a support and small
particles of a dispersed phase. Bottom:
Idealization of this structure as seen by the
electron energy analyzer.

Fig. 2. Normalized depth profiles in function of
the sputtered thickness, calculated for several
sizes of the particles of the dispersed phase
for Rh/CeO$_2$ catalysts.

metal particles from d=1.0 to 7.0 nm. For the samples with
small particle sizes (d<2.0 nm) the calculated profiles are
characterized by a sharp decrease in intensity to reach the
steady state when a=d. This sharp decrease is in agreement
with a readily removal by sputtering of these small particles
at the surface of the sample (fig.1). However, the profiles
of the samples with particles of bigger sizes show an initial
slow increase in intensity followed by a fast decrease to
reach the steady state for a=d. This behaviour is the result
of a practically constant contribution of the particles at
the surface for a<<d (i.e. I$_s$) and a sharp decrease to become
zero for a=d and of an increase in the contribution of those
embedded particles which are near the surface and are
progressively exposed by sputtering up to reach the constant
value for a=d. It is worthy of note that in the

representation, the steady state intensity decreases as the
particle size increases, a feature that can be used as an
additional test for particle size assessment.

RESULTS AND DISCUSSION

Fig. 3 shows the experimental depth profiles obtained
for the two catalyst samples in their precursor, calcined and
reduced states. A common feature of the three profiles for
each sample is that they become less sharp when going from
the precursor to the its reduced states. According to the
theoretical profiles in Fig. 2, this change could be
explained if a progressive agglomeration of the rhodium
occurs with the activation treatment. For the sample

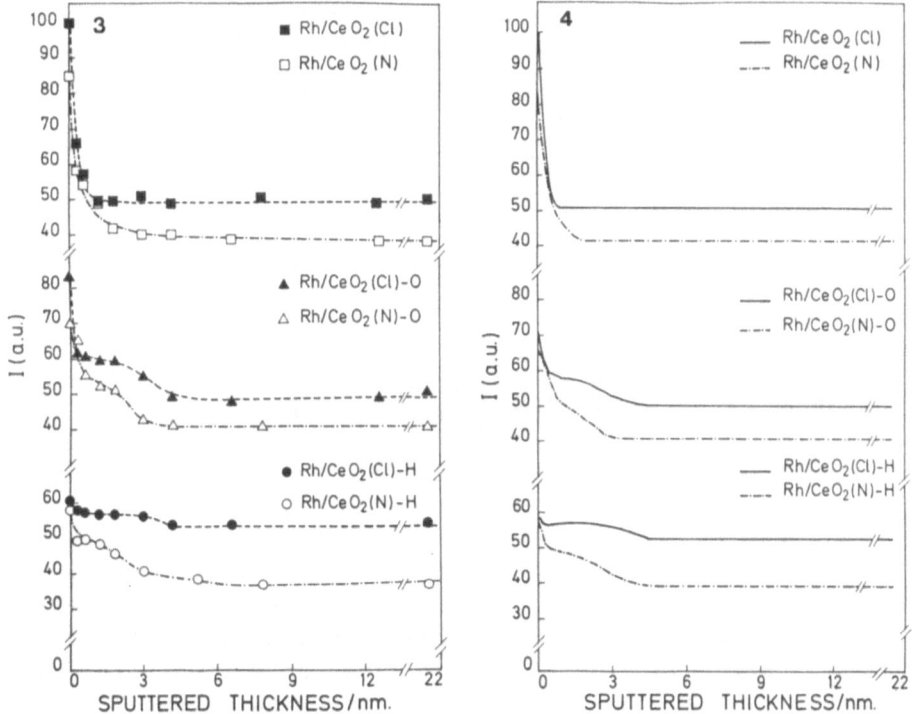

Fig. 3.- Experimental depth profiles of Rh for
Rh/CeO$_2$(N), Rh/CeO$_2$(N)-O, Rh/CeO$_2$(N)-H and
Rh/CeO$_2$(Cl), Rh/CeO$_2$(Cl)-O, Rh/CeO$_2$(Cl)-H
samples.

Fig. 4.- Calculated depth profiles of Rh for
Rh/CeO$_2$(N), Rh/CeO$_2$(N)-O, Rh/CeO$_2$(N)-H and
Rh/CeO$_2$(Cl), Rh/CeO$_2$(Cl)-O, Rh/CeO$_2$(Cl)-H
samples.

Rh/CeO$_2$(N), prepared from the nitrate salt, this is further suggested by the lower intensity of the steady state after reduction. However, the opposite is found for the sample Rh/CeO$_2$(Cl), prepared from the chloride, where the flattening of the profile in the reduced state is accompanied by an increase in the steady state intensity.

To get a deeper insight in the evolution of the profiles of the two samples, a simulation has been made combining a series of equations [1] and [2] with different particle sizes. The best simulations obtained for the two samples are shown in Fig. 4, using the S$_{BET}$ values measured for each sample and the relative particle sizes reported in Table 1.

Table 1. Percentage of particles in each interval of sizes (nm) used for the simulation of the depth profiles of rhodium in different samples

| Sample | D | Interval of particles sizes / nm | | | | | |
		0-1	1-2	2-3	3-4	4-5	Atomic solution
Rh/CeO2(N)	.015	69	31	–	–	–	–
Rh/CeO2(N)-O	.016	31	23	34	12	–	–
Rh/CeO2(N)-H	.016	21	13	25	31	10	–
Rh/CeO2(Cl)	.015	100	–	–	–	–	–
Rh/CeO2(Cl)-O	.017	34	8	17	23	8	–
Rh/CeO2(Cl)-H	.017	9	6	11	11	10	53

It must be stressed that these particles sizes have not an absolute meaning since they are the result of using the sputtering rate of a Ta$_2$O$_5$ standard to make the conversion for our samples of sputtering time into sputtered thickness. However, although these values are submitted to this uncertainty, they can be taken as approximate and, in any case, be used to compare the two samples. From table 1, it is interesting that, for the two precursors, the best simulations have been obtained with a value of D=0.015 (for convenience we have always assumed the same calculated value of A for all the samples), which is smaller than that calculated assuming spherical particles. This difference might be related to the actual morphology of the original CeO$_2$ support which, as it is shown in the micrograph in Fig. 5, consists of big needle-shape particles (>1 nm) fully covered by very small pores. It is likely that for this structure, preferential deposition of rhodium might occur on the external surface of the support, which according to the idealization in Fig. 1, would result in values of D smaller than those expected for small spherically-shaped particles with an homogeneous distribution of the dispersed phase on their surface.

It is known that direct information on the dispersion of a supported phase can be obtained from the values of its XPS intensity[4]. These values for our samples correspond to the points for a sputtered thickness equal zero in the profiles in fig. 3 (i.e. for a sputtering time t=0), and show that the dispersion degree of rhodium varies in the sense precursors > oxidized forms > reduced forms. In addition, simulation of the sputtering profiles provide a semiquantitative information on the particle size distributions. Thus, from the values in Table 1 it appears that the rhodium should be slightly less dispersed in $Rh/CeO_2(N)$ than in $Rh/CeO_2(Cl)$. However, in the two cases the calcining treatment produces a considerable agglomeration as indicated by the need of a higher percentage of particles of bigger sizes to simulate the profiles. It is worthy of note that according to table 1, in the two calcined samples there is a heterogeneous distribution of Rh_2O_3 (Rh^{3+} B.E. ca. 309.1 eV) which, at least for the $Rh/CeO_2(Cl)-O$ sample, has a bimodal distribution with most of the rhodium in the form of small (34% , <1 nm) and much bigger particles (23% , 2-4 nm).

Further agglomeration of the rhodium seems to be produced by the reduction treatment which also leads to changes affecting the texture of the support, as stated in fig. 5 for the $Rh/CeO_2(Cl)-H$ sample showing a great disruption of the CeO_2 particles. This effect can be monitored by changes in the D parameter (table 1), producing the presence of chlorine an enhancement of it (D=0.017 v.s. 0.0 16). Meanwhile a high percentage of the rhodium in "atomic form" is required to obtain a good simulation of the flat profile of this sample (see Table 1). The existence of an "atomic solution" required to reproduce this profile, suggests a rather homogeneous distribution of rhodium atoms, or very small clusters, within the bulk of the support. From the point of view of the sputtering behaviour this would be equivalent to the formation of a solid solution of the metallic phase into within the support matrix. Such diffusion of metallic atoms or clusters within the oxide support has been claimed by several authors [5], even at room temperature [6], in other metal-oxide catalysts showing SMSI phenomena. In the case of the $Rh/CeO_2(Cl)-H$ sample this diffusion would produce the observed effect of a flattening of the profile with a net increase of its steady state intensity the opposite to that expected from our model for the presence of big particles. However, an equivalent effect also should be observed it incorporation of small clusters of rhodium occurs into the pores of the CeO_2 support during the activation treatments of this sample (see Fig. 5) which, in addition, could produce the observed breaking and reordering of the CeO_2 matrix . A more detailed study on the influence of chlorine in these processes for chlorine containing samples is now in progress in our laboratory.

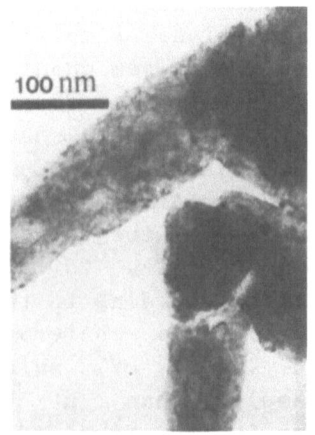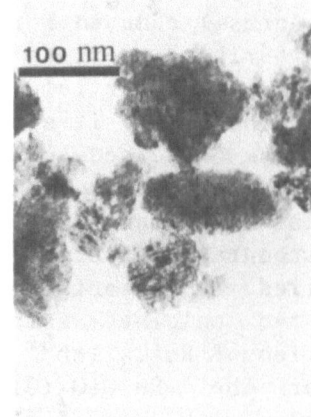

Fig. 5. TEM micrographs of the CeO_2 support (left) and $Rh/CeO_2(Cl)-H$ sample (right).

ACKNOWLEDGMENTS

We thank the CICYT (project MAT88-223) for the financial support and the "Servicio de Microscopía Electrónica de la Universidad de Sevilla" for the TEM analysis of samples.

REFERENCES

1. S. Hofmann, "Quantitative depth profiling in Surface Analysis: A review", Surf. Interf. Anal. 2:148 (1980).
2. A. R. González-Elipe, J. P. Espinós, A. Fernández, and G. Munuera, "Depth profiling of catalyst samples: A XPS based model for the sputtering behaviour of powder materials", J. Catal. accepted for publication.

3 . G. Betz and G. K. Wehner, "Sputtering by Particle Bombardment II" , Ed. R. Behrisch (Springer, Berlin 1984) .

4 . H. P. C. E. Kuipers, H. C. E. van Leuven, W.M. Visser, "The characterization of heterogeneous catalysts by XPS based on geometrical probability. 1 : Monometallic catalysts" , Surf. Interf. Anal. 8 : 235 (1986) .

5 . M. G. Sánchez and J. L. Gázquez, " Oxygen vacancy model in Strong Metal – Support Interaction " , J. Catal. 104 : 120 (1987) .

6 . S. Bourgeois, D. Diakite , F. Jomard , M. Perdereau , and R. Poirault, " A SIMS and AES study of nickel deposition on TiO_2 (100) . Influence of the stoichiometry of the support ", Studies in Surf . Sci. Catal. 48 : 191 (1989).

ELECTRON STIMULATED DESORPTION AND OTHER METHODS FOR THE STUDY OF SURFACE PHENOMENA RELATED TO ATOMIC LEVEL ASPECTS OF HETEROGENEOUS CATALYSIS

John T. Yates, Jr.

Surface Science Center
Department of Chemistry
University of Pittsburgh
Pittsburgh, PA 15260 USA

INTRODUCTION

The three lectures to be presented at this workshop illustrate the use of modern physical methods of research for understanding the details of the molecular behavior of chemisorbed species on transition metal surfaces. The use of electron stimulated desorption as an investigative tool for the study of small molecules adsorbed on single crystal surfaces will be emphasized in the first two lectures. In the third lecture both infrared spectroscopy and high resolution electron energy loss spectroscopy will be employed to study the behavior and modification of model supported catalyst systems.

It is intended that the lectures will emphasize the unique power of modern physical and chemical methods for the study of the surface science of model catalytic systems, where the generation of transferrable concepts about the molecular behavior of surface species is a major goal.

Fundamental Aspects of Heterogeneous Catalysis Studied by Particle Beams,
Edited by H.H. Brongersma and R.A. van Santen, Plenum Press, New York, 1991

LECTURE I.   IMAGING CHEMICAL BOND DIRECTIONS IN ADSORBATES ON SINGLE CRYSTAL METAL SURFACES

In this lecture we will explore electron stimulated desorption as a tool for the investigation of the directionalty of chemical bonds on surfaces.

When an oriented chemisorbed molecule is electronically excited it may choose to eject an ion or other excited species. It has been found that this electron stimulated desorption process often produces sharp beams of desorbing particles, and the directionality of the emission of these beams may be measured. The emission directions are closely related to the original chemical bond direction which is broken by the electronic excitation process. The technique to image ion emission directions is termed "electron stimulated desorption ion angular distribution" or ESDIAD. The ESDIAD process involves an electronic excitation of the adsorbed species to a repulsive electronic state, where rapid desorption along the chemical bond direction occurs. ESDIAD is therefore a sensitive method for imaging the chemical bond directions which exist in an ensemble of adsorbed molecules on a single crystal surface. A more general phenomenon involving electron bombardment of surfaces with the production of new surface and gas phase species is termed "electron stimulated desorption" or ESD. Table I lists the current review articles which have been written on ESD and ESDIAD.

To begin our study of surface species with ESDIAD, the adsorption of the NH_3 molecule on Ni(110) has been chosen for illustrative purposes.[25-28] NH_3 chemisorbs non-dissociatively on Ni(110) and presents its three N-H bonds in an upward-pointing direction. These H moieties can be ionized by electron impact, yielding H^+ species. The observation of a ring pattern for H^+ ESDIAD indicates that the adsorption potential experienced by the NH_3 molecule on the Ni(110)

238

Table I- Reviews on Electron Stimulated Desorption.

Year	Author(s)	References
1970	Redhead	[1]
	Menzel	[2]
1971	Madey and Yates	[3]
1972	Leck and Stimpson	[4]
1974	Ageev and Ionov	[5]
1975	Menzel	[6]
1976	Drinkwine, Shapira and Lichtman	[7]
1977	Madey and Yates	[8]
	Drinkwine and Lichtman	[9]
1979	Bauer	[10]
1981	Taglauer and Heiland (Eds.)	[11]
1982	Menzel	[12]
1983	Tolk, Traum, Tully, Madey (Eds.)	[13]
1984	Madey, Ramaker and Stockbauer	[14]
	Knotek	[15]
1985	Brenig and Menzel (Eds.)	[16]
	Madey and Stockbauer	[17]
1986	Menzel	[18]
	Menzel	[19]
1988	Stulen and Knotek (Eds.)	[20]
1989	Stulen	[21]
	Avouris and Walkup	[22]
1990	DIET IV (in press)	[23]
	Ramsier and Yates (in preparation)	[24]

surface (2 fold symmetry) does not prevent it from establishing an isotropic azimuthal distribution of N-H bond orientations, and it is believed that the molecule is rotating about its vertical C_{3v} axis as it sits on atop Ni sites.

Studies of the ESD-degradative chemistry of NH_3 have been carried out using several methods, and it has been found that the first new species to be produced by ESD is NH_2 which is captured as a chemisorbed species by the Ni(110) surface. The orientation of the H-N-H plane of NH_2 is perpendicular to the rows and channels of the Ni(110) surface, and it is strongly bound.

The chemisorption of the PF_3 molecule has been studied on Ni(111) using ESDIAD and other methods.[29] Here it has been possible to directly observe the thermal excitation of the molecular rotation of chemisorbed PF_3 at its atop binding site on Ni(111). Rotation occurs about the vertical C_{3v} axis. A measurement of the barrier to rotation has been made using ESDIAD. It has been found that rotation will only occur in very dilute layers of PF_3. When the coverage becomes high, mutual interactions between PF_3 molecules cause the rotational behavior to freeze out. This work represents the first <u>direct</u> observation of the thermal activation of the rotation of a chemisorbed molecule.

Chemisorbed PF_3 molecules may be degraded by ESD to PF_2 and PF species which are trapped by chemisorption on the surface at two-fold bridge and three-fold hole sites respectively. The site symmetries can be deduced from the observation of unique P-F bond directions in the species. Recent theoretical calculations[30] have demonstrated that the observed sites and molecular orientation for $PF_3(a)$, $PF_2(a)$ and PF(a) agree with the experimental findings.

In this lecture we have seen that ESDIAD may be used to observe chemical bond directions of species chemisorbed on single crystal surfaces. In addition ESDIAD may be used to observe and measure the dynamical behavior of adsorbates as they swing to-and-fro on their adsorption sites. Finally, the ability of ESD to produce high surface concentrations of coordinatively unsaturated species is demonstrated. These species may be the same as the intermediates which occur in heterogeneous catalytic chemistry, and if so, ESD provides a route to their preparation on surfaces for investigative purposes.

LECTURE II. APPLICATION OF ESDIAD TO STUDIES OF BONDING AND CHEMISTRY AT DEFECT SITES AND NEAR FOREIGN ATOMS ON SURFACES

In this lecture we investigate the use of ESDIAD as a probe of the chemical behavior of the CO and NO molecules on the smooth Pt(111) surface and the stepped Pt(112) surface.[31-34]

When CO chemisorbs on Pt(111), both terminal-CO and two-fold bridged-CO species are produced. The terminal-CO species pack together in domain boundaries at coverages above 0.5, and in these closely spaced domain boundary regions, intermolecular forces between the neighboring CO molecules cause them to tilt away from each other. We measure a tilt angle of 6 degrees from the normal, while Monte Carlo calculations estimate this tilt angle to be 10 degrees.[35] These tilted CO molecules exhibit very soft vibrational modes perpendicular to the plane of tilt, as observed by the thermal broadening of the ESDIAD beams.

In contrast to the behavior of CO molecules on Pt(111), the behavior on a stepped Pt(111) surface offers a fascinating view of the role of steps in preferentially adsorbing molecules while permitting the terraces to remain clean at low average CO coverages. It is possible to arrange to fill the steps to high coverages before the terraces fill, and hence to observe the dynamical behavior of one-dimensional arrays of chemisorbed CO molecules which decorate the step sites. It is found that close packing of the CO molecules on the step sites leads to tilting, as on Pt(111). Initially, for CO coverages between 1/2 and 3/4 filling of the steps, tilting occurs in directions parallel to the step edge. As the remainder of the step sites fill, a new pair of orthogonal tilt directions become favored for the closely-packed CO molecules, up and down stairs. This tilting behavior is due to steric interactional effects, since the dipole-dipole forces between the neighboring CO molecules are far too small to cause the observed tilt angles.

The adsorption of CO to low coverages on the stepped sites of Pt(112) reveals that the isolated molecules are oriented about 20 degrees away from the macroscopic plane in the downstairs direction. It has been shown that the librational potentials experienced by these tilted CO molecules differ in the various directions from the steps. A measure of the character of the librational potential energy surface may be obtained through the study of the vibrational amplitude of the adsorbed molecule, using ESDIAD. Thus, the vibrational amplitude of step site-CO molecules is largest along the step edge, intermediate in the upstairs direction, and smallest in the downstairs direction. This might suggest that the favored migration direction of a CO molecule on a step site would be along the step edge if the amplitude of vibration is an indicator of preferential direction for surface migration. This postulate is not unreasonable, since the vibrational amplitude being observed is a measure of the molecular excursions of the molecule as it undergoes frustrated translational motions.

The adsorption of NO has also been investigated on P(112) using ESDIAD.[36] It has been found through observations of the amplitude of frustrated translational vibration of NO adsorbed on step sites that the librational potential perpendicular to the step edge is very soft compared to the librational potential parallel to the step edge. This is consistent with the fact that NO bridge bonds to two Pt atoms on the step edge and that in this configuration librational motion perpendicular to the plane of the molecule would be expected to be very free compared to motion in the plane of the molecule on its adsorption site. Thus ESDIAD has been used to study the hybridization of an adsorbed molecule.

Adsorption of CO on a Pt(111) surface poisoned to various coverages with Se has been studied using the width of the normally directed ESDIAD pattern as a measure of the librational freedom of the CO molecule on its adsorption site. It has been found that a surface phase transition occurs in which the CO molecule becomes more confined in its interaction with neighboring Se adatoms.[37] This effect causes the angular width of the CO ESDIAD pattern to diminish. Thus, Se, a catalytic poison, has been observed to diminish the librational mode amplitude of chemisorbed CO which is squeezed between Se neighbors. This observation suggests that a steric effect may be involved in the action of the poison atom in reducing catalytic activity.

Finally, it will be shown in this lecture how the reaction between chemisorbed CO and chemisorbed O occurs on a stepped P(112) surface. Using ESDIAD, and the ability to observe preferential adsorption of molecules on step sites, it was possible to place and to maintain different isotopic CO species on the steps and terraces of Pt(112). Reaction of these species with chemisorbed O produces CO_2 at low temperatures, with the isotopic label characteristic only of the terrace-CO. The stronger binding of CO at the step sites is probably responsible for the lower reactivity of these species compared to those on the terrace sites.[38] This represents the first direct observation of the site of preferential catalytic activity, although many spectroscopic studies of catalytic site character have been reported.

In this lecture, the power of the ESDIAD method for observing the surface dynamical behavior of adsorbates and the details of adsorbate-adsorbate intermolecular interactions on single crystal surfaces has been demonstrated. This capability has been used to measure the site preference for a simple catalytic reaction where step sites and terrace sites are compared. The distinction between the catalytic activity of different adsorption sites has been a major goal of surface chemists since the time of H.S. Taylor in 1925 who first postulated that active sites exist on surfaces.[39]

LECTURE III. ELECTRON AND PHOTON VIBRATIONAL SPECTROSCOPIES TO STUDY MODEL CATALYST BEHAVIOR

In this lecture both infrared spectroscopy and high resolution electron energy loss spectroscopy are employed to learn about the behavior of an important catalyst system, Rh supported on Al_2O_3.[40-48] This work is timely because the Rh/Al_2O_3 catalyst is of central importance in automotive environmental catalysis.

It has been found that in the presence of chemisorbed CO, $Rh_x(0)$ catalyst sites are converted to Rh(I) sites. This is postulated to occur via the action of isolated surface hydroxyl groups which are reduced.[49,50] Such a loss of metallic Rh sites would be expected to reduce the catalytic effectiveness of the Rh catalyst. Evidence for this surface redox process will be presented, and the use of chemical methods which remove the isolated hydroxyl groups and hence prevent the $Rh_x(0)$ ---> Rh(I) conversion process will be described.[48] The chemical method employed involves the use of a silation agent, $(CH_3)_3Si-Cl$ which preferentially reacts with the isolated hydroxyl groups. This procedure is effective therefore in preventing the conversion of active metallic Rh sites to catalytically inactive Rh(I) sites, and recent studies have found that this protective behavior extends to at least 600 K catalyst temperature.[51]

Another problem to be described involves the study of the interaction between $Rh_x(0)$ sites and the Al_2O_3 support. In order to investigate this metal-support interaction, we have deposited metallic Rh onto an Al_2O_3 film in ultrahigh vacuum. Using high resolution electron energy loss spectroscopy and Auger spectroscopy it is possible to observe the high temperature behavior of the surface Rh

244

atoms. We observe that Rh enters into the subsurface region of the Al_2O_3 support at temperatures near 1100 K. The chemisorptive capability for CO diminishes to zero as this penetration process occurs in vacuum. To compare this behavior with that of a real Rh/Al_2O_3 catalyst, similar experiments have been conducted in a high-temperature transmission infrared spectroscopy cell,[52] using chemically prepared Rh/Al_2O_3 catalysts. Here it has been found that heating in vacuum initially causes sintering of the supported Rh catalyst. When the structural transformation from γ-Al_2O_3 to α-Al_2O_3 occurs near 1400 K, all Rh catalytic sites disappear as determined by infrared studies of CO chemisorption.[53]

This lecture has attempted to illustrate the power of combining modern surface science experimental methods with more conventional methods employed by the catalytic chemist for understanding the molecular details of catalyst behavior. It has been shown that concepts from the more physical approach of surface science readily transfer to the study of catalyst surfaces. It has also been shown that controlled surface functionalization methods may be used to radically alter the behavior of supported Rh catalysts under extreme environmental conditions.

ACKNOWLEDGEMENTS

We acknowledge with thanks the support received for this work from the Air Force Office of Scientific Research, The Department of Energy (Basic Energy Sciences), and the General Motors Research Laboratory.

REFERENCES

1. P.A. Redhead, J. Vac. Sci. Technol. 7, 182 (1970).
2. D. Menzel, Angew, Chem. Internal. Edit. 9, 255 (1970).

3. T.E. Madey and J.T. Yates, Jr., J. Vac. Sci. Technol. 8, 525 (1971).

4. J.H. Leck and B.P. Stimpson, J. Vac. Sci. Technol. 9, 293 (1972).

5. V.N. Ageev and N.I. Ionov, Prog. Surf. Sci. 5, 1 (1974).

6. D. Menzel, Surf. Sci. 47, 370 (1975).

7. M.J. Drinkwine, Y. Shapira and D.L. Lichtman, in Radiation Effects on Solid Surfaces, edited by M. Kaminsky, (ACS, Washington, DC 1976).

8. T.E. Madey and J.T. Yates, Jr., Surf. Sci. 63, 203 (1977).

9. M.J. Drinkwine and D. Lichtman, Prog. Surf. Sci. 8, 123 (1977).

10. E. Bauer, J. Electron Spectrosc. Relat. Phenom. 15, 119 (1979).

11. E. Taglauer and W. Heiland (Eds), Inelastic Particle-Surface Collisions, (Springer, Berlin, 1981).

12. D. Menzel, J. Vac. Sci. Technol. 20, 538 (1982).

13. N.H. Tolk, M.M. Traum, J.C. Tully and T.E. Madey (Eds), Desorption Induced by Electronic Transitions, DIET I, (Springer, Berlin, 1983).

14. T.E. Madey, D.E. Ramaker and R. Stockbauer, Ann. Rev. Phys. Chem. 35, 215 (1984).

15. M.L. Knotek, Rep. Prog. Phys. 47, 1499 (1984).

16. W. Brenig and D. Menzel (Eds.), Desorption Induced by Electronic Transitions, DIET II, (Springer, Berlin, 1985).

17. T.E. Madey and R. Stockbauer, Methods of Experimental Physics, 22, 465 (1985).

18. D. Menzel, Nucl. Instrum. Methods, B13, 507 (1986).

19. D. Menzel, Ber. Bunsenges. Phys. Chem. 72, 591 (1986).

20. R.H. Stulen and M.L. Knotek (Eds.), Desorption Induced by Electronic Transitions, DIET III, (Springer, Berlin, 1988).

21. R.H. Stulen, Prog. Surf. Sci. 32, 1 (1989).

22. P. Avouris and R.E. Walkup, Ann. Rev. Phys. Chem. 40, 173 (1989).

23. <u>Desorption</u> <u>Induced</u> <u>by</u> <u>Electronic</u> <u>Transitions</u>, DIET IV, in press (Springer, Berlin, 1990).

24. R.D. Ramsier and J.T. Yates, Jr., in preparation (1990).

25. C. Klauber, M.D. Alvey and J.T. Yates, Jr., Chem. Phys. Letters 106, 47 (1984).

26. C. Klauber, M.D. Alvey and J.T. Yates, Jr., Surf. Sci. 154, 139 (1985).

27. M.D. Alvey, C. Klauber and J.T. Yates, Jr., J. Vac. Sci. & Technol. A3(3), 1631 (1985).

28. M.J. Dresser, M.D. Alvey and J.T. Yates, Jr., Surf. Sci. 169, 91 (1986).

29. M.D. Alvey, J.T. Yates, Jr. and K.J. Uram, J. Chem. Phys. 87(12), 7221 (1987).

30. A.W.E. Chan and R. Hoffmann, J. Chem. Phys. 92, 699 (1990).

31. M. Kiskinova, A. Szabo and J.T. Yates, Jr., Surf. Sci. 205, 215 (1988).

32. M.A. Henderson, A. Szabo and J.T. Yates, Jr., J. Chem. Phys. 91, 7245 (1989).

33. M.A. Henderson, A. Szabo and J.T. Yates, Jr., J. Chem. Phys. 91, 7255 (1989).

34. M.A. Henderson, A. Szabo and J.T. Yates, Jr., Chem. Phys. Lett. 162, 51 (1990).

35. B.N.J. Persson, M. Tushaus and A.M. Bradshaw, to be published; B.N.J. Persson and J.E. Müller, Surf. Sci. 171, 219 (1986).

36. A. Szabo, M.A. Henderson and J.T. Yates, Jr., J. Chem. Phys. 92, 2208 (1990).

37. M. Kiskinova, A. Szabo and J.T. Yates, Jr., Phys. Rev. Lett. 61, 2875 (1988).

38. J.T. Yates, Jr., A. Szabo and M.A. Henderson, <u>The</u> <u>Influence</u> <u>of</u> <u>Surface</u> <u>Defect</u> <u>Sites</u> <u>on</u> <u>Chemisorption</u> <u>and</u> <u>Catalysis</u>, Division

of Petroleum Chemistry, American Chemical Society Symposium, "Structure in Heterogeneous Catalysis," Boston, MA, April 1990.

39. H.S. Taylor, Proc. Roy. Soc. A108, 105 (1925).

40. J.G. Chen, M.L. Colaianni, P. Chen, J.T. Yates, Jr. and G.B. Fisher, J. Phys. Chem. 94, 5059 (1990).

41. J.T. Yates, Jr., T.M. Duncan, S.D. Worley and R.W. Vaughan, J. Chem. Phys. 70, 1219 (1979).

42. T.M. Duncan, J.T. Yates, Jr. and R.W. Vaughan, J. Chem. Phys. 71, 3129 (1979).

43. T.M. Duncan, J.T. Yates, Jr. and R. W. Vaughan, J. Chem. Phys. 73, 975 (1980).

44. R.R. Cavanagh and J.T. Yates, Jr., J. Chem. Phys. 74, 4150 (1981).

45. J.T. Yates, Jr. and K. Kolasinski, J. Chem. Phys. 79, 1026 (1983).

46. D. Panayotov, P. Basu and J.T. Yates, Jr., J. Phys. Chem. 92, 6066 (1988).

47. T.H. Ballinger, P. Basu and J.T. Yates, Jr., J. Phys. Chem. 93, 6758 (1989).

48. D.K. Paul, T.H. Ballinger and J.T. Yates, Jr., J. Phys. Chem. 94, 4617 (1990).

49. P. Basu, D. Panayotov and J.T. Yates, Jr., J. Phys. Chem. 91, 3133 (1987).

50. P. Basu, D. Panayotov and J.T. Yates, Jr., J. Am. Chem. Soc. 110, 2074 (1988).

51. D.K. Paul and J.T. Yates, Jr., J. Phys. Chem., submitted.

52. P. Basu, T.H. Ballinger and J.T. Yates, Jr., Rev. Sci. Inst. 59(8), 1321 (1988).

53. T.H. Ballinger and J.T. Yates, Jr., J. Phys. Chem., submitted.

HREELS AND TDS STUDIES OF NO+H$_2$ AND NH$_3$+O$_2$ REACTIONS ON Pt(111)

M.Yu. Smirnov, V.V. Gorodetskii, A.R. Cholach

Institute of Catalysis
Novosibirsk
USSR

Abstract

The reactions of NO+H$_2$ and NH$_3$+0$_2$ on Pt(111) surface were studied at low pressures in the temperature range of 300-350 K. The data obtained indicate that both reactions proceed through the same surface intermediates, namely N$_{ads}$ and HNO$_{ads}$. HREELS spectra give evidence that the HNO intermediates are adsorbed in the side-on state. These species react with hydrogen at 350 K to form N$_{ads}$. The latter may be easily removed from the surface by exposure in hydrogen at 400 K. The catalytic reactions of ammonia oxidation and nitric oxide hydrogenation on platinum group metals are of practical importance. Study of their mechanisms is also of interest from the viewpoint of fundamental metal. The reaction pathway and selectivity of these reactions depend on the properties of surface species formed during the catalytic reaction. The known kinetic schemes include such intermediates as NH$_{ads}$, HNO$_{ads}$ etc., but experimental evidence is scarce. This paper reports further on our studies of the adsorption and catalytic properties of Pt(111) surface by means of High Resolution Electron Energy Loss Spectroscopy (HREELS) and Thermal Desorption Spectroscopy (TDS) [1,2]. The aim of the present work is to identify and to characterise the intermediates of NO+H$_2$ and NH$_3$+O$_2$ reactions on Pt(111). The experiments were performed in a VG ADES-400 Spectrometer with a residual pressure of $2 \cdot 10^{-9}$ Pa. A monochromatic electron beam of 2.3 eV energy, the primary current of $5 \cdot 10^{-11}$ A and an incidence angle of 45° with respect to the surface normal were used. Energy resolution of the elastically reflected beam was ca. 80 cm^{-1} at an intensity of 105 cps. Spectra were recorded in the specular direction. The TD spectra were registered by means of QMS VG QXK-400, a heating rate of 10K/s (DC) was used. Characteristics of the Pt(111) crystal and the procedure of surface cleaning were described elsewhere [1,2].

Results and Discussion

NO+H$_2$ reaction was studied over the temperature range of 300-350 K and the reactant pressure ratio range P$_{NO}$:P$_{H_2}$ = κ = 0.1-5 at a total pressure of $1.3 \cdot 10^{-6}$ Pa. After some reaction time surface reaction intermediates were studied by abrupt cooling of the crystal down to 140 K in order to stop the reaction. Then the crystal was evacuated to P $\leq 10^{-8}$ Pa and again flashed up to the reaction temperature to remove the surface species adsorbed in the course of the cooling procedure. Finally the spectrum was registered. The invariance of this loss spectrum versus reaction time indicated that reaction had reached its steady state.

Fundamental Aspects of Heterogeneous Catalysis Studied by Particle Beams,
Edited by H.H. Brongersma and R.A. van Santen, Plenum Press, New York, 1991

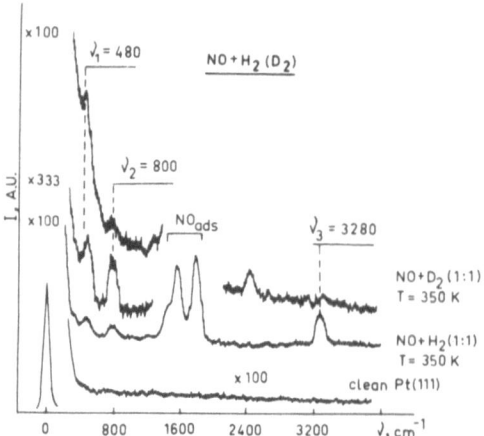

Figure 1
HREEL spectra obtained after steady
state reaction of NO+H$_2$ and NO+D$_2$
on Pt(111) at 350 K and total reactant pressure of 1.3·10^{-6} Pa.

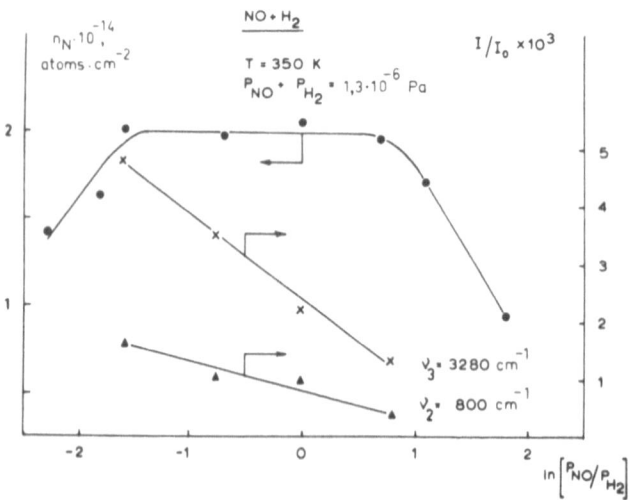

Figure 2
Steady state nitrogen coverage η_N (TDS data) and ν_2, ν_3 loss intensities versus NO:H$_2$ ratio
in the reaction mixture.

Figure 1 shows a HREEL spectrum obtained at 350 K and $\kappa = 1$. It reveals a number of new features at ν_1=480, ν_2=800 and ν_3=3280 cm^{-1} besides that of valence bands
ν(NO) of adsorbed NO [1]. Removal of any reactant resulted in the disappearance of
the new losses mentioned above. Therefore, these bands correspond to new surface
species due to the interaction of NO+H$_2$. The H$_2$ to D$_2$ substitution in the reaction
mixture results in ν_3 frequency shift from 3280 to 2430 cm^{-1}, whereas ν_1 and ν_2 get
slightly changed values at 455 and 770 cm^{-1}, respectively. Heating of the crystal leads
to gradually and simultaneously decreasing ν_2 and ν_3 band intensities down to their
complete disappearance at 420 K, but the ν_1 band intensity remains almost constant.
The synchronous change of ν_2 and ν_3 mode intensities is also observed under the κ value
variation (Fig. 2). The ν_1 band dissapears at 470-500 K, and nitrogen desorption is

detected in the same temperature range by the mass-spectrometer. The data obtained permit us to conclude that the ν_2 and ν_3 losses are due to the same particle. The ν_1 band at 480 cm^{-1} may be attributed to the Pt-N stretching mode of dissociatively adsorbed nitrogen [1]. The band at 3280 cm^{-1} is in the region of stretching N-H vibrations 3240-3375 cm^{-1}, so it probably corresponds to a NH$_x$-fragment (x=1;2) of an adsorbed surface species. However, the ν_2 band is not characteristic of a species as (NH$_x$)$_{ads}$. The absence of losses corresponding to δ_s(NH$_3$) vibrations excludes the value of x=3. An overlap of δ(NH$_x$) bending mode region and ν(NO) stretchings [2] prevents determination of x. However, known NH$_2$-containing molecules also show a wagging loss w(NH$_2$) mode at 1270 cm^{-1} [3]. The absence of the latter leads to the conclusion that the new surface species has to be NH. The use of D$_2$ has shown that the ν_2 band does not correspond to bending modes as δ(NH), δ(OH). This band can be assigned to N-O stretching modes with a bond order of about 1. For instance, low NO frequencies in the frequency regime 900-1150 cm^{-1} were measured for ν(NO) modes of NO adsorbed in the side-on state [4]. The ν_2 band probably derives from HNO$_{ads}$, adsorbed in a μ(N,O) state:

$$\overset{\displaystyle H}{\diagdown}\underset{\overset{|}{\text{\textbar}}}{\overset{|}{N}}\text{--}\underset{\overset{|}{\text{\textbar}}}{O}$$

The slower accumulation of the same species, N$_{ads}$ and HNO$_{ads}$, was observed at lower temperatures. TDS data show much more that nitrogen (η_N) desorbs after reaction than after NO adsorption. Figure 2 indicates that η_N is nearly constant in the κ range of 0.2 - 2.5 and decreases abruptly when a large excess of any reactant is present. Increase of the hydrogen content in the reaction mixture induces HNO-to-N ratio increase at constant total nitrogen surface coverage.

The HNO$_{ads}$+H$_2$ interaction was studied by exposure of the adlayer, obtained as in Fig. 1, to hydrogen at 350 K. Additional HNO$_{ads}$ formation by hydrogenation of residual NO$_{ads}$ is found to occur at H$_2$ exposure of ϵ_{H_2}=30L (1 L=1.33·10^{-4} Pa·s)(Fig. 3). The bridge α_1-state of NO$_{ads}$ on Pt(111) is the only reactive one with respect to H$_2$ [3]. Upon further H$_2$ exposure HNO$_{ads}$ coverage is decreased at constant η_N (Fig. 3). This indicates HNO$_{ads}$ to N$_{ads}$ conversion by hydrogen. Moreover, the band of δ_s(NH$_3$) at 1225 cm^{-1} appeared in the loss spectra after $\epsilon_{H_2} \sim 100$ L exposure. This loss is due to the formation of NH$_3$. According to [5], the present shift provides information on such complex formation as [NH$_3$·NO]$_{ads}$. As will be shown below, the same mode of adsorbed ammonia is displaced to 1160 cm^{-1}.

To study the N$_{ads}$+H$_2$ reaction the adlayer was formed as in Fig. 1 and then heated for a short time at 410 K. The initial N$_{ads}$ coverage was estimated to be $\theta_N = 0.125$ ML (monolayer). This layer was exposed to hydrogen at 400 K to prevent the background NO adsorption. An exposure of 0.6 L depletes θ_N down to 0.06 ML. A value of ϵ =1.5 L was enough to remove N$_{ads}$ completely. The usual mechanism of the NO+H$_2$ reaction on platinum group metals is NO dissociation followed by N$_{ads}$ and O$_{ads}$ hydrogenation to the final products. The present paper gives experimental evidence for another NO activation mechanism - hydrogenation of molecular NO$_{ads}$ to form HNO$_{ads}$ species. This kind of mechanism appears to be characteristic for metal surfaces with respect to dissociative NO adsorption like Pt(111).

Data on the NH$_3$+O$_2$ interaction are presented in Fig. 4. The HREELS spectrum (B) of adsorbed ammonia reveals the bands of (NH$_3$)=680, δ_s(NH$_3$)=1160, δd(NH$_3$)=1615 and ν(NH)=3320 cm^{-1}. O$_2$ exposure results in loss of all the bending mode intensities, where as that of ν(NH) enhances up to 2.5 times (Spm C). Moreover, new features are found at 470, 830 and 1530 cm^{-1}. Two strong bands at 830 and 3320

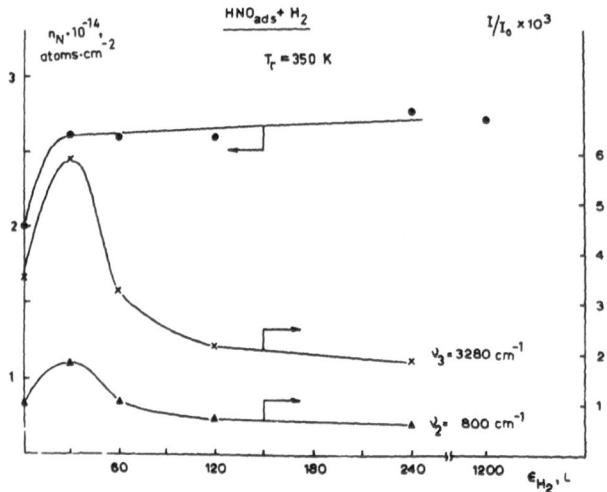

Figure 3

Variation of nitrogen coverage η_N (TDS data) and ν_2, ν_3 loss intensities as a function of hydrogen exposure at 350 K.

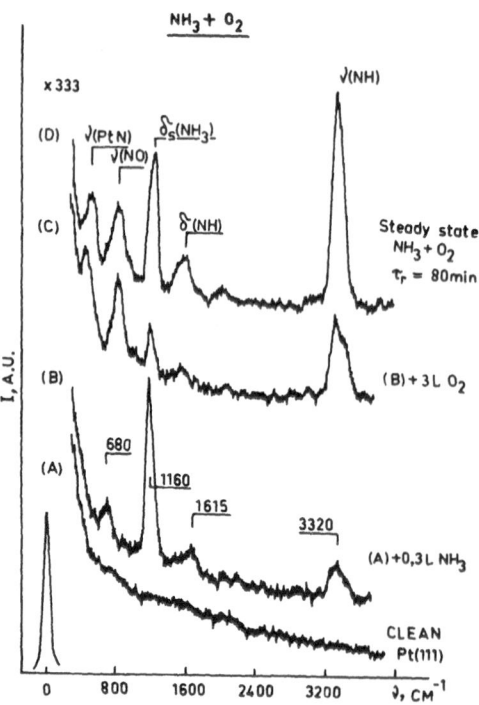

Figure 4

HREELS spectrum of clean Pt(111) surface (A) and spectra obtained after: (B) - NH_3 adsorption over (A) at 300 K; (C)- O_2 adsorption over (B) at 300 K; - (D) the reaction of NH_3+O_2 was allowed to proceed for 80 minutes at 300 K and $P_{NH_3}=P_{O_2}=3.8\cdot10^{-8}$ Pa.

cm^{-1} indicate the formation of HNO_{ads} species, again adsorbed in the side-on state: there are $\nu(NO)$ and $\nu(NH)$ stretchings, respectively. The loss at 470 cm^{-1} is probably due to N_{ads}. The weak band at 1530 cm^{-1} may corresponds also to NO_{ads} stretching modes. However, no NO was detected by TDS. Therefore this is assigned to the bending mode $\delta(NH)$ of the HNO_{ads} species that was mentioned above. The steady state regime of NH_3+O_2 interaction at 300 K and $P_{NH_3}=P_{O_2} = 3.8 \cdot 10^{-8}$ Pa established after the reaction was allowed to proceed for 60 min. The HREEL spectrum reveals the same loss positions and some changes of band intensities (Fig. 4). The weak band of $\nu(CO)$ at 2040 cm^{-1} indicates CO_{ads} accumulation in the spectrum of the experiment up to $\theta_{CO} \sim 5 \cdot 10^{-3}$ ML. The dynamics of the change of spectrum (D) as a function of sample heating is very simular to that of the case of the NO+H$_2$ interaction.

The data obtained allow the conclusion that both reactions, NO+H$_2$ and NH$_3$+O$_2$ on Pt(111), proceed through the same intermediates, namely N_{ads} and HNO_{ads}.

References

1. M.Yu. Smirnov, V.V. Gorodetskii, Poverkhnost(USSR) **9**, (1989) 68.
2. M.Yu. Smirnov, V.V. Gorodetskii, A.R. Cholach, Izv. AN SSSR ser.phys.(USSR) **52**, (1988) 1558.
3. G.W. Watt, B.B. Hutchinson, D.S. Klett, J. Am. Chem. Soc. **89**, (1967) 2007.
4. U. Schwalke, W.H. Weinberg, J.Vac.Sci.Technol. **A5**, (1987) 459.
5. J.L. Gland, B.A. Sexton, J. Catal. **68**, (1981) 286.

USE OF NEUTRON BEAMS IN STUDYING VIBRATIONAL MODES OF MOLECULES

ADSORBED ON CATALYSTS

H. Jobic

Institut de Recherches sur la Catalyse
2 avenue A. Einstein
69626 – Villeurbanne, France

1. INTRODUCTION

When one looks over the different types of instruments available around a high flux reactor (e.g. ref. 1), it immediately appears that neutron based techniques cover a wide range of applications. There are two main classes of instruments : elastic scattering instruments which are used for determining the structure of materials, and inelastic scattering spectrometers which, by measuring energy transfers, give information on atomic and molecular motions. Therefore, leaving aside fundamental and nuclear physics, the applications of neutron scattering to condensed matter include crystal and magnetic structures, molecular dynamics, disordered materials, colloids, polymers, structural excitations, biology, etc.

Several of these neutron based techniques have been applied in catalysis research. The structure of hydrogeneous layers and of molecules adsorbed in zeolites has been determined by neutron diffraction. Textural or clustering studies have been performed by small angle neutron scattering (SANS). The rotational and translational motions of molecules have been characterized by quasi–elastic neutron scattering (QENS). Finally, the vibrational modes of catalysts and of adsorbed molecules have been measured by inelastic neutron scattering (INS). It is this last application which will be mainly discussed here.

Vibrational spectroscopic techniques have proved to be very useful for the understanding of surface and catalysis phenomena. It has been possible with these techniques to identify adsorbed species and to characterize their structure and bonding with the surface. If transmission infrared has long been the only vibrational spectroscopy to be used for surface and catalysis studies, in the last 15 years or so other spectroscopic methods have emerged, e.g. electron energy loss (EELS), inelastic electron tunneling, Raman, reflection–absorption infrared and INS. Since none of these techniques can answer to all questions, they should be viewed as complementary. Each has its advantages for a particular system, i.e. spectral domain, resolution, sensitivity and experimental conditions.

Although neutrons are not be themselves a surface probe, it is possible to obtain vibrational spectra of surfaces provided that the adsorbate has a larger cross

Fundamental Aspects of Heterogeneous Catalysis Studied by Particle Beams,
Edited by H.H. Brongersma and R.A. van Santen, Plenum Press, New York, 1991

section than the substrate. Large surface area materials are required but such conditions usually prevail in catalytic systems. There are a number of other features which can be associated with INS :

(1)   there are no "selection rules" so that the whole vibrational density of states can be measured,
(2)   with present neutron spectrometers, the whole spectral range $1-4000$ cm^{-1} can be observed,
(3)   optically opaque materials can be almost transparent for neutrons,
(4)   the large incoherent cross section of protons makes INS very sensitive to hydrogen motions, but the scattering from other atoms can be measured,
(5)   it is possible to compute both the frequencies **and** the intensities of the vibrational transitions, whereas other spectroscopies tend to rely only upon frequencies (this compensates in a way for the poorer resolution of INS).

Several reviews on inelastic neutron scattering from adsorbates have been written, e.g. refs. 2-6. Therefore only a few selected examples will be described here, with some emphasis on the role of the Debye-Waller factor.

2. THEORY OF NEUTRON SCATTERING

2.1. Introduction

The neutron wavelength is given by the de Broglie relationship :

$$\lambda = \frac{h}{mv} = \frac{h}{p} = \frac{2\pi}{|\underline{k}|} \tag{1}$$

where h is Plank's constant, m the neutron mass, v its velocity, p its momentum and $\underline{k}$ the wavevector.

The neutron energy is given by :

$$E = \frac{1}{2} mv^2 = \frac{\hbar^2 k^2}{2m} \tag{2}$$

The two measured quantities in an INS experiment, apart from the intensity, can then be defined. These are the neutron energy transfer

$$\hbar\omega = E_o - E' = \frac{\hbar^2}{2m} (k_o^2 - k'^2) \tag{3}$$

and the neutron momentum transfer

$$\hbar\underline{Q} = \hbar(\underline{k}_o - \underline{k}') \tag{4}$$

where E_o ($\underline{k}_o$) and $E'(\underline{k}')$ are respectively the incident and scattered neutron energies (wavevectors).

Some useful neutron properties :

mass 1.67×10^{-24}g
charge   0
spin   1/2
magnetic moment $\mu = -1.913$ nuclear magnetons

The unit which is the most used in INS experiments is the meV, it corresponds to 8.065 cm^{-1}.

Table 1. Bound coherent scattering lengths and coherent, incoherent and absorption cross sections (1 barn = $10^{-28}m^2$) for some common elements (σ_{abs} is proportional to λ, it corresponds here to $\lambda = 1$ Å).

element	$<b>$ $(10^{-14}m)$	σ_{coh} (barns)	σ_{inc} (barns)	σ_{abs} (barns)
H	– 0.37	1.76	79.9	0.19
D	0.67	5.6	2.04	0
^{3}He	0.57–0.15i	4.42	1.2	2960.
B	0.53–0.02i	3.54	1.7	426.
C	0.66	5.55	0	0
N	0.94	11.01	0.49	1.1
O	0.58	4.23	0	0
Na	0.36	1.66	1.62	0.3
Mg	0.54	3.63	0.08	0.04
Al	0.35	1.5	0	0.13
Si	0.42	2.16	0	0.1
S	0.28	1.02	0	0.3
Cl	0.96	11.53	5.2	18.6
Ti	– 0.33	1.37	2.67	3.4
V	– 0.04	0.02	5.18	2.82
Cr	0.36	1.66	1.83	1.7
Fe	0.95	11.44	0.39	1.4
Co	0.25	0.79	4.8	20.6
Ni	1.03	13.3	5.2	2.5
Zn	0.57	4.05	0.08	0.6
Mo	0.7	6.07	0.28	1.4
Pd	0.59	4.39	0.09	3.8
Cd	0.51–0.07i	3.3	2.4	1400.
Pt	0.96	11.65	0.13	5.7

2.2. Scattering cross section, coherent and incoherent scattering

In INS experiments, the quantity which is measured is the double–differential cross section, $d^2\sigma/d\Omega dE$, which represents the number of neutrons scattered into a solid angle $d\Omega$ and with energy in the range dE. This quantity can be calculated from scattering theory for different scattering processes. A number of text–books and review articles describe well the theory of neutron scattering (7–10), only the most important features will be described here.

The interaction of neutrons with matter occurs via nuclear forces (magnetic interactions with unpaired electrons will be ignored). When a beam of neutrons impinges on a sample, the neutrons can be scattered or absorbed or transmitted without interaction. The scattering can be either coherent or incoherent, depending on whether there is a fixed phase relationship between the scattered waves or not.

The neutron–nucleus interaction potential V(r) is known to be rather strong and of very short range ($\approx 10^{-15}$ m). For neutron scattering, the Fermi pseudo-potential is used, it consists of a delta function and of a single parameter, b, the scattering length :

$$V(\underline{r}) = \frac{2\pi\hbar^2}{m} \, b \, \delta(\underline{r} - \underline{R}) \tag{5}$$

where r and R are the neutron and nucleus positions.

This potential allows the scattered wave to be isotropic (which is not the case with X–rays), the phase is determined by the sign of the scattering length which is usually real and positive but b can be complex if there is neutron absorption.

In a real experiment, the sample contains different elements, averages have then to be performed over all isotopes and spin distributions. The total cross section per scatterer corresponds to :

$$\sigma = \int \frac{d\sigma}{d\Omega} \, d\Omega = \int \int \frac{d^2\sigma}{d\Omega dE} \, d\Omega dE = 4\pi \, \langle b^2 \rangle \tag{6}$$

where the bracket <> denotes an average value.

If we consider first the case of a single isotope of nuclear spin i, it may combine with a neutron of spin 1/2 to give two possible spin states : $i + 1/2$ and $i - 1/2$. There are two scattering lengths associated with the two states : b^+ and b^-. Since for a total spin quantum number t there are $2t + 1$ states, there will be $2(i + 1/2) + 1 = 2(i + 1)$ states for b^+ and $2(i - 1/2) + 1 = 2i$ states for b^-. The total number of states is $2(i + 1) + 2i = 2(2i + 1)$ so that the probability of b^+ occuring is $p^+ = (i + 1)/(2i + 1)$ and of b^- is $p^- = i/(2i + 1)$. This verifies the normalization property :

$$p^+ + p^- = 1 \tag{7}$$

The average over spin states is :

$$\langle b \rangle = p^+ \, b^+ + p^- \, b^- \tag{8}$$

This gives rise to coherent scattering with a cross section

$$\sigma_{coh} = 4\pi \, \langle b \rangle^2 = 4\pi \, |p^+ \, b^+ + p^- \, b^-|^2 \tag{9}$$

The incoherent scattering corresponds to the mean square deviation from the mean potential, it is therefore given by

$$\sigma_{inc} = 4\pi \, (\langle b^2 \rangle - \langle b \rangle^2) \tag{10}$$

with $\quad \langle b^2 \rangle = p^+ \, |b^+|^2 + p^- \, |b^-|^2 \tag{11}$

The scattering of neutrons by hydrogen atoms ($i = 1/2$) is a useful example. The scattering lengths which have been determined experimentally are $b^+ = 1.082 \times 10^{-14}$ m and $b^- = -4.742 \times 10^{-14}$ m so that
$$\langle b \rangle = 3/4 \, b^+ + 1/4 \, b^- = -0.374 \times 10^{-14} \text{ m}$$
which gives $\quad \sigma_{coh} = 1.76$ barn $\quad$ (1 barn $= 10^{-28}$ m^2)
and $\quad \sigma_{inc} = 79.9$ barn

When the sample consists of several isotopes, more general expressions are calculated by taking into account the concentration c_p of each isotope p, with scattering length b_p :

$$\langle b \rangle = \sum_p \frac{c_p}{2i_p + 1} \left((i_p + 1)b_p^+ + i_p \, b_p^- \right) \tag{12a}$$

$$\langle b^2 \rangle = \sum_p \frac{c_p}{2i_p + 1} \left((i_p + 1)(b_p^+)^2 + i_p \, (b_p^-)^2 \right) \tag{12b}$$

Some scattering lengths and cross sections are given in Table 1 for common elements. For inelastic scattering studies of adsorbates, it appears that hydrogen or hydrogen containing molecules are the best candidates because of the high incoherent cross section of hydrogen, but other gases adsorbed on different substrates (mainly physisorbed layers) have been studied (3).

Note that at the high Q values often involved in INS, coherent effects can be neglected and the coherent cross section can be added to the incoherent one, this is called the incoherent approximation.

2.3. Differential cross section corresponding to the vibrational motions

The neutron can exchange energy with the sample and excite transitions between vibrational levels, this produces a change in energy of the incident neutron wave. The double differential cross section can be derived from perturbation theory and it is related to matrix elements of the form $\langle f| \exp(i\underline{Q}.\underline{R}_d)|i \rangle$, where $|i\rangle$ and $|f\rangle$ are the initial and final states and $\underline{R}_d$ the position of the d th nucleus (9). Just as the scattering cross section can be split into coherent and incoherent contributions, the double-differential cross section can be written :

$$\frac{d^2\sigma}{d\Omega dE} = \frac{k'}{k_o} \frac{N}{4\pi} \left[\sigma_{inc} \, S_{inc}(\underline{Q},\omega) + \sigma_{coh} \, S_{coh}(\underline{Q},\omega) \right] \tag{13}$$

where $S(\underline{Q},\omega)$ is called the scattering law. Since hydrogenated compounds are often studied in chemisorption, the main term in Eq. (13) will be the incoherent one. The incoherent scattering law then depends only on the self-motion of the atoms :

$$S_{inc}(\underline{Q},\omega) = \frac{1}{N} \sum_{i,f} P_i \sum_d |\langle f|\exp(i\underline{Q}.\underline{R}_d)|i\rangle|^2 \delta(\hbar\omega + E_i - E_f) \tag{14}$$

where P_i is the probability of occupation of the initial state i, the delta function representing the condition of conservation of energy.

This formalism is also found in other spectroscopies, the main difference lying in the matrix elements. In infrared, the matrix elements are $\langle f|\underline{e}.\underline{\mu}|i\rangle$ where $\underline{e}$ is a unit vector of the incident electromagnetic field and μ the electric dipole moment. In Raman, the matrix elements are $\langle f|\underline{e}_o.\underline{\underline{\alpha}}.\underline{e}'|i\rangle$ where $\underline{\underline{\alpha}}$ is the polarisability tensor. This gives rise to selection rules in optical spectroscopies. In neutron scattering, a method to compute the matrix elements is through correlation functions, using in Eq. (14) time-dependent Heisenberg operators for the nuclei position vectors and the Fourier representation of the delta function :

$$S_{inc}(\underline{Q},\omega) = \frac{1}{2\pi\hbar N} \int_{-\infty}^{\infty} dt \, \exp(-i\omega t) \sum_d \langle \exp\{-i\underline{Q}.\underline{R}_d(0)\} \exp\{ i\underline{Q}.\underline{R}_d(t)\} \rangle \tag{15}$$

where <> denotes a thermal average. The correlation function in Eq. (15) is called the intermediate scattering function, $I_s(Q, t)$, which is defined as :

$$I_s(\underline{Q}, t) = \frac{1}{N} \sum_d \langle \exp\{-i\underline{Q}\cdot\underline{R}_d(0)\} \ \exp\{ i\underline{Q}\cdot\underline{R}_d(t)\}\rangle \qquad (16)$$

This correlation function describes the decay in time of single particle fluctuations of momentum transfer $\underline{Q}$ (hence the subscript s which stands for self).

2.3.1. Scattering by a nucleus in a harmonic oscillator potential.

In the case of a nucleus of mass m bound harmonically to a fixed point, the function $I(Q, t)$ is found to correspond to (8) :

$$I(Q, t) = \exp\{-\frac{\hbar Q^2}{2m\omega_o} \coth(\frac{\hbar\omega_o}{2k_B T})\}$$

$$x \ \exp\left[\frac{\hbar Q^2}{4m\omega_o \sinh(\frac{\hbar\omega_o}{2k_B T})} \{\exp(i\omega_o t + \frac{\hbar\omega_o}{2k_B T})+\exp(-i\omega_o t - \frac{\hbar\omega_o}{2k_B T})\}\right] \qquad (17)$$

The potential is isotropic and ω_o is the frequency of vibration. In order to perform the Fourier transform in Eq. (15), the second exponential in Eq. (17) can be replaced by the modified Bessel function of the first kind, $I_n(y)$, since it has the form of the generating function :

$$\exp\{\frac{y}{2}(x + \frac{1}{x})\} = \sum_{n=-\infty}^{\infty} x^n I_n(y) \qquad (18)$$

The scattering law corresponds finally to :

$$S(Q, \omega) = \exp\left[-2W(Q)\right] \sum_{n=-\infty}^{\infty} \exp(\frac{n\hbar\omega_o}{2k_B T}) \ I_n(y) \ \delta(\hbar\omega - n\hbar\omega_o) \qquad (19)$$

$$\text{where} \qquad W(Q) = \frac{\hbar Q^2}{4m\omega_o} \coth(\frac{\hbar\omega_o}{2k_B T}) \qquad (20)$$

$$\text{and} \qquad y = \frac{\hbar Q^2}{2m\omega_o \sinh(\frac{\hbar\omega_o}{2k_B T})} \qquad (21)$$

The spectrum consists of a series of δ functions corresponding to transitions of n quanta of energy $\hbar\omega_o$. The system can be excited (n >0, i.e. neutron energy loss) or deexcited (n<0) in the scattering process. The factor $\exp(n\hbar\omega_o/2k_B T)$ ensures that at a given temperature, the processes in which the neutron loses energy have a higher intensity than those for which it gains energy (note that $I_n(y) = I_{-n}(y)$).

The elastic intensity (n = 0) is governed by the Debye–Waller factor $\exp[-2W(Q)]$. This factor was first derived for X–ray scattering, its influence was later studied in the case of electron scattering and atomic scattering (e.g. ref. 11).

2.3.2. Scattering law for intramolecular vibrations

The calculation performed for the harmonic oscillator can be extended to polyatomic molecules to derive the intensity of the normal modes of vibration.

If one considers a molecular crystal, the low frequency intermolecular modes (lattice modes) are usually decoupled from the high frequency intramolecular modes so that the position vector $\underline{R}_d$ of the atom d in the Lth molecule can be written as :

$$\underline{R}_d = \underline{u}_L + \underline{u}_d \qquad (22)$$

where $\underline{u}_L$ describes the rigid–body motion of the molecule and $\underline{u}_d$ the displacement of a hydrogen atom d from the equilibrium position under the effect of vibrations. Since these two motions are uncoupled, the thermal averages in Eq. (16) can be performed separately and the intermediate scattering function can be written :

$$I_s(\underline{Q},t) = I_s^L(\underline{Q},t)\ I_s^M(\underline{Q},t) \qquad (23)$$

$$\text{with } I_s^L(\underline{Q},t) = \frac{1}{N_L} \sum_L\ \langle \exp\{-i\underline{Q}\cdot\underline{u}_L(0)\}\ \exp\{\ i\underline{Q}\cdot\underline{u}_L(t)\}\rangle \qquad (24a)$$

$$\text{and } I_s^M(\underline{Q},t) = \frac{1}{n_H} \sum_d\ \langle \exp\{-i\underline{Q}\cdot\underline{u}_d(0)\}\ \exp\{\ i\underline{Q}\cdot\underline{u}_d(t)\}\rangle \qquad (24b)$$

where N_L defines the number of molecules and n_H the number of hydrogen atoms in each molecule. Only the translational and rotational degrees of freedom of a molecule are involved in $I_s^L(\underline{Q},t)$.

The molecular part, $I_s^M(\underline{Q},t)$, has been discussed in detail by Zemach and Glauber (12). The molecular vibrations are resolved into normal modes and the displacement vector $\underline{u}_d$ is expressed in terms of the normal coordinates q_λ :

$$\underline{u}_d(t) = \sum_\lambda\ \underline{C}_d^\lambda\ q_\lambda(t) \qquad (25)$$

where $\underline{C}_d^\lambda$ is the amplitude vector for the d th proton in the λ th normal mode, in mass–weighted cartesian coordinates. Since for a harmonic system the normal modes are dynamically independent, exponential functions of $\underline{u}_d(t)$ can be factored into products of exponentials :

$$\exp\{i\underline{Q}\cdot\underline{u}_d(t)\} = \prod_\lambda \exp\{i\underline{Q}\cdot\underline{C}_d^\lambda\ q_\lambda(t)\} \qquad (26)$$

The derivation of the scattering cross section then follows the same calculation as for the harmonic oscillator : the intermediate scattering function, Eq. (24b), is separated into a product of mean values and the Fourier transform yields the following scattering law :

$$S_s^M(\underline{Q},\omega) = \sum_d \exp\left[-2W_M^d(\underline{Q})\right] \prod_\lambda \left[\sum_{n_\lambda} \exp\left(\frac{n_\lambda \hbar\omega_\lambda}{2k_B T}\right) I_{n_\lambda}\left(\frac{\hbar|\underline{Q}\cdot\underline{C}_d^\lambda|^2}{2m_d\omega_\lambda \sinh\left(\frac{\hbar\omega_\lambda}{2k_B T}\right)}\right) \delta\left(\hbar\omega - \sum_\lambda n_\lambda \hbar\omega_\lambda\right) \right] \qquad (27)$$

The total spectrum is made up of series of delta functions, which indicates that transitions can occur in the sample whenever the neutron energy transfer equals some combination of quanta of energy $n_\lambda \hbar\omega_\lambda$.

The Debye–Waller factor in Eq. (27) includes only the mean–square displacement $<u^2>$ due to the molecular modes :

$$\exp\left[-2W_M^d\,(\underline{Q})\right]=\exp\left[-Q^2\,<\underline{u}_d^2>_M\right]\ =\ \exp\left\{-\ \frac{\hbar}{2\ m_d}\ \sum_\lambda\ \frac{|\underline{Q}.\underline{C}_d^\lambda|^2}{\omega_\lambda}\ \coth(\ \frac{\hbar\omega_\lambda}{2k_BT}\ )\ \right\} \qquad (28)$$

This factor is expected to reduce the intensity at large Q values involved in high energy transfers. Note that $<u_d^2>$ is non zero at $T = 0$ because of zero–point motion.

In most experiments, the argument of the Bessel function in Eq. (27) is small compared with unity, so that only the first term in its power expansion has to be considered :

$$I_{n_\lambda}(y)\ =\ \frac{1}{|n_\lambda|!}\ (\frac{y}{2})^{|n_\lambda|} \qquad (29)$$

2.3.3. Scattering for a fundamental mode

The intensity of a fundamental in neutron energy loss ($n_\lambda = 1$) will be given by :

$$S_s^M\,(\underline{Q},\omega)=\sum_d\exp\left[-2W_M^d\,(\underline{Q})\right]\ \frac{1}{1-\exp(\ -\dfrac{\hbar\omega_\lambda}{k_BT}\ )}\ \frac{\hbar\,|\,\underline{Q}.\underline{C}_d^\lambda\,|^2}{2m_d\,\omega_\lambda}\ \delta\,(\hbar\omega-\hbar\omega_\lambda) \qquad (30)$$

Most of the characteristics associated with INS can be understood from this equation :
- there are no considerations on the symmetry of the vibrations so that all modes can be a priori observed,
- neutrons   are very sensitive to hydrogen because of its large incoherent cross section, σ_H, and its low mass, m_H. Therefore when the concentration of hydrogen in the sample is not too low, the sum over the atoms $\sum_d$ can be restricted to hydrogen atoms,
- the vibrational modes which involve large amplitude hydrogen motions, $\underline{C}_d^\lambda$, will have high intensities,
- for a powder sample, an average over all possible orientations of $\underline{C}$ relative to $\underline{Q}$ must be performed. In anisotropic systems, this can be calculated using analytical methods (13,14), but for most samples one can replace $|\underline{Q}.\underline{C}_d^\lambda|^2$ by $(Q^2|\underline{C}_d^\lambda|^2)/3$ to a good approximation. For a crystal however, polarization effects can be measured (15),
- even if the largest intensity is obtained with hydrogen, the scattering from other atoms can be measured. For example, selective deuteration will not only shift the vibrational frequencies but also decrease the intensity of the modes involved because deuterium has a smaller cross section,
- information on the vibrational modes can be obtained from the frequencies and from the intensities. For a molecular compound, these two quantities can be obtained from a normal coordinate analysis.

2.3.4. Scattering for overtones and combination bands

In general, the fundamentals will be the most intense features of the spectrum but overtones and combination bands can also be found in the INS spectrum because of the product $\prod\limits_{\lambda}$ in Eq. (27). Their intensity will be governed by factors of the form :

$$\left(\frac{\hbar \, |\underline{Q} \cdot \underline{C}_d^\lambda|^2}{4 m_d \, \omega_\lambda \, \sinh(\frac{\hbar \omega_\lambda}{2 k_B T})} \right)^{|n_\lambda|} \tag{31}$$

The intensity due to these features will be minimized at low temperatures and small momentum transfers. These terms usually give rise to a structureless background when the number of quanta, n_λ , is large.

2.3.5. Total incoherent scattering law for molecular systems, influence of the Debye–Waller factor

The total incoherent scattering law can be written under the form of a convolution product, since it is the Fourier transform of a product of two terms (Eq. 23) :

$$S_{inc}(\underline{Q},\omega) = S_S^M(\underline{Q},\omega) \otimes S_S^L(\underline{Q},\omega) \tag{32}$$

If one is studying the intensity around a fundamental at frequency ω_λ , the convolution yields :

$$S_{inc}(\underline{Q},\omega) = \sum_d \frac{\exp\left[-2W_M^d(\underline{Q})\right]}{1-\exp(-\frac{\hbar\omega_\lambda}{k_B T})} \frac{\hbar |\underline{Q} \cdot \underline{C}_d^\lambda|^2}{2m_d \omega_\lambda} S_S^L(\underline{Q}, \hbar\omega - \hbar\omega_\lambda) \tag{33}$$

Therefore the line shape around the frequency ω_λ is given by the lattice scattering law $S_S^L(\underline{Q},\omega)$. For high Q, at room temperature, and with intermediate resolution, a Gaussian shape for $S_S^L(\underline{Q},\omega)$ was a good approximation (16,17). However with the higher resolution spectrometers which have been recently built (18–20) several side bands due to the lattice modes can be resolved from the fundamentals, even for intermediate values of Q or T. In this case, the measured phonon density of states can be used instead of actually computing the lattice scattering law through a phonon expansion of $I_S^L(\underline{Q},t)$, Eq. (24a). Fourier transform has been performed numerically for a simple system (17) but this method is complicated and cannot be applied to large systems. This lattice scattering law contains another Debye–Waller factor, $\exp(-2W_L)$, which drains further intensity from the internal modes. The intensity which is taken from the fundamentals is redistributed into side bands (or phonon wings) as a function of Q and T. For stiff lattices, i.e. for large Debye–Waller factors, the phonon wings can be simulated with one and two-phonon contributions (21). However, when the lattice is softer, it has been found that multiphonon lattice modes up to the eighth term needed to be considered (22).

To illustrate the redistribution of the intensity of a fundamental as a function of Q and T, a simulation is shown in Fig. 1 for two modes ω_1 and ω_2 . The simulation is performed at two temperatures : 5 K and 100 K, as a function of energy (for most INS spectrometers ω is proportional to Q^2). A typical lattice density of states and Debye–Waller factor, $\exp(-2W_L)$, corresponding to a stiff lattice has been considered, and multiphonon terms are calculated up to the second order. This simulation clearly shows the necessity to work at very low temperture to observe the fundamental vibrations of molecular crystals. At 100 K, phonon wings can be found on both sides of the zero–phonon peak because both phonon creation and annihilation can occur. The lattice Debye–Waller factor governs the intensity of the zero–phonon peak, I_z, to the total intensity, $I_z + I_L$, where I_L is the intensity of the side bands :

$$\exp(-2W_L) = I_z / (I_z + I_L) \tag{34}$$

When the total Debye–Waller factor is larger, e.g. for an hydrogen atom bonded to a "rigid" framework (metal or zeolite), it is not necessary to go down to very low temperatures, measurements can be made at room temperature.

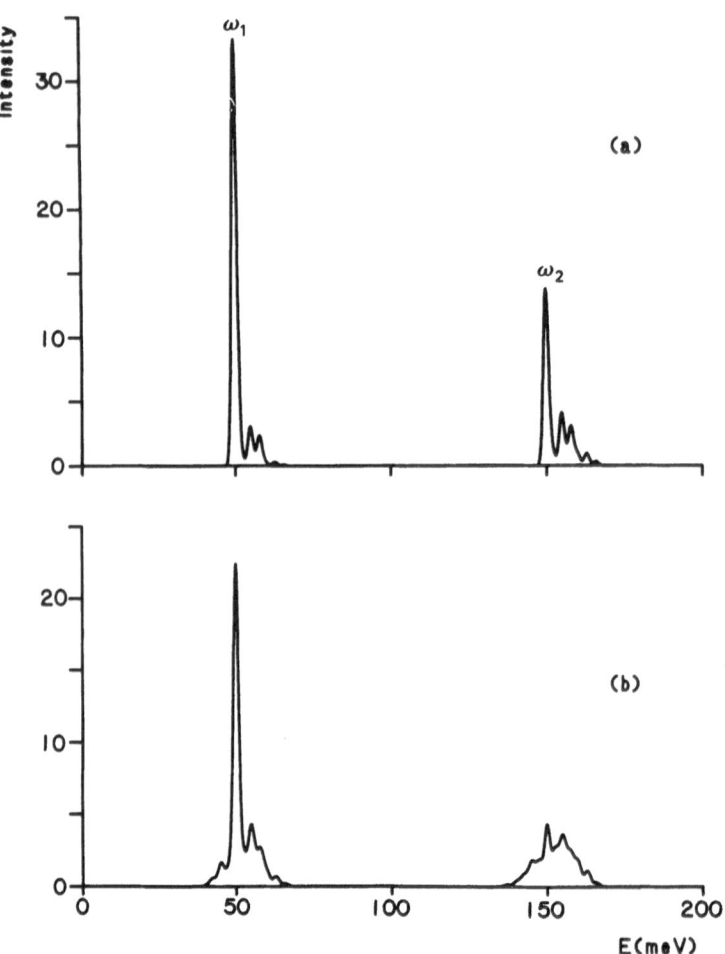

Fig. 1 . Simulation, in the case of a stiff lattice, of the redistribution of the intensity for two fundamentals ω_1 and ω_2, at two temperatures (a) 5 K and (b) 100 K

2.3.6. Recoil scattering

If the Debye–Waller factor due to the external modes is very small, which corresponds to large lattice motion, recoil scattering can be observed. This may happen either for a soft lattice at low temperature or for a stiff lattice at high Q and high T. In this case, there is no intensity left at the frequency of the internal modes, a large number of multiphonon are excited and produce a broad band shifted to higher frequency by the recoil energy, E_R, which corresponds to :

$$E_R = \frac{\hbar Q^2}{2\overline{M}} \tag{35}$$

where $\overline{M}$ is the effective mass of the molecule which combines the translational and rotational motions of a molecule in a perfect gas, when the neutron energy is much larger than the difference in rotational levels. This effective mass can be obtained after averaging the Sachs–Teller inverse mass tensor (23, 24). The effective mass for a given scattering atom corresponds to :

$$\frac{1}{\overline{M}} = \frac{1}{M} + \frac{1}{3} \left(\frac{r_j^2 + r_k^2}{I_i} + \frac{r_i^2 + r_k^2}{I_j} + \frac{r_i^2 + r_j^2}{I_k} \right) \tag{36}$$

where M is the molecular mass and (r_i, r_j, r_k) are the coordinates of the atom in the system of the principal axes of inertia (I).

The line shape of the recoiled band corresponds to a Gaussian whose width is related to the average kinetic energy, $<E_{kin}>$, of the recoiling molecule (25) :

$$\Gamma = \left(\frac{4}{3} <E_{kin}> E_R \right)^{1/2} \tag{37}$$

Such measurements can therefore provide a direct measure of the rotational and translational kinetic energies of various systems (26–30).

2.3.7. Derivation of the displacement vectors from a force field calculation

Several methods can be used to compute the displacement vectors c_d^λ which govern the vibrational intensities, Eq. (30). When the molecule contains more than 10 atoms, empirical methods are used most often.

A force field calculation, following the treatment of Wilson et al. (31), can be used to derive the vibrational frequencies and the atomic displacements by solving the secular equation :

$$\underline{\underline{G}}\,\underline{\underline{F}}\,\underline{\underline{L}} = \underline{\underline{L}}\ \underline{\underline{\Lambda}} \tag{38}$$

where $\underline{\underline{G}}$ is the inverse kinetic energy matrix, $\underline{\underline{F}}$ the matrix of the force constants, $\underline{\underline{L}}$ the eigenvector matrix and $\underline{\underline{\Lambda}}$ is a diagonal matrix of eigenvalues directly related with the vibrational frequencies.

Therefore, knowing the geometry of the molecule and its force field, the vibrational frequencies and normal modes can be derived.

The problem is usually set up in internal coordinates, r, e.g. bond stretching, angle bending and torsion. The internal and cartesian coordinates are related by a linear transformation :

$$\underline{r} = \underline{\underline{B}}\,\underline{x} \tag{39}$$

The displacement vectors, in mass–weighted cartesian coordinates, are calculated through :

$$\underline{\underline{C}} = \underline{\underline{M}}^{-1/2}\,\underline{\underline{B}}^T\,(\underline{\underline{L}}^{-1})^T \tag{40}$$

where $\underline{\underline{M}}$ is a diagonal matrix containing the masses of the atoms, ^{T} denotes the transpose of a matrix and $\underline{\underline{L}}^{-1}$ is the inverse of $\underline{\underline{L}}$.

The displacement vectors are normalized for each normal mode :

$$\sum_d (\underline{c}_d^\lambda)^2 = 1 \tag{41}$$

The vibrational analysis usually consists in measuring vibrational frequencies and assigning these frequencies to particular modes of vibration. Then the $\underline{\underline{G}}$ matrix is calculated and the initial force field, which contains several adjustable parameters, is refined to give the best fit to the experimental frequencies. Since for a molecule with N atoms, there are 3N−6 vibrational degrees of freedom, for a non-linear molecule, the force constant matrix will have 1/2 (3N−6)(3N−5) different terms ($\underline{\underline{F}}$ is symmetric). However there are only 3N−6 experimental frequencies so that the final force field will not be unique. The number of parameters can be reduced using the symmetry elements of the molecule, Coriolis coupling constants (for gaseous substances), or isotopic derivatives. More commonly, approximations are made by fixing certain force constants arbitrarily, by making the assumption that the high frequencies do not mix at all with the low frequencies or by transferring force constants between different molecules for identical groups.

Several types of force fields have been proposed :
(i) the valence force field, VFF, where the forces considered are those which resist the extension or compression of valence bonds, together with those which oppose the bending or torsion of bonds (32),
(ii) the Urey−Bradley force field, UBFF, which contains terms representing non-bonding interactions between atoms (33),
(iii) the consistent force field, CFF, which is aimed at reproducing vibrational, structural and thermodynamic data (34), etc...

3. EXPERIMENTAL

Neutrons can be produced from steady−state reactors (e.g. the high flux reactor at the Institut Laue−Langevin, Grenoble, France, ref. 1) and also from pulsed sources (e.g. the spallation source ISIS at the Rutherford Appleton Laboratory, Harwell, U.K., ref. 35). Neutrons issued from the source are slowed down in moderators whose temperatures can vary between 25 and 2000 K. This gives neutron wavelengths typically ranging from 0.2 to 20 Å. The number and diversity of neutron instruments being very large, only two types of spectrometers which are used to measure vibrational spectra will be described here.

3.1. Filter detector spectrometer

The intramolecular vibrations are often measured on a beryllium filter detector spectrometer. For example, the spectrometer INFB, at the Institut Laue−Langevin, is used for measuring energy transfers between 12 and 500 meV (100−4000 cm^{-1}). A schematic view of this instrument is shown in Fig. 2. On this spectrometer, the incident energy is made monochromatic by Bragg reflection from copper crystals. Three different planes are available : (200), (220) and (331), according to the desired range of energy transfer and resolution. A spectrum is obtained by varying stepwise the incident energy (typically by 1 meV). The spectrum is normalized by counting at each energy for a constant number of incident neutrons detected by a monitor placed before the sample. The sample and analyser tables slide on air cushions over a marble floor during the energy scan, to follow the neutron beam coming from the monochromator (2 θ_M). The beryllium filter, placed before the detector, transmits only those scattered neutrons which have lost almost all their energy, due to a vibrational mode. Neutrons which have energies higher than the cut-off energy of beryllium, $E_C \approx 5$ meV ($\equiv 4$ Å), are scattered out of the beam. The beryllium block is cooled to 80 K to decrease the inelastic scattering and thus increase the transmission of the filter.

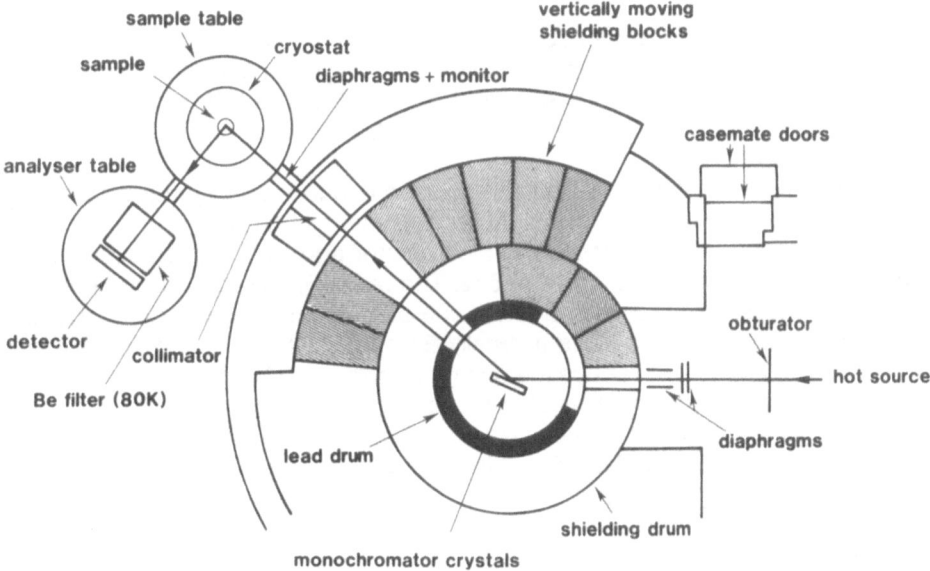

Fig. 2 . Schematic view of the beryllium filter detector spectrometer INFB (ILL).

This corresponds to a neutron energy loss technique. The scattered neutrons have very small energies compared to the incident energy so that the energy transfer, Eq. (3), is to a good approximation :

$$\hbar\omega = \frac{\hbar Q^2}{2m} \qquad (42)$$

Therefore, the neutron momentum transfer is not constant on this type of spectrometer, large energy transfers will imply large momentum transfers. There are however a few simplifications which can be made in the double-differential cross section. Firstly, the scattered wavevector $|\underline{k}'|$ is constant. Secondly, the monitor efficiency being inversely proportional to the incident neutron velocity, this eliminates the k_o dependence in Eq. (13). Finally, the Q^2 and ω terms cancel in the scattering law for fundamentals because of Eq. (42), and the exponential term, $\exp(-\hbar\omega_\lambda/k_B T)$, is very small for energy transfers large compared to $k_B T$.

If the influence due to the lattice Debye–Waller factor can be neglected, i.e. if most of the intensity is due to fundamentals, and introducing the atomic amplitude weighted density of states (the modes are assumed to have no dispersion) :

$$G_d(\omega) = \frac{1}{3N} \sum_\lambda |c_d^\lambda|^2 \, \delta(\omega - \omega_\lambda) \tag{43}$$

the measured spectrum will then correspond to :

$$\frac{d^2\sigma}{d\Omega dE} \propto \sum_d \frac{\sigma_d}{m_d} \exp(-2\,W_d)\;G_d(\omega) \tag{44}$$

In the case of a hydrogeneous compound and neglecting the Debye–Waller factor, it appears that the spectrum measured on this type of spectrometer is closely related to the hydrogen amplitude weighted density of states. The vibrational peaks will however be shifted up in energy by $\approx 4\,\mathrm{meV}$, because of the beryllium filter response function.

Below 100 meV, the instrumental resolution is dominated by the bandpass of the Be filter (the FWHM is $\approx 4\,\mathrm{meV}$). At higher energy transfers, the resolution is limited by the mosaic spread of the monochromator and by the collimation ; in this energy range $\Delta E/E$ is of the order of 5 %. The resolution of this spectrometer can be improved by using a filter with a lower cut-off (18). Another solution consists in detecting neutrons using two filters (typically Be and BeO) and taking the difference between the two spectra, to reduce the energy bandpass (36).

This type of spectrometer is also found on pulsed neutron sources (e.g. ref. 19). In this case, there are no movable parts, one employs a white incident beam and the scattered neutrons are selected at a fixed energy by an analyser crystal, a time-of-flight analysis giving the energy transfer.

3.2. Time–of–flight spectrometer

Time–of–flight spectrometers are best suited for low-energy transfers (typically $0.01-100\,\mathrm{cm}^{-1}$), they can be used in both neutron energy gain and energy loss.

On a steady–state reactor, a primary spectrometer (choppers, or rotating crystal, or crystal + chopper) is used to produce a pulsed monochromatic beam of neutrons. Monitors are placed before and after the sample in order to determine its transmission. After scattering by the sample, the neutrons are analysed in a secondary spectrometer as a function of time and angle. The time–of–flight of the scattered neutrons is related to the energy transfer and the scattering angle to the momentum transfer. Since a large number of detectors is used to cover a wide range of scattering angles, these detectors have to be normalized from the scattering of a vanadium plate because the cross section of this element is almost totally incoherent. The neutron time–of–flight, τ, is defined as the inverse of the velocity ($\tau_{\mu s/m} = 252.77\ \lambda_{\text{Å}}$).

The spectra, which are recorded as a function of time–of–flight, are transformed into an energy scale using

$$\frac{d^2\sigma}{d\Omega dE} = -\frac{\tau^3}{m}\frac{d^2\sigma}{d\Omega d\tau} \tag{45}$$

This type of spectrometer is well suited to study tunneling transitions and diffusive motions, in addition to low-frequency modes.

3.3. Samples

In order to define the quantity of sample required for an INS experiment, the following relation can be used :

$$\frac{I}{I_o} = \exp\left(-\sum_i N\sigma_i \ \ell\right) \tag{46}$$

where I/I_o gives the transmission of the sample perpendicular to beam, N is the number of scatterers per unit volume, σ_i their total cross section and ℓ the thickness of the sample.

For quasi-elastic experiments, which concern the study of diffusive motions occuring at small energy transfers ($\leq$ 2 meV), the percentage of scattering is kept below 10 % otherwise multiple scattering effects must be taken into account. If we consider a hydrogenated molecule in the liquid phase, for example benzene, its density is 0.8765 g/cm^3, its molecular weight 78.12 g/mol and one finds using Table 1 : $\Sigma\sigma_i$ = 529. x 10^{-24} cm^2 (for λ = 5 Å). This gives a transmission of 90 % for a thickness of 0.03 cm. The beam area at the sample position being typically of 10 cm^2, it appears that a quantity of 0.263 g is sufficient for quasi-elastic experiments.

In order to put in the beam the same number of protons for surface studies, the quantity of substrate has to be much larger. If we consider for example an unsupported metal powder of 50 m^2/g of surface area, the average number of H atoms per square meter is about 1.5 x 10^{19} at saturation. It results from these numbers that a quantity of 20 g is required. If one wants to study smaller coverages, then the quantity of powder should be augmented.

For vibrational studies, it has been found that the percentage of scattering can be increased up to 25 % without degrading the spectral quality (37). It is therefore a frequent practise to use large quantities of sample (up to 100 g for metal powders).

The substrates which can be studied in INS have usually large surface areas : unsupported metals (Raney or colloidal black), or small metallic particles supported on silica or alumina. Different oxides have also been studied, as well as sulphides or charcoal. Zeolites, which are well-crystallized materials, offer very high surface areas. All these powders are widely used in catalysis and cannot be easily studied by other vibrational techniques.

The preparation of such large amounts of substrate is a crucial point. The powder has to be cleaned from any residual water or hydrocarbons and kept in a sealed cell. The neutron cells can be cylindrical or slab-shaped, depending on the neutron beam size and scattering strength. They are usually made from aluminium but stainless-steel or quartz cells have been used.

For quasi-elastic studies, experiments can be performed from a few Kelvin to several hundreds of Kelvin. For inelastic studies, it has been shown in previous paragraphs that the vibrational spectrum was sharpened at low temperature so that the sample is usually placed in a liquid nitrogen or helium cryostat. However INS experiments on chemisorbed hydrogen have been reported at room temperature (38,39). It can also be noted that INS and QENS experiments can be performed with pressures of adsorbates up to a few atmospheres.

4. EXAMPLES

4.1. Hydrogen chemisorption

Hydrogen chemisorbed on large surface area materials is difficult to characterize by most of the vibrational techniques. With INS, all the local modes of hydrogen adsorbed on a given site can be observed, but the assignment can be difficult if several species contribute to the spectrum.

INS has been used to study the adsorption of hydrogen on different transition metals : Ni (38–43), Pd (44–48) and Pt (49–51). Adsorption on non–metallic catalysts has also been investigated, mainly on sulphides : MoS_2 (52–54), WS_2 (54,55), TaS_2 (56–57) and RuS_2 (58), but also on oxides such as MoO_3 and WO_3 (59,60) or ZnO (61). The assignment of the vibrational features of these inhomogeneous surfaces has been controversial (62–66) and several questions remain open.

If we consider for example the system hydrogen on nickel, all the recent INS studies conclude that hydrogen is preferentially adsorbed in sites of C_{3v} symmetry, sites which are found on the (111) planes (39,42,43,64,66). This species gives rise to strong peaks at $\approx$ 940 and 1120 cm^{-1}, the lowest frequency peak corresponding to the antisymmetric stretch, of E symmetry, since it is twice as intense as the other mode of A symmetry.

However, other species whose proportion depends on the sample preparation can be found. At low coverages, peaks have been observed at 750, 815 and 1080 cm^{-1} (43), suggesting the presence of sites of nearly three–fold symmetry, occuring on the (110) planes. Other peaks measured in the range 500–650 cm^{-1} have been previously assigned to hydrogen adsorbed at four–coordinated sites, on the (100) planes. Therefore the picture of hydrogen adsorption on polycrystalline materials is much more complicated than on single crystals. If all the vibrational peaks due to the different species could be resolved, the proportion of planes exposed at the surface could be derived ; but this is not the case experimentally, even with the recent INS spectrometers, because the intrinsic width of the modes is larger than the instrumental resolution.

Even on single crystal surfaces, the situation is sometimes complicated. The recent EELS results obtained for hydrogen adsorption on the unreconstructed as well as reconstructed Ni(110) surface have revealed many vibrational modes corresponding to several species (67). In particular, a peak was observed at $\approx$ 460 cm^{-1}, which is close in frequency to a vibrational feature reported by Stockmeyer et al. at 480 cm^{-1} on supported nickel (38). It was suggested in ref. 38 that this peak could be due to absorbed hydrogen, but it has been shown more recently that no subsurface or hydride species are formed up to 80 bars (39) ; this peak could therefore originate from (110) planes. Another peak was observed by EELS on Ni(110) at 600–640 cm^{-1}, and the INS intensity which is measured in this range on powder samples could thus be due to hydrogen adsorbed on (110) planes and not on (100) planes as previously proposed. Another point is still unclear about the (100) plane : it is believed that hydrogen populates the four–fold hollow positions (68) but only the perpendicular mode could be measured by EELS so that the frequency of the other modes is still unknown.

The proportion of terminal hydrogen on polycrystalline Ni samples is another important question. This species is still considered by some authors to be the predominant one (e.g. ref. 69), in contradiction with most experimental work obtained on powders or single crystals. The proportion of terminal hydrogen, derived from INS experiments on Raney nickel, was previoulsy estimated to be less than 10 % (43). However, recent studies indicate that this proportion could be higher on some samples, up to 20 % (70). Terminal hydrogen is difficult to measure with INS because the stretching mode is situated at ≈ 1800 cm^{-1} near multiphonon features, and the bending modes are found at about 1000 cm^{-1} and are thus hidden by the intense peaks due to hydrogen adsorbed in three–coordinated sites.

These spectroscopic measurements do not indicate which species is active in catalytic reactions, it could well be the small proportion of terminal hydrogen which continues to be populated, with C_{3v} sites, at high pressures (70).

All these contradictory results explain why this system is still being studied by different techniques. For INS, other samples with a better definition of surface planes should be prepared and a reassignment of the weaker features can be expected. The presence of molecular hydrogen, which has been observed by EELS on a stepped nickel surface (71), has not been yet characterized on Ni powders (it has however been measured on sulphides, ref. 54).

The adsorption of hydrogen on other metals has also been studied by INS. On palladium, hydrogen appears to be preferentially adsorbed in sites of three–fold symmetry, which indicates that a majority of (111) planes are found on the surface, as on Ni. On the other hand, the INS spectra of hydrogen adsorbed on Pt/SiO$_2$ and on Raney Pt are completely different (51). The most intense features are found at 500–600 cm^{-1} and these were assigned to hydrogen adsorbed on (100) planes which would be predominant on these samples.

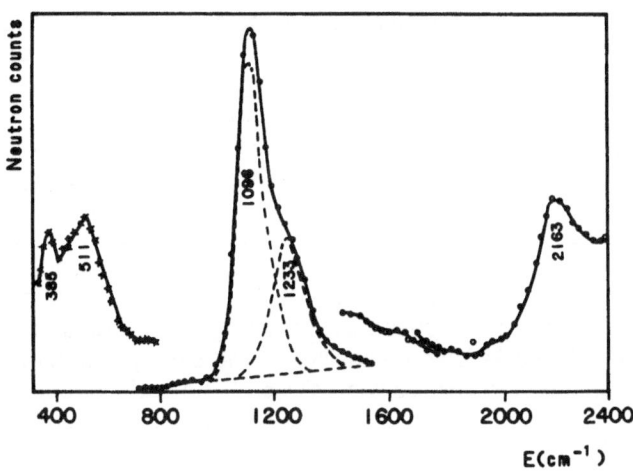

Fig. 3 .   INS spectrum of HFeCo$_3$(CO)$_{12}$ (74).

4.2. Organometallic model compounds

Very often, the assignment of the observed bands of chemisorbed species is made through comparison with data obtained on model compounds. Since there are difficulties in using optical spectroscopies for these compounds, several studies have been performed with INS (72–80). With this technique, one can measure vibrational modes which are inactive in infrared or Raman (because of selection rules) or which have small optical intensity because of small dipole moment or polarisability changes. INS is also useful to study samples which decompose or fluoresce in a laser beam or which have intense bands masking the vibrations involving hydrogen atoms (e.g. the CO vibrations).

In the case of hydrogen, which was discussed in the previous paragraph, the vibrational data obtained on transition metal hydridocarbonyls have proved to be very useful. Hydrogen, like CO, is able to bind at points, along edges, or on faces of metal clusters.

For example, the $[HCo_6(CO)_{15}]^-$ ion is a hexanuclear metal cluster in which the hydrogen is lying at the centre of an octahedral hole (75). The triply degenerate stretching mode of hydrogen was assigned at 1056 cm^{-1} in this compound.

In $HFeCo_3(CO)_{12}$, which contains a (μ_3–H) species (74), two intense bands are found at 1096 and 1233 cm^{-1} (see Fig. 3). It is straightforward to assign the strongest band at 1096 cm^{-1} to v_{as}, of E symmetry, and the other one at 1233 cm^{-1} to v_s. The peak at 2163 cm^{-1} in Fig. 3 is due to overtones and the bands at 385 and 511 cm^{-1} correspond to deformations of the cluster. There is a close similarity between these peaks and those which were described for hydrogen chemisorbed on Raney nickel.

The INS spectra of organometallic molecules have also been useful to assign the vibrational modes of adsorbed molecules. One example is Zeise's salt, $K[Pt(C_2H_4)Cl_3]$, which is the prototype of a transition metal–olefin complex. The bonding between a metal atom and an olefin was originally described by Dewar (81) and latter by Chatt and Duncanson in the particular case of Zeise's salt (82). The main features of the model, σ bonding and π back–bonding, are often invoked in the case of chemisorbed ethylene.

The vibrational spectrum of Zeise's salt has been much studied and the assignment of Hiraishi (83) is still used in recent papers on chemisorbed ethylene (e.g. ref. 84) even though important modifications in the assignment of the A_2 and B_2 species have appeared since (79). The new assignment in ref. 79 was supported by a calculation of the vibrational frequencies and of the neutron intensities. The experimental and calculated INS spectra of Zeise's salt are shown in Fig. 4, in the range 350–1650 cm^{-1}. The out-of-phase CH_2 twisting mode $v_4(A_2)$, which had not been observed by optical spectroscopies, is an intense feature of the INS spectrum and is located at 1020 cm^{-1}. The lowest internal mode is found at 845 cm^{-1}, it corresponds to the CH_2 rocking mode v_{10} (B_2) ; the other rocking mode v_6 (A_2) is placed at 1180 cm^{-1} instead of 841 cm^{-1} in ref. 83. The band measured at 720 cm^{-1} is assigned to the in–phase CH_2 twisting mode (rotation about the C–C bond) v_{13} (B_2), this mode is very sensitive to the strength of coordination and its frequency has been found to shift down to 418 cm^{-1} in zeolites (85).

In most studies, like for Zeise's salt, the refinement is based on the vibrational frequencies and the validity of the assignment is assessed by a comparison of the observed and calculated spectra. Another method of interpretation of the INS spectra is based on a refinement of the force constants directly to the observed spectral profile (80). This method has been illustrated by an analysis of the INS spectrum of $CH_3CCo_3(CO)_9$. The CH_3 torsional mode, which could not be observed by optical spectroscopies, was located at 383 cm^{-1}. This approach appears to be promising for small systems, but its application to larger molecules has to be tested.

The study of the vibrational spectrum of these model compounds is useful because not only the assignment but also the force constants can be transferred to the surface complex, examples are given in the next paragraph.

4.3. Hydrocarbons on metals

The INS spectra of benzene chemisorbed on Ni and Pt were the first systems to be quantitatively interpreted with a normal coordinate analysis (86–88). It was found that benzene was adsorbed flat on the surfaces and a significant weakening of the C–C stretching force constant was observed, the perturbation being larger on

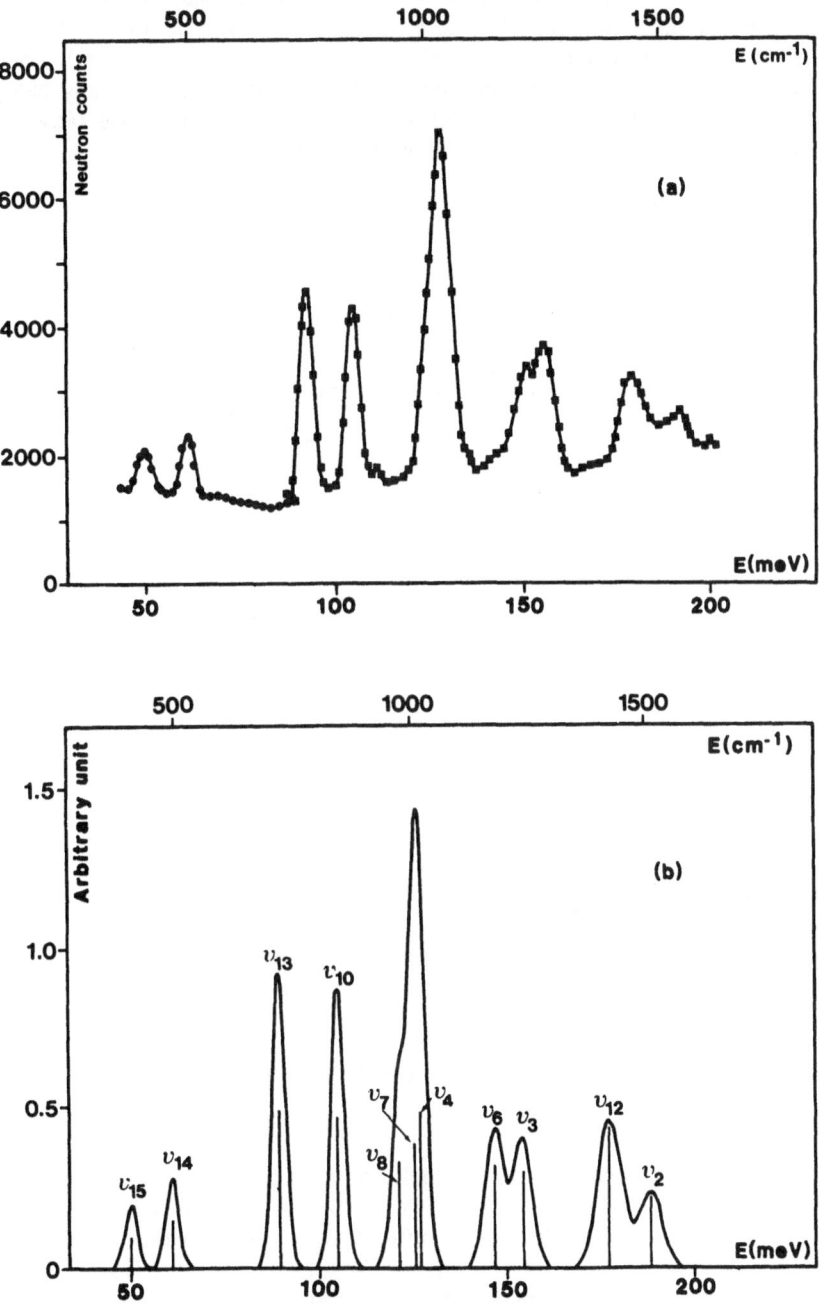

Fig. 4 . INS spectra of Zeise's salt, K[Pt(C$_2$H$_4$)Cl$_3$], (a) experimental spectrum
recorded at 5 K, (b) calculated (79).

bare Ni than on Pt or Ni precovered with H_2. On bare nickel, a small proportion (≈ 15 %) of benzene was found to be totally dehydrogenated but not on platinum. The force field of a model molecule (η^6-C_6H_6)$Cr(CO)_3$ (76), was transferred to the case of adsorbed benzene by changing a few force constants. In the force field calculation, the benzene molecule was taken bound to one metal atom because the EELS spectra obtained on Ni(111) and (100) were almost identical (89). However, even if one considers that the internal modes of benzene are reasonably assigned, all the vibrational modes of benzene relative to the surface cannot be observed so that the bonding geometry cannot be established definitely.

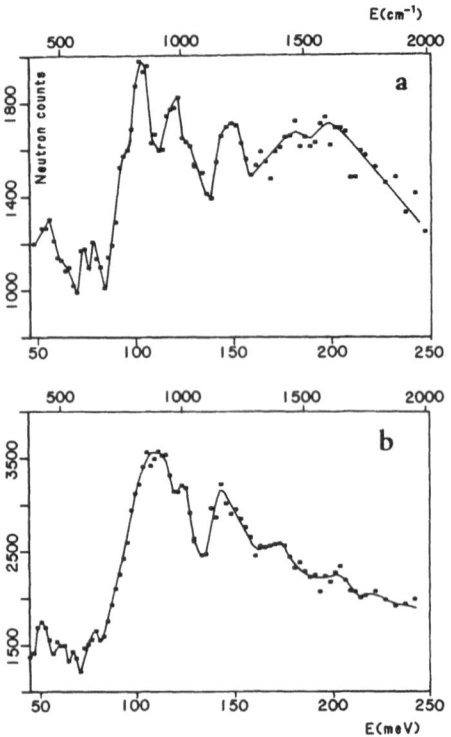

Fig. 5 . INS spectra of acetylene on Raney platinum (a) after adsorption at 200 K, (b) after adsorption at 300 K (the spectra were recorded at 80 K).

The adsorption of benzene on Pt black has also been studied by INS (90) but good cleaning of this material is very difficult (91) and contributions in the INS spectra from OH groups or Pt–Cl vibrations (90) have been admitted.

The adsorption of other hydrocarbons, mainly C_2H_2 and C_2H_4, has been studied by INS on different metals (92-95). On Ni, the adsorption at room temperature of C_2H_2 and C_2H_4 results in extensive decomposition and rearrangement of the molecular fragments. If the adsorption is performed at low temperature, this process is slowed down but the interpretation of the INS spectra in terms of one molecular species seems dubious.

On platinum, the situation is more favourable since no dehydrogenation is observed at low temperature. Cavanagh et al. have studied the adsorption of C_2H_2 and C_2H_4 on Pt black (95). The INS spectrum of C_2H_2 adsorbed at 150 K was assigned to a bent acetylene molecule lying parallel to the surface. A force field calculation was carried out starting with the force constants of a cobalt model compound, $(C_2H_2)Co_2(CO)_6$. When the sample was warmed at room temperature, a quite different spectrum was obtained. On Raney Pt, similar results have been obtained after adsorption of C_2H_2 at 200 K (96), the INS spectrum is shown in Fig. 5a. It appears that the intensity is small around 70–80 meV, where the modes of atomic hydrogen are expected, so that this species is negligible. The peaks observed at 104, 120 and 148 meV can be assigned to $\delta(CH)$ vibrations of a bent C_2H_2 molecule, but no force field calculation was attempted because of the bad resolution of the molecule–metal vibrations, and because it has been found by EELS that C_2H_2 and C_2H_4 decomposition on single crystals is structure sensitive. When the adsorption is performed at room temperature, the resulting spectrum is shown in Fig. 5b ; it appears that the overall shape of the spectrum is preserved, at least below 150 meV, which was not the case with Pt black.

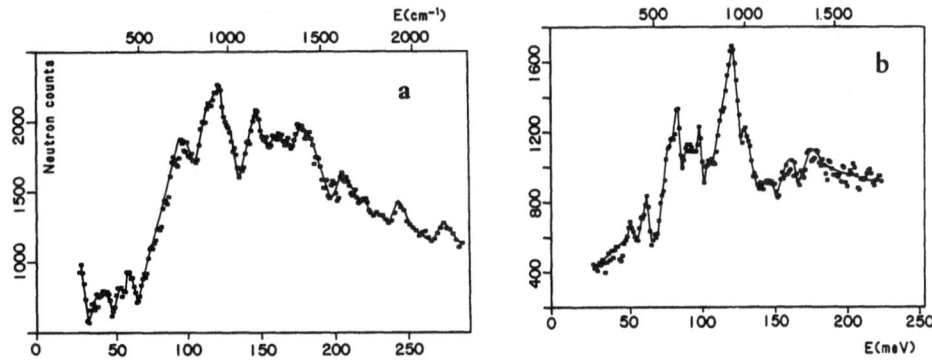

Fig. 6. INS spectra of hydrocarbons adsorbed at 300 K on Raney palladium, (a) acetylene, (b) ethylene (the spectra were recorded at 10 K).

The number of studies of hydrocarbon adsorptions on Pd is less extensive. The adsorption of C_2H_2 on Raney Pd has been studied at room temperature (96), the INS spectrum is shown in Fig. 6a. Similarities are found with C_2H_2 adsorbed on Pt, but the large number of vibrational peaks suggests that other species are present on the surface. The INS spectrum of C_2H_4 adsorbed at 300 K on Raney Pd is shown in Fig. 6b, this spectrum is very similar to the one of Zeise's salt (Fig. 4a). This indicates the presence on the surface of π–bound ethylene, but again, the apparition of extra peaks did not allow a quantitative interpretation of the INS intensities.

4.4. Zeolitic systems

There are several applications of neutron scattering to zeolitic systems. Neutron diffraction is used to locate molecules adsorbed in zeolites (97–100). Information on the clustering of molecules can be obtained from small–angle neutron scattering, SANS (101). There is a growing interest in the characterization of the intracrystalline mobility with quasi-elastic neutron scattering, QENS (102–110). Finally, the vibrational modes of the zeolite and of adsorbed molecules can be measured with INS (111).

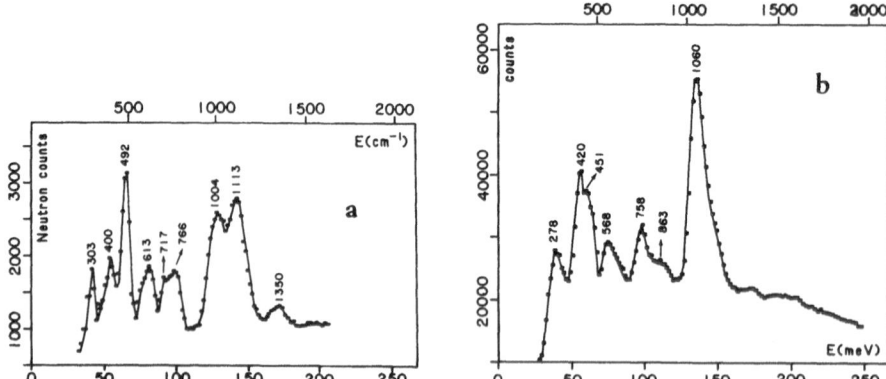

Fig. 7 . INS spectra of (a) NaY and (b) HNaY. The values within the figures give the peak frequencies in cm^{-1} (115).

4.4.1. Fundamental bending vibrations of hydroxyl groups in zeolites

Since bridged hydroxyl groups are at the origin of the Brönsted acidity of zeolitic catalysts, much work has been done to characterize these sites. Direct information on the local geometry of bridged hydroxyls is difficult to obtain with X–ray or neutron diffraction techniques. Therefore several spectroscopic methods, mainly infrared and NMR, have been extensively used to identify the different types of hydroxyl groups in zeolites. Only the ν(OH) stretching vibrations can be easily observed with infrared, the in–plane (δ) and out–of–plane (γ) bending modes cannot be measured because of strong absorption bands of the framework in the range 200–1300 cm^{-1}. The δ(OH) vibrations of some zeolites have however been derived from overtones and combination bands by diffuse reflectance spectroscopy (112), these bending modes were found to be more sensitive to the structure and composition of the zeolites than the ν(OH) modes.

With INS, the ν(OH) vibrations are difficult to observe because of the effect of the Debye–Waller factor and because of overlap with multiphonon. On the other hand, the bending fundamentals, which occur at lower frequency, can be easily observed.

Wax et al. have studied zeolite Rho under the sodium and acid forms (113). Peaks were measured in the INS spectra at 1150, 1060 and 360 cm^{-1}. They assigned the peaks at 1150 and 1060 cm^{-1} to δ(OH) and γ(OH), respectively, and left the peak at 360 cm^{-1} unassigned. It was latter suggested that the peak at 360 cm^{-1} could correspond to γ(OH) (114), so that the peaks at 1150 and 1060 cm^{-1} would be due to δ(OH) modes.

The bending vibrations of hydroxyl groups in Y zeolite have also been studied (115). The INS spectra of NaY and HNaY, recorded at 20 K, are shown in Fig. 7a and 7b, respectively. Comparison between the two spectra evidences the apparition of new peaks in NHaY at 1060, 863 and 420 cm^{-1}. The peak at 1060 cm^{-1} is assigned to the δ(OH) modes of bridged hydroxyls. The peak at 863 cm^{-1} is also assigned to δ(OH) modes, but for silanol groups (these groups are produced during the thermal treatment of the decationized zeolite). The peak at 420 cm^{-1} is assigned to the γ(OH) modes of the bridged hydroxyls, in good agreement with quantum chemical calculations (114). The framework vibrations also appear in the spectrum of HNaY because for these framework modes, the hydrogen atom will "ride" on the oxygen atoms, receiving an appreciable displacement with respect to the centre of mass. Comparison between Figs. 7a and 7b shows that in HNaY the vibrations of the framework are shifted to lower frequency, compared to NaY. This is in agreement with the computed weakening of the [Si–O–Al] bond upon bonding by a proton (116).

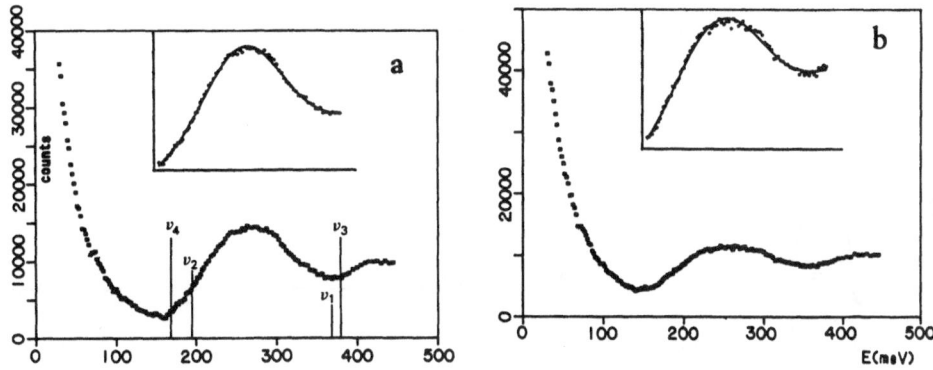

Fig. 8 .   INS spectra of methane adsorbed in NaZSM–5, for a loading of 2.2 mol./u.c. (a) at 10 K and (b) at 100 K (30). The experimental points are fitted with Gaussians between 150 and 380 meV (continuous line).

4.4.2. Recoil scattering from methane adsorbed in NaZSM–5

Molecular dynamics and Monte–Carlo calculations are now frequently used to study the mobility of small molecules inside zeolites (117–121). Some simplifications are made in these calculations : the framework is treated as rigid and classical mechanics is used. For a light molecule like CH$_4$, at low temperature, these approximations could be not justified.

Neutron recoil scattering from CH_4 adsorbed in NaZSM–5 has been measured at 100 K and 10 K, in order to determine the average kinetic energy of CH_4 in the zeolite (30). The INS spectra of CH_4 adsorbed at 10 K and 100 K, for a loading of 2.2 mol./u.c., are shown in Fig. 8.

The frequencies of the CH bending modes, v_2 and v_4, and of the CH stretching modes, v_3 and v_1, of the CH_4 molecule in the gas phase are also indicated in Fig. 8a. It appears that there is no intensity left at the frequencies of the internal modes, the modes have been shifted up in frequency by the recoil energy E_R, Eq. (35). The value of the effective mass which is derived : 3.5 amu at 100 K (3 amu at 10 K) corresponds well to the value calculated with Eq. (36). The line shapes were fitted with Gaussians and from the widths, Eq. (37), average kinetic energy values were obtained. It was found that the state of the adsorbed molecule was far from the perfect gas. At 100 K, the total kinetic energy (translation + rotation) is 14 meV compared to 25.9 meV for the ideal gas. At 10 K, the translational kinetic energy is much higher than in the gas : 7.3 meV compared to 1.3 meV. This implies that zero–point motion makes an important contribution to the total energy at low temperature.

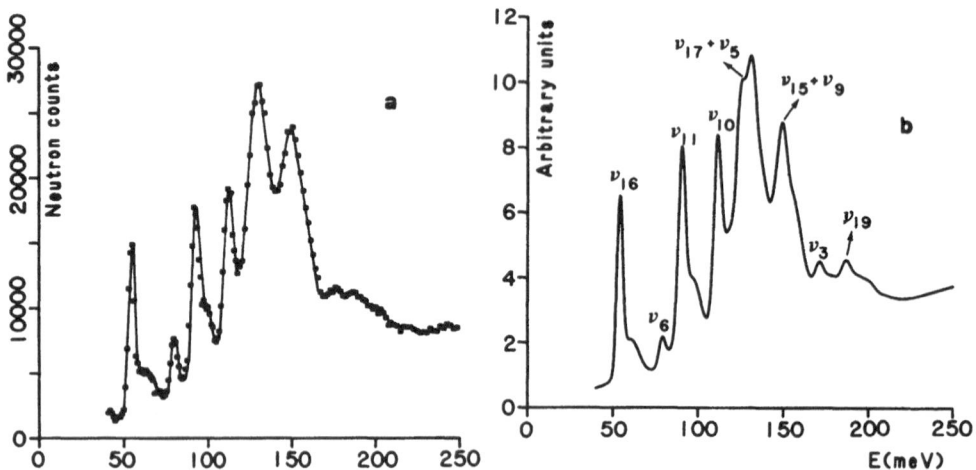

Fig. 9 :   INS spectra of benzene adsorbed in NaY, for a loading of 1 mol./supercage (a) experimental spectrum recorded at 20 K, (b) calculated with multiphonon processes (122).

4.4.3. Benzene adsorbed in Na–Y

The vibrational modes of benzene adsorbed in different zeolites have been studied by several vibrational techniques : infrared, Raman and neutron.

The INS spectrum of benzene adsorbed in NaY, for a loading of 1 mol./u.c., is shown in Fig. 9a   in the range 300–2000 cm^{-1} (122). This spectrum was recorded at 20 K with the spectrometer INFB, at the ILL. Such low temperatures are necessary for these systems because of the high mean – square amplitudes of the benzene molecules in the zeolite. The INS spectrum of bare NaY has been subtracted (it is much less intense anyway as can be seen from a comparison of the intensity scales in Figs. 7a and 9a). This implies that all the fundamental modes of adsorbed benzene can be observed in that range, unlike in infrared. The experimental spectrum was simulated by computing the intensities of the fundamentals and of multiphonon (Fig. 9b). The frequencies of the modes involving the hydrogen atoms : the in–plane and out–of–plane bending modes, could thus be derived. It was found that the in–plane and torsional modes vary little in energy but the γ(CH) modes are shifted to higher frequencies, up to 25 cm^{-1} compared to the gas phase.

5. CONCLUSION

The application of inelastic neutron scattering to the study of vibrational modes of atoms or molecules adsorbed on catalysts has been reviewed. The small number of high resolution spectrometers available in the world (only four !) limits the number of applications to selected systems. However this technique is useful when the information cannot be obtained from the more traditional vibrational techniques (because of selection rules, absorption, fluorescence, etc.).

Since not much improvement in the instrumental resolution can be expected for the near future, one should concentrate on the information contained in the peak intensities. Recent theoretical work shows that the calculations have to include the contributions from multiphonon, overtones and combinations. Such simulations or direct fitting of the INS spectra give a much higher confidence in the assignments.

REFERENCES

1   Neutron Research Facilities at the ILL High Flux Reactor (Internal Scientific Report, Institut Laue Langevin, 1986).
2   J.W. White, in "Dynamics of Solids and Liquids by Neutron Scattering," (S.W. Lovesey and T. Springer Eds.) p. 197, Springer–Verlag, New York (1977).
3   R.K. Thomas, Prog. Solid St. Chem. 14, 1 (1982).
4   C.J. Wright and C.M. Sayers, Rep. Prog. Phys. 46, 773 (1983).
5   R.R. Cavanagh, J.J. Rush and R.D. Kelley, in "Vibrational Spectroscopy of Molecules on surfaces", (J.T. Yates Jr. and T.E. Madey, Eds.) p. 183, Plenum, New York (1987).
6   H. Jobic, in "Physical Techniques in the Study of Catalysts", (B. Imelik and J.C. Védrine, Eds.), Plenum, in press.
7   I.I. Gurevich and L.V. Tarasov, Low–Energy Neutron Physics, North–Holland, Amsterdam (1968).
8   H. Boutin and S. Yip, Molecular Spectroscopy with Neutrons, M.I.T. Press, Cambridge (1968).
9   W. Marshall and S. Lovesey, Theory of Thermal Neutron Scattering, Oxford University Press (1971).
10   G.L. Squires, Introduction to the Theory of Thermal Neutron Scattering, Cambridge University Press (1978).
11   A.C. Levi and H. Suhl, Surface Sci. 88, 221 (1979).
12   A.C. Zemach and R.J. Glauber, Phys. Rev. 101, 118 (1956).
13   J. Tomkinson, M. Warner and A.D. Taylor, Molec. Phys. 51, 381 (1984).
14   J. Howard, J.M. Nicol, B.C. Boland, J. Tomkinson, J. Eckert, J.A. Goldstone and A.D. Taylor, Molec. Phys. 53, 323 (1984).
15   H. Jobic, J. Chem. Phys. 76, 2693 (1982).

16   A. Griffin and H. Jobic, J. Chem. Phys. 75, 5940 (1981).

17   M. Warner, S.W. Lovesey and J. Smith, Z. Physik B 51, 109 (1983).

18   H.J. Lauter and H. Jobic, Chem. Phys. Lett. 108, 393 (1984).

19   J. Penfold and J. Tomkinson, Rutherford Appleton Lab. Rep. RAL-86-019 (1986).

20   W.B. Nelligan, D.J. Lepoire, C.K. Loong, T.O. Brun and S.H. Chen, Nucl. Instrum. Meth. Phys. Res. A 254, 563 (1987).

21   H. Jobic and H.J. Lauter, J. Chem. Phys. 88, 5450 (1988).

22   J. Tomkinson and G.J. Kearley, J. Chem. Phys. 91, 5164 (1989).

23   R.G. Sachs and E. Teller, Phys. Rev. 60, 18 (1941).

24   T.J. Krieger and M.S. Nelkin, Phys. Rev. 106, 290 (1957).

25   A. Sjölander, Ark. Fys. 14, 315 (1958).

26   W. Langel, D.L. Price, R.O. Simmons and P.E. Sokol, Phys. Rev. B 38, 11275 (1988).

27   J. Tomkinson, Chem. Phys. 127, 445 (1988).

28   R.S. Holt, L.M. Needham and M.P. Paoli, Phys. Letters, 126 A, 373 (1988).

29   R. Rauh and N. Watanabe, Phys. Letters 100 A, 244 (1984).

30   H. Jobic, Chem. Phys. Lett. 170, 217 (1990).

31   E.B. Wilson, J.C. Decius and P.C. Cross, Molecular Vibrations, Mc Graw-Hill, New York (1955).

32   N. Bjerrum, Verhandl-deut. physik. Ges. 16, 737 (1914) ; R.G. Snyder, J.H. Schachtschneider, Spectrochim. Acta 21, 169 (1965).

33   H.C. Urey and C.A. Bradley, Phys. Rev. 38, 1969 (1931).

34   S. Lifson and A. Warshel, J. Chem. Phys. 49, 5116 (1968).

35   ISIS Annual Report, Rutherford Appleton Lab. (1989).

36   A.D. Taylor, E.J. Wood, J.A. Goldstone and J. Eckert, Nucl. Instrum. Meth. Phys. Res. 221, 408 (1984).

37   P.S. Goyal, J. Penfold and J. Tomkinson, Rutherford Appleton Lab. Rep. RAL-86-070 (1986).

38   R. Stockmeyer, H. Stortnik, I. Natkaniec and J. Mayer, Ber. Bunsenges. Phys. Chem. 84, 79 (1980).

39   H. Jobic, G. Clugnet and A. Renouprez, J. Elec. Spect. Relat. Phenom. 45, 281 (1987).

40   R. Stockmeyer, H. Conrad, A. Renouprez and P. Fouilloux, Surf. Sci. 49, 549 (1975).

41   A.J. Renouprez, P. Fouilloux, G. Coudurier, D. Tocchetti and R. Stockmeyer, J. Chem. Soc., Faraday Trans. 1, 73, 1 (1977).

42   R.R. Cavanagh, R.D. Kelley and J.J. Rush, J. Chem. Phys. 77, 1540 (1982).

43   H. Jobic and A. Renouprez, J. Chem. Soc., Faraday Trans. 1, 80, 1991 (1984).

44   J. Howard, T.C. Waddington and C.J. Wright, Chem. Phys. Lett. 56, 258 (1978).

45   I.J. Braid, J. Howard and J. Tomkinson, J. Chem. Soc., Faraday Trans. 2, 79, 253 (1983).

46   H. Jobic, J.P. Candy, V. Perrichon and A. Renouprez, J. Chem. Soc., Faraday Trans. 1, 81, 1955 (1985).

47   H. Jobic and A.J. Renouprez, J. Less-Common Met. 129, 311 (1987).

48   J.M. Nicol, J.J. Rush and R.D. Kelley, Surf. Sci. 197, 67 (1988).

49   J. Howard, T.C. Waddington and C.J. Wright, in "Neutron Inelastic Scattering", Vol. II (IAEA, Vienna), p. 499 (1978).

50   J.J. Rush, R.R. Cavanagh, R.D. Kelley and J.M. Rowe, J. Chem. Phys. 83, 5339 (1985).

51   A.J. Renouprez and H. Jobic, J. Catal. 113, 509 (1988).

52 C.J. Wright, C. Sampson, D. Fraser, R.B. Moyes, P.B. Wells and C. Riekel, J. Chem. Soc., Faraday Trans. 1, 76, 1585 (1980).

53 C.F. Sampson, J.M. Thomas, S. Vasudevan and C.J. Wright, Bull. Soc. Chim. Belges, 90, 1215 (1981).

54 P.N. Jones, E. Knözinger, W. Langel, R.B. Moyes and J. Tomkinson, Surf. Sci. 207, 159 (1988).

55 C.J. Wright, D. Fraser, R.B. Moyes and P.B. Wells, Appl. Catal. 1, 49 (1981).

56 C.J. Wright, C. Riekel, R. Schöllhorn and B.C. Tofield, J. Solid State Chem. 24, 219 (1978).

57 C. Riekel, H.G. Reznik, R. Schöllhorn and C.J. Wright, J. Chem. Phys. 70, 5203 (1979).

58 W.H. Heise, K.Lu, Y.J. Kuo, T..J. Udovic, J.J. Rush and B.J. Tatarchuk, J. Phys. Chem. 92, 5184 (1988).

59 C.J. Wright, J. Solid State Chem. 20, 89 (1977).

60 P.G. Dickens, J.J. Birtill and C.J. Wright, J. Solid State Chem. 28, 185 (1979).

61 J. Howard, I.J. Braid and J. Tomkinson, J. Chem. Soc., Faraday Trans. 1, 80, 225 (1984).

62 C.J. Wright, J. Chem. Soc., Faraday Trans. 2, 73, 1497 (1977).

63 D. Graham, J. Howard and T.C. Waddington, J. Chem. Soc., Faraday Trans. 1, 79, 1281 (1983).

64 C.M. Sayers, J. Phys. C, Solid State Phys. 16, 2381 (1983).

65 C.M. Sayers, Surf. Sci., 143, 411 (1984).

66 C.M. Sayers and C.J. Wright, J. Chem. Soc., Faraday Trans. 1, 80, 1217 (1984).

67 B. Voigtlander, S. Lehwald and H. Ibach, Surf. Sci. 208, 113 (1989).

68 P.A. Karlsson, A.S. Martensson, S. Anderson and P. Nordlander, Surf. Sci. 175, L 759 (1986).

69 J.T. Richardson and T.S. Cale, J. Catal. 102, 419 (1986).

70 H. Jobic et al. (unpublished results).

71 A.S. Martensson, C. Nyberg and S. Anderson, Surf. Sci. 205, 12 (1988).

72 J.W. White and C.J. Wright, J. Chem. Soc. A, 2843 (1971).

73 J.W. White and C.J. Wright, J. Chem. Soc., Faraday Trans. 2, 68, 1423 (1972).

74 D. Graham, J. Howard and T.C. Waddington, J. Chem. Soc., Faraday Trans. 1, 79, 1281 (1983).

75 D. Graham, J. Howard, T.C. Waddington and J. Tomkinson, J. Chem. Soc., Faraday Trans. 2, 79, 1713 (1983).

76 H. Jobic, J. Tomkinson and A. Renouprez, Mol. Phys. 39, 989 (1980).

77 J. Howard, K. Robson and T.C. Waddington, J. Chem. Soc., Dalton Trans. 967 (1982).

78 J. Howard, T.C. Waddington and C.J. Wright, J. Chem. Soc., Faraday Trans. 2, 73, 1768 (1977).

79 H. Jobic, J. Mol. Struct. 131, 167 (1985).

80 G.J. Kearley, J. Chem. Soc., Faraday Trans. 2, 82, 41 (1986).

81 M.J.S. Dewar, Bull. Soc. Chim. Fr. 18, C71 (1951) ;
    M.J.S. Dewar and G.F. Ford, J. Am. Chem. Soc. 101, 783 (1979).

82 J. Chatt and L.A. Duncanson, J. Chem. Soc. 2339 (1953).

83 J. Hiraishi, Spectrochim. Acta, Part A, 25, 749 (1969).

84 E. Yagasaki and R.I. Masel, Surf. Sci. 222, 430 (1989).

85 J. Howard, T.C. Waddington and C.J. Wright, J. Chem. Soc., Faraday Trans. 2, 73, 1768 (1977).

86 H. Jobic, J. Tomkinson, J.P. Candy, P. Fouilloux and A. Renouprez, Surf. Sci. 95, 496 (1980).

87 H. Jobic and A. Renouprez, Surf. Sci. 111, 53 (1981).

88   A. Renouprez, G. Clugnet and H. Jobic, J. Catal. 74, 296 (1982).

89   J.C. Bertolini and J. Rousseau, Surf. Sci. 78, 577 (1978).

90   D. Graham and J. Howard, J. Chem. Soc., Faraday Trans. 1, 80, 3365 (1984).

91   Z. Paal and D. Marton, Appl. Surf. Sci. 26, 161 (1986).

92   H. Jobic and A. Renouprez, Proc. ICSS4 and ECOSS3, Suppl. "Le Vide, Les Couches Minces", n° 201, Vol II, p. 746 (1980).

93   J.P. Candy, H. Jobic and A. Renouprez, J. Phys. Chem. 87, 1227 (1983).

94   R.D. Kelley, R.R. Cavanagh, J.J. Rush and T.E. Madey, Surf. Sci. 155, 480 (1985).

95   R.R. Cavanagh, J.J. Rush, R.D. Kelley and T.J. Udovic, J. Chem. Phys. 80, 3478 (1984).

96   H. Jobic et al. (unpublished results).

97   A.N. Fitch, H. Jobic and A. Renouprez, J. Chem. Soc., Chem. Commun. 284 (1985) ; J. Phys. Chem. 90, 1311 (1986).

98   P.A. Wright, J.M. Thomas, A.K. Cheetham and A.K. Nowak, Nature 318, 611 (1985).

99   J.M. Newsam, J. Phys. Chem. 93, 7689 (1989).

100   M. Czjzek, T. Vogt and H. Fuess, Angew. Chem., 28, 770 (1989).

101   A. Renouprez, H. Jobic and R.C. Oberthur, Zeolites 5, 222 (1985).

102   C.J. Wright and C. Riekel, Mol. Phys. 36, 695 (1978).

103   E. Cohen de Lara and R. Kahn, J. Physique 42, 1029 (1981).

104   H. Jobic, M. Bée and A. Renouprez, Surf. Sci. 140, 307 (1984).

105   H. Jobic, A. Renouprez, M. Bée and C. Poinsignon, J. Phys. Chem. 90, 1059 (1986).

106   H. Jobic, M. Bée and G.J. Kearley, Zeolites, 9, 312 (1989).

107   H. Jobic, M. Bée, J. Caro, M. Bülow and J. Kärger, J. Chem. Soc., Faraday Trans. 1, 85, 4201 (1989).

108   R. Kahn, E. Cohen de Lara and E. Viennet, J. Chem. Phys. 91, 5097 (1989).

109   E. Cohen de Lara, R. Kahn and F. Mezei, J. Chem. Soc., Faraday Trans 1, 79, 1911 (1983).

110   H. Jobic, M. Bée, J. Kärger, H. Pfeifer and J. Caro, J. Chem. Soc., Chem. Commun., 341 (1990).

111   H. Jobic, Spectrochim. Acta (in press).

112   L.M. Kustov, V.Yu. Borovkov and V.B. Kazansky, J. Catal. 72, 149 (1981).

113   M.J. Wax, R.R. Cavanagh, J.J. Rush, G.D. Stucky, L.Abrams and D.R. Corbin, J. Phys. Chem. 90, 532 (1986).

114   J. Sauer, J. Molec. Catalysis 54, 312 (1989).

115   H. Jobic, J. Catal. (submitted).

116   R.A. van Santen, B.W.H. van Beest and A.J.M. de Man, in "Physicochemical Properties of Zeolitic Systems and their Low Dimensionality", NATO workshop (in press).

117   S. Yashonath, J.M. Thomas, A.K. Nowak and A.K. Cheetham, Nature, 331, 601 (1988).

118   P. Demontis, G.B. Suffritti, A. Alberti, S. Quartieri, E.S. Fois and A. Gamba, Gazz. Chim. Ital. 116, 459 (1986).

119   L. Leherte, D.P. Vercauteren, E.G. Derouane, G.C. Lie, E. Clementi and J.M. André, in "Zeolites : Facts, Figures, Future", (Ed. P.A. Jacobs and R.A. van Santen), Elsevier, Amsterdam p. 773 (1989).

120   P. Demontis, S. Yashonath and M.L. Klein, J. Phys. Chem. 93, 5016 (1989).

121   B. Smit and C.J.J. den Ouden, J. Phys. Chem. 92, 7169 (1988).

122   H. Jobic, A.N. Fitch, H.J. Lauter and J. Tomkinson (in preparation).

LOW–ENERGY ION SCATTERING INVESTIGATIONS OF CATALYSTS I

H.H. Brongersma and G.C. van Leerdam

Department of Physics and Schuit Institute of Catalysis
Eindhoven University of Technology, P.O. Box 513
5600 MB Eindhoven, The Netherlands

ABSTRACT

Low–energy ion scattering (LEIS) provides unique information on the composition and structure of the outermost layers of a solid. Recent progress in both theory and experiment have led to the development of new methods of catalyst characterization.

The fundamental principles of LEIS are first discussed. These principles are then applied to investigate the surface structure of oxides and La/γ–Al$_2$O$_3$, Mo/γ–Al$_2$O$_3$ and Mo/SiO$_2$. It is shown that quantification of the LEIS results seriously effects earlier models for these supports and catalysts.

1. INTRODUCTION

Low–energy ion scattering (LEIS) is a surface analysis technique that selectively probes the outermost atomic layer of a solid. Since it is precisely this layer that largely determines the catalytic activity of a solid, LEIS should be an ideal technique for characterizing catalysts. The experimental set–up, however, requires that the sample under investigations is kept in a vacuum chamber. This limits the application of LEIS to surfaces that are stable in such an environment. Also, the ions can only probe the external and not the internal surface of a catalysts. For zeolites, where the precise internal pore structure plays a very important role, LEIS will be of little value. Since the processes determining surface segregation, relaxation and spreading will be similar for the internal and external surface, some limited applications are known.

In section 2 a description is given of the physical properties and features of LEIS. At the end a summary is given of the essentials that are necessary to understand and interpret the applications of LEIS that are described in section 3.

Both the description of the principles involved and that of the applications are highly complementary to the paper by E. Taglauer [1]. Since the applications of LEIS to catalyst characterization have also been reviewed by Horrell and Cocke [2], emphasis is put in the present paper on the latest developments and possibilities.

2. LOW–ENERGY ION SCATTERING

2a. Introduction

In a LEIS experiment (fig. 1) a beam of low–energy E_i (0.1 – 10keV) ions is directed onto the sample. The energy E_f of the ions scattered over an angle Θ is

Fundamental Aspects of Heterogeneous Catalysis Studied by Particle Beams,
Edited by H.H. Brongersma and R.A. van Santen, Plenum Press, New York, 1991

283

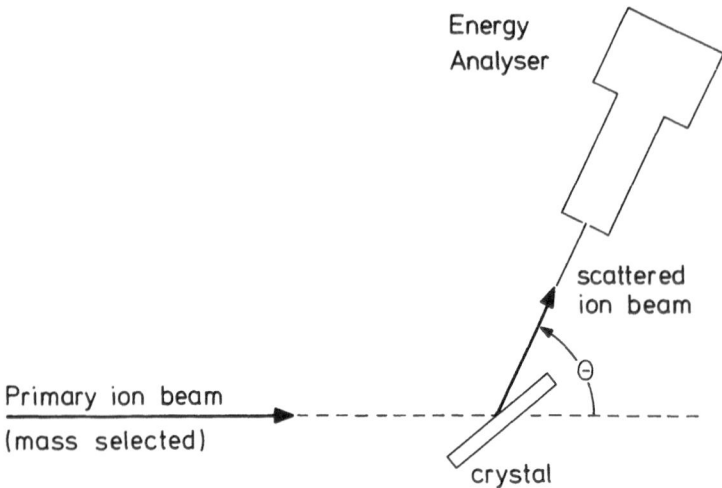

Fig. 1.   *Set–up of a LEIS experiment*

measured. Very often the experiments are carried out with inert gas ions such as He⁺. The high neutralization probability for these ions (due to the high electron affinity) ensures that only ions that made a single collision with an atom in the outermost atomic layer have a finite probability to be scattered in an ionized state. The analyzer is such that only <u>ions</u> are energy analyzed and detected. For such ions one can derive from the laws of conservation of momentum and energy that

$$E_f/E_i = \left[\frac{\cos \Theta \pm \sqrt{q^2 - \sin^2\Theta}}{1 + q} \right]^2 \qquad (1)$$

where $q = M_{at}/M_{ion}$ is the ratio of the masses of the surface atom and the ion. The minus sign (second solution) only exists if $1 \geq q \geq |\sin \Theta|$.

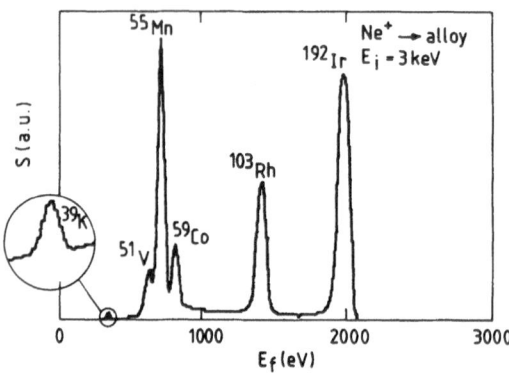

Fig. 2   *Energy spectrum of 3 keV Ne⁺ ions scattered by an alloy.*

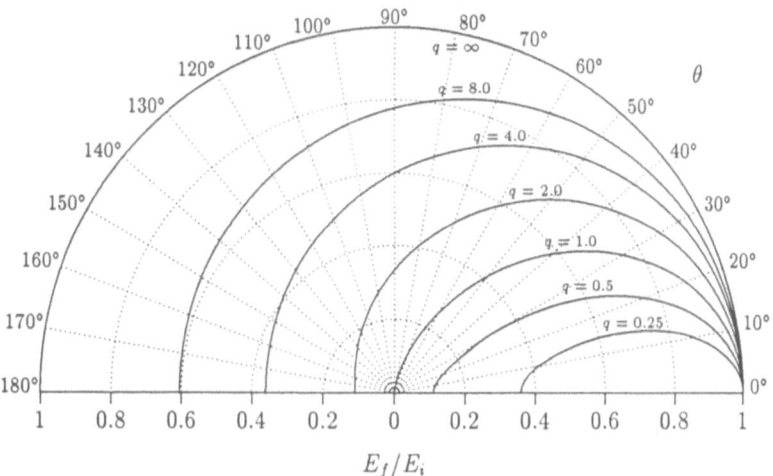

Fig. 3 Polar diagram of E_f/E_i vs the scattering angle Θ for a number of mass ratios q $= M_{at}/M_{ion}$.

The equation shows that for given experimental conditions (fixed E_i, Θ and M_{ion}), the energy of the scattered ions is fully determined by the mass of the surface atom. An energy spectrum of these ions is thus equivalent to a mass spectrum of the surface atoms. A typical spectrum is given in fig. 2. In fig. 3 eq. 1 is illustrated as a polar diagram. The energy ratio E_f/E_i is plotted for several values of q as a function of the scattering angle Θ. The diagram is helpful in selecting suitable experimental conditions. It shows, for instance, that the mass separation increases with increasing Θ and that an ill–defined scattering angle gives a much larger energy spread for small q values than for large ones.

Eq. 1 is based on the implicit assumptions that only elastic scattering (no electronic excitation of the ion or atom) occurs and that the surface atom can be regarded as a free atom at rest. In practice, these assumptions are very well fulfilled. Deviations give rise, however, to peak broadening and small energy shifts.

The scattered ion signal S depends on the scattering probability (differential scattering cross–section $\frac{d\alpha}{d\Omega}$) and the ion fraction P^+ of the scattered particles

$$S \sim P^+ \frac{d\alpha}{d\Omega}. \tag{2}$$

The scattering is caused by the screened Coulomb repulsion of the nuclei of the ion and atom. This process is well understood and can be accurately calculated (see e.g. [3]). In fig. 4 this is illustrated with a computer simulation for a homogeneous beam of 1 keV He^+ ions scattered by a Si atom. The shadowing behind the atom and the flux peaking at the side of the shadowcone are nowadays very successfully used to determine the precise surface structure of single crystals and its adsorbates [4].

For He^+ ions the ion fraction is typically $10^{-2} - 10^{-4}$ after a single collision. This reduces, of course, the sensitivity of the technique. However, the probability of He^+ ions to penetrate beyond the first atomic layer and still scatter back in an ionized state, is generally very small. The contribution to the signal from deeper layers is then negligible. LEIS can thus be used to selectively analyze the outermost atomic layer.

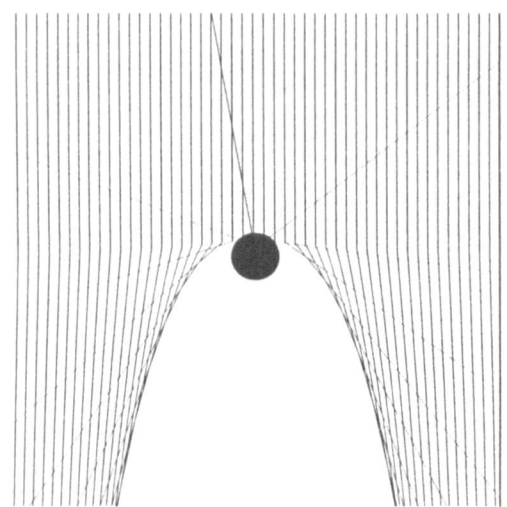

Fig. 4   Shadow cone for a homogeneous beam of 1 keV He⁺ ions scattered by a Si–atom.

2b. Experimental

Low–energy ion scattering has now been used for more than 20 years. Many experimental set–ups have been described. The scattering angles are generally large (around 140^0) to obtain a good mass resolution. Although the set–ups are generally adequate for the analysis of metals and semiconductors, the study of supported catalysts requires special precautions. In those studies it is particularly important to minimize damage and surface charging by the incident ion beam.

Damage

The ion beams cause sputtering and damage (change of the local structure). At low ion doses this is not serious since the ions will be scattered from the undamaged surface. Although the light He⁺ ions are generally used (to enable/facilitate the detection of light elements such as C, O, Al and Si), the damage is still significant. Through the years more and more efficient energy analyzers have, therefore, been developed, thus enabling analysis at lower and lower ion doses. The use of a cylindrical mirror analyzer [5] and of a hemispherical analyzer with a special input lens [6] proved to be major improvements.

Typical ion beams are now of the order of 10–100 nA ($10 - 10^3$ nA/cm²) for an analysis in 1–10 min.

Recently we have developed ([7], fig. 5) a new instrument ("EARISS") where the scattered ions are imaged in energy and angle on a position sensitive 2–dimensional detector. Although this instrument has not yet been used for supported catalysts, the gain in sensitivity that EARISS offers is already very clear. Amongst others, it was possible to analyze with this instrument the composition of a Pd/Au alloy with a 3 keV He⁺ ion beam of 3.4 pA in 30 s ! The surface damage is especially low, if one realizes that the sputtering action of He⁺ ions is about a factor of 10 lower than that of Ar⁺ ions.

The energy of the primary ions is generally between 0.5 and 3 keV. The lower energies having the clear advantage of lower damage per ion. However, quantification of the results becomes more problematic. At energies as high as 2 keV, the presence of hydrogen will seriously effect the scattered ion signals. This complicates the quantification of catalysts containing a large amount of water or hydroxyl–groups, especially since the presence of hydrogen ($M_{at} < M_{ion}$) cannot be detected by LEIS.

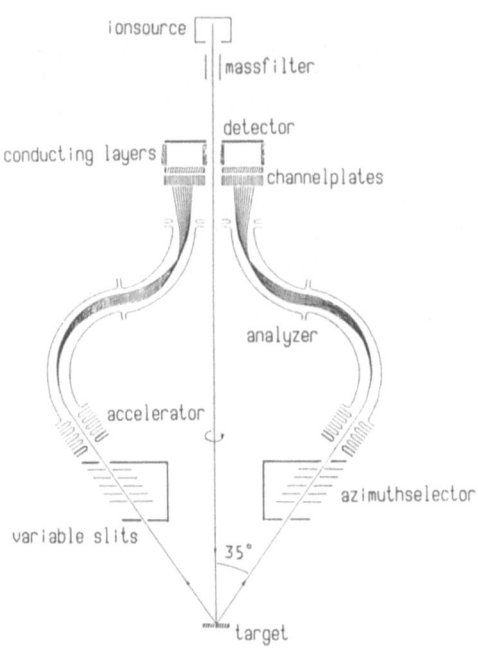

ionsource

massfilter

detector

conducting layers

channelplates

analyzer

accelerator

azimuthselector

variable slits

35°

target

Fig. 5 *Schematic diagram of EARISS. The ion trajectories are shown for several backscattered energies.*

Surface charging

Surface charging is a serious problem in the analysis of supported catalysts, since they are generally very good insulators. Although, in principle, one can compensate the charging by spraying electrons from a flood gun, this is not very effective for a rough surface. Charging will lead to a shift in the energy (the effective E_i in eq. 1 is lower) and a strong reduction in the scattered ion signal. In the past the reduction was generally attributed to the geometry of the surface: incident ions that penetrate into a pore, are blocked from scattering into the detector. Although this geometric factor certainly contributes to the loss in signal, computer simulations suggest that it can only account for a reduction by a factor of 2–3. The much larger reduction, that is generally found experimentally (easily an order of magnitude), is due to the electrostatic influence of the surface charge on the trajectories of the incoming and scattered ions. If the charging is not homogeneous, which is very likely for rough surfaces, the influence will not be the same for all ions. This will lead to a strong decrease in the number of ions that are accepted by the analyzer.

Adjusting the target position to try and compensate this effect does not lead to a major improvement.

Due to the presence of the inhomogeneous electrostatic field the scattered ion trajectories do not seem to originate from a point source anymore. For quantitative measurements it is thus crucial to effectively compensate such charging. In our own set–up this is accomplished using a kind of ring–shaped filament that surrounds the primary ion beam. The filament sprays electrons from all sides onto the target. The difference between signals of atomically flat targets and those of high surface area targets is then reduced to what is expected on the basis of geometric consideration.

We observe e.g. even a somewhat lower overall signal for α–Al_2O_3 than for γ–Al_2O_3, having specific surface areas of 5.5 and 280 m^2/g respectively. This is in contrast to findings of Margraf et al. [8] who found a 5 times larger signal for single crystalline α–Al_2O_3 than for γ–Al_2O_3 powder. The difference may, at least partially, be due to the lower primary energy of 500 eV they use (more sensitive to shielding by hydrogen) and to differences in surface charging of the aluminas.

Target preparation

Supported catalysts are pressed into Ta holders and then introduced in a pretreatment chamber. They are generally heated in a specific atmosphere to remove contaminations as water, hydrocarbons, etc. For oxidic catalysts heating at 300^0C at a pressure of $5 \cdot 10^3$ Pa of oxygen during 30 min. is generally adequate.

After the pretreatment, the catalysts are transferred to a carousel in the analysis chamber via a load lock. The carousel enables one to make a direct comparison with other catalysts having different loadings or with reference targets, etc.

2c. Neutralization and reionization processes

When an ion with kinetic energy E interacts with an atom or passes through a solid, it may be neutralized. The ion fraction is

$$P^+ = e^{-ct}, \tag{3}$$

where c is a constant and t the interaction time (proportional to $1/\sqrt{E}$). Although eq. 3 gives a good overall description of the charge state of the particle after scattering, it has in recent years become very clear that the situation is much more complex. Thomas et al. [10] and Aono et al. [10] have shown, for instance, that neutral He atoms can even be reionized by a close interaction with certain elements (amongst others Si, Al) at kinetic energies below 500 eV. For the He^+ ions in a LEIS experiment this implies that ions that are first neutralized while approaching the atom, are then reionized at close separation but have a high probability to be neutralized again upon leaving. The overall effect of the reionization is a higher ion fraction and thus a higher signal for that particular element. Quantum mechanical calculations of Muda and Newns [11] and of Verbist et al. [12] for He^+ ion scattering from Si provide more quantitative information: while most He^+ ions are first neutralized, about half of them is reionized at close separation. It is interesting that this number is rather insensitive to the precise energy. The scattered ion fraction, however, depends strongly on energy (10^{-3} at 500 eV to 10^{-1} at 10 kV).

2d. Quantification of LEIS

In the past LEIS has generally been described as a qualitative method. Quantification on an ab–initio basis proved difficult, since there was no good theory for calculating P^+. Also, experiments were generally not very reproducible. The main reason being that LEIS is extremely surface sensitive. A monolayer of contamination may completely shield the underlying substrate. The presence of hydrogen is particularly dangerous: since the mass of hydrogen is lower than that of the lightest inert gas ion, backscattering is impossible; however, hydrogen can neutralize the ions and thus strongly reduce the signal from deeper layers. For catalysts, which are in general non–conducting materials, surface charging is an additional complication (see section 2b).

Notwithstanding all these problems, it has become clear that a reliable quantification is possible under well–defined conditions.

A good procedure to check the absence of matrix effects and the reliability of a quantification is the following.

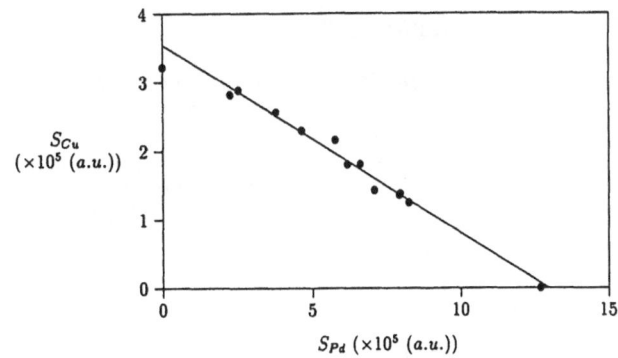

Fig. 6.   *Plot of the Cu signals vs Pd signals for a number of Cu/Pd alloys, Cu and Pd.*
The surface of the Cu target had not been properly cleaned; the data point is,
therefore, too low.

Suppose that the elements A and B, covering the fractions ν_A and ν_B of the surface,
give rise to the ion scattering signals S_A and S_B, then

$$S_A = \eta_A \cdot \nu_A \qquad (4a)$$
$$S_B = \eta_B \cdot \nu_B, \qquad (4b)$$

where η_A and η_B are the sensitivity factors for A and B. These equations are so simple,
since LEIS only samples the outermost atomic layer. The eqs. 4 only hold if the
sensitivity factors are not influenced by the identity of the neighbouring atoms ("no
matrix effects").

If the surface consists only of A and B atoms,

$$\nu_A + \nu_B = 1. \qquad (5)$$

Combining these eqs. gives

$$S_A = \eta_A - \frac{\eta_A}{\eta_B} \cdot S_B. \qquad (6)$$

A plot of S_A vs S_B for alloys A/B should be a linear function. It has been shown before
[13] that, under well-chosen experimental conditions, the linear relationship holds. In
fig. 6 this is illustrated for Cu/Pd alloys of various compositions and after various
treatments (sputtering, annealing). The fact that the data point for pure Cu does not fall
on the straight line is attributed to improper cleaning of this target.

For oxides the situation is not as simple as for metals. The open structure makes it
possible that atoms can be adsorbed without covering an atom in the layer just below.
The results suggest, however, that equations such as eq. 4 also hold for oxidic materials.
Some examples will be given in section 3.

Recently, another possibility to quantify the surface composition has been given by
Ackermans et al. [14]. By comparing the signal intensities for scattering of $^3He^+$ and $^4He^+$
ions it was shown how one can obtain the ion fractions P^+ in situ. Although it is
necessary to carefully select the energy at which the measurements are carried out
[15,16], it has the advantage that it circumvents the problem of reference targets and
open structures.

3. CHARACTERIZATION OF CATALYSTS

LEIS is a very powerful technique for determining the elemental composition of the outermost atomic layer of·a solid. For single crystals the surface structure can also be derived from angular dependent measurements. Unfortunately, this method is not applicable to polycrystalline compounds. Several authors have, therefore, developed variations of the LEIS technique to infer the atomic arrangements at the surface of such solids. In these techniques the scattered ion signals are determined as a function of:

–a. incident energy of the ions ("the energy method")
–b. loading of the catalyst
–c. time (or depth) in a sputtering experiment.

ad a. Neutralization and shadowing or blocking reduce the information depth. At higher energies neutralization (shorter interaction time) and scattering (trajectory of the ion less easily affected) become less important. It is thus expected that the information depth increases with energy.

ad b. If the signal of a metal on a support increases linearly with its loading, this is a strong indication that the metal is in the same type of site at all loadings.

ad c. When an element is only present in the outermost atomic layer, its signal will drop exponentially during sputtering. For double and thicker layers the signal will be a constant at first. Once the layers are sputtered through, it will also decrease in time. Preferential sputtering may change the rate of sputtering, but the initial slope of the signal remains characteristic.

In practice, it is not always easy to obtain well–defined reproducible signals from catalysts. As indicated in section 2b the measurements may be hampered by surface charging or contaminations. It has, therefore, become common practice to determine signal ratio's rather than absolute signals. It will be shown below that this is very dangerous. Probably all conclusions based on the "energy method" (method a) and many conclusions based on method b have to be revoked.

After a discussion of the energy method, the dangers and possibilities of the other two methods will be illustrated with some recent examples.

The energy method

It has been found by many authors that the energy dependence of the cation/anion signal ratio is very characteristic for a compound. McCune [17,18] classified the oxidic compounds into two groups. One group in which the cation is strongly shielded by the oxygen (e.g. in SiO_2, Al_2O_3, TiO_2 and Ta_2O_5) and another group for which the cation is not shielded (e.g. $Na\,NO_3$, $Ca\,CO_3$ and ZnO). In agreement with this partition, it was found that the cation/anion signal ratio increased strongly with energy for the first and not for the second group. The increasing coordination by oxygen in the series MgO, Al_2O_3, SiO_2 also nicely correlated with the observed shielding.

Nelson [19] and Martin and Netterfield [20] also interpreted their energy dependent results for Ta_2O_5 and ZrO_2, respectively, in terms of a strong shielding of the cation by oxygen. Similar arguments have been used to show that Mo is shielded by S in MoS_2 [21].

Apart from shielding effects, the energy dependence of the signals will depend on the differential scattering cross–section and on the ion fractions (eq. 2). It is possible, that the observed energy dependence is a consequence of the intrinsic atomic properties and not of the local atomic arrangement. It is generally accepted that the energy dependence of the scattering cross–sections is not enough to explain the results. For instance, Martin and Netterfield found for the 0.5–2 keV He^+ ion scattering by ZrO_2, that the Zr/O peak height ratio increased by a factor of about six, while the ratio of the scattering cross–sections increased only by a factor of approximately 1.5.

Much less is known about the ion fractions. A straight forward way to check the importance of atomic shielding in a compound is compare its cation/anion signal ratio

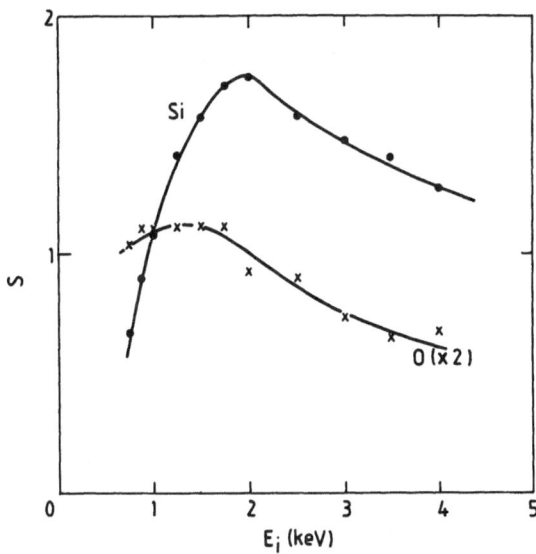

Fig. 7 *Energy dependencies of the Si and O signals S, as derived from the spectra of the reference compounds Si and B_2O_3.*

with that of the pure elements. Van Leerdam and Brongersma [22, 23] have carried out this analysis for SiO_2, Al_2O_3 (group I, according to McCune) and ZnO (group II). As reference targets pure Si, Al and Zn, were taken for the cations. Since pure oxygen is difficult to handle as a target, B_2O_3 was used instead. Due to the layered structure of B_2O_3 [24] the surface is expected to consist of a two–dimensional network of B and O atoms. This, together with the small size of B, will make the shielding of O by B of minor importance.

In fig. 7 the Si and O signals of the reference targets are given as a function of the incident energy. In fig. 8 the Si/O ratio is presented for SiO_2 and for the reference targets. The ratio for SiO_2 has been multiplied by 3.5 to correct for atomic densities. The results show that, within experimental error, the same energy dependent behaviour is found for the Si/O ratios. The influence of shielding on the energy dependence of the Si/O ratio is thus negligible ! In a similar way the cation/anion ratio was obtained for Al/O and Zn/O from the reference targets (fig. 9).

The observed energy dependences agree nicely with those of McCune for the compounds. The very different behaviour of the Zn/O ratio (decrease with increasing energy) thus does not result from a different shielding by oxygen. Van Leerdam and Brongersma [22, 23] could show that the effect is a consequence of differences in neutralization processes, which are intrinsic properties of the atoms. Maschhoff et al. [25] reach a similar conclusion for He^+ ion scattering by TiO_2. Understanding of the neutralization mechanism also suggest that the same type of explanation holds for all oxides studied by Nelson [19], McCune [18] and by Martin and Netterfield [20] and also for MoS_2 studied by Carver et al.[21].

In fig. 10 trajectory calculations are presented for 1 keV He^+ ion scattering by a SiO_2 surface. In low–energy ion scattering a scattering cross–section is more or less proportional to the atomic number of the atom. The diameter of the shadow cone for Si is thus larger than that for O. This is in sharp contrast to chemistry, where the radius of O^{2-} ions (0.14 nm) is much larger that that of Si^{4+} ions (0.04 nm). In agreement with the present findings, the shielding of cations by oxygen is not very effective. In studies of oxides the energy method only has a chance to be successful if much lower energies are used.

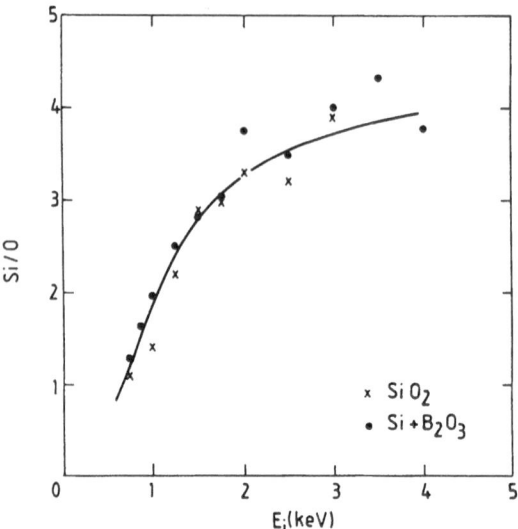

Fig. 8 Energy dependences of the Si/O ratio R for SiO_2 (x) and the reference compounds (•). R has been multiplied for SiO_2 by 3.5 to correct for atomic densities.

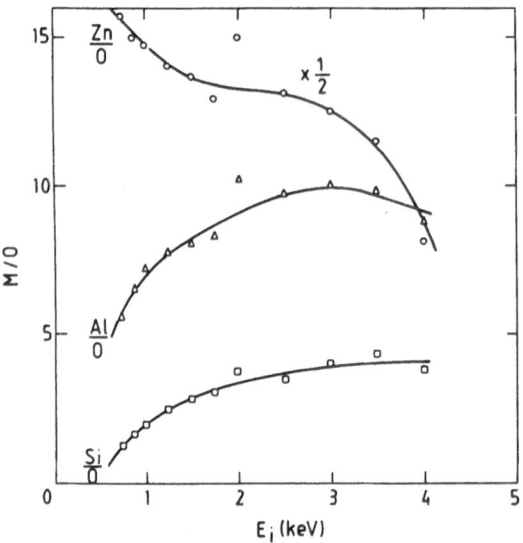

Fig. 9 Energy dependences of the cation/anion signal ratios as derived from the reference compounds Al, Si, Zn and B_2O_3. The exceptionally high ratios at 2 keV are probably due to a too low value for oxygen at this energy.

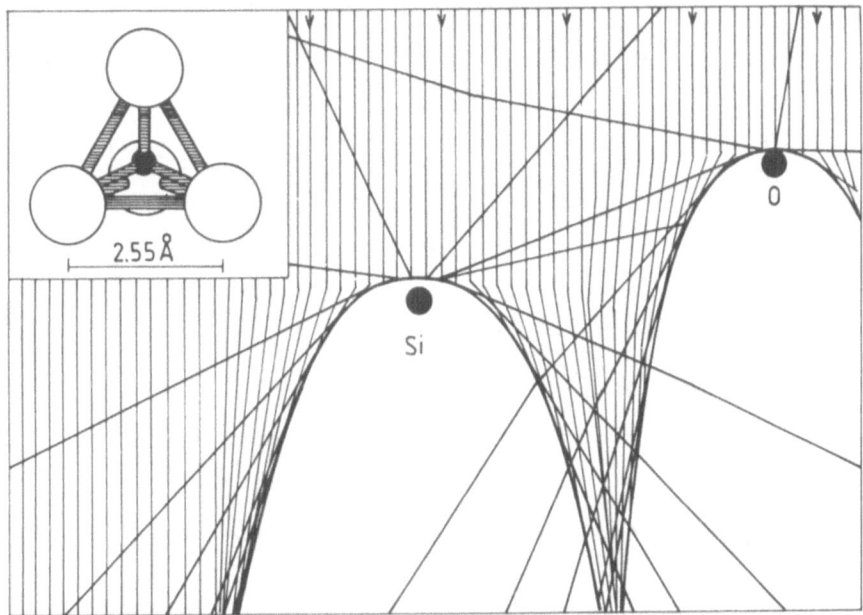

Fig. 10   Trajectory calculations for the scattering of a parallel beam of 1 keV He⁺ ions by SiO_2. The inset shows the SiO_4 tetrahedron.

Since the energy method does not work, the following examples of analyzing catalysts are based on studies as a function of loading and on depth profiling.

$La_2O_3/\gamma-Al_2O_3$

In search for thermostable catalysts for combustion catalysis, $\gamma-Al_2O_3$ has been doped with various elements. In the group of J.W. Geus it has been demonstrated that low loadings of lanthanum are already very effective in preventing sintering, provided the lanthanum has been deposited onto the support by specific adsorption of the [La (EDTA)]⁻ complex [26]. An X-ray Photoelectron Spectroscopy (XPS) study indicated that La remains uniformly distributed over the support after several stages of the sintering process. Although XPS is a surface sensitive technique, it averages the information over a number of atomic layers. We have, therefore, used LEIS to assess more accurately the position of the La species. A more detailed discussion is given elsewhere [15,16]. The samples studied are $\gamma-Al_2O_3$ (Al 4172, Engelhard Chemie) loaded with 1.6, 2.3 and 2.6 wt % La.

Fig. 11 shows the La/Al ratio of the samples after drying overnight at 60°C, after calcination for 5 hours at 550°C and after subsequent heating in stagnant air for 23 and 145 hours. The Al signal decreases in general, somewhat at higher calcination temperatures or longer times.

From the figure it follows that the La/Al ratio of the dried sample increases linearly with the La content. After calcination at 550°C the La/Al ratio has decreased and is no longer proportional to the La loading. After sintering at 1050°C the La/Al ratio has increased and is again proportional to the La content. During sintering the BET surface area has decreased from 265 to 90 m²/g [26].

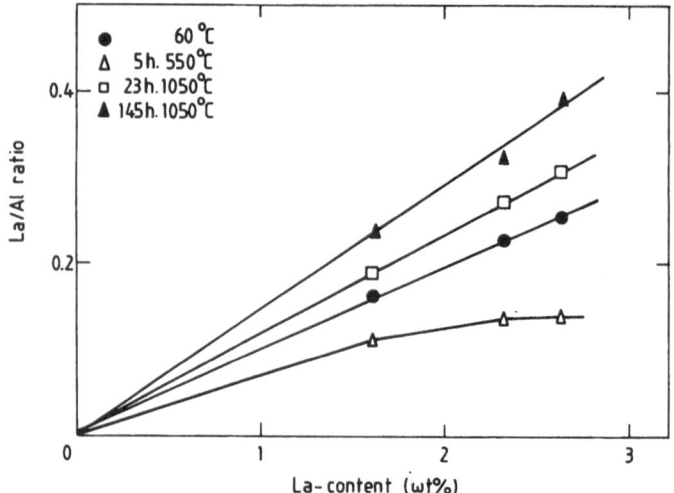

Fig. 11   *Dependence of the La/Al signal ratio on the bulk La–content for several thermal treatments.*

The proportionality between the La/Al ratio and the La loading for the dried samples implies that the La (still embedded in the complex) is fully dispersed or contains equi–sized La particles. An exponential decrease of the La signal as a function of depth profiling by sputtering has been observed. This proves that the La is present in a monolayer.

The considerable drop in the La intensity after calcination at 550°C indicates that La coagulation has occurred, most probably to form La_2O_3 particles which are 2 or 3 layers thick. This has been verified by depth profiling (the La signal hardly decreases in time; see fig. 12).

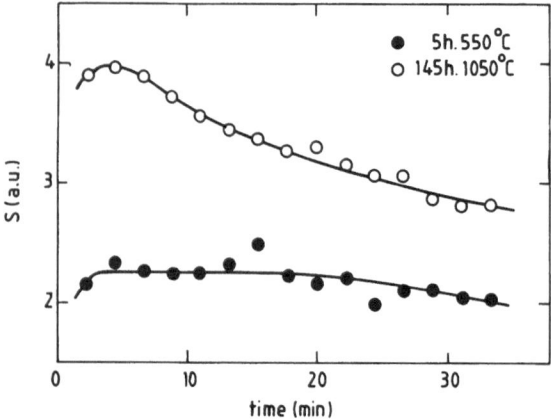

Fig. 12.   *Lanthanum sputter profile for samples containing 2.3 wt % La after calcination at 550°C and sintering at 1050°C.*

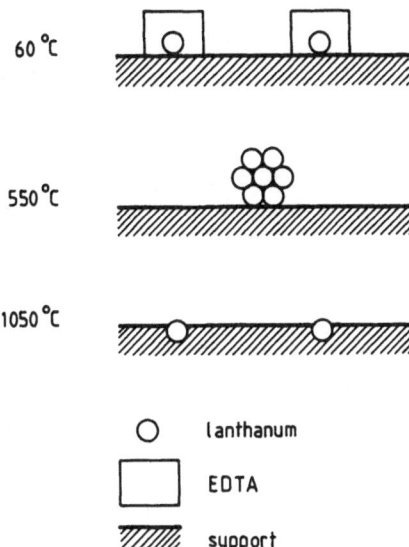

Fig. 13. Schematic picture of the distribution of lanthanum over the $\gamma-Al_2O_3$ after the different thermal treatments.

The increased La/Al ratio for the samples sintered at 1050°C shows that a redispersion of the La occurs under these conditions. In the depth profile the La signal decreases again exponentially (fig. 12). At 1050°C the La has probably reacted with the support and is now present in rather than on the surface of the support. The formation of this stable lanthanum–aluminate surface blocks the transformation to $\alpha-Al_2O_3$. In fig. 13 a schematic picture is given for the distribution of lanthanum after the various treatments.

$MoO_3/\gamma-Al_2O_3$

In the petroleum industry molybdenum based catalysts are extensively used to remove sulfur, nitrogen and metals from petroleum feedstocks or coal products. The activity of these hydrotreating catalysts has often been discussed in relation to their atomic structure and composition. Although these catalysts are usually sulfided under operation conditions, many studies have been carried out on the oxidic precursor. In particular, much insight has been gained by the work of Zingg et al. [27], Margraf et al. [28], Leyrer et al. [29] and by Kasztelan et al. [30]. Zingg et al. found, by plotting the Mo/Al signal ratio as a function of loading, a significant increase in the slope for loading above 8 wt % MoO_3. They concluded that at low loadings Mo occupies tetrahedral positions, whereas at loadings exceeding 8 wt % MoO_3 the octahedral sites are occupied. The distinction was based on the assumption that the tetrahedral positions are located just below the surface, and hence invisible to LEIS, while the octahedral positions are in the surface plane.

Recently, we have made an extensive study of molybdenum oxide on γ–alumina. These catalysts have been prepared by AKZO Chemical Division (Research Centre Amsterdam) by impregnation of the $\gamma-Al_2O_3$ support (surface area 270 m²/g) with ammonium di–molybdate. The LEIS results will be described elsewhere in more detail [15,16]. In this study the same change in the slope of the Mo/Al signal ratio versus loading was observed as by Zingg et al. However, the experimental improvements, as described in section 2, enabled the use of the signals rather than relative signals. This led to a very different picture of the surface structure. Some of the results are here summarized. They illustrate the importance of quantification and show the detailed information that can be obtained by LEIS.

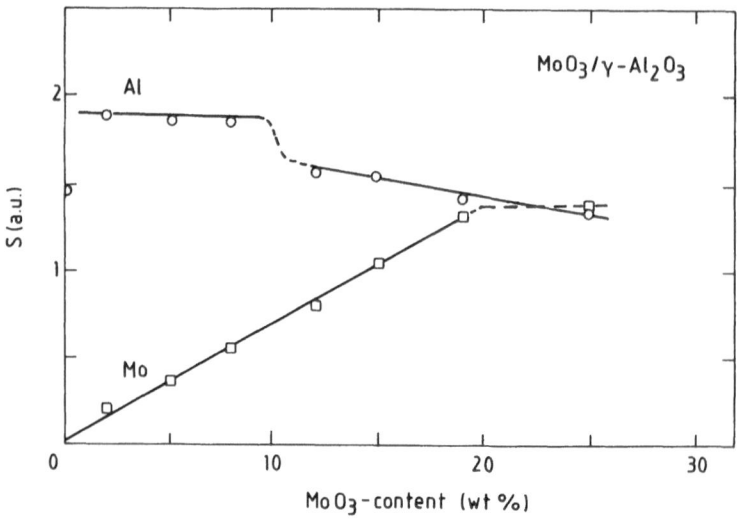

Fig. 14. *The signals S of Mo (squares) and Al (circles) as a function of MoO₃ content.*

In fig. 14 the Al and Mo signals are plotted as a function of the MoO_3 loading. The Mo signal increases perfectly linear as a function of loading. The increase in the slope of the Mo/Al signal ratio, as observed by Zingg et al., is thus due to the sudden decrease of the Al signal!

It is generally assumed that the surface of γ–Al_2O_3 is composed of a mixture of the low–index planes (111), (110) and (100), each having two types of layers containing different cation distributions. The layers are denoted as A and B, C and D, E and F respectively. It is also assumed that there is an equal distribution between the layers of a given low–index plane (e.g. the same amount of C and D surfaces). The present results strongly suggest that this is not the case. It seems that the D or E layer is predominantly present. The occurrence of D or E depends on the history and purity of the surface. The fact that the Al signal is about constant for loadings up to 10 wt% MoO_3 implies that the Al is not shielded by the Mo. This seriously restricts the posssible locations of the Mo. If one assumes that the surface stucture of γ–Al_2O_3 closely resembles that of the bulk, the following (somewhat tentative) picture is obtained [15, 16].

During calcination at a temperature of 500°C a phase transition at the surface of the Mo–loaded support takes place. This transition has been explained in terms of plane specificity of the individual support crystallites. Before calcination these crystallites most probably preferentially expose the D–layer of the (110)–face, while during calcination at 500°C a Mo–induced transition to the E–layer of the (100)–face occurs. Although the models for the surface structure may have been idealized, the use of them proves to be very useful.

During calcination MoO_3 tends to form a monolayer on the surface of the support. At low loadings ($\leq$ 8 wt%) Mo preferentially occupies tetrahedral sites on the E–layer of the support, which may be envisaged as epitaxial Mo–growth. At higher loadings ($\leq$ 19 wt%) the Mo atoms change from position on the substrate, and the formation of a polymeric oxomolybdenum species in the tetrahedral rows of the E–layer of the support is favoured. At loadings exceeding 19 wt% MoO_3 crystallites are observed on the surface.

Although a detailed picture of the present MoO_3/γ–Al_2O_3 catalyst has been attained, one should be careful with generalization of the results. The rather simple picture of predominantly one or two structures for the surface of γ–alumina seems to be in contrast to the infrared absorption measurements of Knözinger [31], where 7 OH–absorption bands can be distinguished. Although there is no doubt about the presence of OH–groups on the surface (they do not influence our results under the present experimental condition), the number of differently bonded groups is difficult to bring into line with a simple surface. Perhaps the fact that infrared absorption probes the bulk as well as the surface and the difficulty to quantify the occurrence of the various OH–groups is partially responsible for the discrepancy with the LEIS results. Also, the plane specificity of the support crystallite for example, may be different for support materials which have been differently synthesized and thermally treated. We have indications, for instance, that some contaminants induce already the formation of the E–layer in unloaded supports. The addition of V_2O_5, however, seems to favour the D–layer (Jacobs et al. [32]). It is clear that further research, using all available analysis techniques, is required to obtain a comprehensive picture of these catalysts.

MoO_3/SiO_2

Silica–supported molybdena catalysts have gained renewed interest by the finding that they can be effective oxidation catalysts. In the preparation of MoO_3 on silica catalysts various problems have been encountered. Part of these problems are due to the relatively weak interaction between the active oxide and the support [33,34]. To obtain highly dispersed catalysts, molybdenum has recently been introduced by homogeneous deposition precipitation using electrochemically reduced precursors [35]. By means of transmission electron microscopy it was confirmed that small Mo–oxide particles were homogeneously distributed on the support of the calcined catalyst [35]. Remarkably, the catalyst appeared to be active in the selective oxidation of ammonia only at loadings exceeding 15 wt% MoO_3. LEIS has been used to investigate whether crystallite size or dispersion effects could be responsible for this effect. A more extensive description will be given elsewhere [15,36].

The surfaces of 5, 10 and 21 wt% MoO_3/SiO_2 catalysts have been studied. After a calcination at 300°C, the samples were introduced into the analysis chamber (UHV) and annealed in steps of 50°C (0.5 hours–step). Fig. 15 shows the Mo and Si signals of the sample containing 21 wt% MoO_3 as a function of the annealing temperature. For all catalysts a considerable increase in the Mo signal is observed at annealing temperatures above 400°C. At the same time the Si signal decreases. Apparently the Mo–oxide spreads over the support. At 650°C a full coverage is reached. The full Mo dispersion is not obtained after prolonged annealing at 500°C. The present results do not give any information on the Mo/O stoichiometry.

Recently, the spreading of an active oxide over a support oxide has been discussed in terms of solid–solid wetting by Leyrer et al. [29]. Wetting will occur when the surface free energy of the active oxide plus the free interfacial energy are lower than the surface free energy of the support. The surface free energy of hydrated SiO_2 (130 mJ/m^2) is somewhat higher than that of MoO_3 (50–70 mJ/m^2). Unfortunately, free interface energies are not known. The moderate tendency of MoO_3 to spread over the support (in contrast to the spreading over Al_2O_3) suggests, however, a slightly positive interface energy. In contrast to our results, Leyrer et al. [29] did not observed any spreading of Mo–oxide at 450°C.

Additional information on the spreading can be obtained by plotting the data of fig. 15 together with the signal intensities for pure SiO_2 and MoO_3 in the form of a calibration curve (fig. 16). This is analogous to the procedure for Cu/Pd in section 2d. Since the data below 550°C fall on the straight line connecting the pure MoO_3 and SiO_2, this implies that upon heating the thickness of the Mo–oxide particles decreases (from about 4 to 2 layers, on the average), while spreading over the SiO_2 support.

At temperatures above 550°C the Mo–oxide becomes more effective in shielding the silica (data points below the line). This suggests that the structure of the Mo–oxide is changing to a monolayer structure which is dictated by the support. The number of Mo atoms per surface area decreases by about a factor of two.

The change in the surface structure around 550°C is probably induced by a dehydroxylation of the SiO_2. The much higher free energy of pure SiO_2 (600 mJ/m^2) will lead to spreading of the Mo–oxide.

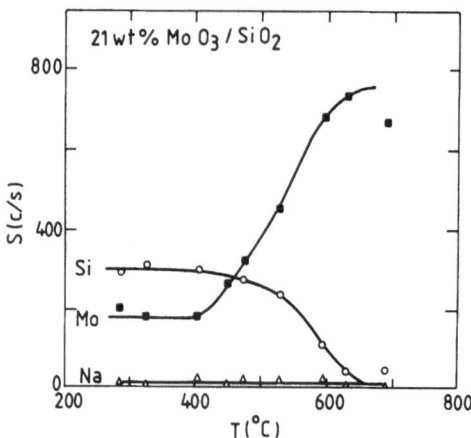

Fig. 15 *Signals of the 21 wt% MoO₃ on SiO₂ catalyst as a function of the annealing temperature.*

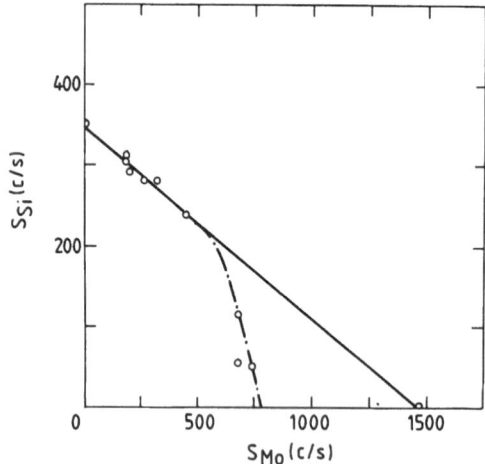

Fig. 16 *Calibration curve for the 21 wt% MoO₃ on SiO₂ catalyst, as derived from the data in fig. 15, together with the signals for pure MoO₃ and SiO₂.*

A reason why Leyrer et al. [29] do not observe any spreading may be their relatively low annealing temperature at which no dehydroxylation of the support takes place. The discrepancy may also be due to differences in particle size. Grove and Schmidt [37], using transmission electron microscopy, found that small (< 5 nm) Mo particles wet the SiO₂ under oxidizing conditions, while larger particles may also combine or vaporize. Since Leyrer et al. start from a physical mixture of MoO₃ and SiO₂, their particles are much larger than those in our experiments (only about 4 layers thick). Also, while spreading in their experiments should occur over macroscopic distances to be detectable, the spreading in our homogeneously deposited Mo takes place at a scale of a few nm's.

In how far do the above results explain the remarkable behaviour of the catalysts to the selective oxidation of ammonia? Since the sudden switch in activity around 15 wt% MoO₃ is very close to the estimated monolayer coverage (about 13 wt%), the activity may well be related to the presence of multilayers or MoO_3 particles on a Mo–covered support. The model of Tatibouet et al. [38] for the selective oxidation of methanol nicely corresponds with this observation, since they assume that two closely related molybdenum sites are necessary for the selective oxidation of methanol. The formation of a monomolecular layer, however, was achieved at annealing temperatures of about 550°C, which is much higher than the 400°C calcination of the catalysts prior to the reaction. One thus has to assume that the hydroxyl groups of the support can be removed chemically at 400°C, while 550°C is necessary during annealing. This assumption is supported by the observation that the colour of the catalyst changes during the initial part of the catalytic reaction [39], as has also been observed during the present annealing procedure.

Note added in proof: Directly after the Advanced Study Institute the group of Prof. J.W. Geus has confirmed by infra red absorption measurements that Mo–oxide starts to spread on the homogeneously deposited MoO_3/SiO_2 catalysts at temperatures around 550°C.

References

1. E. Taglauer, these proceedings.
2. B.A. Horrell and D.L. Cocke, Catal. Rev. Sci. Eng. **29** (1987) 447.
3. M.T. Robinson and I.M. Torrens, Phys. Rev. B9 (1974) 5008.
4. J.W. Rabalais, these proceedings.
5. H.H. Brongersma, N. Hazewindus, J.M. van Nieuwland, A.M.M. Otten and A.J. Smets, Rev. Sci. Instr. **49** (1978) 707.
6. M.P. Seah in: Methods of surface analysis, J.M. Walls ed., Cambridge Univ. Press (1990) p. 67.
7. P.A.J. Ackermans, P.F.H.M. van der Meulen, H. Ottevanger, F.E. van Straten and H.H. Brongersma, Nucl. Instr. Meth. Phys. Rev. **B35** (1988) 541.
8. R. Margraf, H. Knözinger and E. Taglauer, Surf. Sci. **211/212** (1989) 1083.
9. T.M. Thomas, H. Neumann, A.W. Czanderna and J.W. Pitts, Surf. Sci. **175** (1986) L737.
10. M. Aono and R. Souda, Nucl. Instr. Meth. **B27** (1987) 55.
11. Y. Muda and D.M. Newns, Phys. Rev. **B37** (1988) 7048.
12. G. Verbist, H.H. Brongersma and J.T. Devreese, these proceedings.
13. H.H. Brongersma, G.C.J. van der Ligt and G. Rouweler, Philips J. Res. **36** (1981) 1.
14. P.A.J. Ackermans, G.C.R. Krutzen and H.H. Brongersma, Nucl. Instr. Meth. Phys. Res. **B45** (1990) 384.
15. G.C. van Leerdam, Ph.D. Thesis, Eindhoven University, Netherlands (1991).
16. G.C. van Leerdam and H.H. Brongersma, to be published.
17. R.C. McCune, J. Vac. Sci. Technol. **16** (1979) 1569.
18. R.C. McCune, J. Vac. Sci. Technol. **18** (1981) 700.
19. G.C. Nelson, J. Vac. Sci. Technol. **15** (1978) 702.
20. P.J. Martin and R.L.P. Netterfield, Surf. Interf. Anal. **10** (1987) 13.
21. J.C. Carver, S.M. Davis and D.A. Goetsch, in Catalyst Characterization Science, Chapter 12, Ed. Publ. Co. American Chemical Society (1985).
22. G.C. van Leerdam and H.H. Brongersma, Proc. Symp. Surface Science, La Plagne (1990) 161.
23. G.C. van Leerdam and H.H. Brongersma, submitted to Surf. Sci.
24. C.R. Peters and M.E. Milberg, Acta Cryst. **17** (1964) 229.
25. B.L. Maschhoff, J.M. Pan, R.A. Baragiola and T.E. Madey, these proceedings.
26. I.I.M. Tijburg, Ph.D. Thesis, Utrecht University, Netherlands (1989).
27. D.S. Zingg, L.E. Makovsky, R.E. Tischer, F.R. Brown and D.M. Hercules, J. Phys. Chem. **84** (1980) 2898.
28. R. Margraf, J. Leyrer, H. Knözinger and E. Taglauer, Surf. Sci. **189/190** (1987) 842.
29. J. Leyrer, R. Margraf, E. Taglauer and H. Knözinger, Surf. Sci. **201** (1988) 603.
30. S. Kasztelan, J. Grimblot and J.P. Bonnelle, J. Phys. Chem. **91** (1987) 1503.
31. H. Knözinger, these proceedings.
32. J.-P. Jacobs, G.C. van Leerdam and H.H. Brongersma, these proceedings.
33. T. Ono, M. Anpo and Y. Kubokawa, J. Phys. Chem. **90** (1986) 4780.

34. Y. Barbaux, A.R. Elamzani, E. Payen, L. Gengembre, J.P. Bonnelle and B. Grzybowska, Appl. Catal. **44** (1988) 117.
35. J.J.P. Bierman, F.J.J.G. Janssen, M. de Boer, A.J. van Dillen, J.W. Geus and E.T.C. Vogt, J. Mol. Catal. **60** (1990) 229.
36. G.C. van Leerdam, H.H. Brongersma, J.J.P. Bierman and J.W. Geus, to be published.
37. C.L. Grove and L.D. Schmidt, Appl. Surf. Sci. **35** (1988/1989) 199.
38. J.M. Tatibouet, J.E. Germain and J.C. Volta, J. Catal. **82** (1983) 240.
39. E.T.C. Vogt, Ph.D. Thesis, University of Utrecht, Netherlands (1988).

LOW-ENERGY ION SCATTERING INVESTIGATIONS OF CATALYSTS II

Edmund Taglauer

Max-Planck-Institut für Plasmaphysik
EURATOM Association, D-8046 Garching, FRG

INVESTIGATION OF ADSORPTION LAYERS BY ION SCATTERING

Real catalysts are multi-component systems. There is at least one major component finely dispersed on a high surface area support and frequently there are additional components added which act as promotor substances, conferring advantages to the catalyst with respect to activity, selectivity, or stability. Hence, catalysts are adsorption systems that generally are very complex in composition and structure. Since ion scattering spectroscopy (ISS) is specifically sensitive to the outermost atomic layer of a surface /1,2/ and since the catalytic properties may be mainly determined by the outermost chemical species, ISS is well suited to gain useful information about catalytic systems /3/.

The scattering signal intensity from an adsorbed species, I_A, with a surface density N_A can be expressed as

$$I_A = I_o \cdot N_A \cdot d\sigma/d\Omega \quad \cdot \Delta\Omega \cdot K \cdot P_A \tag{1}$$

where I_o is the primary ion current, $d\sigma/d\Omega$ the cross section for scattering into a solid angle $\Delta\Omega$; K is a constant factor characterizing the spectrometer and P_A the ion survival probability. Up to a coverage of one monolayer the signal should therefore increase linearly with the surface density of the adsorbed species. Since the substrate atoms are covered by the adsorbing species, its signal decreases accordingly and this is given by an expression complementary to eq. (1):

$$I_S = I_o \cdot d\sigma_S/d\Omega \cdot \Delta\Omega \cdot K \cdot P_S \cdot (N_S - \alpha N_A) \tag{2}$$

The shadowing factor α indicates how many substrate atoms are shadowed by one adsorbate in a particular scattering geometry. Thus α is a function of the angle of incidence, ψ, and the scattering angle ϑ. In various studies values for α of the order of 1, up to 4 have been determined. For coverages close to one monolayer the shadows of the adsorbates overlap and the apparent shadowing factor decreases. A

Fundamental Aspects of Heterogeneous Catalysis Studied by Particle Beams,
Edited by H.H. Brongersma and R.A. van Santen, Plenum Press, New York, 1991

critical point in eqs. 1 and 2 is the assumption of a constant ion escape probability P. This factor can in principle change during adsorption because of concurring changes in surface structure, work function etc. Also for grazing incidence and exit angles trajectory dependent neutralization has been observed. But there are several cases reported in the literature which show that the linear relations given in eqs. (1) and (2) are justified. As an example we can consider the adsorption of CO on Ni(100), a system which has been extensively studied in view of fundamental catalytic processes. Figure 1 shows the ion scattering intensities as a function of CO exposure /4/. The oxygen and nickel signals follow the relations given by eqs. 1 and 2. Saturation is reached after an exposure of about 5 L, where a c(2x2) structure can be osbserved. CO adsorption on Ni(100) causes a large work function change of 0.9 eV, but a linear dependence of the He^+ scattering signal on CO coverage is still observed. This is probably due to the fact that a much larger energy difference of about 20 eV is available for the Auger neutralization process.

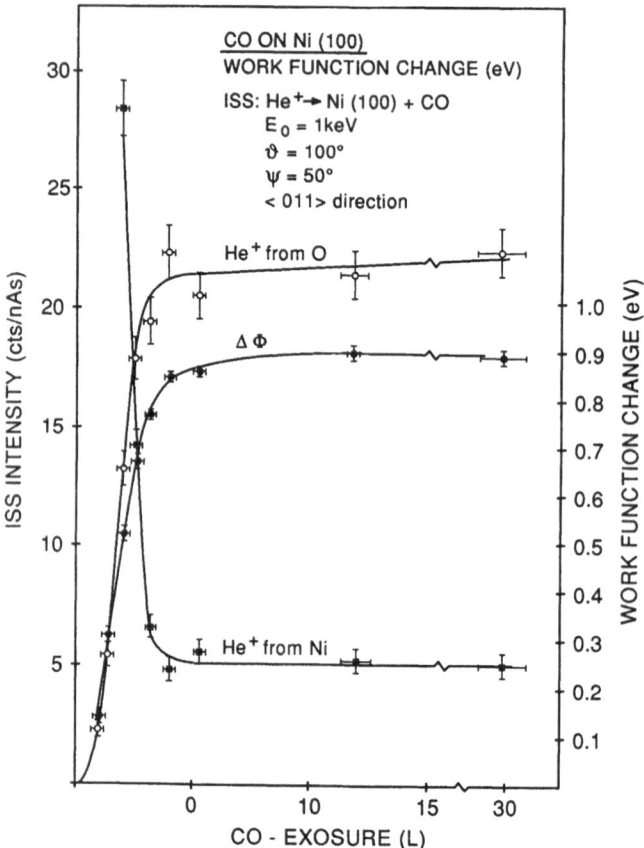

Fig. 1. He^+ ion scattering intensities and work function change for the adsorption of CO on Ni(100) at 300 K /4/.

In catalyst studies additional problems can arise from charging effects, surface roughness etc. Therefore absolute scattering intensities are sometimes difficult to compare and relative measurements, i.e. intensity ratios, are taken instead. For an adsorbate and substrate signal this yields

$$I_A/I_S = d\sigma_A/d\sigma_S \cdot P_A/P_S \cdot \theta/(1-\alpha\theta) \quad \sim \text{const } \theta \ (1+\alpha\theta +...) \quad (3)$$

showing that the ratio is linear in θ only for $\alpha\theta << 1$. For site competing adsorbates the same expression holds, using $\alpha = 1$. Here θ is the coverage N_A/N_S. However, intensity ratios can well represent surface density ratios N_1/N_2 of two species on the surface

$$I_1/I_2 = d\sigma_1/d\sigma_2 \cdot P_1/P_2 \cdot N_1/N_2 \quad (4)$$

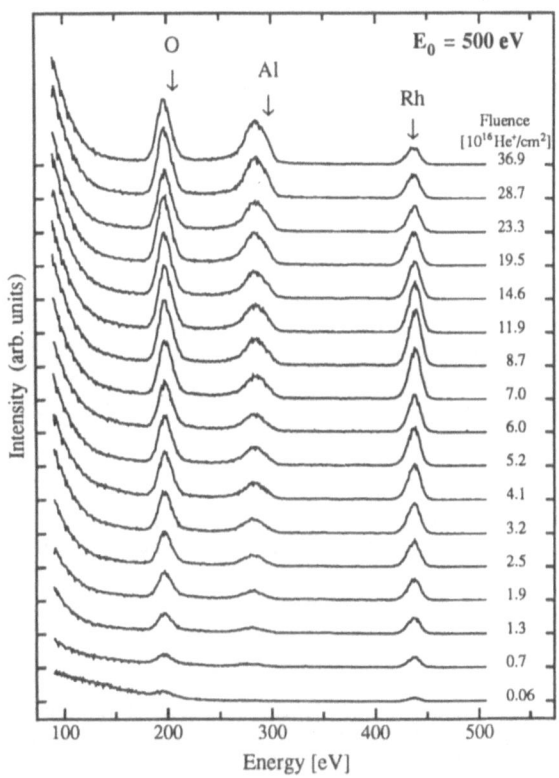

Fig. 2. ISS spectra of an alumina film, formed by oxidation in air, with an overlayer of rhodium. To the right, the fluence at the beginning of the spectra is noted. The arrows indicate the peak positions, calculated with the binary collision model /5/.

Within the concept presented above, I_{SS} can determine the elemental species on the surface, the species which are covered by the adsorbate and also the surface density of the adsorbate.

Intensity ratios are particularly useful for not atomically "clean" samples, because contaminants which consist of light elements (H, CH) do not contribute to the scattering intensity in most scattering geometries. That is, signal conservation is not generally granted in ISS. During sputter etching of such a sample the total scattering signal increases. An example for that behaviour is shown in Fig. 2 /5/. In such cases quantification is possible to some extent by using intensity ratios as mentioned above. Among others, successful data interpretation using intensity ratios was possible for such catalytic systems as MoO_3 /Al_2O_3 /6/, Cu/Al_2O_3 /7/ etc.

If light adsorbates such as hydrogen isotopes have to be detected, this can be done by using fairly small scattering angles /8/, or probably more favourably by using direct recoil detection /9, 10/. A spectrum demonstrating scattering and recoil peaks from H_2O adsorbed on Ni is shown in Fig. 3.

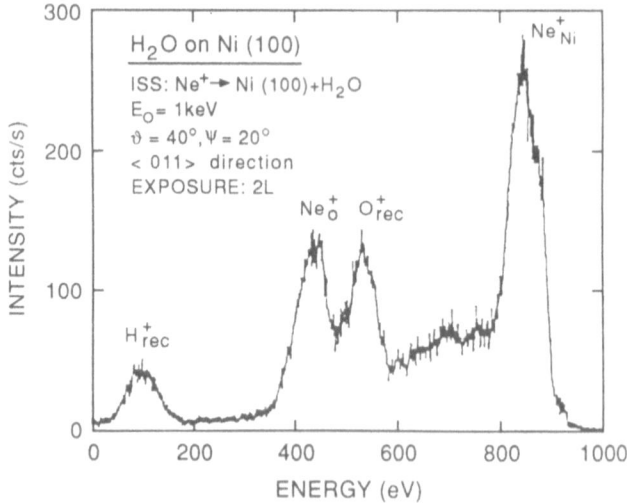

Fig. 3. Ion energy spectrum for Ne^+ bombardment of a Ni(100) surface with H_2O adsorbed at 104 K. Ne^+ scattering and O^+ and H^+ recoil peaks are indicated /4/.

Structural information in terms of next neighbour positions and distances can be obtained from ion scattering by making use of the shadow cone concept. This is explained in part I /1/. Due to flux peaking an intensity maximum is obtained when the edge of the shadow cone from a scattering atom hits a neighbouring atom. The corresponding critical angle of incidence ψ_c is related to the shadow cone radius r_s by

$$\psi_c = \sin^{-1} (r_s/d) \tag{5}$$

where d is the distance between the atoms under consideration.

If the shadow cone from a substrate atom is directed onto an adsorbate which is located at a distance δ above the substrate surface, the relation is

$$\psi_c^A = \sin^{-1}(r_s/(d^2 + \delta^2)^{1/2}) - \tan^{-1}(\delta/d) \qquad (6)$$

Evaluation of eq. 6 shows, that by this means the height δ can be determined quite accurately (with an uncertainty of about 0.1 Å), and less sensitivity is obtained for the substrate-adsorbate distance in that case. Using this concept, e.g. the position of H adsorbed on Ru(001) could be determined by recoil spectroscopy /11/. The concept of recoil detection has also been successfully applied for a number of structure determinations by using a time-of-flight technique /12/.

SPUTTERING EFFECTS

The ion beam used for analysis of a surface in ISS unavoidably causes sputtering of the bombarded area. However, this effect can be minimized such that a quasi-static analysis is possible. For conventional LEIS analysis a fluence of the order of 10^{13} ions/cm² is necessary to take a complete energy spectrum. Using time-of-flight and neutral particle detection techniques, this fluence can be reduced by at least a factor of 100 /13/. On the other hand, the sputtering action of the beam is frequently found serendipituous because it can be used for sputter depth profilng during the analysis. In particular the near-surface "depth" analysis in the monolayer range turned out to be an extremely useful application of ISS analysis of supported catalyst systems.

The efficiency of the sputtering process is characterized by the sputtering yield Y (ejected atoms/incoming ion). Y is determined by the energy deposition of the ion in the surface region and the binding energy of the surface atoms /14/. Consequently, the yield can be expressed by a mass factor and an energy factor, as was proposed for an empirical formula /15/:

$$Y = 6.4 \times 10^{-3} M_2 \times \gamma^{5/3} E'^{1/4}(1 - 1/E')^{7/2} \qquad (7)$$

This universal expression for light ions and normal incidence agrees with experimental sputtering yields within a factor of 2 or better in the parameter range $M_{ion}/M_{target} < 0.4$ and $1 < E' < 20$. Here γ is the energy transfer factor between two atomic masses

$$\gamma = 4 M_1 M_2/(M_1+M_2)^2$$

and E' is the kinetic energy measured by the sputtering threshold energy, $E' = E/E_{th}$. For light ions the threshold energy can be expressed in terms of the surface binding energy E_B and γ :

$$E_{th} = E_B/\gamma(1-\gamma) \qquad (8)$$

For ion-target combinations of interest in our context, e.g. He$^+$ on Al$_2$O$_3$, Y has a maximum value of 0.2 around 1 keV /15/ (normal incidence). Generally compounds such as oxides show a higher empirical threshold energy than elements: for Al E$_{th}$ = 20.5 eV, for Al$_2$O$_3$ E$_{th}$ = 85 eV. It can be concluded that sputtering of elements is sufficiently well understood and yields can be calculated from relations such as eq. 7. The situation is more complex for compounds for which the sputtering yields for different components are different and therefore the ion bombardment results in a change of the stoichiometry of the surface layer /16/. This phenomenon, called preferential sputtering, has to be taken into account for the interpretation of sputter depth profiles combined with surface analysis /17/.

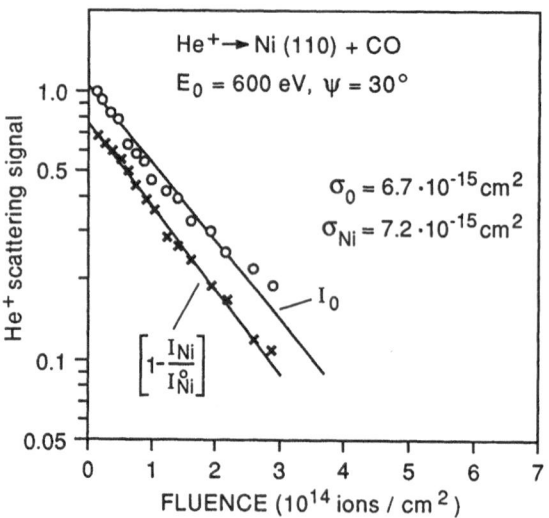

Fig. 4. Sputtering of CO from a Ni(110) surface as monitored by He$^+$ ion scattering from O and Ni, respectively /18/.

A case of special interest is the sputtering of adsorption layers. It is often used in surface cleaning procedures /18/ and in our context it is applied for obtaining information about the layering structure of supported catalyst systems. The idealised concept is to remove layer-by-layer by the sputtering action of the beam. This is of course a simplification of a very complex process governed by physical processes in the collision cascade, by ion beam mixing, recoil implantation etc. Some of these aspects are discussed in ref. 18.

If we deal with a spectroscopy like ISS whose signal is proportional to the surface density N_A of an adsorbed species (eq.1), then the signal from this adsorbate varies with the fluence i.t like

$$\ln(I_A(t)/I_A(o)) = -\sigma \cdot i \cdot t + const \qquad (9)$$

and a similar expression holds for the substrate signal I_s if we plot $\ln (1-I_s/I_s^0)$. This is shown in Fig. 4 demonstrating the applicability of the simple concept for adsorption layers. From the slope of the straight lines the sputtering or desorption cross section σ can be determined for coverages of a monolayer or less (whereby the knowledge of the absolute coverage is not necessary). σ is related to the sputtering yield through the monolayer density N

$$Y = \sigma \cdot N \qquad\qquad (10)$$

Experimental values for σ are in the range of 10^{-14} cm^2 to 10^{-16} cm^2 and the corresponding yields are often significantly higher than those for substrates (metals) /18/, particularly at low ion energies. Concomitantly, the energy dependence is generally weaker than for substrate materials.

Sputtering cross sections of adsorption layers can be calculated by using the theory of Winters and Sigmund /19/ with an accuracy which is usually sufficient for practical purposes. In this concept the cross section is calculated as the sum of three contributions to the sputtering of the adsorbed species: direct knock-off processes by the incoming ions, sputtering by reflected primary ions and sputtering due to momentum transfer from sputtered substrate atoms. Since the first two contributions decrease with increasing ion kinetic energy while contribution three increases, the energy dependence is weak and the yield can even increase towards lower energies in the case of light primary particles.

The removal of adsorbed species from the surface by ion impact can of course not only lead to removal into the vacuum (sputtering or ion impact desorption) but also to relocation into the bulk material, i.e. recoil implantation. This process is particularly effective for heavy adsorbates on a relatively light substrate, and has been studied for such cases /20/ which are also of relevance for supported catalyst systems. Here in addition to the desorption cross section σ_D the recoil cross section σ_R has to be considered, and σ_R can amount to 10 percent of σ_D or more /18,20/. The surface coverage $N(\Phi)$ as a function of fluence Φ = i.t then no longer follows one simple exponential decay but has to be described by an expression like

$$N(\Phi) = N_0 \cdot \sigma_R/(\sigma_D + \sigma_R) + N_0 \cdot \sigma_D/(\sigma_D+\sigma_R) \exp-(\sigma_D+\sigma_R)\cdot\Phi \qquad (11)$$

Thus the frequently observed deviations from an exponential law of the scattering intensity decrease with higher fluences can be interpreted as significant recoil implantation.

STUDIES OF SUPPORTED CATALYST SYSTEMS

Although real catalyst systems generally exhibit very rough surfaces and supports such as Al_2O_3 or SiO_2 are highly insulating, it is possible to perform valuable ISS investigations on such samples by pressing wafers from the powder material and by flooding the specimen with electrons during the analysis. As an example, we consider the case of molybdena catalysts which contain nickel as promotor and are supported on high surface area (160 m^2/g) γ-alumina /21/. For a

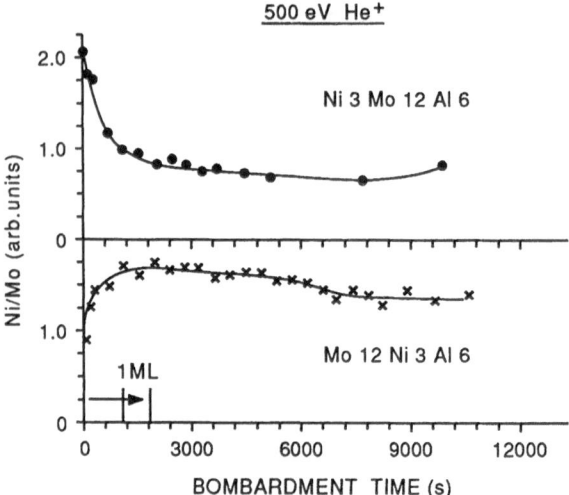

Fig. 5. "Depth profile" of Ni/Mo atomic ratios for catalysts first impregnated with Mo and subsequently with Ni (top) and vice versa (bottom) /21/.

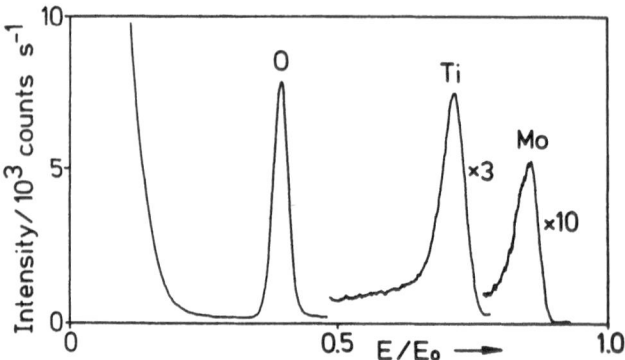

Fig. 6. Typical ^{4}He$^+$ ion scattering spectrum of a physical mixture of TiO$_2$ and MoO$_3$ (4.1 wt.%). Primary energy E$_0$=500eV, scattering angle ϑ = 137^0 /25/.

sequential impregnation to a coverage of about one monolayer, the ISS analysis clearly shows that either the Mo or the Ni species were more abundant on the top atomic layer, depending on the sequence of impregnation, see Fig. 5. It could further be shown that a redistribution of these species occurs at calcination temperatures of 870 K, the ISS results being supported by other spectroscopies /21/. For a similar catalyst, the CO-Mo/Al$_2$O$_3$ system, the formation of a bilayer was deduced from ISS data /22/, whose formation required first Co and subsequent Mo impregnation. Again, a redistribution upon calcination at 900 K was observed /23/. Investigations of that kind have by now proven in many cases the usefulness of ISS analysis of catalytic systems which is connected with the sensitivity of the technique to the outermost atomic layer. Also a strong correlation between thiophene hydrodesulfurisation (HDS) activity and ISS results of such HDS catalysts has been demonstrated /24/.

An interesting pheonomenon is the formation of "monolayers" by solid-solid wetting in supported catalyst systems, as e.g. MoO$_3$ on TiO$_2$ /25/. Figure 6 shows an energy spectrum of a physical mixture of the components and Figs. 7 "depth" profiles, i.e. the He$^+$ fluence dependence of Mo/Ti intensity ratio. It is clearly seen that calcination in the oxygen stream with or without water vapor leads to a steep increase of the Mo/Ti ratio close to the surface, i.e. a spreading of the active Mo compound on the TiO$_2$ support. Similar results are found for WO$_3$ and V$_2$O$_5$ as active components and Al$_2$O$_3$ as support /25/. On SiO$_2$ no spreading is observed. It could also be shown that the formation of the surface polymolybdate, which is considered a necessary precursor state for active catalysts, is a process that has a quite different dependence on calcination time compared to spreading, see Fig. 8. For a calcination temperature of 723 K the spreading is completed after about 5 hours, whereas the polymolybdate formation requires about 30 hours to be completed as shown by Laser Raman Spectroscopy. Therefore it can be concluded that spreading and polymolybdate formation occur in separate processes. Moreover, polymolybdate formation requires the presence of water vapor, while the ISS results for spreading are similar for H$_2$O-saturated and dry conditions.

As mentioned above, surface roughness is an important feature of catalyst supports. Therefore the question arises about its influence on ion scattering results. Generally speaking, a reduction of scattered ion intensities can be expected due to blocking of parts of the bombarded area in the scattered beam direction. For a uniform surface structure, (e.g. hemispherical shape) this effect should be independent of the amplitude (radius) of the structure. In realitiy a distribution of step heights and inclination angles exists. For a two-dimensional model the blocked fraction has been calculated and compared to experiments /26/. For Au on ceramic Al$_2$O$_3$ as the rough surface (rms roughness 0.25 μm, rms slope 21°) compared to Au on sapphire as the smooth surface, an intensity reduction of about 20 % has been obtained from calculations and experiments (1500 eV He$^+$ and Ne$^+$ scattering). Much larger effects were observed with Al$_2$O$_3$ supported systems, where intensities differ by more than a factor of 6 betweeen smooth alumina films and powder wafers /27/. However, comparatively small differences were found for intensity ratios (e.g. O/Al) and also the depth profiles after molybdate impregnation of these systems were very similar. It can therefore be concluded that relative measurements are significant also for systems with strong variations in surface roughness. This supports the intention that studies on plane model catalysts bear relevance for the description of real catalyst systems.

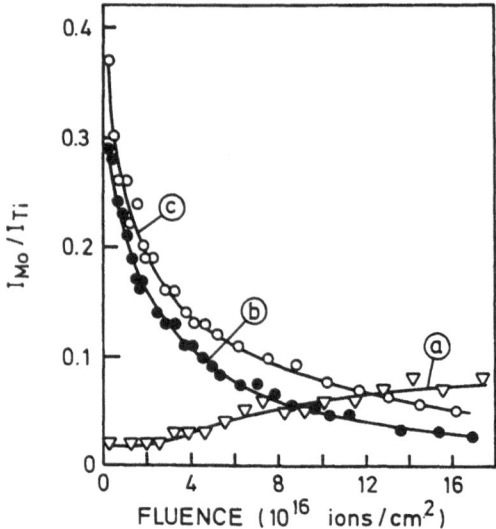

Fig. 7. Scattered He$^+$ ion intensity ratios Mo/Ti as a function of He$^+$ fluence yield "depth profiles" of MoO_3/TiO_2: (a) physical mixture. (b) after calcination at 720 K for 24 h in dry O_2, and (c) in H_2O-saturated O_2 /25/.

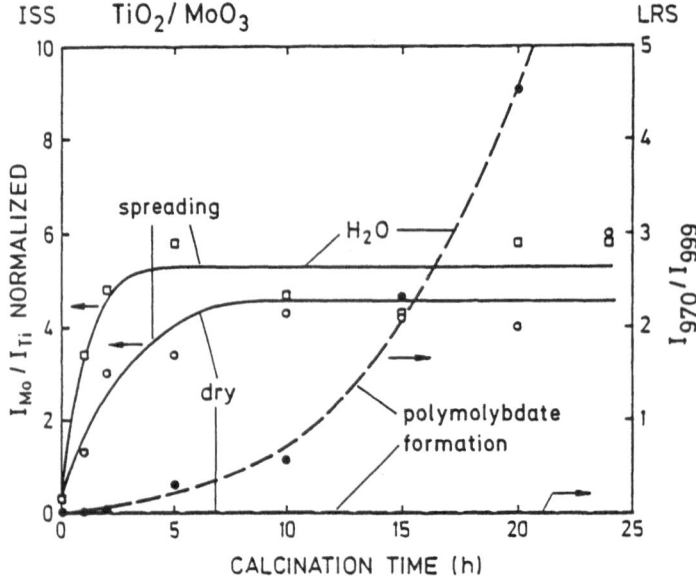

Fig. 8. Dependence of the normalized Mo/Ti He$^+$ ion scattering intensity ratios on calcination time. In order to obtain a normalized measure for the degree of spreading the ratios of the initial to final intensity values from fig. 7 are taken. The laser Raman signal ratio for the 970 cm^{-1} and 999 cm^{-1} bands, indicative of polymolybdate formation, is also shown.

REFERENCES

/1/   H. H. Brongersma, LEIS Investigations of Catalysts I, this volume.

/2/   E. Taglauer, Appl. Phys. A38, 161 (1985).

/3/   H. Jeziorowski, H. Knözinger, E. Taglauer and
C. Vogdt,
J. Catal. 80, 286 (1983).
B.A. Horrell and D.L. Cocke, Catal. Rev. - Sci. Eng. 29, 447 (1987).

/4/   M. Beckschulte, D. Mehl and E. Taglauer, Vacuum 41, 67 (1990).

/5/   Ch. Linsmeier, E. Taglauer and H. Knözinger, contribution in this volume.

/6/   F.R. Brown and D.M. Hercules, J. Phys. Chem. 84, 2898 (1980).

/7/   B.R. Strohmeier and D.M. Hercules, J. Phys. Chem. 88, 4922 (1984).

/8/   R. Bastasz, T.E. Felter and W.P. Ellis, Phys. Rev. Lett. 63, 558 (1989).

/9/   R.J. Schneider, W. Eckstein and H. Verbeek, Nucl. Instr. Meth. 218, 713 (1983).
J.W. Rabalais, J.A. Schultz and R. Kurnov, Nucl. Instr. Meth. Phys. Res. 218, 719 (1983).

/10/   G.N. van Wyk, W. Englert and E. Taglauer, Nucl. Instr. Meth. in Phys. Res. B35, 504 (1988).

/11/   J. Schulz, E. Taglauer, P. Feulner and D. Menzel, Verhandl. DPG (VI) 25, 1166 (1990).

/12/   J.W. Rabalais, this volume.

/13/   T.M. Buck, G. H. Wheatley, G.L. Miller,
D.A.H. Robinson and Y.-S. Chen, Nucl. Instr. Meth. 149, 591 (1978).

/14/   P. Sigmund in "Sputtering by Particle Bombardment I", R. Behrisch, ed., Springer Berlin, Heidelberg (1981).

/15/   J. Bohdansky, J. Roth and F. Brossa, J. Nucl. Mater. 85&86, 1145 (1979).

/16/   G. Betz and G.K. Wehner in "Sputtering by Particle Bombardment II", R. Behrisch, ed., Springer, Berlin, Heidelberg (1983).

/17/   E. Taglauer, Appl. Surf. Sci. 13, 80 (1982).

/18/   E. Taglauer, Appl. Phys. A51, 238 (1990).

/19/   H.F. Winters and P. Sigmund, J. Appl. Phys. 45, 4760 (1974).

/20/   K. Morita and H. Morii, Nucl. Instr. Meth. in Phys. Res. B13, 457 (1986).

/21/   J. Abart, E. Delgado, G. Ertl, H. Jeziorowski,
H. Knözinger, N. Thiele, X. Zh. Wang and
E. Taglauer, Appl. Catal. 2, 155 (19882).

/22/   F. Delannay, E.N. Haeussler and B. Delmon, J. Catal. 66, 469 (1980).

/23/   B.A. Horrell, D.L. Cocke, G.R. Sparrow and
J. Murray, J. Catal. 95, 309 (1985).

/24/   M. Houalla, C.L. Kibby, E.L. Eddy, L. Petrakis and D.M. Hercules, J. Catal. 83, 50 (1983).

/25/   J. Leyrer, R. Margraf, E. Taglauer and H. Knözinger, Surf. Sci. 201, 603 (1988).

/26/   G.C. Nelson, J. Appl. Phys. 47, 1253 (1976).

/27/   R. Margraf, H. Knözinger and E. Taglauer, Surf. Sci. 211/212, 1083 (1989).

TIME-OF-FLIGHT SCATTERING AND RECOILING SPECTROMETRY (TOF-SARS) FOR SURFACE STRUCTURE DETERMINATIONS

J. Wayne Rabalais

Department of Chemistry
University of Houston
Houston, Texas   77204-5641

INTRODUCTION

Low energy (<10 keV) ion scattering spectrometry[1] is becoming increasingly important as a surface analysis technique in three specific areas, i.e. surface elemental analysis,[2] probing surface structure,[3] and studying electronic transition probabilities[4] between ions or atoms and surfaces. This is largely due to the following recent advances: (i) impact collision ion scattering spectrometry[3] (ICISS) in which the scattering angle is close to 180°, thus simplifying the scattering geometry and allowing experimental determination of the shadow cone radii, (ii) the use of alkali primary ions[5] which have low neutralization probabilities, leading to higher scattered ion fluxes, (iii) time-of-flight (TOF) techniques[6] with detection of both neutrals and ions in a multichannel mode in order to enhance sensitivity, (iv) scattered ion fractions[4] to probe the spatial distributions of electrons, and (v) the use of recoiling[7] atoms to determine the structure of light adsorbates on surfaces.

TOF-SARS

The technique of time-of-flight ion scattering and recoiling spectrometry (TOF-SARS) is described herein. It is a nondestructive method of analysis due to collection of both neutrals plus ions in a multichannel TOF mode, it is sensitive to all elements (including hydrogen) due to the combination of scattering and recoiling with continuous angular variation, it can determine surface and adsorbate structures to an accuracy of <0.1 Å, and it can delineate

Fundamental Aspects of Heterogeneous Catalysis Studied by Particle Beams,
Edited by H.H. Brongersma and R.A. van Santen, Plenum Press, New York, 1991

313

between structural effects and electronic exchange (neutralization) effects by collection of neutrals + ions and neutrals only.

TOF-SARS has the following characteristics for surface analysis: (i) <u>Elemental Analysis</u> - All elements can be analyzed by either scattering, recoiling, or both techniques, TOF peak identification is straightforward using classical scattering expressions, and collection of neutrals plus ions results in scattering and recoiling intensities that are determined by elemental concentrations, shadowing and blocking effects, and classical cross sections. (ii) <u>Structural Analysis</u> - It provides "real space" information on the relative positions of all atoms in the surface region, including hydrogen, based on simple classical concepts, i.e. shadowing and blocking cones. The cone dimensions can be calculated and calibrated from known interatomic spacings and the analyses are not complicated by ion neutralization effects. (iii) <u>Ion-Surface Electronic Transition Probabilities</u> - Since the scattered and recoiled ion fractions can be directly measured, electron exchange probabilities as a function of ion energy and type, surface type, crystallographic orientation and chemical state, and scattering geometry can be determined.

HISTORICAL PROSPECTIVE

The technique of TOF-SARS is an outgrowth of conventional ion scattering spectrometry (ISS). In 1967, Smith[8] recognized that the scattering of low energy ions from surfaces was well-described by the binary collision approximation and that analysis of such spectra provided surface elemental analysis. Detection of only the scattered ions made the technique very surface sensitive, although difficult to quantitate. In 1982, Aono et al.[3] proposed a technique called impact collision ion scattering spectroscopy (ICISS) in which θ (defined in Fig. 1) is close to 180°, ca. 165°. This greatly simplified the scattering geometry and allowed experimental determination of shadow cone radii. There are two difficulties with this technique: (i) It analyzes only the scattered ions; these are typically only a very small fraction (<5%) of the total scattered flux. Thus, high primary ion doses are required for spectral acquisition. (ii) Neutralization probabilities tend to change as the ion-beam incidence angle with respect to the surface changes. As a result, this can attenuate the features that arise from specific geometrical arrangements. In 1984 Niehus[5] proposed the use of alkali primary ions which have low neutralization probabilities, leading to higher intensities. The contamination of the sample surface by the reactive alkali ions is a potential problem with this method. Buck et al.[9] used TOF methods for surface structure analysis in 1984 and demonstrated the capabilities and high sensitivity of the technique when both neutrals and ions are detected. In 1987 van Zoest, et al.[10] demonstrated that TOF analysis of both the scattered and recoiled neutrals and ions provided unique information about surface structure. In 1989, Rabalais, et al.[11-15] presented a TOF spectrometer system with sufficient resolution to separate scattered and recoiled particles.

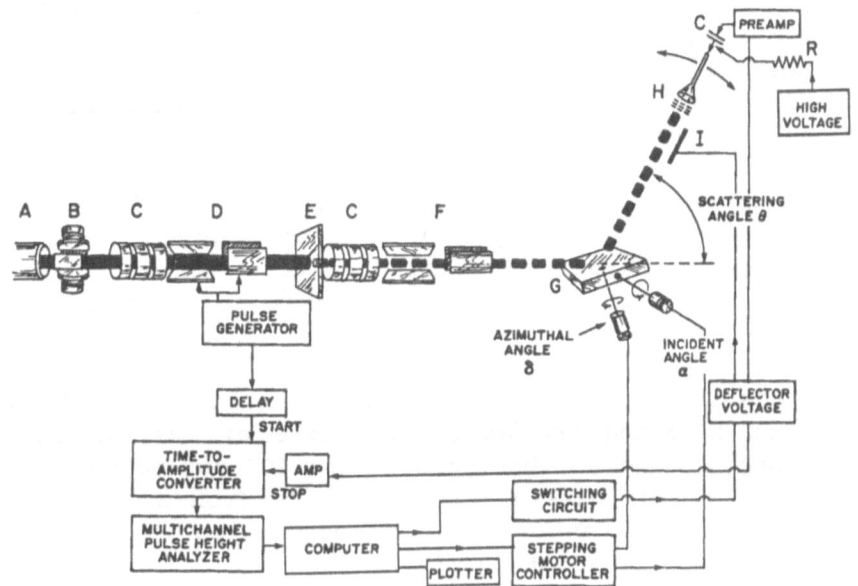

Fig. 1. Schematic drawing of pulsed ion beam line, sample, detector, and associated electronics for TOF-SARS.

EXPERIMENTAL METHOD

A schematic drawing of a pulsed ion beam and timing electronics for TOF-SARS is shown in Fig. 1. It consists of the following components: (i) large flat vacuum chamber (radius of 1 m), (ii) pulsed primary ion beam, (iii) detector (channel electron multiplier) rotatable through continuous scattering angles over the range θ = 0-165°, (iv) 1 m TOF drift region, (v) pulse generating, timing, detection, and control electronics, (vi) and conventional surface analysis techniques (AES, XPS, and LEED). The instrument is described in detail elsewhere.[16] Typical experimental parameters are as follows: 2-5 keV He[+], Ne[+], or Ar[+] pulsed primary ion beam; pulse width 20-50 ns; pulse rate 10-50 kHz; average current density 0.05-0.1 nA/mm[2]; signal detection rate up to ~30,000 c/s. A TOF spectrum can be acquired in 20 s with a dose of <10^{-3} ions/target atoms.

TOF SPECTRA

Example TOF spectra for Ar[+] scattering from a Si sample which has some residual oxygen and hydrogen adsorbed on the surface are shown in Fig. 2 for several values of the polar beam incident angle α and scattering angle θ. These angles are defined in the inset of Fig. 2; for demonstrative purposes they have been adjusted to specular conditions in the range θ = 4.5° - 85°. The sharp, intense peak observed at low values of α and θ corresponds to Ar projectiles scattered with minor energy loss from combinations of single and multiple collisions with target atoms. In this angular region and for atomically flat surfaces, projectiles moving along azimuths of high symmetry collide with several target atoms before leaving the surface with exit angles similar to the incident angle. The surface channeling effects are useful for aligning the azimuthal angle of the sample, as will be shown later in this paper. The TOF for scattering of an incident ion of mass M_1 and energy E_o from a target atom of mass M_2 into an angle θ can be calculated by the binary collision (BC) approximation[7,17] as

$$t_{SS}=1(M_1+M_2)/(2M_1E_o)^{1/2}\{\cos\theta+[(M_2/M_1)^2-\sin^2\theta]^{1/2}\}, \quad (1)$$

where 1 is the distance from the target to detector. For cases where $M_1>M_2$, there is a critical angle $\theta_c = \sin^{-1}(M_2/M_1)$ above which only multiple scattering can occur. Since θ_c = 44.8° for Ar\Si collisions, a broad and featureless structure results at large θ as shown in Fig. 2. Recoils ejected in single collisions with the projectile into an angle ϕ, i.e. direct recoils (DR), have a TOF given by[7]

$$t_{DR} = 1(M_1+M_2)/(8M_1E_o)^{1/2} \cos\phi. \quad (2)$$

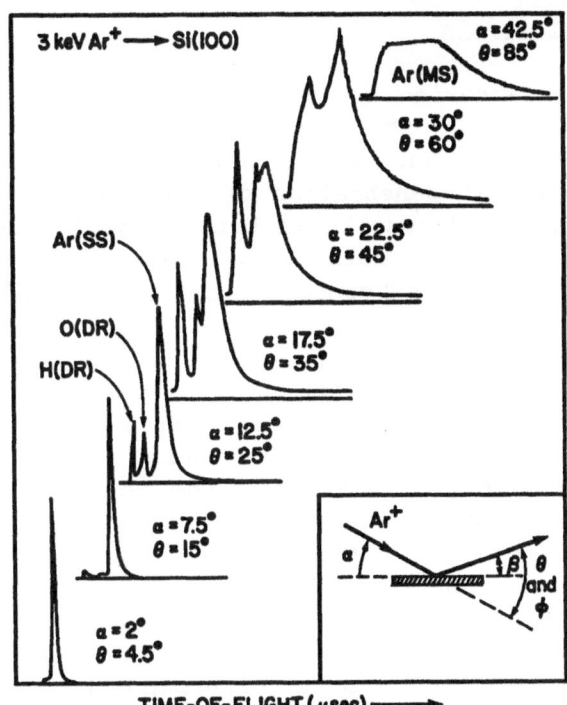

Fig. 2. TOF spectra for 3 keV Ar⁺
scattering from a Si{100}
surface with hydrogen and
oxygen adsorbates at
several α and θ values.

Both H and O DR peaks are present in the spectra of Fig. 2.
For low α they appear at the short TOF side of the scattering
peak, while for α > 35° the O DR and the scattering peak
overlap. The angular range where this overlap occurs is
determined by Eqs. 1 and 2 and by the width of the peaks,
which has a nonlinear relation with the recoiled energy E_r
(it is proportional to $E_r^{-3/2}$).

SURFACE STRUCTURE ANALYSIS

The ability to obtain direct information on surface
structures from TOF-SARS is one of the most important
applications of the technique. In this section the technique
will be applied to analysis of a W{211} surface, oxygen and
hydrogen chemisorbed on W{211}, and reconstruction of a
Pt{110} surface.

Interatomic Potentials

Although scattering in the keV range is dominated by repulsive potentials, it is not simply a hard sphere or billiard ball collision where there is a clean "hit" or "miss". The partial penetration of the ion into the target atom's electron cloud results in bent trajectories even when there is not a "head-on" collision. This type of interaction is well-described by a screened Coulomb potential[17] such as

$$V(r) = [Z_1 Z_2 e^2 / R] \Phi (R/CR_s) \tag{3}$$

where R is the internuclear separation and the Z_i are the atomic numbers of the collision partners. Φ is a screening function which is determined by R, the screening radius R_s, and a scaling parameter C; there are several good approximations for Φ.[17] Using such a potential, one can determine the relationship between the scattering angle θ and the impact parameter p. The p is defined as the minimum perpendicular distance from the target atom to the ion trajectory. A small value of p corresponds to a near head-on collision and backscattering and a large value corresponds to a glancing collision and forward scattering. Similarly, the recoiling angle ϕ is also determined by p.

Backscattering

When an atomic projectile approaches a target atom, the trajectories are bent such that an excluded volume, i.e. a shadow cone, in the shape of a paraboloid is formed behind the target atom (Fig. 3). Since the radii of the cones are of the order of 1 Å, the technique is very sensitive to the outermost atomic layers. When an ion beam is incident on an atomically flat surface at grazing angles, each surface atom is shadowed by its neighboring atom such that only forward scattering (large impact parameter (p) collisions) is possible. As α increases, a critical value $\alpha_{c,sh}^i$ is reached each time the ith-layer of target atoms move out of the shadow cone allowing for large angle backscattering (BS), i.e. small p collisions, from the specific ith-layer. If the BS intensity I(BS) is monitored as a function of α, steep rises with well defined maxima are observed when the focused trajectories at the edge of the shadow cone pass close to the center of neighboring atoms. These effects are shown in Fig. 3 for Ar$^+$ scattering at θ = 163° from a W{211} surface along the [1Ī1] azimuth. A schematic drawing of the surface with possible adsorption sites is shown in Fig. 4. For this particular azimuth, the 1st- and 2nd-atomic layers are

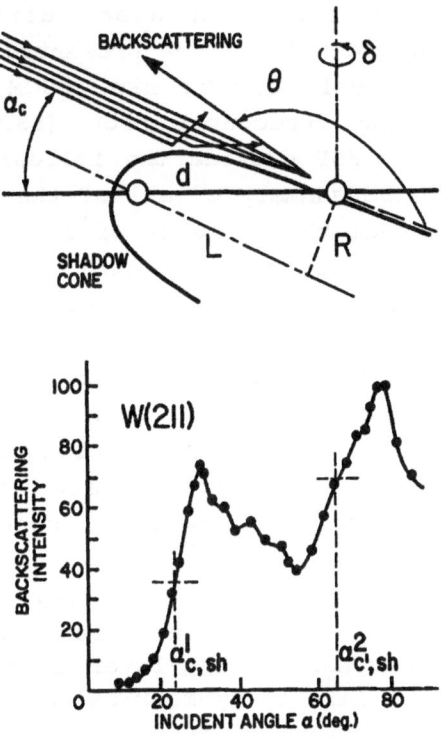

Fig. 3. Schematic illustration
of backscattering events
observed in TOF-SARS along
with a representative plot
of I(S) versus α along the
[Ī11] azimuth of W{211}
using 4 keV Ar⁺.

exposed to the beam at the same incident angle (see Fig. 4)
giving rise to the peak observed at $\alpha = 29°$ with $\alpha^1_{c,sh} = 23°$;
the underlying 3rd- and 4th-layers start to be accessible to
the beam at $\alpha = 60°$, giving rise to the intense peak observed
at $\alpha = 79°$. The $\alpha^i_{c,sh}$ values are taken at 50% of the peak
height.

If the shape of the cone,[17-19] i.e. the radius (R) as a
function of distance (L) behind the target atom, is known,
the interatomic spacing (d) can be directly determined from
the I(BS) vs. α plots. For example, by measuring $\alpha^1_{c,sh}$ along
directions for which specific crystal azimuths are aligned
with the projectile direction and using $d = R/\sin \alpha^1_{c,sh}$,
interatomic spacings in the 1st-atomic layer can be determined. The 1st-2nd-layer spacing can be obtained in a similar
manner from $\alpha^2_{c,sh}$ measured along directions for which the
1st- and 2nd-layer atoms are aligned. An example of the use
of the shadow cone in this determination is shown in Fig. 5.

I(BS) as a function of α along different crystal azimuths for $-90° < \delta < +90°$ can be presented in the form of a scattering structural contour map (SSCM) and a three-dimensional scattering structural plot (SSP) as shown in Fig. 6. The SSCM and SSP provide the following information. (i) They are a concise summary of all of the experimental BS data. (ii) They reveal the symmetry of the I(BS) data in α, δ space, thereby providing a fingerprint for a specific crystal face and type with minor perturbations due to relaxation and possible major perturbations due to reconstruction. (iii) They show what general regions of α, δ-space contain interesting structures for more detailed investigation. (iv) Comparison of the clean surface SSCM and SSP to those of the adsorbate covered surface allows determination of adsorbate induced reconstruction or relaxation.

Consider details of the SSCM and SSP of Fig. 6. The thick line at low α gives the value of $\alpha^1_{C,sh}$ vs. δ corresponding to shadowing of 1st-layer atoms by their 1st-layer neighbors; this line is symmetrical about $\delta = 0°$, as is the 1st-atomic layer. Intense structures are observed as α increases above 20° due to subsurface-layer scattering; the high intensities are due to focusing and channeling of ion trajectories by 1st- and 2nd-layer W atoms onto 3rd- and 4th-layer W atoms and back out again. The asymmetry about $\delta = 0°$ is a result of the lack of symmetry between the 1st- and underlying layers, i.e. there is no mirror plane through the $\delta = 0°$ azimuth. The diagonal orientation of the line of intense peaks observed from $\alpha = 30°$, $\delta = -70°$ towards $\alpha = 75°$, $\delta = +80°$ results from the fact that focusing onto subsurface layers for $\delta < 0°$ occurs mainly at low α, while for $\delta > 0°$, this focusing occurs only at high α. The values of $\alpha^1_{C,sh}$ vs. δ are consistent with 1st-layer interatomic spacings corresponding to the bulk truncated W{211} structure showing that the surface is not reconstructed. The 1st-layer is, however, relaxed vertically and laterally, as discussed elsewhere[11,13] together with other details of the structures observed in the SSCM and SSP of Fig. 6.

Recoil of Oxygen and Hydrogen

Light adsorbates can be efficiently detected[7] by recoiling them into forward scattering angles ϕ (Fig. 7). As α increases, the adsorbate atoms move out of their neighboring atom shadow cones so that direct collisions from

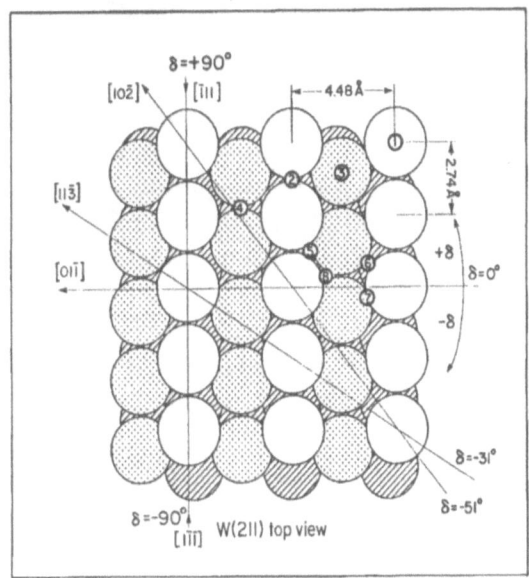

Fig. 4. Structure of the bulk
truncated W{211} surface.
Open circles - 1st-layer
atoms; Dotted circles -
2nd-layer atoms; Dashed
circles - 3rd- and 4th-
layer atoms. Geometrically
different adsorbate sites
are indicated.

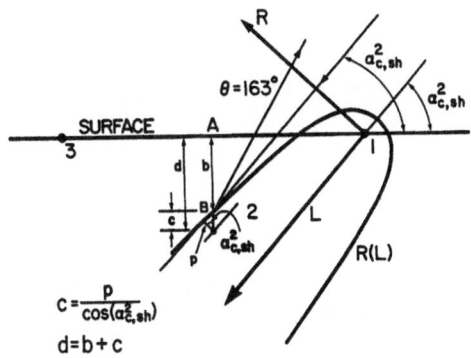

Fig. 5. Schematic diagram showing the use of
the calculated shadow cone and experimental $\alpha_{c,sh}^2$ value measured for second-
layer scattering to determine the first-
to second-layer spacing. Atoms 1 and 3
are in the first-layer, atom 2 is in
the second layer, c is the correction
induced by the impact parameter p, and
the first- to second-layer spacing is
d = b+c. The shadow cone coordinates
are labeled R and L.

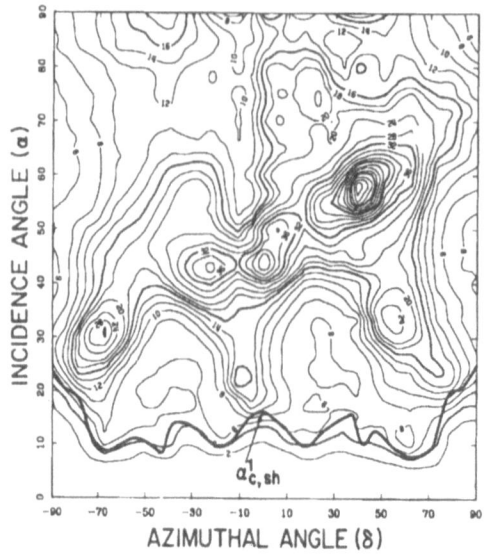

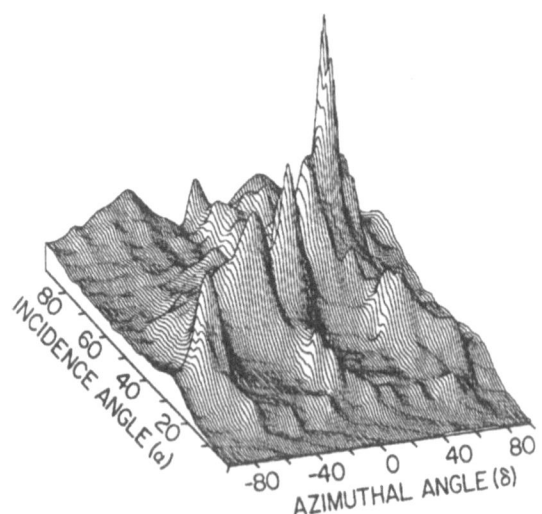

Fig. 6. (Top) Scattering structural
contour map (SSCM) for the
clean W(211) surface. Primary ion: 4 keV Ar$^+$;
$\theta = 163°$; $\delta = 0°$ is
the [01$\bar{1}$] azimuth; $\delta = -90°$
is the [1$\bar{1}\bar{1}$] azimuth;
$\delta = +90°$ is the [$\bar{1}$11] azimuth.
The critical value of α at low
angles, $\alpha_{c,sh}$, is plotted as a
heavy line. (Bottom) Three-
dimensional scattering structural plot (SSP) for the clean
W(211) surface. Viewing direc-
tions are $\alpha = 35°$ and $\delta = -20°$.

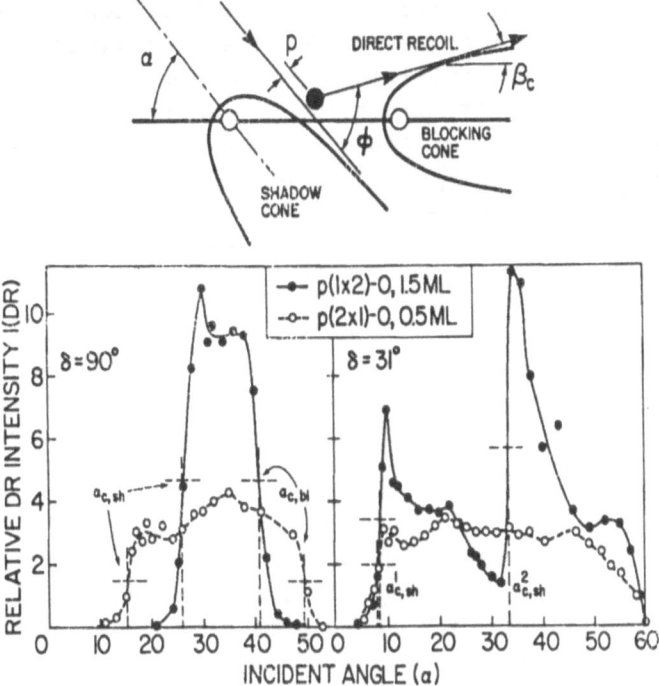

Fig. 7. Schematic illustration of direct recoiling events observed in TOF-SARS along with representative plots of I(R) versus α for two different azimuths and two different oxygen coverages.

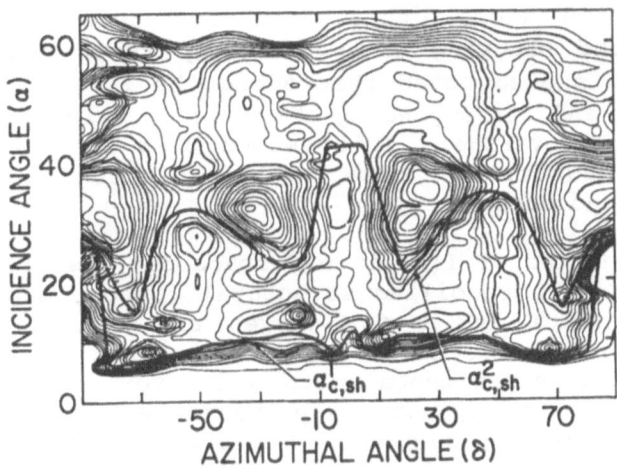

Fig. 8. Recoiling structural contour map (RSCM) for the W{211}-p(1x2)-O surface. Contours of equal O recoil intensity I(R) are plotted as a function of α and δ. The two critical shadowing angles, $\alpha_{c,sh}$ and $\alpha^2_{c,sh}$, are plotted on the map.

incident ions are possible. When the p values necessary for recoiling of the adsorbate atom into a specific ϕ becomes possible in a single collision, adsorbate direct recoils are observed. Focusing at the edge of the shadow cone produces sharp rises in the recoiling intensity I(R) as a function of α. By measuring $\alpha_{c,sh}$ corresponding to the direct recoil event, the interatomic distance of the adsorbate atom relative to its nearest neighbors along the direction of the projectile can be directly determined from p and the shape of the shadow cone.

Example plots of oxygen recoiling intensities as a function of α obtained by bombardment of an oxygen covered W{211} surface with 4 keV Ar$^+$ ions are shown in Fig. 7 for two different azimuths and two different oxygen coverages. Besides the sharp rises of I(R) determined by shadowing effects, sharp decreases in I(R) are observed at high α values. These result from the deflection of the recoiling trajectories by the blocking cones cast by neighboring atoms. The $\alpha_{c,bl}$, defined in Fig. 7, provide a measure of the size of the blocking cones and can also be used to determine interatomic distances. Along both of the $\delta = 90°$ azimuths, $\alpha_{c,sh} = 24°$ and $\alpha_{c,bl} = 42°$ at high coverage, which is considerably higher and lower, respectively, than the values $\alpha_{c,sh} = 16°$ and $\alpha_{c,bl} = 48°$ obtained at low coverage. Along other azimuths, such as $\delta = 31°$, the $\alpha_{c,sh}^1$ and $\alpha_{c,bl}$ values are nearly identical for different coverages. These data indicate that as coverage increases, both the shadowing and blocking effects become more severe along the $\delta = 90°$ azimuths than along other azimuths; this results from shadowing and blocking of O atoms by their neighboring O atoms. It is also noted that along the $\delta = 31°$ azimuth, a second peak and corresponding sharp $\alpha_{c,sh}^2$ value are observed only at high coverage. This second peak indicates that an additional adsorption site is being occupied which was previously unoccupied.

The features of I(R) in α, δ space can also be presented in the form of a recoiling structural contour map (RSCM) as shown in Figs. 8 and 9 for oxygen and hydrogen saturated W{211} surfaces, respectively. These maps are a concise summary of the experimental data and reveal the symmetry of the recoil data in α, δ space, providing a fingerprint for O and H on the W{211} surface. For example, the symmetry of the O RSCM about $\delta = 0°$ indicates that the adsorption sites are symmetrical about this azimuth, thus eliminating the asymmetrical sites 3,4,6,7, and 8 shown in Fig. 4. Comparison of the BS intensities for the clean and the O covered surface indicates that only site 5, i.e. the three-fold site in which the O atom is bound to two 1st- and one 2nd-layer W atoms, is consistent with all of the experimental data. The O-W bond lengths are determined from detailed

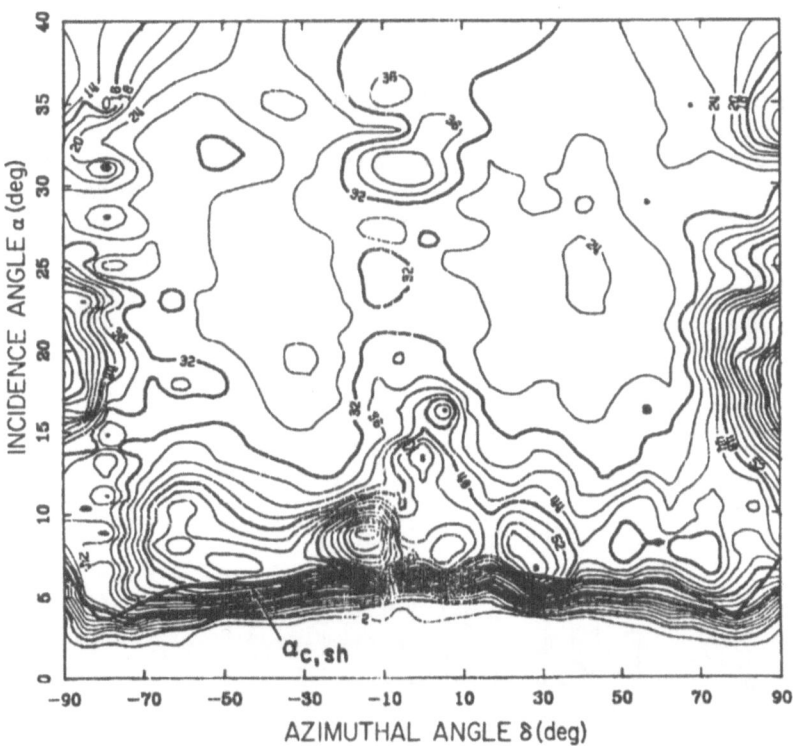

Fig. 9.   Recoiling structural contour map (RSCM)
for the W{211}-H surface.   Contours of
equal H recoil intensity I(R) are
plotted as a function of α and δ. The
critical value of α, $\alpha_{c,sh}$, is plotted
on the map.

analysis[14] as 1.83 Å and 2.17 Å to the 1st- and 2nd-layer W 1atoms, respectively. In contradiction with the O case, the H RSCM shows only one $\alpha_{c,sh}$ at low α, which is approximately constant at 4° for −85° < δ < +85° and increases to $\alpha_{c,sh}$ = 10° for δ = 90°; a relatively flat region is observed in the center and background of the H RSCM. This indicates that close packing along the δ = 90° direction occurs and that there is no H buried in subsurface layers that is accessible for recoiling. A detailed analysis presented elsewhere[15] allows determination of the hydrogen position as 0.58 ± 0.2 Å above the 1st-layer W plane and confined within a band that is centered above the [1Ī1] troughs.

Surface Periodicity

When the scattering angle θ is decreased to a forward angle (<90°), both shadowing effects along the incoming trajectory and blocking effects along the outgoing trajectory contribute to I(FS) patterns. The blocking effects arise because the exit angle $\beta=\theta-\alpha$ is small at high α angles. The

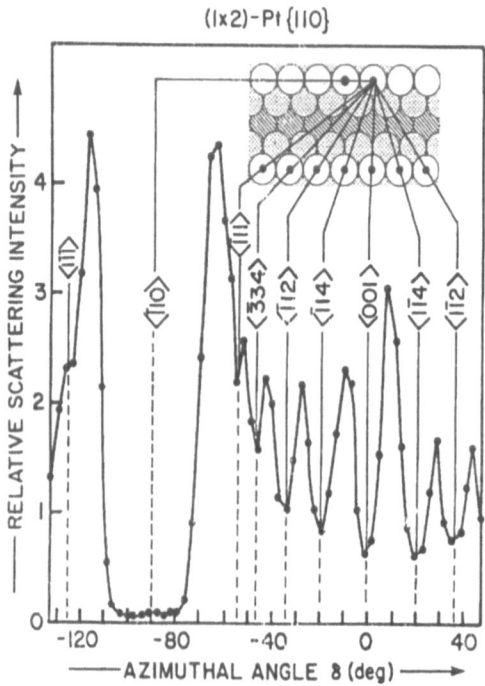

Fig. 10. Forward scattering intensity I(FS) versus azimuthal angle δ for Pt{110} in the (1x2) missing row reconstruction using 4 keV Ne⁺ at α = 6° and θ = 28° illustrating surface periodicity.

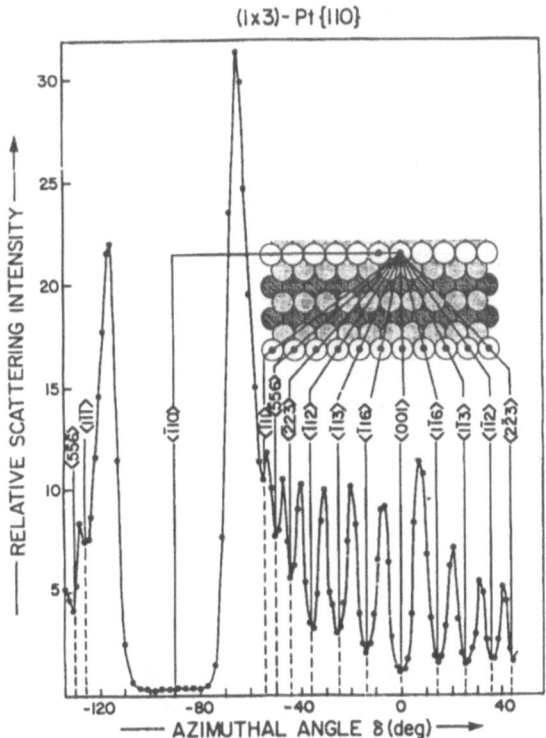

Fig. 11. Forward scattering intensity I(FS) versus azimuthal angle δ for Pt{110} in the (1x3) missing row reconstruction using 2 keV Ne⁺ at $\alpha = 4°$ and $\theta = 28°$ illustrating surface periodicity.

features of I(FS) vs. δ plots, with constant α and θ, are determined by shadowing and blocking along the close-packed azimuths. Surface periodicity can be read directly from these features as shown in Figs. 10 and 11 for Pt{110} in both the (1x2) and (1x3) reconstructed surface phases. Minima are observed at the δ positions corresponding to alignment of the beam along specific azimuths. These minima are a result of shadowing and blocking along the close-packed directions, thus providing a direct reading of the surface periodicity.

Sputtering Induced Damage

TOF-SARS is capable of monitoring the damage in the outermost atomic layers caused by ion collisions during sputtering. Such an experiment was carried out by bombarding a clean W{211} surface with 3 keV Ar⁺ ions for a specific

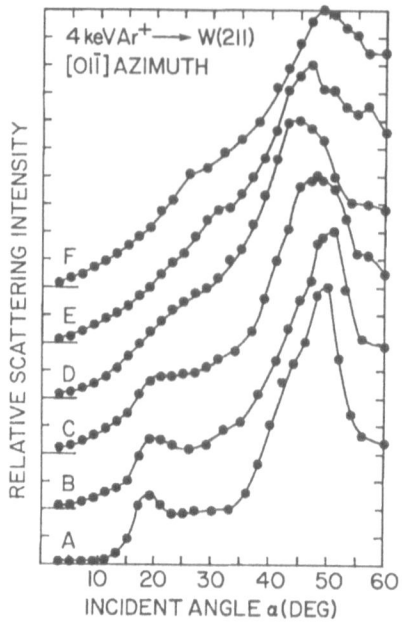

Fig. 12. Relative scattering
intensity I(S)
versus α after
bombardment with
different 4 keV
Ar$^+$ ion doses.
(A) clean W{211},
(B) 9.4x10^{14},
(C) 2.8x10^{15},
(D) 7.5x10^{15},
(E) 3.8x10^{16},
and (F) 7.9x10^{16}
ions/cm^2.

dose and then collecting backscattering (θ = 163°) spectral
intensities as a function of incident angle α. After every
such incident angle scan, the crystal was annealed to 2000°C
for 10-15 min. in order to eliminate damage caused by the
previous sputtering dose and to clean off any impurities that
may have accumulated on the surface. The primary pulsed ion
beam used for scattering was 4 keV Ar$^+$ at a flux density of
0.1 nA/mm^2; a dose of ≈10^{13} ions/cm^2 was used for each
incident angle scan. This analysis dose is negligible
compared to the doses used to create damage, i.e. > 9x10^{14}
ions/cm^2.

The results of this experiment are shown in Fig. 12 as
incident angle scans along the [01Ī] azimuth for the clean
surface and the clean surface after bombardment with specific

Ar$^+$ ion doses. The clean surface scan exhibits two peaks due to shadowing and focusing effects of the 1st-layer atoms with their 1st-layer neighbors (low α) and 1st-layer atoms with their 2nd-layer neighbors (high α). The atomic density of the W(211) surface is 8x10^{14} atoms/cm^2. At the lowest Ar$^+$ dose of 9x10^{14} ions/cm^2, i.e. ≈1 ion/surface atom, the initial slope of the low α peak begins to decrease and the intensity at α < 15° begins to increase. The 1st-layer scattering peak is reduced to a minor inflection as the dose increases. The intensity of the valley between the 1st- and 2nd-peaks increases, the height of the plateau following the 2nd-peak increases, and the 2nd-peak broadens as the dose increases. This signifies the destruction of the 1st-atomic layer, i.e. the fixed interatomic spacings between 1st-layer atoms are replaced by a 1st-layer structure with random interatomic distances. The effects observed at higher α values associated with the 2nd-peak signify randomization of the 1st-2nd-layer interatomic spacings.

Surface Semichanneling

Sharp spatial anisotropy of scattered ion intensity has been observed[17] when an ion beam is incident at a glancing angle on a surface which has so-called "surface semichannels". These semichannels are formed by close-packed rows in the 1st-atomic layer which serve as the "walls" of the channel and similar rows in the 2nd-atomic layer which serve as the "base" of the channel. Such surface semichannels are able to focus scattered ions effectively under certain conditions when the incident plane of the projectile is parallel to the channels. The W{211} surface has semichannels along the [1Ī1] azimuthal direction of width 4.48 Å and depth 1.17 Å, considering the clean relaxed[13] {211} structure. The measurements were performed by directing a pulsed 5 keV Ne$^+$ ion beam along the [1Ī1] direction and collecting a series of TOF spectra as a function of surface azimuthal angle δ in a region ≈±25° about the [1Ī1] axis (labeled δ = 90°). Specific beam incident α and exit β angles were chosen and I(S) was monitored as a function of δ. As an example of the results, Fig. 13 shows plots of I(S) vs. δ at α = 10° and different β values. The scans are symmetric about the [1Ī1] troughs and their structure is very sensitive to β. A sharp peak (FWHM 4°-5°) is observed at low β due to surface semichanneling. As β increases above the specular condition, i.e. α = β = 10°, the intensity of the focusing peak decreases until a minimum is finally observed.

If the angles between the direction of motion of the

scattering particles and the atomic rows are small enough, the scattering potential can be approximated as a multi-atom potential due to the chains of atoms along the close-packed rows. For such a case, focusing is enhanced at the specular condition, i.e. $\alpha = \beta$. Lindhard has defined a critical angle α_L above which the concept of continuous chains of atoms is no longer valid; for $\alpha > \alpha_L$, focusing is lost due to scattering from individual atomic centers rather than continuous multi-atom potentials. In order to estimate α_L, the expressions[17] for the Lindhard inverse-square

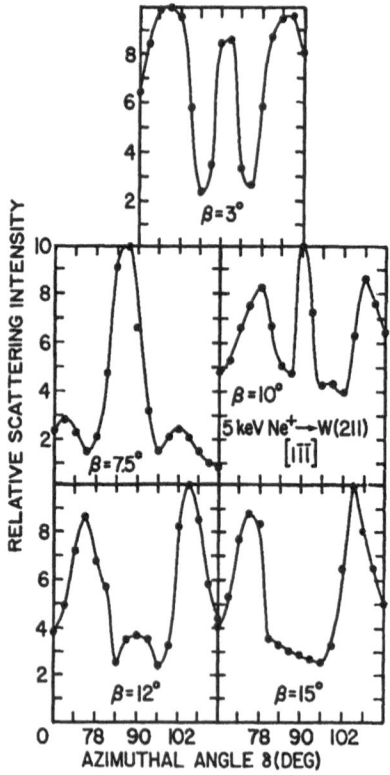

Fig. 13. Surface semichanneling illustrated by plots of I(S) versus δ in the region near the W{211} δ = 90° [1$\bar{1}\bar{1}$] azimuth with α = 10° and different β values using 5 keV Ne$^+$.

approximation to the Thomas-Fermi continuum potential, $(V(r))$, and the critical angle, given by

$$V(r) = \pi A/rd \quad \text{and} \quad \alpha_L = (\pi A/E_o d^2)^{1/3} \tag{3}$$

were evaluated. In these expressions, r is the distance of closest approach of the projectiles to the rows, d is the lattice distance along the rows, $A = \epsilon_v Z_1 Z_2 e^2 a/2$ (the potential constant), $\epsilon_v = 2/(2.7183 \times 0.8853)$, Z_1 and Z_2 are the atomic numbers of the projectile and target atoms, and the screening length $a = 0.8853 a_o (Z_1^{2/3} + Z_2^{2/3})^{-1/2}$ where a_o is the Bohr radius. The resulting calculated α_L's are 12.3°, 19.0°, and 23.3° for 4 keV He^+, 5 keV Ne^+, and 5 keV Ar^+, respectively. These α_L's exhibit the same qualitative ordering as observed for the experimental α_{max} values, however quantitatively they are all slightly larger than the α_{max} values; this is because α_L provides the upper limit for the breakdown of the continuous chain approximation.

ION-SURFACE ELECTRON EXCHANGE

Ion-surface electronic transition probabilities are determined by electron tunneling between the valence bands of the surface and the atomic orbitals of the ion. Such transition probabilities are highest for close distances of approach. An example of this is shown in Fig. 14 as a plot of the Ne^+ ion fraction, that is $F = $ ions/(neutrals + ions), scattered from a Ni{100} surface. The F is highest along the [100] azimuth where scattering is only from the first-layer (second-layer atoms are shadowed). The F is lowest along the [110] azimuth where both first- and second-layer scattering occur. The degree of neutralization is higher for second-layer scattering because the ion trajectories are close to several other atoms (both first- and second-layer ones). The symmetrical nature of the plot reflects the lattice periodicity. In general, low scattered ion fractions F are observed when the ion trajectory goes through a region of high electron density near several atoms. Monitoring F at selected scattering geometries can serve as a probe surface electron density.

ROLE OF TOF-SARS IN CATALYSIS

TOF-SARS has applications to catalysis as an elemental and structural analysis technique.[1] The unique features for elemental analysis are direct monitoring of surface hydrogen and the extreme sensitivity to the outermost atomic layers.

However for general surface elemental analyses, x-ray photoelectron spectroscopy (XPS) and Auger electron spectroscopy (AES) remain the techniques of choice. The major role of TOF-SARS is as a surface structural analysis technique which is capable of probing the positions of all elements, including hydrogen. It is sensitive to short-range order, i.e. individual interatomic spacings along azimuths. It is complementary to low energy electron diffraction (LEED) which probes long-range order, i.e. minimum domain size of 100-200 Å.

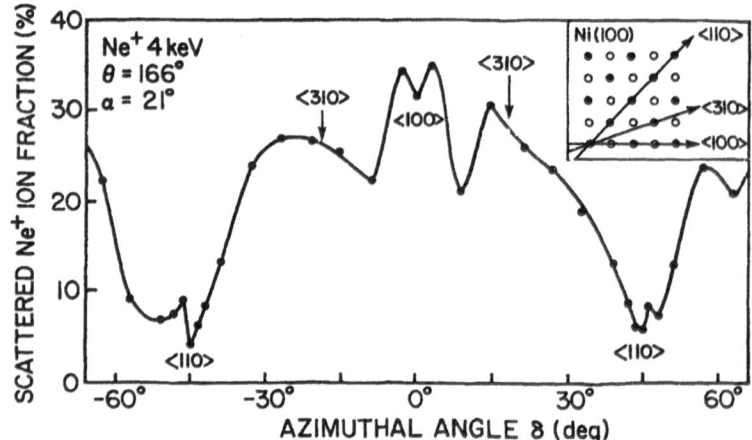

Fig. 14. Scattered Ne[+] ion fractions for 4 keV Ne[+] impinging on a Ni{100} surface as a function of azimuthal angle δ.

Thus far, the structural studies by TOF-SARS have been carried out on monocrystalline surfaces. Adsorption on such well-defined surfaces provides excellent models for more complicated practical catalysts. These studies allow one to determine the active adsorption sites, reactivity of different crystal faces, and relationship between surface reconstruction and adsorption. TOF-SARS also makes it possible to probe the role of surface species which are poisons or promoters for catalytic reactions and their influence on the dynamical properties of catalytically interesting surface species. It is now possible to investigate the influence of surface defect step sites on adsorption and the dynamics of adsorbed species at these defect sites. Active sites for simple catalytic processes, reactions at these active sites and artificial production of high coverages of catalytic reaction intermediates are examples of details of surface reactions that are accessible by TOF-SARS.

The near-surface layers of practical supported catalysts have been studied through low energy ion scattering[20] by making use of the high surface sensitivity and ion beam sputtering to obtain monolayer "depth-profiles". This can be used to determine the outermost or reactive species and the near-surface concentration distribution. Charging problems from the insulating support material, which can be serious for some forms of spectroscopy, can be overcome in scattering by neutralizing with thermal electrons. Some successful applications of scattering have been in studies of the spreading behavior of oxides,[21,22] such as MoO_3 on Al_2O_3, the formation of monolayers by solid-solid wetting,[23,24] such as the spreading of MoO_3, WO_3, and V_2O_5 as active components on an Al_2O_3 support, and the conditions for formation of surface polymolybdate, which is considered[25] to be a precursor state for the active catalyst.

The research described in this article has shown that TOF-SARS has significant applications in catalysis and can provide unique information about surface reactions. Advances in all aspects of low energy ion scattering are occurring at such a rapid rate that the breadth of these applications will certainly be expanded in the near future.

ACKNOWLEDGMENTS

This work was supported by the U.S. National Science Foundation Grant CHE-8814337 and the R. A. Welch Foundation Grant E-0656. I am grateful to all of the members of my research group and particularly H. Bu, O. Grizzi, C. C. Hsu, F. Masson, M. Shi, and R. Trehan for their contributions to this work.

REFERENCES

1.   T. Fauster, Surface Geometry Determination by Large-Angle Ion Scattering, Vacuum 38:119 (1988).

2. Y. S. Jo, J. A. Schultz, S. Tachi, S. Contarini and J. W. Rabalais, Surface Stoichiometry, Structure, and Chemisorption on Silicon Nitride Studied by Direct Recoiling, XPS, and AES, <u>J. Appl. Phys</u>. 60:2564 (1986).

3. M. Aono, Y. Hou, C. Oshima, and Y. Ishizawa, Low-Energy Ion Scattering From the Si(001) Surface, <u>Phys. Rev. Lett</u>. 49:567 (1982).

4. J. W. Rabalais, J. A. Schultz, R. Kumar, and P. T. Murray, Simultaneous TOF Spectra of Sputtered and Scattered Neutrals and Ions From 3 keV Ne$^+$, Ar$^+$ and Ar^{+2} Bombardment of CsBr, <u>J. Chem. Phys.</u> 78:5250 (1983).

5. H. Niehus, Analysis of the Pt(110)-(1x2) Surface Reconstruction, <u>Surface Sci</u>. 145:407 (1984).

6. J. W. Rabalais, J. A. Schultz, and R. Kumar, Surface Analysis Using Scattered Primary and Recoiled Secondary Neutrals and Ions by TOF and ESA Techniques, <u>Nucl. Inst. Meth</u>. 218:719 (1983).

7. J. W. Rabalais, Direct Recoil Spectrometry, <u>CRC Critical Rev. Sol. St. Mat. Sci</u>. 14:319 (1988).

8. D. P. Smith, Scattering of Low-Energy Noble Gas Ions From Metal Surfaces, <u>J. Appl. Phys</u>. 38:340 (1967).

9. L. Marchut, T. M. Buck, G. H. Wheatley, and C. J. McMahon, Jr., Surface Structure Analysis Using Low Energy Ion Scattering I. Clean Fe(001), <u>Surface Sci</u>. 141:549 (1984).

10. J. M. van Zoest, J. M. Fluit, T. J. Vink, and B. A. van Hassel, Surface Structure Analysis of Oxidized Fe(100) by Low Energy Ion Scattering, <u>Surface Sci</u>. 182:179 (1987).

11. J. W. Rabalais, O. Grizzi, M. Shi, and H. Bu, Surface Structure Determination From Scattering and Recoiling: W(211) and W(211)-p(1x2)-O, <u>Phys. Rev. Lett</u>. 63:51 (1989).

12. O. Grizzi, M. Shi, H. Bu, J. W. Rabalais, R. R. Rye, and P. Norlander, Determination of the Structure of Hydrogen on a W(211) Surface, <u>Phys. Rev. Lett</u>. 63:1408 (1989).

13. O. Grizzi, M. Shi, H. Bu, J. W. Rabalais, and P. Hochmann, Time-Of-Flight Scattering and Recoiling. I. Structure of the W(211) Surface, <u>Phys. Rev. B</u> 40:10127 (1989).

14. H. Bu, O. Grizzi, M. Shi, and J. W. Rabalais, Time-Of-Flight Scattering and Recoiling. II. The Structure of Oxygen on the W(211) Surface, <u>Phys. Rev. B</u> 40:10147 (1989).

15. M. Shi, O. Grizzi, H. Bu, J. W. Rabalais, R. R. Rye, and P. Nordlander, Time-Of-Flight Scattering and Recoiling Spectrometry. III. The Structure of Hydrogen on the W(211) Surface, <u>Phys. Rev. B</u> 40:10163 (1989).

16. O. Grizzi, M. Shi, H. Bu, and J. W. Rabalais, <u>Rev. Sci. Instrum</u>. 61:740 (1990).

17. E. S. Mashkova and V. A. Molchanov, "Medium Energy Ion Reflection From Solids," North-Holland, Amsterdam (1985).

18. J. F. Ziegler, J. P. Biersack, and U. Littmark, "The Stopping and Range of Ions in Solids," Pergamon, New York (1985).

19. S. R. Kasi, M. A. Kilburn, H. Kang, J. W. Rabalais, L. Tavernini, and P. Hochmann, Interaction of Low Energy Reactive Ions With Surfaces. III. Scattering of 30-200 eV Ne^+, O^+, C^+, and CO^+ From Ni(111), <u>J. Chem. Phys</u>. 88:5902 (1988).

20. E. Taglauer, M. Beckschulte, R. Margraf, and D. Mehl, Recent Developments in the Applications of Ion Scattering Spectroscopy, <u>Nucl. Inst. Meth</u>. B35:404 (1988).

21. H. Knözinger and E. Taglauer, Ion Scattering Spectroscopy and Raman Spectroscopy for Catalyst Characterization, <u>Amer. Che. Soc. Div. Petr. Chem. Prepr</u>. 26:357 (1981).

22. H. Knözinger, H. Jeziorowski, and E. Taglauer, <u>Proc. 7th Int. Congr. Catalysts, Toyko 1980</u>, Elsevier, Amsterdam (1981), Part A, 604.

23. J. Leyrer, R. Margraf, E. Taglauer, and H. Knözinger, Solid-Solid Wetting and Formation of Monolayers in Supported Oxide Systems, <u>Surface Sci.</u> 201:603 (1988).

24. R. Margraf, H. Leyrer, H. Knözinger, and E. Taglauer, Study of Molybdate Dispersion on Supported Catalysts Using Ion Scattering and Raman Spectroscopy, <u>Surface Sci</u>. 189/190:842 (1987).

25. H. Jeziorowski and H. Knözinger, Raman and Ultraviolet Spectroscopic Characterization of Catalysts, <u>J. Phys. Chem</u>. 83:1166 (1979).

HIGH-E ION SCATTERING AND ATOM LOCATION

J.A. Davies

Institute for Materials Research
McMaster University
Hamilton, Ontario

The emphasis on ion scattering techniques at this ASI is a clear indication of their importance in the field of catalysis. Two lecturers (Dr. Brongersma and Dr. Tagelauer) are covering the key low-energy (keV) scattering regime and two additional lecturers (Dr. Rabelais and Dr. Vickerman) will be discussing specific applications of such keV scattering to surface structure and secondary ion mass spectroscopy (SIMS) analysis.

My role is to discuss the high-energy (Rutherford) scattering regime which, although not so widely used in catalysis studies, nevertheless plays an extremely useful complementary role to the low-energy scattering process. Indeed, Rutherford backscattering (RBS) has become firmly established[1] as one of the most quantitative and versatile techniques for surface and thin film analysis. It offers the following distinct advantages: (i) quantitative analysis (typically $\pm 2\%$) for almost all elements, without requiring calibrated standards; (ii) mass (and hence in many cases "elemental") resolution of near-surface impurities; (iii) excellent non-destructive depth resolution, i.e. $\leq 30 \text{Å}$ under optimum conditions; and (iv) information on surface structure and foreign-atom location (by combining RBS with channeling). In the present lecture, we will review briefly each of these features and describe the role of various experimental parameters such as choice of ion beam (Z_1, E_0), and angle of beam incidence/exit with respect to the target surface.

First of all, in order to appreciate the distinction between the low- and high-energy scattering regimes, we start with a brief overview of the two main energy-loss processes involved – namely, nuclear stopping (S_n) and electronic stopping (S_e) – and their respective dependence on the ion energy. In fig. 1, our abscissa scale is the ion energy, ε, expressed in Lindhard's dimensionless Thomas-Fermi (T.F.) units[2]: i.e., it is the ratio of the T.F. screening length, $\underline{a}$, to the distance of closest approach $\underline{b}$ in an unscreened collision, where

$$a = 0.8853 \ a_0 \ (Z_1^{2/3} + Z_2^{2/3})^{1/2} \ ; \tag{1a}$$

$$b = Z_1 Z_2 \ e^2 \ (M_1 + M_2)/M_2 E \ ; \tag{1b}$$

a_0 is the Bohr radius, e is the electronic charge, E is the incident ion energy, and Z_1, Z_2 and M_1, M_2 are the atomic numbers and masses of the ion and target atom, respectively.

The advantages of this T.F. energy scaling are two-fold. Firstly,

Fundamental Aspects of Heterogeneous Catalysis Studied by Particle Beams,
Edited by H.H. Brongersma and R.A. van Santen, Plenum Press, New York, 1991

it enables S_n (and all associated quantities such as the scattering cross section, sputtering yield, damage production) to be expressed in terms of a semi-universal curve (fig. 1) for all combinations of Z_1 and Z_2. Secondly, it provides a clean separation of almost all ion scattering studies into two widely different regimes: namely,

(i) <u>Low-E scattering</u> (ion implantation) regime - $\varepsilon \leq 10$.

This consists of relatively slow moving heavy ions with velocities much smaller than the T.F. velocity, $v_0Z_1^{2/3}$, and hence, is the regime in which electron capture and loss cross sections are large and the scattering occurs in a highly screened coulomb field; also, as seen in fig. 1, it is the regime in which S_n and S_e both contribute significantly to the slowing down. Consequently, theoretical treatment is complex and much less quantitative than in the high-energy regime. Regime (i) covers the energy region in which sputtering and damage cascade effects are dominant; hence, sputter-etching and SIMS fall exclusively within this domain. It is also the energy region of interest in almost all ion implantation studies.

(ii) <u>High-energy</u> (nuclear microanalysis) regime - $\varepsilon \geq 100$.

In this ε region, S_n has fallen to a negligible and roughly constant fraction ($\sim10^{-3}$) of S_e. Hence, the ion trajectories are almost linear, collision cascade effects such as sputtering are generally small and, in contrast to regime (i), a quantitative theoretical framework exists for accurately predicting energy loss, scattering and channeling behaviour. This is the energy region in which quantitative nuclear microanalysis methods such as Rutherford backscattering and nuclear reaction analyses are carried out - and hence is the energy region of interest in the present lecture. Before leaving fig. 1, it should be noted that S_e does not exhibit the convenient T.F. scaling behaviour; consequently, unlike S_n, it cannot be described by a single universal curve. At low energies, S_e increases linearly with $\varepsilon^{1/2}$ (i.e. with ion velocity) with a slope k that depends weakly on both Z_1 and Z_2. The magnitude of $\varepsilon^{1/2}$ for some typical ions used in ion scattering are illustrated on Figure 1 for a medium mass target (copper). Note that the so-called "medium-energy-ion-

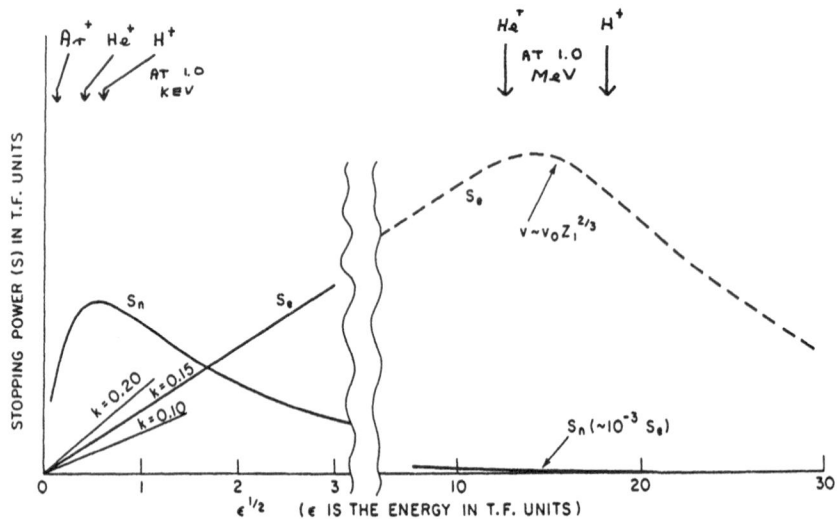

Fig 1. Stopping power versus energy in Lindhard's dimensionless units (see ref. 2).

scattering (MEIS)" regime developed by the FOM group in Amsterdam utilizes a beam of 50-200 kev protons and hence (as shown in figure 1) is really part of this high-energy Rutherford scattering regime.

RBS Technique

A typical RBS energy spectrum for 1.0 MeV ^{4}He$^+$ backscattered through 150° in an Si target is shown in figure 2. The sharp threshold around 0.6 MeV corresponds to backscattering of ^{4}He from near-surface ^{28}Si atoms. The slowly climbing continuum at lower energies (in the "random" spectrum) corresponds to backscattering from progressively larger depths in the crystal; from the known rate of energy loss, S_e (~60 eV/Å in the present case), the appropriate energy-to-depth-scale conversion can then be made.

Heavy surface impurities, such as As and Sb, give rise to backscattered ^{4}He peaks at higher energies than 0.6 E_0; because of the Z^2-dependence of the Rutherford cross section, the detection sensitivity increases strongly with increasing atomic number. For the heaviest impurities (eg., Au, Ta, W) in a low-mass substrate such as Si, a detection sensitivity better than 10^{-3} monolayers is easily achieved.

Surface impurities lighter than the target atoms (eg., ^{12}C and ^{16}O produce backscattered ^{4}He peaks at lower energies than the 0.6 E_0 threshold and hence are usually difficult to detect. Note, however, that under channeling conditions the RBS yield from the substrate can be suppressed almost 100-fold and, as shown in the <111> aligned spectrum of

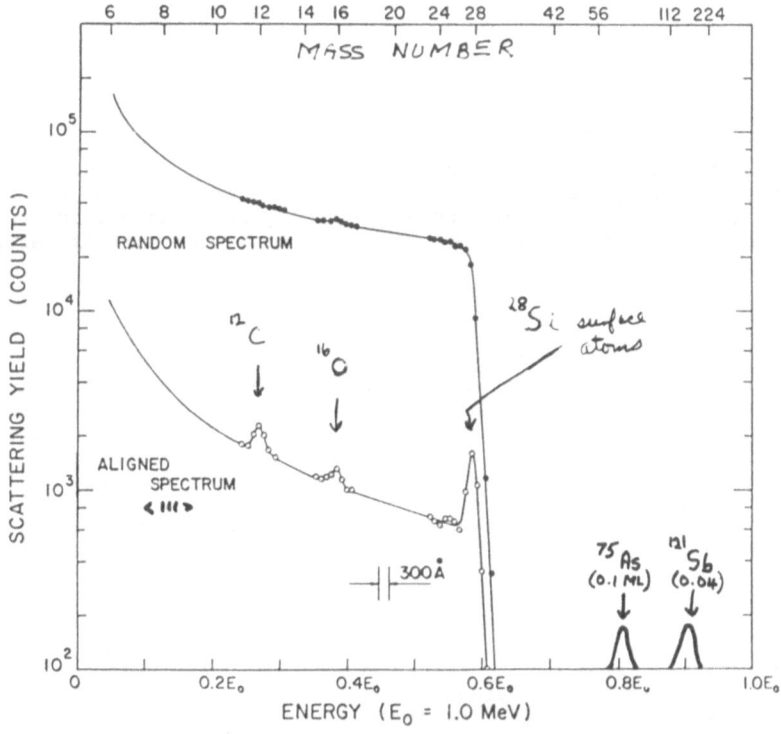

Fig. 2.   Aligned <111> and random RBS spectra for 1.0 MeV ^{4}He in silicon at 150° (from fig. 4.9 in ref. 3).

figure 2, these low-mass impurity peaks can then be clearly resolved at about the monolayer level.

Many of the low-Z atoms (hydrogen, carbon, nitrogen, oxygen) have useful nuclear reactions such as $H(^{15}N,\gamma)^{16}O$; $D(^{3}He,p)^{4}He$; $^{12}C(d,p)^{13}C$; $^{14}N(d,\alpha)^{12}C$; and $^{16}O(d,p)^{17}O$. These reactions can often be used to achieve a detection sensitivity of better than 10^{-2} monolayers for specific low-Z atoms, while still yielding RBS information from the heavier surface atoms.

The peak at mass 28 in the <111> spectrum of fig. 2 is referred to as the surface peak from the substrate atoms; as we shall discuss later, this surface peak provides quantitative information on the number of surface atoms visible to the incident beam.

Quantitative Accuracy

The key components in a typical RBS analysis are summarized in the following relationship:

$$\frac{Y}{Q} = N \cdot \sigma(E_1\Theta_s) \cdot \Delta\Omega \qquad (2)$$
$$\text{(counts/ion)} \quad \text{(atoms/cm}^2) \quad \text{(cm}^2/\text{steradian)} \quad \text{(steradian)}$$

Experimentally, one measures the backscattering yield Y for a specific number of incident ions, Q. Hence, to achieve an absolute accuracy better than 5-10%, proper Faraday cup techniques must be utilized in integrating the incident beam current; also, the beam-line vacuum should be low enough ($\leq 10^{-6}$ torr.) to avoid significant errors due to charge exchange effects.

The Rutherford cross section $\sigma_R(E,\Theta_s)$ gives us a very accurate conversion factor (i.e. better than 1%) for almost all target atoms, provided that the beam energy does not exceed ~1 MeV per atomic mass unit.

Hence, equation 2 can be used to obtain the number N of scattering atoms directly from the measured Y/Q values without the need for calibrated standards. In practice, however, instead of measuring the detector solid angle $\Delta\Omega$, it is usually easier (and more accurate) to calibrate the detector system with a previously calibrated standard, such as the widely used Bi-implanted Si wafers known as the Harwell Series I and II standards.

If the incident beam energy is too high, then the collision distance b (eq. 1b) in the backscattering event becomes smaller than the nuclear radius of the target (r_{nuc} ~1.5 $Z_2^{1/3}$ x 10^{-5}Å); under such circumstances, large and complex deviations from the Rutherford scattering law can occur. With the 1-3 MeV ^{4}He beams commonly used for RBS, such deviations[4] remain negligible for atoms as light as Si, Al and oxygen. In certain cases such as ^{12}C and ^{16}O there are large nuclear resonances in the elastic scattering cross section at somewhat higher energies and these can sometimes be utilized to enhance the detection sensitivity for these low-Z elements. However, the cross section is no longer predictable from the Rutherford law and careful calibration with appropriate standards is necessary.

Within the Rutherford regime, the true scattering cross section σ is always slightly smaller than the Rutherford value σ_R, due to the screening effect of the electrons on the target atom. Fortunately, this is ordinarily a rather small correction and it can be estimated quite accurately[5] via the following expression:

$$\frac{\sigma}{\sigma_R} = 1 - 0.049 \, Z_1 Z_2^{4/3}/E_0\text{(keV)} \qquad (3)$$

Even for the heaviest target atoms, the correction factor for a 2.0 MeV ^{4}He beam is only ~2%. However, if using heavier projectiles such as ^{14}N, this screening correction could become significant unless E_0 is also increased.

<u>Mass Resolution</u>

To achieve optimum mass resolution in the backscattering process, the mass ratio between the target atom and the incident ion should typically be ~5-10. Hence, for a fairly heavy substrate such as gallium arsenide, a 10-20 MeV ^{14}N beam gives much better resolution than a 2.0 MeV He beam, as shown in fig. 3. Unfortunately, however, accelerators for producing 10-20 MeV heavy ion are not widely available and so ^{4}He is almost universally used for backscattering studies.

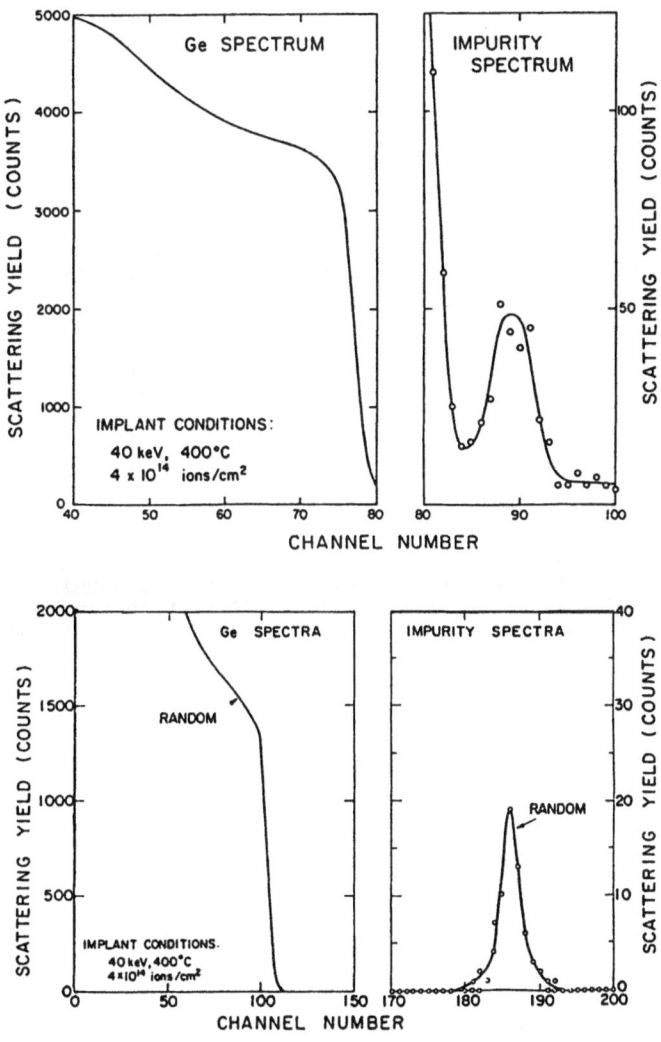

Fig. 3. Comparison of backscattering spectra obtained at random incidence with 1.0 MeV ^{4}He (upper panel) and 20.0 MeV ^{16}O (lower panel) beams. The same Tl-implanted Ge target was used in both cases (from fig. 4.19 in ref. 3).

Depth Resolution

The simplest method of improving the depth resolution is to increase the incident and/or emergent path lengths in the target, i.e., to apply the grazing-angle technique developed several years ago by J.S. Williams[6]. Table I shows that the grazing-exit geometry (i) gives significantly better resolution than grazing-incidence geometry (ii), due to the influence of the kinematic energy loss factor in the latter case. Grazing-exit geometry also permits the "channeling option" to be fully retained, since the target can be tilted in the plane orthogonal to the scattering plane and also rotated around the surface normal, without changing significantly the ratio of incident/emergent path lengths. This use of grazing-exit to enhance the depth resolution in a channeled RBS spectrum is illustrated in fig. 4.

Another useful feature of grazing-angle RBS measurements is in verifying whether or not an impurity peak (figure 2) is due to a specific near-surface impurity or alternatively to a much heavier impurity buried significantly beneath the surface. In the latter case, as α_{out} is reduced, the peak position (relative to the substrate edge) would shift dramatically to lower energies; on the other hand, for a near-surface impurity the peak position would be independent of α_{out}.

One obvious limitation of the glancing-angle technique is that it can only be applied to highly polished, atomically-flat surfaces. Furthermore, even for a completely flat surface, plural scattering effects impose serious constraints on the minimum value of α_{out}, and hence on the ultimate achievable depth resolution. Obviously, any particle undergoing secondary scattering through an angle comparable to α_{out} will have lost its depth/energy correlation. Since we are often interested in not just the mean depth but also the tails of the depth distribution, we must ensure that no more than a few % of the emerging ^{4}He ions have been deflected through an angle comparable to α_{out}. For this reason, in a high-Z target such as Pt (fig. 4), optimum resolution requires α_{out} to be at least 10–12°. In a similar study in Si, because of its lower atomic number, Feldman et al.[8] were able to use a significantly smaller α_{out} value (5°) and thus achieve a better depth resolution.

Quantitative evaluation of plural scattering is extremely complex and hence the objective is to select a sufficiently large α_{out} value that the resulting correction is negligible. Fortunately, the _tail_ of the plural scattering distribution is always dominated by _single_ scattering events, even at fairly large path lengths, as shown in fig. 5. Hence, simple analytical estimates may be used to establish the minimum acceptable values for α_{out}.

Table I. Grazing-Angle Technique (1.0 MeV ^{4}He in Si).

α_{in}	α_{out}	Depth Resolution (Å)
> 60°	> 60°	200
> 60°	4°	22 (i)
4°	> 60°	33 (ii)
6°	4°	17

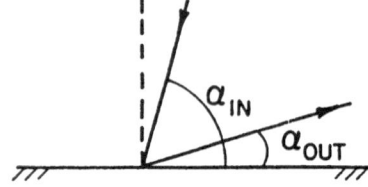

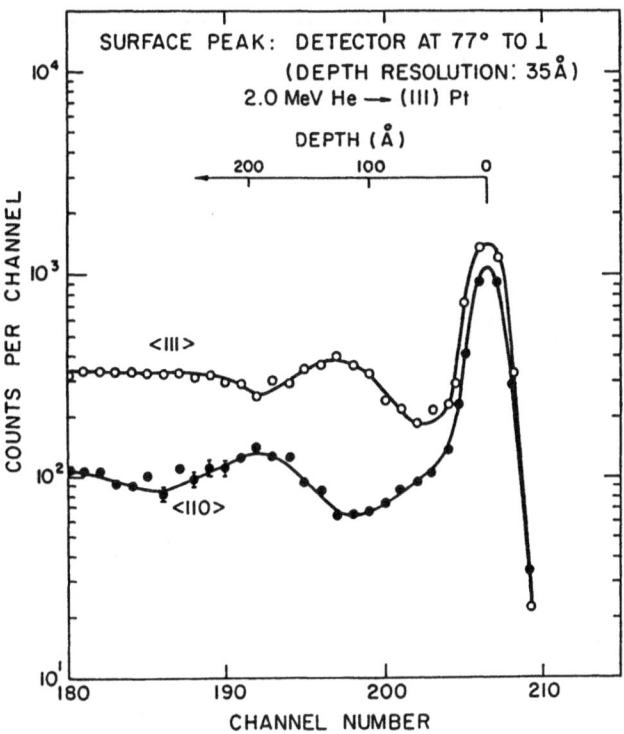

Fig. 4. RBS spectrum for 2.0 MeV ^{4}He$^+$ incident on a Pt (111) surface, with the detector at 13° to the crystal surface to enhance the depth resolution (adapted from fig. 2 in ref. 7).

Thus, for the 1.0 MeV ^{4}He → Ag case of fig. 5 and a path length of 6 x 10^{17} atoms cm^{-2} (ie. 100Å depth at α_{out} = 6°), 10% of the emerging beam undergoes at least one scattering event greater than 3°. Hence, the minimum acceptable α_{out} (in this case) should be somewhat larger than 3°. Note that the corresponding multiple-scattering angle (0.8°) would have indicated a much lower limit, since it drastically underestimates the tail of the scattering distribution.

Disadvantages of RBS

One of the main limitations is the radiation damage induced by such a high-E beam. Insulators such as oxides, ceramics, alkali halides, polymers and other organic targets are strongly affected by both the electronic (S_e) and nuclear (S_n) energy loss components in fig. 1 and hence may undergo significant damage under the bombardment doses involved in RBS analyses.

Metals and most semiconductor targets, on the other hand, are relatively unaffected by electronic energy loss processes (S_e in fig. 1) and hence it is only the extremely small S_n component that can produce any permanent damage. However, in surface adsorption studies, the metal-adsorbate bond may be quite sensitive also to S_e, and hence significant beam-stimulated desorption effects can be expected to occur. In Dr. Yates' lecture the topic of adsorbate desorption under electron bombardment is discussed in detail. Here, it is sufficient to point out that protons and electrons at the same velocity have identical S_e values and hence produce the same density of secondary electrons and other electronic excitation along their track. Consequently, a 250-keV H$^+$ beam and a 130-eV electron beam should exhibit rather similar rates of beam-

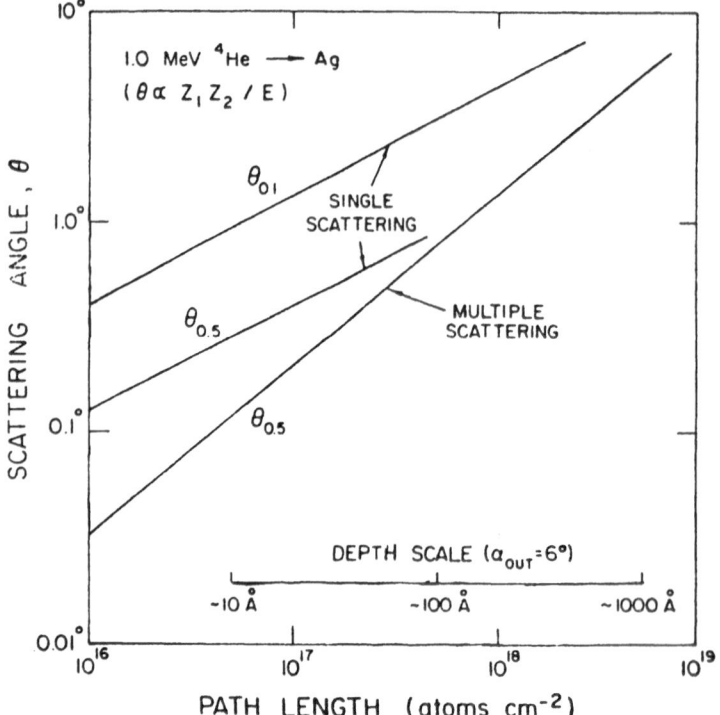

Fig. 5. Comparison of single- and multiple-scattering estimates as a function of path length. Curves labelled $\theta_{0.1}$ and $\theta_{0.5}$ represent angles through which 10% and 50% of the beam have been scattered.

induced desorption. Since a typical RBS measurement involves 1-10 μC on a 1 mm² spot, it is roughly equivalent to an electron bombardment of 0.4-4 mC/cm² at ~130 eV.

Another major limitation is that RBS requires a relatively flat target surface over distances of at least 0.1 μm and hence it is of little use on rough surfaces or powders.

Channeling

Channeling is a gentle, coulombic steering process that occurs whenever a positively charged particle (positron, proton, helium, heavy

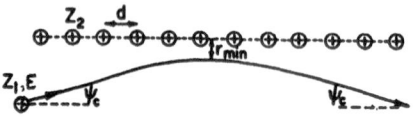

Fig. 6. Schematic of a channeled ion trajectory, illustrating conservation of transverse motion.

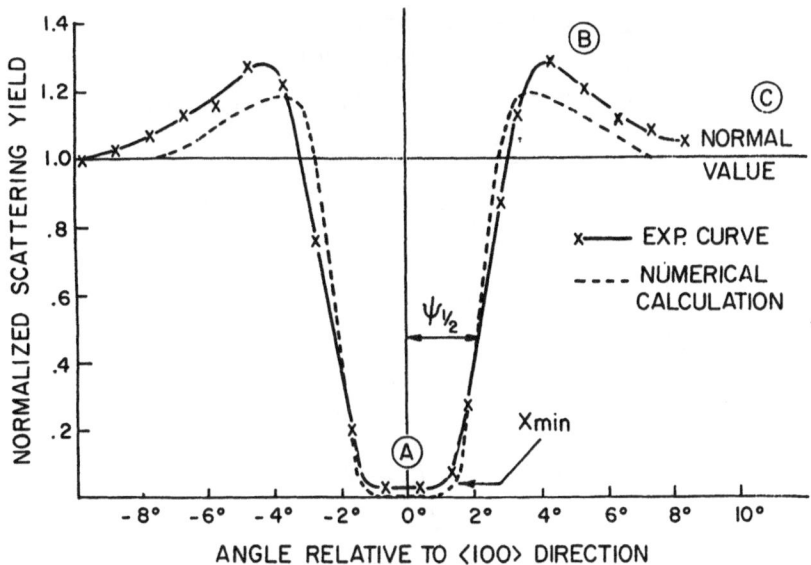

Fig. 7. Angular scan through the <100> axis for an 0.5 MeV ⁴He⁺ beam
incident on (100) W at room temperature (ref. 10).

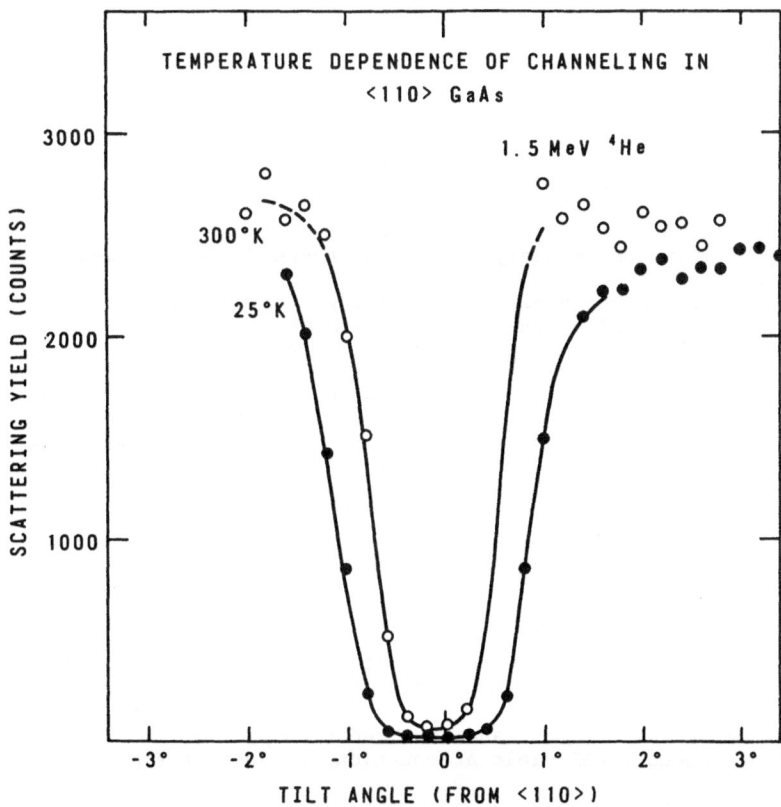

Fig. 8. Temperature dependence of the angular scan through the <110>
axis in GaAs.

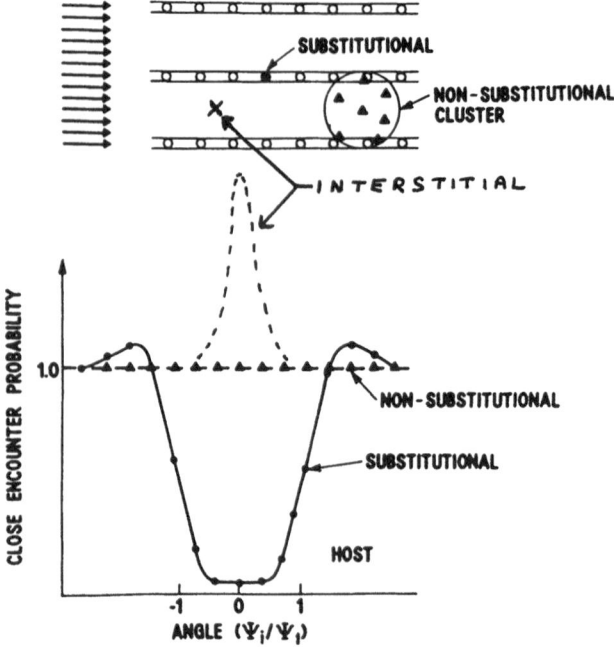

Fig. 9. Schematic of the close-encounter probability curve for substitutional and interstitial impurities and also for a non-substitutional cluster (adapted from Fig. 5.4 in ref. 11).

ion) travels through a crystal lattice within a small, predictable angle (ψ_c) of a low-index axis or plane (fig. 6). It has been studied experimentally at energies as low as 100 eV, but it is only in the high-E regime (fig. 1) that an accurate theoretical framework[9] exists to describe the channeled trajectories. Hence, channeling plays an extremely useful supplementary role in high-energy RBS studies.

The main consequence of the steering process in fig. 6 is that impact parameters smaller than r_{min} are prohibited; hence all violent collisions such as RBS and nuclear reactions are strongly attenuated. The resulting RBS yield curve versus angle of incidence is illustrated in fig. 7. By assuming conservation of transverse energy, $E_\perp = E \psi_c^2$, and equating r_{min} (fig. 6) to the (2-dimensional) vibrational amplitude ($\rho_\perp$) of the lattice atoms, we obtain a good fit to the experimentally observed angular dependence. The resulting expressions for the angular half-width ($\psi_{1/2}$) and the minimum aligned yield (χ_{min}) are:

$$\psi_{1/2} = \left(\frac{Z_1 Z_2 e^2}{E\ d} \right)^{1/2} \cdot \left[\ln \left\{ \left(\frac{Ca}{\rho_\perp} \right)^2 + 1 \right\} \right]^{1/2} \quad (4)$$

and

$$\chi_{min} = N\ d\ \pi \rho_\perp^2 \quad (5)$$

where d is the spacing along the axis (fig. 6), N is the atomic density (atoms cm^{-3}) and C is a constant of magnitude ~3.

By cooling the crystal, and thus reducing $\rho_\perp$, one obtains a wider and significantly deeper RBS yield attenuation, as seen in fig. 8.

Foreign Atom Location

The existence of such large, predictable yield attenuations has led to several major applications in near-surface lattice studies. One of

346

the most useful and widespread of these is in locating the lattice site
of specific foreign atoms. The principle is illustrated in fig. 9.
Foreign atoms located on lattice sites (i.e. substitutional) will exhibit
an identical angular scan to that of the host atoms. Randomly-
distributed atoms (such as the non-substitutional cluster in fig. 9) will
exhibit no orientation dependence at all. Mid-channel interstitials will
exhibit a narrow peak (instead of a dip) due to a pronounced flux-peaking
that occurs in the mid-channel region.

By making similar measurements down two or more axes or planes, and
invoking triangulation and lattice symmetry arguments (fig. 10), it is
relatively simple to pinpoint the exact location of the foreign atom
within the crystal lattice; in favourable cases, a precision of 0.02Å can
be readily attained. The RBS/channeling "signatures" for various lattice
sites and combinations of sites are shown in fig. 11.

Surface Structure

So far, we have been considering channeling within the bulk of the
crystal lattice, i.e., at sufficiently large depths (≥ 500Å) that
equilibrium within the transverse plane has already occurred. Let us now
turn our attention to the surface and sub-surface regions shown
schematically in fig. 12.

At the surface, the first atom in each atomic row is obviously <u>not</u>
<u>shadowed</u> and hence must exhibit the normal backscattering yield. This is
the main cause of the prominent surface peak (Mass 28) in fig. 2. Note
that the second and subsequent atoms in the row fall within a 'shadow
cone' created by the surface atom and hence will not contribute to the
observed 'surface peak' unless their $\rho_\perp$ value is comparable to the radius
R of the shadow cone. In practice, at the 1-3 MeV ^{4}He energies normally
used, $\rho_\perp$ and R are comparable in magnitude for at least the first 2-3
atomic planes beneath the surface and so the observed 'surface peak'
contains a significant (but predictable) sub-surface contribution as
shown in fig. 13. At slightly larger depths (A), the RBS yield falls

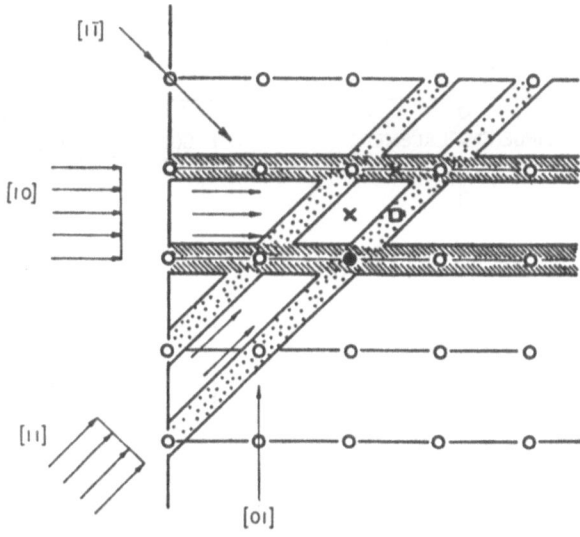

	SHADOWED FRACTION		
	●	x	□
⟨10⟩	1.0	0.5	0.0
⟨11⟩	1.0	0.0	1.0

Fig. 10. A 2-d model illustrating how the channeling effect may be used
to locate foreign atoms in the crystal (fig. 2 in ref. 12).

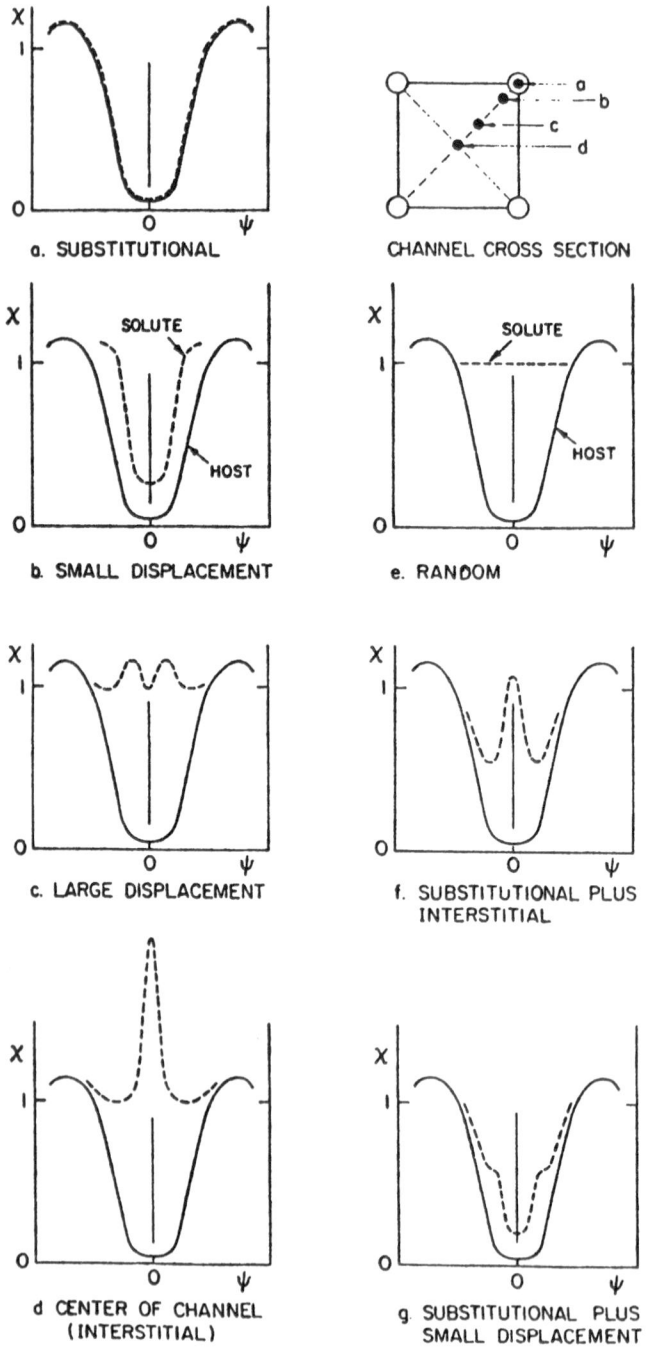

a. SUBSTITUTIONAL

CHANNEL CROSS SECTION

b. SMALL DISPLACEMENT

e. RANDOM

c. LARGE DISPLACEMENT

f. SUBSTITUTIONAL PLUS
INTERSTITIAL

d. CENTER OF CHANNEL
(INTERSTITIAL)

g. SUBSTITUTIONAL PLUS
SMALL DISPLACEMENT

Fig. 11. Angular scan RBS yield profiles for different projections of a solute atom into an axial channel (fig. 30 in ref. 12).

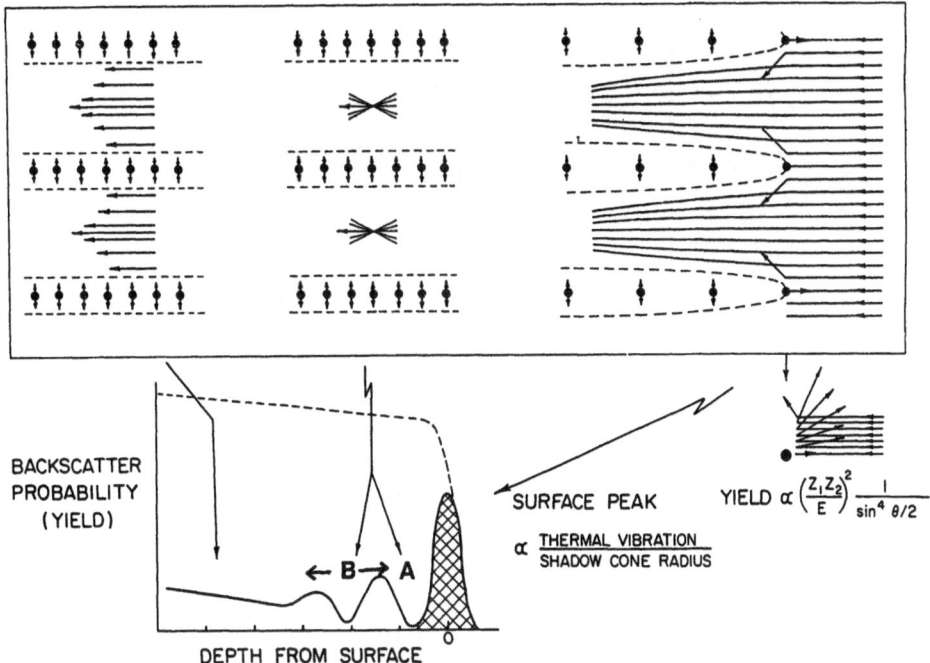

Fig. 12. Schematic illustration of the surface, subsurface and bulk channeling behaviour (ref. 13).

essentially to zero and then exhibits a rapidly damped oscillatory depth dependence (B), with a period of approximately $(2d/\psi_{1/2})$, before eventually levelling off at the 1-2% value characteristic of χ_{min} in the bulk crystal. This oscillatory sub-surface behaviour arises because the non-channeled (1-2%) component of the beam is determined initially by the point of impact at the surface and hence is not randomly distributed across the channel.

The magnitude of the surface peak provides a quantitative measure of the number of near surface atoms visible to the incident beam. As noted above, this number can be predicted quite accurately for an ideal crystal as a function of the vibrational amplitude of the surface atoms. Deviations from the ideal bulk structure (e.g., reconstruction, relaxation, enhanced $\rho_{\perp}$, etc.) will ordinarily increase the observed surface peak, as shown schematically in fig. 14. Hence, surface peak measurements provide a powerful tool for studying surface structure.

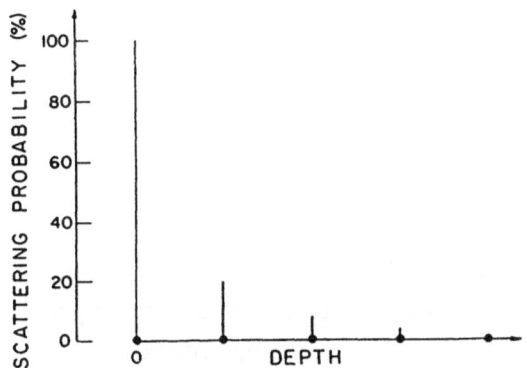

Fig. 13. Calculated RBS contribution from successive planes in an Au (011) crystal, using 0.5 MeV ${}^{4}He^{+}$ ions at 300K (ref. 14).

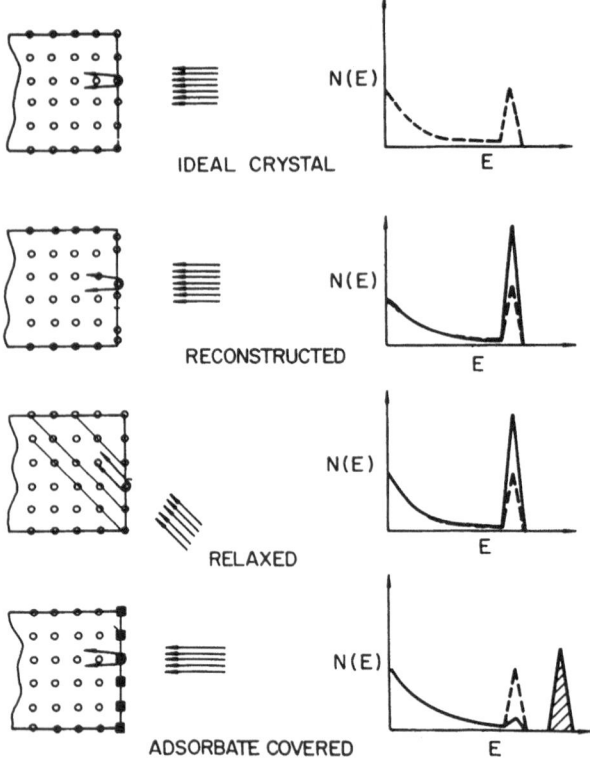

IDEAL CRYSTAL

RECONSTRUCTED

RELAXED

ADSORBATE COVERED

Fig. 14. Representations of different surface effects on a simple cubic lattice and the resulting channeled RBS spectra. The dashed line represents the signal from the ideal "bulk-like" crystal (fig. 5.15 in ref. 11).

Figure 15 shows an example of how such RBS surface peak measurements can be used to follow the reversible surface phase transition (reconstruction) between the cubic (1x1) and the hexagonal (5x20) structures of Pt(100) as a function of target temperature and pressure of CO. In this case, the fractional coverage θ_{co} of CO on the surface was also monitored in situ by switching to a 1.0 MeV deuterium beam (and the

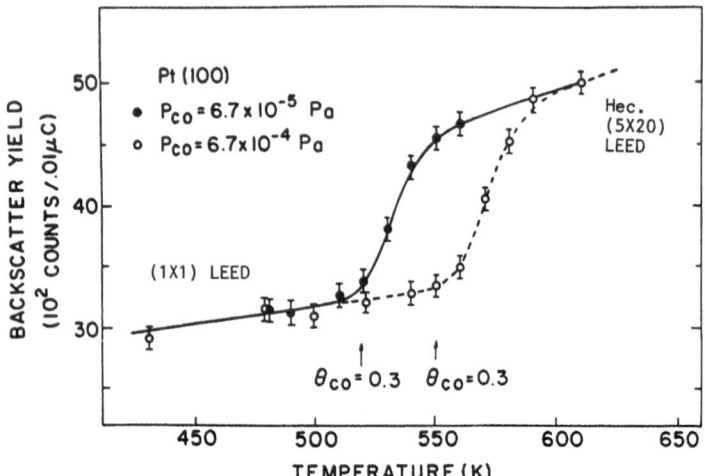

Fig. 15. Pt <100> surface peak yield at increasing sample temperature for two different CO pressures (ref. 15).

350

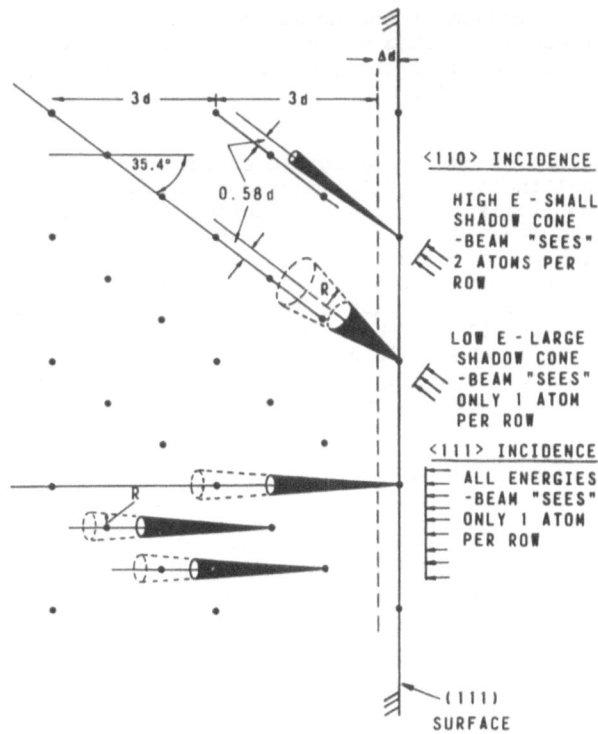

Fig. 16. Atomic configuration in the (110) plane of Pt, illustrating the use of channeling to measure the surface relaxation, Δd (ref. 16).

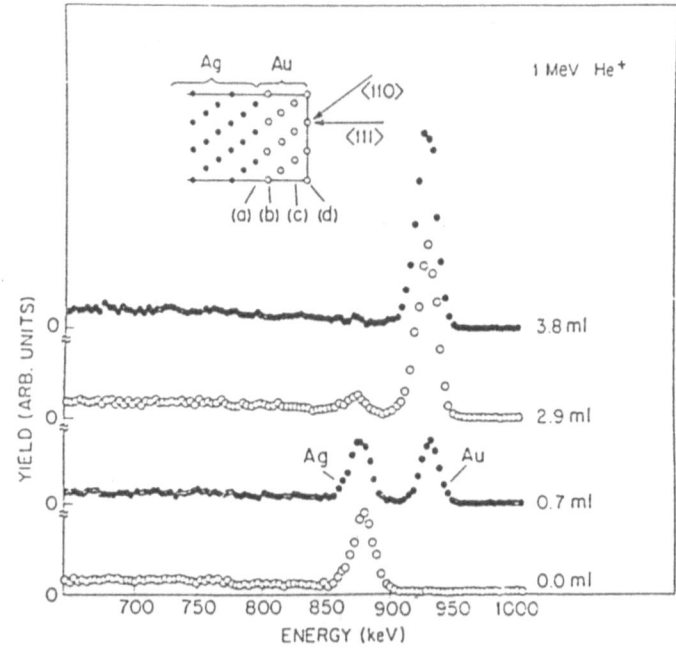

Fig. 17. RBS spectra for ⁴He⁺ ions incident along the <110> axis of a) clean Ag (111) surface, and b), c) and d) after various coverages of Au had been deposited (fig. 5.17 in ref. 11).

^{12}C (d,p) reaction) instead of the normal ^{4}He beam. It should be noted that high-energy backscattering measurements are relatively insensitive to the ambient pressure in the target chamber and hence can be used to follow surface catalysis effects at working pressures up to about 10^{-3} atmospheres.

Surface relaxation has a negligible effect on the surface peak measured at normal incidence (<111> incidence in fig. 16). However, by using a non-perpendicular channeling direction (eg. <110>) the effect of the outward displacement Δd of the surface plane is immediately visible, provided the shadow cone radius R is comparable in magnitude to the projected relaxation, $\Delta d \sin 35°$, normal to the <110> direction. Since Δd values are usually quite small (≤ 0.05Å), the optimum beam energies for such studies are typically in the 0.1-3.0 MeV range. (Dr. Rabelais' lecture also discusses the topic of shadow cones, but at keV energies where their much larger radii are well suited for studying the large lateral displacements involved in surface reconstructions such as that in fig. 15).

Low-Z surface adsorbates, such as the CO in fig. 15, produce such small shadow cones that their effect on the substrate surface peak is usually quite small. However, with a high-Z adsorbate such as Au on Ag (fig. 17), the attenuation of the substrate surface peak and its replacement by a Au peak at higher RBS energy is a powerful and quantitative tool for studying the initial stages of epitaxial growth.

Double-Alignment Technique

A closely related phenomenon to channeling, known as blocking, occurs whenever the outgoing (backscattered) trajectory is aligned with a crystal axis or plane. The same steering process that prevents a channeled incident ion from hitting a lattice atom also prevents any ion that is backscattered from a lattice atom from emerging along a channeling direction. By invoking the principle of reversibility, Lindhard [9] has shown that channeling and blocking involve exactly the same steering process and hence the same $\psi_{1/2}$ and χ_{min} values apply to both.

This has led to the development of the so-called "double-alignment" technique by Saris et al.[17] in which both the incident and scattered

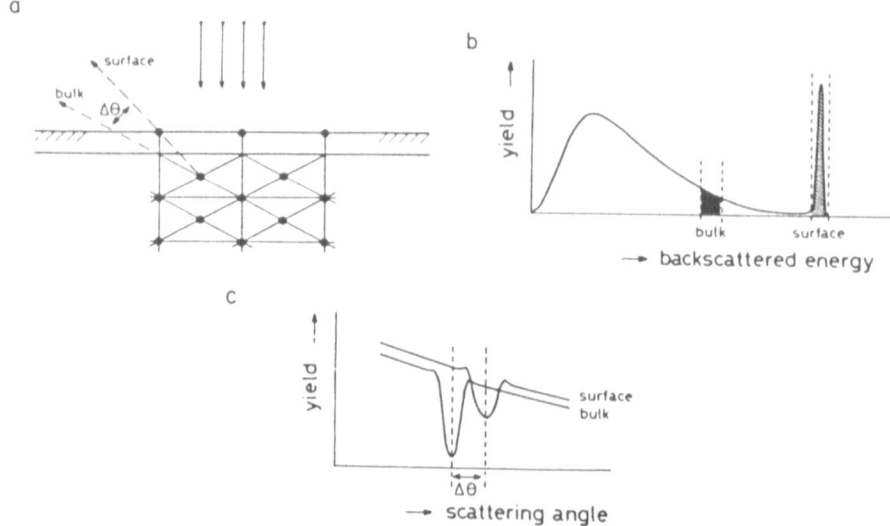

Fig. 18. (a) Schematic diagram of a combined channeling-blocking experiment; (b) Channeled RBS spectrum showing a clearly resolvable surface peak; and (c) Detector scan showing the blocking dips for the two shaded regions in (b) (ref.17).

trajectories are aligned with low-index crystal axes or planes, so that channeling and blocking effects occur simultaneously. The result is a very powerful tool for surface structure analysis, as illustrated schematically in fig. 18. First, the incident beam is aligned with an appropriate axis in order to produce a well-resolved surface peak involving atoms in the first 2-3 atomic planes of the target (fig. 18b). Note that, along the incident beam direction in fig. 18a, only 50% of the atomic rows originate in the surface plane; the other 50% originate in the second plane. Consequently, at least 50% of the backscattered ions contributing to the surface peak can be blocked along certain emission directions by the surface plane of atoms, thus producing a blocking dip in the surface peak yield. Any relaxation Δd of the surface plane will rotate the position of this blocking dip through an angle $\Delta\theta$ relative to the underlying bulk blocking direction, as shown in fig. 18c. Hence the magnitude and sign of Δd may be obtained from purely geometrical considerations. Another big advantage of this blocking technique is that it can be readily adapted to provide information on the location of adsorbate atoms since these too may produce observable blocking dips in the backscattered yield from the underlying crystal.

An excellent example of this is the study of sulfur on Ni(110) shown in fig. 19. The shadowing effect causes the backscattering to occur mainly from the first Ni atom in each row (and also from the adsorbed S atoms, but these give a peak at much lower backscattered energy). For perfect shadowing, the Ni surface peak intensity would be exactly 1 atom per row; in fig. 19, it is considerably higher (i.e., ~1.5 atoms per row), indicating that significant scattering contributions also come from the second and even the third monolayers. Note that the clean surface exhibits a prominent blocking dip at 59° - i.e., displaced ~1° from the bulk [011] blocking direction - thus indicating a relaxation of -4% (a small contraction) for the surface plane.

Comparison of the clean and sulfur-covered spectra in fig. 19 reveals two significant differences. First, there is a small dip in yield around 52° in the sulfur-covered case; this dip is attributed to

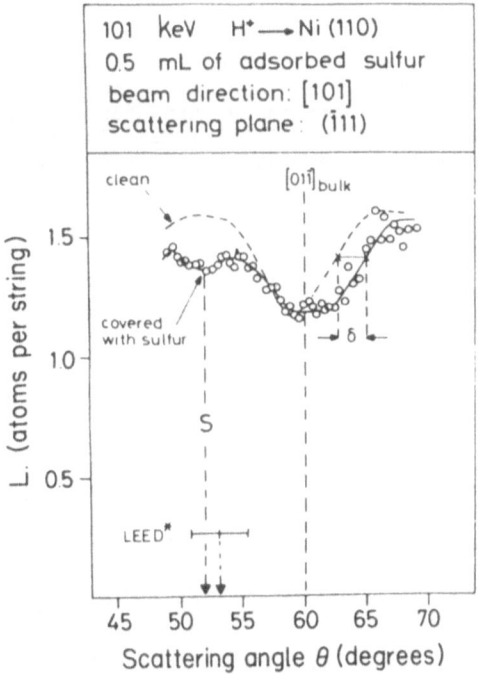

Fig. 19. Blocking angular scan for 101 keV H⁺ scattered from clean and sulfur-covered Ni (011) surfaces (ref. 18).

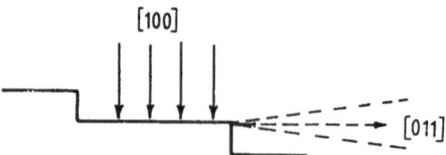

Fig. 20. Double-alignment configuration in Cohen et al's stepped crystal experiment[19].

those particles backscattered from Ni surface atoms and subsequently blocked along their outgoing path by the sulfur overlayer. From the angular position of the dip (52°), the authors deduced that the sulfur atoms lie in the 4-fold coordinated hollow site, 0.087 ± 0.003 nm above the Ni surface, in good agreement with the LEED analysis value of 0.093 ± 0.01 nm.

Second, the sulfur-covered surface exhibits a much broader dip around the [011] blocking direction, with the centroid being displaced to angles slightly *larger* than 60°. This shift to larger angles suggests that the Ni surface undergoes an ~6% expansion in the sulfur-covered case, compared to the 4% contraction of the clean surface.

Another potential application of double-alignment, developed by C. Cohen et al. at the Universite de Paris, is in studying the surface of a stepped crystal such as Cu (511). The concept is illustrated in fig. 20. Here, the incident beam direction (<100>) is perpendicular to the terrace and the detector scans through a low-index direction (eg. <110>) located in the step surface, at 90° to the incident beam. Since each terrace on such a stepped surface is only 4-5 atoms wide, any relaxation or enhanced vibration of step-edge atoms produces a considerable weakening and broadening of the observed blocking dips. Hence, double-alignment measurements yield quantitative structural information on step-edge atoms. In the case of the Cu (511) surface, they observed a readily measureable effect with an anomalously large temperature dependence. Unfortunately, the resulting experimental data indicate that a very complex surface behaviour is involved and they have not yet found a reasonable model for interpreting their observations.

One major disadvantage of the double-alignment technique is that the acceptance angle to the detector must be very small, i.e., less than the blocking angle R/d. Hence rather large analyzing beam fluences are used and considerable care is required in order to minimize radiation damage and beam-induced desorption effects.

Summary

High energy ion scattering provides a quantitative analysis technique for surface and thin film studies that is completely independent of matrix effects and hence does not require the use of precalibrated standards. Probably its most useful role in heterocatalysis and related surface studies is in providing a quantitative cross calibration for some of the more widely used surface techniques (eg. Auger, SIMS and low energy ion scattering) where matrix effects are often dominant.

On single crystal surfaces, channeling and blocking add an extra 'dimension' to the RBS analysis in that quantitative structural information on surface relaxation, reconstruction, enhanced vibration, and foreign atom (adsorbate) location is also obtained.

References

1. W.K. Chu, J.W. Mayer and M.A. Nicolet, Backscattering Spectrometry, Academic Press, New York (1978).
2. J. Lindhard, M. Scharff and H.E. Schiott, Kgl. Danske Vid. Selsk., Mat. fys. Medd. $\underline{33}$, No. 14 (1963).
3. J.W. Mayer, L. Eriksson and J.A. Davies, Ion Implantation in Semiconductors, Academic Press, New York (1970).
4. J.R. MacDonald, J.A. Davies, T.E. Jackman and L.C. Feldman, J. Appl. Phys. $\underline{54}$, 1800 (1983).
5. J. L'Ecuyer, J.A. Davies and N. Matsunami, Nucl. Instr. Meth. $\underline{160}$, 337 (1979). (See also H.H. Andersen et al., Phys. Rev. $\underline{A21}$, 1891 (1980) and M. Hautala and M. Luomagarvi, Rad. Eff. $\underline{45}$, 159 (1980).
6. J.S. Williams, Nucl. Instr. Meth. $\underline{126}$, 205 (1975).
7. J.A. Davies, D.P. Jackson, N. Matsunami, P.R. Norton and J.U. Andersen, Surf. Sci. $\underline{78}$, 274 (1978).
8. T.E. Jackman, J.R. MacDonald, L.C. Feldman, P.J. Silverman and I. Stensgaard, Surf. Sci. $\underline{100}$, 35 (1980).
9. J. Lindhard, Kgl. Danske Vid. Selsk., Mat. fys. Medd. $\underline{34}$, No. 14 (1965).
10. J.U. Andersen and E. Uggerhoj, Can. J. Phys. $\underline{46}$, 517 (1968).
11. L.C. Feldman and J.W. Mayer, Fundamentals of Surface and Thin Film Analysis, North Holland (1986).
12. L.M. Howe, M.L. Swanson and J.A. Davies, Methods of Experimental Physics, $\underline{21}$, 275 (1983).
13. T.E. Jackman, private communication (1984).
14. E. Bogh in Channeling: Theory, Observation, and Applications, ed. D.V. Morgan, Wiley (Interscience) New York (1973) p. 435.
15. T.E. Jackman, K. Griffiths, J.A. Davies and P.R. Norton, J. Chem. Phys. $\underline{79}$, 3529 (1983).
16. J.A. Davies, D.P. Jackson, J.B. Mitchell, P.R. Norton and R.L. Tapping, Physics Letters $\underline{54A}$, 239 (1975).
17. W.C. Turkenburg, W. Soszka, F.W. Saris, H.H. Kersten and B.G. Colenbrander, Nucl. Instr. Meth. $\underline{132}$, 587 (1976).
18. J.F. Van der Veen, R.M. Tromp, R.G. Smeenk and F.W. Saris, Surf. Sci. $\underline{82}$, 468 (1979).
19. C. Cohen, private communication (1988).

SECONDARY ION MASS SPECTROMETRY - FUNDAMENTALS AND APPLICATION TO HETEROGENEOUS CATALYSIS

N. M. Reed and J. C. Vickerman

Centre for Surface and Materials Analysis
Department of Chemistry
University of Manchester Institute of Science and Technology
Manchester M60 1QD, UK

1. INTRODUCTION

Secondary ion mass spectrometry, SIMS, is the mass spectrometry of atomic or molecular particles which are emitted (a process known as sputtering) when a surface, usually a solid although it may be a liquid, is bombarded by energetic primary particles. The primary particles may be electrons, ions, neutrals or photons. The secondary ions which are detected may be emitted from the surface in the ionised state or they may be initially emitted as neutrals to be post-ionised before analysis.

It was Sir J J Thomson who first observed and identified in 1910 the emission of positive secondary ions when primary ions bombarded a metal surface in a discharge tube[1]. However it wasn't until the period 1948-58 that instrumentation was developed by Herzog and then Honig to exploit the phenomenon in analysis[2,3]. In the 60-70's there were rapid advances in the use of SIMS in high sensitivity elemental analysis. In this mode of operation, sometimes known as ion probe or dynamic SIMS, a high flux of primary ions is directed at the material surface in order to obtain a very high yield of secondary ions. The surface is eroded away very rapidly and it is possible to monitor changes of elemental composition with depth and thus a depth profile may be generated.

The SIMS technique may be thought to be destructive since it relies on particle removal from a surface. The true power of SIMS in surface analysis only emerged when it became possible in the early 70's to reduce the primary bombarding beam density to a very low level and yet retain high sensitivity in a modification of the technique, known as static SIMS, demonstrated by Benninghoven[4]. In this mode secondary ions are emitted from areas not previously damaged and the surface monolayer lifetime is well in excess of the time required for analysis. The importance of static SIMS for surface analysis arises from the fact that a surface mass spectrum is generated and thus in principle the power associated with analytical organic mass spectrometry for chemical structure characterisation is available to surface analysis.
Surface analysts frequently require high spatial resolution and this is dependent on the focussing characteristics of the probe beam. In the 1980's sub-micron spatial resolution has become possible with the advent of highly focussed

Fundamental Aspects of Heterogeneous Catalysis Studied by Particle Beams,
Edited by H.H. Brongersma and R.A. van Santen, Plenum Press, New York, 1991

357

liquid metal primary beam sources[5]. By raster scanning the beam across an area of surface and collecting the secondary ions at each point a chemical image can be generated. This variant of the technique is known as <u>imaging SIMS.</u> In common with many techniques of surface analysis which utilise a charged particle as the probe species, ion bombardment of a poorly conducting surface can result in surface charging which in the SIMS experiment can result in spectral loss or instability. Flooding the surface with low energy electrons can be a solution although stable neutralisation of surface charge is not always easy. In addition electron bombardment can have a deleterious effect on delicate surfaces. Although neutral beams had been used by Devienne to generate secondary ions[6] in the period 1979-81 Vickerman and co-workers showed that the charging problem could be almost completely eliminated by using a neutral primary beam instead of an ion beam and, in consequence, insulator materials could be routinely analysed[7,8]. This modification of static SIMS is sometimes known as fast atom bombardment (FAB) SIMS.

2 The SIMS Phenomenon

SIMS analysis is fundamentally concerned with the yield of secondary ions. The basic equation is

$$i_s M = i_p \, YR^+ \theta M \eta \qquad 1$$

where $i_s M$ is the secondary ion current of an element or species M, i_p is the primary particle flux, Y is the sputter rate for M, R^+ is the ionisation probability to positive secondary ions (if negative ions are to be collected R would be appropriate) θM is the fractional coverage of M and η is the transmission of the analysis system. Clearly sensitivity to M is fundamentally controlled by S, R and η. The transmission of the analysis system is determined by the ion collection/mass spectrometer system used.

2.1. <u>Sputter yield.</u> The magnitude of Y is a function of the primary beam parameters, see fig. 1 The sputter yield generally increases with beam energy and with the primary particle mass although the increase is not linear in either parameter and at high energies (> 10 keV) a maximum between 1 and 10 is reached[9]. Y is also a function of impact angle and maximises at about 70° to the surface normal as a consequence of depositing most of the impact energy in the surface layers. It is frequently assumed that the sputter rate is independent of the charge state of the primary particles. Whilst this is probably so in the case of metal targets where the charge from ion bombardment can be dissipated easily, as we have seen above it is unlikely to be the case with poor conductors or insulators.

Sputter rate is also sensitive to the properties of the material being analysed. In the case of pure elements there is a variation across the periodic table which suggests the influence not only of atomic mass but also solid state bonding. The sputter rate of organic materials in terms of the structural damage done is considerably higher (probably as high as 100 to 1000 times) than for metals or other inorganic solids[10,11]. The sputter yields of molecular ions are certainly influenced by the type of bonds to be broken. This is not surprising since some organic materials are molecular solids held together with weak van der Waals forces, whereas the molecular units in polymers are bound together by strong chemical bonds.

2. 2. <u>Ionisation probability</u>. Since ionisation occurs as the secondary particles leave the surface, the electronic state of the surface has an important influence on the probability that the particle will leave the surface as an ion. There is a dependence on the ionisation potential and electron affinity of the particle but it

is observed that the electronic state of the material being analysed has a major influence on the ion yield. In common with dynamic SIMS it has been found that[12]:

(a) positive ion yields of different elements sputtered from a <u>common matrix</u> exhibit an inverse exponential dependence on the ionisation potential of the sputtered atom, (there is some evidence for a similar dependence of negative ion yield on electron affinity)[13];

(b) positive ion yields are greatly enhanced in the presence of oxygen or other electronegative species at the surface[14]

(c) positive ion yields are sensitive to the electronic state of the matrix from which they are sputtered[15].

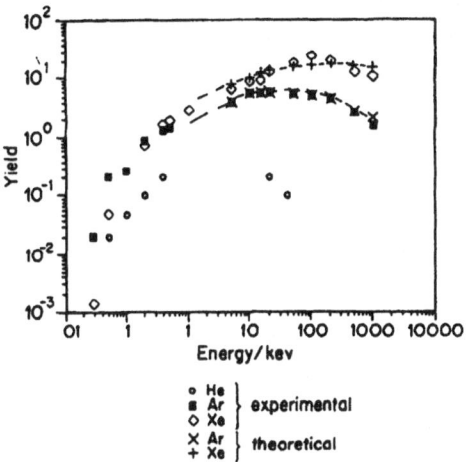

Fig. 1 Variation of sputter yield of copper atoms from a copper matrix as a function of bombarding energy for helium, argon and xenon primary ions

Table 1 highlights the large variation in secondary ion yield between a range of metals whose crystallographic and electronic state vary very considerably, and it illustrates the tremendous increase in secondary ion yield observed when they are exposed to oxygen. Thus sensitivity to a particular element or species will vary from matrix to matrix. It is this which makes quantitative analysis by SIMS a matter of very careful calibration. In Sputtered Neutral Mass Spectrometry, SNMS, where ionisation is accomplished after emission, the elemental ion yields are only dependent on the ionisation potential of the element whose values vary by less than an order of magnitude through the periodic table[16].

The above observations refer to the emission of atomic ions. Little in the way of detailed studies on the parameters which affect the yield of molecular cluster ions have been carried out. A start has been made in seeking an understanding of the fragmentation pathways for polymers[17]. Clearly there will be significant involvement of chemical structure parameters in addition to the electronic factors outlined above.

TABLE 1

Metal	Clean elements (M^+ yield)	Oxide (M^+ yield)
Al	0.007	0.7
Ti	0.0013	0.4
Cr	0.0012	1.2
Fe	0.0015	0.35
Cu	0.0003	0.007
Mo	0.00065	0.4
Ba	0.0002	0.03
W	0.00009	0.035
Si	0.0084	0.58
Ge	0.0044	0.02

3. THE SIMS EXPERIMENT

There are two main components in the SIMS experiment - the primary particle beam and the mass spectrometer, fig 2. Their specifications and how they are used depends on the information required.

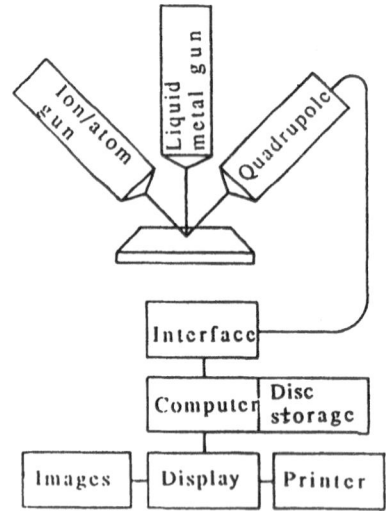

Fig. 2 Schematic diagram of the components of a typical SIMS system

3.1 Primary particle beams for static SIMS. From equation 1 it is clear that if in dynamic SIMS we require very high elemental sensitivity and surface is not important then we require a high primary beam current and a high ionisation probability. For elemental positive ions the latter can be realised using an oxygen primary ion beam. As the surface is eroded it is oxidised and R is high. For negative ion detection a caesium beam may be used. For depth profiling the beam must be rastered over a precise area to form an accurately defined crater. However in static SIMS we are concerned to have high surface sensitivity with minimal surface damage. To a first approximation surface damage is dependent solely on the sputter rate, although particle bombardment may also generate additional surface chemical modification which can also be regarded as damage. Considering sputtering alone, a crucial parameter is the lifetime of the surface monolayer

$$t_m = 10^{15} / Y\, i_p \qquad 2$$

where the density of atoms in the surface is taken as 10^{15} cm^{-2}; Y is the sputter rate which usually lies between 0.1 and 10; [18] and i_p is the number of primary particles incident at the surface cm^{-2}. Thus when i_p is 1 nA (10^{10} particles cm^{-2} s^{-1}) and Y=l, then $t_m = 10^5$ s, good conditions for static SIMS.

However surface chemical modification must also be minimised, consequently inert gas beams, argon or xenon, are normally used . Particle bombardment disrupts surface chemistry, it roughens metal surfaces, it can reduce oxide surfaces and there is good evidence that it can cause extensive chemical fragmentation of organic materials. Furthermore there is evidence that the charge deposited by ion bombardment of organic materials generates further damage so the use of atom beams is beneficial. Organic materials are most sensitive to bombardment induced damage. The molecular identity is rapidly lost as a function of primary beam dose. Static conditions for these materials are thus more rigourous than for inorganic. It is estimated that each primary particle impact generates collisional effects within the surface layer extending out to a region of about 10 nm diameter. Calculation would therefore suggest that only 10^{13} impacts per cm^2 would affect all 10^{15} atoms in the 1 cm^2 of surface. Thus the monolayer lifetimes should be divided by 100 to yield analysis lifetimes. Even at 0.1 nA cm^{-2} the analysis lifetime reduces to only 5 h, or 30 min if the sputter rate is 10. The problem becomes more important when we consider analysis at high spatial resolution, either the spectral investigation of small surface features or the generation of chemical images. As the area of interest decreases, to generate sufficient signal an increasingly large proportion of the atoms in the surface will be removed for analysis thus destroying what is to be analysed. Table 2 shows the monolayer removal rate as a function of magnification and primary beam current assuming a sputter yield of 1 and an analysis time of 100 s. If we remember that damage begins to become evident after 0.01 monolayers, it is clear that *static* SIMS analysis at high magnification will be very difficult.

TABLE 2

Analysis Mode	Primary beam parameters			Data
	i_p	d_p	t_m	
Static SIMS	Ar$^+$/Xe$^+$;Ar•/Xe•			
	<10 nA cm^{-2}	2-3 mm	>10^3 s	Spectra
Dynamic SIMS	O$_2^+$; Cs$^+$			
	>10 µA cm^{-2}	≥10 µm	<1 s	Depth Profiles
Imaging SIMS	Ga$^+$; In$^+$			
	≈ 10 nA cm^{-2}	100 nm	≤10^3 s	Images or maps

3.1.1 <u>Ion beams</u> Since low beam currents (10^{-11} to 10^{-8} A cm^{-2}) are required most static SIMS systems use electron bombardment of gaseous molecules to produce the ion beam. The beam is then accelerated to energies between 0.5 and 4 keV. To maximise the secondary ion yield an impact angle of about 70^0 to the surface normal is used and the impact area is usually large (ca. 0.1 cm^2) compared to the field of view of the analyser. To reduce the possibility of sample contamination by impurity ions from the source filament or residual gases in the vacuum system the beam should be mass filtered by a Wien filter or a small RF quadrupole.

Where the analysis of electrically insulating materials is of significant interest, it is necessary to overcome the surface charging which results from ion bombardment. This can be accomplished by simultaneous irradiation of the sample surface with a low energy <u>electron beam.</u> The beam flux has to be carefully matched with the ion beam density otherwise further charging can occur. It is also important to ensure that the conditions are such that electron stimulated desorption does not occur.

The requirements for positive ion detection are somewhat different from those for the detection of negative ions. Particle bombardment always leads to the emission of vast numbers of secondary electrons as well as ions and neutrals. Hence positive charging of the surface almost always occurs. Under ion bombardment the material can charge to the potential of the primary beam due to the emission of electrons plus the injection of positive ions. Clearly the charging potential reached will be dependent on the capacitance of the material. Such charging will cause positive ions to be ejected with considerably increased energy such that they cannot be analysed by the mass spectrometer. The function of electron flooding is to depress the surface potential to a low level, although still positive, such that the positive ions can be analysed by the mass spectrometer.
It is frequently easier to use an atom beam for these analyses because although charging still occurs due to electron emission it is very much slower and to a much lower potential. It is frequently possible to tune the ion collection optics to accept the emitted positive ions without any electron flooding[19]. To detect negative ions it is necessary to make the surface potential negative otherwise most of the negative ions will not be released from the surface. Thus electron flooding is usually required for both ion and atom primary beams. Clearly the fluence of electrons has to be higher for negative ion SIMS than for positive ion SIMS[20].

3.1.2 Atom beams are usually produced by passing a high energy ion beam through a chamber containing a high pressure (10^{-4} mbar) of inert gas. A proportion of the fast ions (up to ca. 20% depending on the chamber length and the gas pressure) lose their charge by capturing an electron from the atoms or molecules randomly moving in the chamber. Although these charge exchanged particles have lost their charge their kinetic energy and direction of motion are essentially unchanged, consequently a neutral high energy beam emerges from the charge exchange chamber. The ion component of the beam is removed using electrostatic deflection plates.

It is frequently advantageous to have both an ion and atom beam source available because whilst atom beams have advantages over ion beams for static SIMS, the range of beam fluences available from an ion beam source is usually larger. Van den Berg and Vickerman have described a combined ion-atom source which can be rapidly switched from ion beam to atom beam operation and is mass filtered in both modes[21], it is ideally suited for SSIMS. Beam energy can be varied between 500 eV and 2000 eV. The beam flux densities are controllable from 10 nA cm^{-2} in ion mode and 1 nA cm^{-2} in atom mode down to 1 pA cm^{-2}, and the beam diameter at the target surface is 3 mm in both modes.

3.1.3 Liquid metal ion beams. Where there is a need to investigate small surface features or to produce chemical maps of a surface a highly focussed beam is necessary. The beam is rastered over the surface and the secondary ions generated at each point are collected and stored on a storage scope or in a computer to form a chemical map of the surface. It is not possible to generate gas phase based ion beams with adequate intensity having beam diameters less than 1 mm. Due to their high brightness, the recently developed liquid metal ion beams are capable of delivering adequate beam current into a 50 nm beam diameter. These sources are typically metals which are liquid near room temperature, for example gallium, indium and caesium. Ion beam formation is generally as follows: The liquid metal wets and flows over a fine needle, a high positive field is established between the needle tip and an extractor and an intense positive ion emission of metal ions occurs. The result is a very high brightness, highly collimated ion source[7,22,23].

3.2 Mass Analysers

Today the majority of SSIMS systems utilize quadrupole mass spectrometers and their use is critically reviewed[24]. The requirement of good vacua because surface studies are involved is probably the main reason for this. It is much easier to enclose the compact quadrupole filter in a UHV system than to provide good pumping for a magnetic spectrometer. They are small but the latest large rod systems are very sensitive and have mass ranges well in excess of 1000 dalton. Where high mass resolution is required a magnetic sector instrument has to be used and many dynamic SIMS systems are of this type. If sensitivity with minimum damage is of primary concern the time-of-flight analyser is rapidly becoming popular.

3.2.1 The Quadrupole Analyser.
Mass analysis is performed by superimposing a varying DC field on a radio frequency (RF) electric field which has been established between the four parallel circular rods which form the analyser[25]. Despite its popularity there are a number of features of the quadrupole filter which have to be carefully borne in mind when using it as a SSIMS analyser[26]. Although mass spectra can be obtained rapidly and easily the relative intensities of detected ions depend rather critically on the operational parameters of the quadrupole. This is because the transmission and mass resolution of the analyser are very sensitive to the axial and transverse velocities of the incoming ions. When comparing absolute ion abundances from different analysers care has to be exercised that standard calibrants are compared first[27]. Large fringing fields exist at the entrance and exit of this type of mass filter, their presence can severely impair the performance of the analyser as a consequence of ion trapping and the reflection of slow moving ions resulting in severe mass discrimination with increasing mass. This problem is alleviated by the addition of a separate set of short quadrupole rods at the entrance to the main analyser. Only the RF component is applied to these rods. Many modern SSIMS quadrupoles are now equipped with these segmented rods and their performance can often match magnetic sector instruments[28].

A further complication is the large spread of kinetic energy of the secondary ions produced in the sputtering process. Ions having high kinetic energies cannot be effectively mass filtered by the analyser, consequently the mass resolution is degraded. This problem can be alleviated by using an energy filter to select a small energy "window" of ions. However this procedure can modify the spectrum since secondary ions are characterised not only by mass and charge state but also by the shape of their kinetic energy distribution, fig 3. The more complex the secondary ion cluster, the sharper this distribution has been found to be. The energy analyser is the electrostatic device at the entrance to the quadrupole which selects the required energy window of ions, however it can perform other functions. To maintain the mass resolution performance of the analyser it should select an energy window of about 5 to 10 eV; it should also prevent direct line of sight into the analyser to reduce the background intensity due to sputtered neutrals etc; it is desirable to have some arrangement which will also collect ions from a large area and from a wide range of emission angles in order that adequate sensitivity can be obtained with a small primary beam current.

To fulfil all these requirements is not straight-forward and most energy analysers currently in use owe their design parameters mainly to practical considerations and experience rather than any theoretical ion trajectory calculations. The designs include cylindrical electrostatic sectors, cylindrical mirror, parallel plate analysers, and retarding-accelerating lens systems with a central stop.

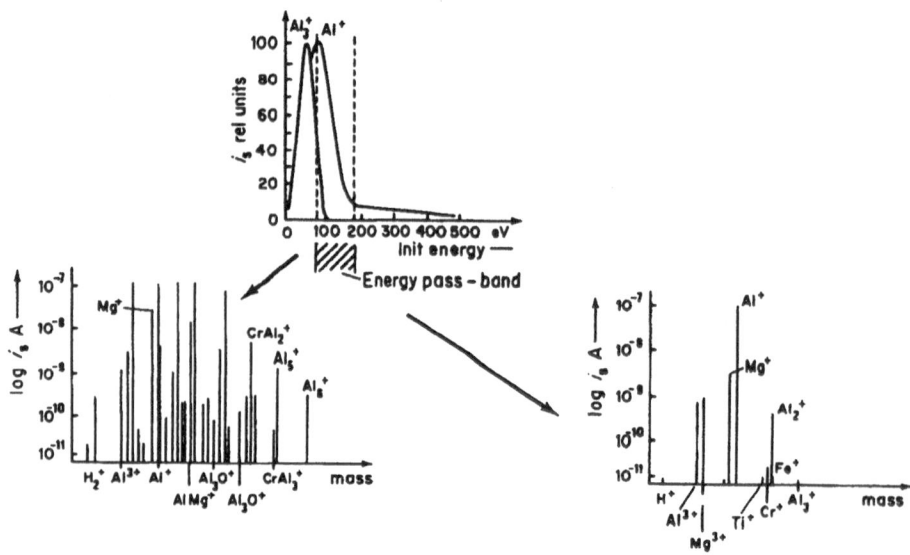

Fig. 3 Effect of position and band pass of an energy filter
on the detected ion spectra

As the use of static SIMS for chemical structure analysis grows the benefits
of MS /MS will be recognised as they are in organic and bio-organic mass
spectrometry[29]. This spectrometry can be carried out using the triple quadrupole
analyser. The arrangement of the three quadrupoles allows a cluster ion of
uncertain origin to be selected in the first mass analyser, the ion then passes into
the second quadrupole region which is filled with a low pressure of inert gas. This
quadrupole only operates with an RF field on the rods, the selected ion collides
with the inert gas atoms and dissociates (collision induced dissociation or CID). The
fragment ions are then analysed in the third quadrupole thus clarifying their
assignment. Basic studies of ion origin can also be carried out. For example it has
been shown that many large organic clusters fragment to yield the smaller
fragments seen lower down the spectrum.

3.2.3 <u>Magnetic Sector Analyser</u>. The double focussing electrostatic
analyser/magnetic sector instrument offers many advantages as far as the mass
spectrometry is concerned. Since ions are accelerated into the spectrometer with
energies of several keV the energy acceptance of the analyser is much greater
and mass discrimination is much less of a problem, consequently total ion
transmission can be 1% or even greater. Ultimate mass resolution ranging from
10, 000 to in excess of 100, 000 is possible. Whilst resolution of this order can be
necessary for trace element analysis due to isotopic interferences, this is also an
important consideration for the complex mass spectra which are beginning to
arise in the surface analysis of organic surfaces. Furthermore double focussing
instruments and tandem instruments involving their combination with another
mass spectrometer, magnetic or quadrupole, are well suited to the MS/MS
experiments mentioned above. However until recently the size of double focussing
spectrometers and their lack of UHV facilities has ruled out their application to
SSIMS in all but a few cases. In the main they are used for dynamic and depth
profiling SIMS.

3.2.4 <u>Time-of-flight analysers.</u> Although SSIMS is very sensitive its ultimate
sensitivity is limited by the proportion of the secondary ions which are actually
received by the detector, <0.1% by quadrupoles and 1% by magnetic analysers. In
the early 1980s the possibility of utilising the time of flight (ToF) mass
spectrometer began to be explored. This type of analyser had been used by
Togerson and Macfarlane for plasma desorption mass spectrometry in which
involatile organics were sputtered by the 50-100 MeV particles resulting from the

fission decay of Cf^{30}. This was a linear ToF analyser and resolution was degraded by the large kinetic energy spread of the sputtered ions. Standing has shown that using a simple pulsed Cs ion source, high sensitivity and high mass resolution can be obtained with a linear ToF^{31}. However the energy compensated ToF mass spectrometer proposed by Poschenrieder[32] or Mamyrin [33] which incorporate an electrostatic section in the flight tube, is ideal for SSIMS. Ions of different energy may take trajectories of different lengths, and with a suitable geometry, the slower speed of the low energy ions will be exactly compensated by the shorter flight path taken. Thus the flight time is then only dependent on the mass of the ions and is independent of their exact energy. The great advantage of this analyser is its enormously greater transmission up to 30% and its almost simultaneous detection of all ions.

In studies of thin organic films on metal substrates, Benninghoven has shown that such analysers can yield increases in sensitivity over the quadrupole analyser of $100-1000^{34}$. However this is obtained at the cost of increased analysis times. Since the ToF system necessarily uses a pulsed primary beam the duty time is of the order of 10^{-4} s/s. Table 3 compares the sensitivity of the two analyser systems assuming a total primary dose of 10^{13} ions cm^{-2}, a sputter yield of 1 and an ionisation probability of 0.01.

The minimum significant ion count for the quadrupole analyser is taken as 10 and for the ToF, 5. If the range m/z = 0 to 250 is scanned the ToF will collect ions at all m/z for the whole analysis time whereas the quadrupole will only collect each m/z for 1/250 th of the total analysis time. It can be seen that whilst considerable improvements are possible in sensitivity with the ToF, these will almost always have to be sacrificed to a degree to reduce the analysis time to reasonable levels. Thus whilst analytical fractions of the order of 5×10^{-8} are in principle possible, ridiculous analysis times are required with the pulsed beam fluxes at present available. Sensitivities in the region of ppm or at best 0.1 ppm are more likely.

TABLE 3

	Resolution	Mass range	Transmission	Mass Detection	Relative Sensitivity
Quadrupole	$10^2 - 10^3$	$\leq 10^3$	0.01 - 0.1	sequential	1
Magnetic Sector	10^4	$>10^4$	0.5 - 1.0	Sequential	10
Time-of Flight	$>10^3$	$10^3 - 10^4$	0.5 - 1.0	Parallel	10^4

It is evident therefore that whilst static SIMS conditions are best maintained by the ToF analyser practical analysis where rapid acquisition is required can be performed very satisfactorily in many cases using a quadrupole filter.

4. The SIMS phenomenon

There are two processes involved in the SIMS phenomenon: the emission of the particle or sputtering, and particle ionisation. There is a good deal of disagreement as to whether they occur simultaneously or consecutively.

4.1 Sputtering

Although there is a wealth of experimental data concerned with sputtering we are still some way from a basic understanding of the mechanism by which the impact of a high energy particle with a solid surface gives rise to the emission of low energy secondary ions. However it is recognised that there are a number of mechanisms which can give rise to sputtering and they can be classified according to the time-scale of their occurrence as outlined below[35]. A primary particle is incident at time t=0.

(a) Prompt collisional sputtering occurs at .t=10^{-15} to 10^{-14} s. This is a direct or almost direct impact process between the incident particle and the sputtered particle involving only surface atoms. The sputtered atoms are removed at high energy.

(b) Slow collisional sputtering occurs due to the internal flux of moving target atoms intersecting the surface in the time scale 10^{-14}s $\leq$ t $\leq$ 10^{-12}s. It is this process with which we are mainly concerned in static SIMS because it refers to the situation when the primary flux density is not large. Although the process has been modelled using Monte-Carlo simulations the quantitative treatment using linear transport theory is due predominantly to Sigmund[36]. The treatment is known as the collision cascade theory of sputtering and the general outline of the process is widely accepted. The primary particle of energy E_0 impacts with the surface atoms, symmetry considerations suggest that at least three surface atoms may be involved. Some energy may be lost by electronic excitation but most is transferred by a nuclear stopping mechanism during hard-sphere or billiard ball type collisions. Knock-on collision cascades between atoms in the near surface region are initiated. The energy transferred, E_t, in a nuclear stopping collision from atom M_1 to atom M_2 varies from 0 to a maximum of $[(4M_1M_2)/(M_1+M_2)]E_0$. Some energy will be dissipated into the bulk by displacement cascades. Here the incident ion can remove bulk atoms out of their regular lattice sites. After displacement, however, a very efficient process of recombination usually occurs such that the number of defects remaining per ion is 1 to 10^{-3}. Some cascades will return to the surface causing the emission of secondary particles or sputtering. This latter process occurs within 1 to 2 nm of the surface whereas displacement or lattice damage can occur down to 10 to 25 nm. The atomic sputter yields and secondary particle energy distributions predicted by Sigmund's treatment are generally in good agreement with experiment.

(c) Thermal sputtering seems to occur at higher primary beam densities and is due to transient vaporisation from the surface of the impact region[37] This is known as the thermal spike and it occurs in the time zone 10^{-13} s $\leq$ t $\leq$ 10^{-10} s.

(d) Electronic sputtering has only fairly recently been recognised as being a significant process. However it is very relevant to the static SIMS analysis of poor conductors since most of the examples where it has been identified are concerned with charged particle bombardment of poorly conducting targets. This process may be slow or fast thus t $\geq$ or $\leq$ 10^{-11} s to 10^{-10} s. Three or four possible mechanisms have been identified involving either direct interaction of the charged particle with the lattice or the creation and diffusion of charge[38,39] vacancies.

Almost all the suggested mechanisms have been primarily concerned with understanding the emission of atomic species. The importance of cluster or molecular ion emission has only been recognised as static SIMS has developed. Classical molecular dynamics and Monte-Carlo calculations of the collisional sputtering process have been very helpful in providing a qualitative picture of the sputtering of molecular species[40,41].

4.2 Ionisation

The question as to whether ionisation occurs at the moment of emission or just after emission involves a complex mixture of collisional dynamics and quantum mechanics and is still a matter of considerable debate. Again a number of theories have been developed which give some insight into the possible processes which occur. Although surface annealing is thought to occur within 10 s, in the region of the primary particle impact zone the crystalline structure of the surface will be considerably disrupted. Models which rely on a normal solid band structure are not likely to be valid. It is probably more satisfactory to consider that the area of emission is amorphous with a continuum of energy states

Many treatments consider that the ionisation probability of a departing atom is determined by an atom-surface interaction within a few tenths of nanometres from the surface. Various mechanisms of the interaction have been analysed and are reviewed by Williams[12]. The perturbation model for example[42] suggests that the ionisation probability is a maximum closest to the surface where atom surface coupling varies most strongly. Of course this coupling also produces efficient neutralisation. As the distance from the surface increases, the coupling becomes weaker and the ionisation probability is less but the ion escape probability begins to be finite. Thus the probability of creating and observing an ion maximises in a region some few Ångstroms from the surface. For metals this model rationalises the well-known dependence of yield on ionisation potential and on adsorption or matrix effects which give rise to work function changes.

The surface excitation model[12] is a development of the above model. Resonant electron transfer between the surface and the departing atom occurs up to a distance of 5-10 Å. For ionisation to occur and the ion to escape it is necessary that a valence level in the departing atom is isoenergetic with a vacant level in the surface below the photo-threshold (Fermi level in a metal or valence band edge in a semiconductor or insulator). The surface excitation resulting from the primary ion impact greatly increases the probability of this occurring. Ion emission from ionic solids is considered in the bond-breaking model[43]. The key assumption is that the ion state is the ground state, and the probability of observing a secondary ion depends simply on whether the ion state remains energetically favoured over the neutral state as the ion moves away from the surface. Static SIMS is far more concerned with the yield of molecular ions, however our understanding is much less advanced in this area.

4.3 Molecular Cluster Sputtering - the theories.

As for atomic secondary ion generation the process can be considered in two stages to make it susceptible to relatively easy theoretical treatment. First there is the sputtering process and then the ionisation of the sputtered cluster. We will consider some of the ideas of the sputtering process first, followed by the the theories of ionisation.

The collision cascade theory of sputtering has not really been applied in any detail to cluster sputtering, indeed by virtue of the lack of structural parameters in its basic suppositions it cannot easily account for sputtering which may be influenced by the structure or chemistry of the solid. Molecular dynamics models attempt to provide a structural framework for sputtering.

4.3.1. Molecular Dynamics Models

Clearly models are required which take account of the full three dimensional structure of the target compound. In principle this can be accomplished using computer methods. Two fundamental assumptions are common to all computer simulation models used to study the SIMS process. First, the constituent particles of the solid are assumed to be atoms or atomic ions interacting with one another through conservative pairwise forces. A volume-dependent interaction term may also be included; however, the assumption means that only atomic properties are studied, not electronic properties. Second, model calculations are based on classical mechanics. Newtonian mechanics will

367

obviously neglect electronic effects, and it is clear that electronic effects can be very important when collisions between atoms are considered. Corrections may be included but not by adding specifically electronic components to the modelling.

In setting up the model one has to specify the number of mutually interacting atoms and the boundary conditions which guarantee the stability of the model and specify its interactions with a surrounding matrix. Usually the computational crystallite consists of between 200 and 5000 atoms which interact according to two body forces derived from a potential $V(r)$. Typical potentials used are the purely repulsive Born-Meyer potential[26].

$$V_{BM}(r) = C_{BM}\exp(-r/a_{BM}) \qquad\qquad 1$$

where $C_{BM} = 52(Z_1 Z_2)^{3/4}$ eV and $a_{BM} = 0.0219$ nm, and the Morse potential

$$V_{MO}(r) = C_{MO}y(y-2) \qquad\qquad 2$$

where $y = \exp[-(r-r_0)/a_{MO}]$ and C_{MO} and a_{MO} are similar constants.

The potential function is usually set so that it vanishes smoothly at, say, the second neighbour separation in the crystal. Thus a relatively small number of atoms contribute to a force at a particular point. If the potential between the atoms is only repulsive the stability of the crystallite is maintained by putting special static external forces on the atoms at the crystallite boundaries. Even when the specified interatomic forces are partially attractive other external boundary forces may be required to define the correct lattice spacing or simulate the cohesive energy or elastic constants. The embedding of the 'crystallite' in its macroscopic medium also has to be specified by a further set of boundary conditions. For example, the atomic displacements in the crystallite may be influenced by the matrix and these could be simulated by a spring force to represent the elastic response of the matrix.
To simulate the impact of an ion on the crystallite the dynamical model selects an atom which is set moving in a particular direction with the desired kinetic energy. The classical equations of motion are integrated for all the particles in the crystallite as long as enough of the original kinetic energy remains. The time steps between integrations are usually 1-3 fs and the whole calculation will require 100-1000 steps. Harrison, Winograd and Garrison et al have been active in the use of this modelling approach to sputtering. In the main only simple inert gas solids or homogeneous metals have been studied. Initially atomic sputtering was investigated. This was rather successful, for example the sputter yield of Cu from Cu(100) was calculated to be 4.7 atoms/ion using 5 keV Ar$^+$ compared to an experimental value of 4.1 atoms/ion[26]. Furthermore they were able to reproduce the Wehner-type spots generated as a consequence of channelling during Ar$^+$ bombardment of the (001) face of Ni.

However it is in the simulation of molecular or cluster sputtering that MD calculations have made the most interesting contribution. The first studies were directed towards the generation of multi-metal atom clusters from clean metal surfaces. These were followed by the study of clusters formed from adsorbed species on metal surfaces. The basic approach used was common for both. The crystallite consisted of four layers with about 60 atoms per layer and the bombardment was normal to the crystallite with 600 eV Ar$^+$. The interaction of the Ar$^+$ with the other species is represented by a purely repulsive term

$$V_{ij} = Ae^{-BR} \qquad\qquad 3$$

Winograd et al found that the character of the sputter yield was not affected if the area or depth of the microcrystallite was increased[26]. *The criterion for the formation of a dimer or multimer is important*, it is based on the balance of the total kinetic energy and potential energy of the apparently close moving dimer of multimer. The relative kinetic energy of the component atoms, T_R, plus the potential energy, V, for all pairs of ejected atoms are first computed. The potential energy between any pair of atoms i and j, V_{ij}, is calculated using a Morse potential. If the total energy of the pair

$$E_{tot} = T_R + V_{ij} \qquad\qquad 4$$

is negative, then the tested dimer is considered bound. Using this approach the yield and origin of multimers from the three faces of copper were monitored. The yields of Cu_2^+ and Cu_3^+ relative to Cu^+ were sensitive to the structure of the surface plane and are in reasonable agreement with experimental data for the emission of Ni_2^+ ions from Ni surface planes (see below)[27]. However it is perhaps more significant to note that the authors found that the sputtered multimers *do not emit as intact sections of the solid, they form above the surface and the component atoms do not necessarily arise from atoms which were nearest neighbours in the surface.* Although in many cases nearest neighbour atoms are components of a multimer, some components arise from as far apart as 7 nm. Thus although sensitivity to atomic geometry was found, the possibility that clusters could be made up from atoms arising from remote positions could reduce the value of SSIMS for surface structure analysis.

The model was applied to the adsorption of oxygen on a number of metals. On Cu are large number of clusters were suggested - Cu_2, Cu_3, O_2, Cu_2O, CuO_3, Cu_2O_3, Cu_3O, Cu_4O Cu_3O_2, Cu_2O_4, Cu_3O_3. They were all found to be formed *above* the surface from atoms after sputtering[28]. However in this case it was found that most clusters, though not all, were formed from nearest neighbour atoms, so a significant relationship with structure was maintained. Other adsorbates have been modelled. CO adsorption on Ni was shown to generate Ni_xCO clusters and although many arise from nearest neighbour atoms they are not formed by the emission of an intact cluster from the surface[28,44,45]. Garrison and Winograd conclude that 'there is apparently no direct relationship between these moities and linear and bridge-bonded surface states'. It will be seen in the next section that this runs counter to much of the experimental data.

In contrast however when large adsorbed molecules such as benzene are sputtered, the molecule itself may be emitted intact[27,28,46]. The attachment of substrate metal atoms however apparently occurs after emission. The reason for the maintenance of the benzene structure during emission is thought to be due to the large number of internal degrees of freedom it has to absorb the excess energy it receives during an energetic collision. Fragmentation does occur in the MD process but this is partly due to direct bombardment by a primary particle and partly due to very energetic substrate atoms striking the molecule.

Whilst the MD modelling is helpful in providing a physical picture of the sputtering process and the mechanism by which secondary atoms and clusters *may* be emitted it clearly has weaknesses associated with the approximations which have to be introduced to make the calculation tractable. The pictorial representations of the process are easily accepted as representative of the actual process occurring in the sputtering of complex materials. This may be dangerous because the simple model may be a far from real representation of bonding in the real materials[27], for example:

1. The same interaction potentials are employed for all atoms of the same type no matter where they are in the solid. Thus atoms at a step in a surface will

be bound more strongly. Indeed even the surface atoms of a smooth surface are known to be bound with somewhat higher energy than those in the bulk.

2. Similarly adsorbed species may also increase the cohesion between the substrate atoms to which they are bound. It can be envisaged that an adsorbed molecule and the surface atoms to which it is attached may be approximated to a molecular cluster embedded in the surface layers. Thus just as organic molecules are predicted to desorb intact so may an adsorbate structure. This type of concept has given rise to the intact cluster emission theory for the emission and ionisation of clusters from inorganic and organic compounds (see below).

3. There is the general problem that most MD models do not take any account of ionisation. The stability of a cluster will be significantly affected once ionised. Garrison and Sroubek have begun to address this omission[47].

4.3.2 Other Cluster emission models

In seeking an understanding of cluster ion emission it is not always helpful to restrict models to classical collisional dynamics. It is evident that other processes contribute especially when insulating materials are analysed. This is highlighted by the observation that the secondary cluster ions generated by both keV and MeV excitation are frequently very similar. The two regimes involve different excitation mechanisms. keV bombardment is thought to mainly involve collisional processes, whereas the MeV regime is primarily electronic. If the material to be analysed is crystalline the primary bombardment process will generate damage and may render the structure amorphous. The energy dispersion processes will excite lattice vibrations to a level where inter-molecular bonds will be broken. If the material is an insulator it can be postulated that the primary deposited energy will be transported away by excitons (quanta of electronic excitation) which can propagate through the solid[48]. These radiationless transitions deposit their energy into vibrational excitation throughout the matrix. Macfarlane points out that the attractive feature of this exciton model is that exciton states can be made either by electronic excitation or by collisional processes and *both* could decay into the same pattern of electronic and vibrational excitations. Clearly this approach can explain the similarities in observed spectra from SSIMS and ^{252}Cf PDMS (Plasma Desorption Mass Spectrometry). It would also explain why the emission of secondary cluster ions is sensitive to the matrix, because the energy dispersion in an exciton-phonon process will be influenced by the physical features of the solid. Thus emission can be described as vibrational and if the energy reaching the molecule is low, intermolecular bonds will be broken and the desorption of an intact molecule will result. Where the energy arriving at the molecule is high, fragmentation will result.

The possibility of an electronic contribution to sputtering, even in the keV regime, has only fairly recently been recognised as being a significant process. Most situations where it has been identified involve charged particle bombardment of insulators. There may be a number of contributing processes. For example direct interaction of the impacting ion with the lattice will deposit significant amounts of electronic energy, for example the electronic stopping power of 10 keV Ar^+ is about 90 eV nm^{-1}. This is sufficient to cause ionisation in surface layers of halides and oxides with consequential loss of bonding.

4.4. Ionisation

As in the case of sputtering the thinking on cluster ionisation is much less advanced but a number of qualitative models have been suggested which do provide insights into how the process may occur. Most theories of ionisation in SIMS suggest that the process occurs in the surface region as the particle is leaving the surface. When considering inorganic or indeed organic solids, in some cases ions may already exist in the solid so the generation of free ions may occur simply by bond-breaking.

4.4.1 Bond-breaking model.

This model was developed primarily to explain *atomic* secondary ion formation from ionic solids however it is the precursor to models of cluster ion emission. The key assumption is that the ion state is the ground state and the probability of observing a secondary ion depends on whether the ion state remains energetically favoured as the ion moves away from the surface[43]. If the atom-surface separation is < 1 nm the ion and neutral states will be strongly mixed. Clearly the ion emission probability will be strongly dependent on the ground state of the associated species and the difference between the ionisation potential and electron affinity of the positive and negative ions.

4.4.2 Nascent ion molecule model.

The electronic transition rates are very fast ($10^{14} - 10^{16}$ s^{-1}), it is therefore unlikely that an ionised species could escape without de-excitation in the time taken to leave the surface region (10^{-13} s). It is the suggestion of this model that ions can only escape if they are formed some distance from the surface. Ions are therefore formed by dissociation of sputtered neutral molecular species[49]. Energy transfer in the collision cascade in the solid gives rise to collision energy transfer between the atoms of the molecule leaving the surface, consequently, dissociative ion formation may occur with charge exchange during dissociation of the molecule remote from the surface. The model has been used to determine the mass dependence of the emission of Me^+ from oxide specimens under Ar^+ bombardment. Me^+ mainly arises from the dissociation of MeO^0 and a dependence of the form

$$S^+_{max}(m_{max}) = \text{const. } m_{metal}^{-2.4}. \qquad 5$$

was obtained. This mirrors the dependence found for Benninghoven's empirical valence model.

The generation of cluster ions is thought to occur by the dissociation of large emitted neutrals $Me_nO_m^0$ although no quantitative predictions have been derived. These ideas appear reasonable because it is certainly known from MS/MS techniques that cluster dissociation does occur during their flight from the surface. However other models suggest that most neutral clusters emitted would have insufficient energy to dissociate in quantities required to give the high cluster ion yields observed.

4.4.3 Desorption ionisation model

Cooks and Busch have proposed this model mainly to provide an understanding of molecular and cluster ion emission from organic materials where they believe vibrational excitation to be important[50]. There are many inorganic materials where this thinking may be appropriate. The model emphasises that the processes of desorption and ionisation may be considered separately, fig. 4.

Whatever the form of the initial excitation process, the energy is transformed into thermal/vibrational motion as far as the molecules are concerned. The only ions desorbed immediately are those which are in the cationic or anionic state in the solid. There is *no net creation of ions during desorption*. Most of these ions will be desorbed without fragmentation because their internal energy will be too low. Neutral molecules will be desorbed in high yield but to be observed require to be ionised. To generate further ions in what is referred to as the selvedge region or in the free vacuum the DI model suggests that desorption is *followed* by chemical reactions of two possible types:

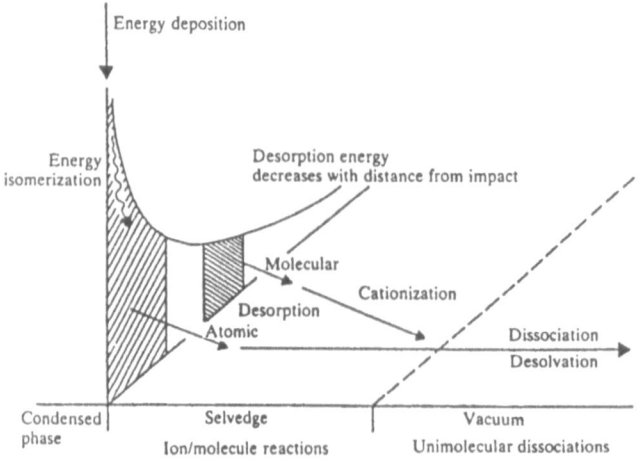

Fig. 4 Summary of processes thought to occur during desorption ionisation

1. in the selvedge region fast ion/molecule reactions or electron ionisation can occur;

2. in free vacuum, unimolecular dissociations may occur, governed by the internal energy of the parent ion giving rise to fragment ions.

The model is consistent with the fact that the SSIMS spectra for many materials show cluster or molecular ions in high yield. However it does not seem entirely appropriate to suggest that the initial collision process does not produce significant numbers of additional ions relative to those existing preformed in the material. It is clear from this discussion that for many materials a combination of the various ideas gives a broad understanding of the processes which give rise to the emission of cluster ions. The MD model of sputtering has strongly suggested that clusters are formed by atom recombination above the surface. On the other hand many of the ionisation models which seek to address cluster ion generation see the dissociation of large emitted neutral clusters, presumably emitted intact, as being an important ionisation mechanism. A definitive statement regarding the relationship between spectra and chemical structure requires experimental evidence. This has been explored using, on the one hand model compounds, from which a library of standard spectra have been assembled which clearly demonstrate the straightforward relationship between spectra and chemical structure. The approach of special relevance to surface science and heterogeneous catalysis has been to investigate model adsorption systems which are well characterised by other surface science techniques.

5. Model Adsorption Studies

One of the earliest studies was to investigate the sensitivity of cluster ion emission to the surface of clean single crystal surfaces.

5.1 Metal surface structure

The surface structure of the low index (100), (110) and (111) planes of fcc metal single crystals have characteristic distributions of surface metal atoms. By considering only the top surface atoms it is possible to calculate the statistical probability of the emission of M_2 and M_3 clusters relative to M, see Table 4. A static SIMS study of the cleaned annealed planes of nickel gave the Ni_x^+/Ni^+ ratios shown in the table. Very close agreement is obtained between the expected relative yields and those observed[51]. The only variation was for Ni_3^+ from the (111) plane where especially stable triangular clusters are probably possible.

When the surfaces were sputter etched for a few minutes with a 1 μA beam, the ratios for all three surfaces were very similar. The top surface structures had been destroyed.

TABLE 4

	(110)	(100)	(111)
No. of ways of producing Ni_2^+	2	4	6
Obs. Ni_2^+/Ni^+	1	1.8	3.2
No. of ways of producing Ni_3^+	4	24	48
Obs. Ni_3^+/Ni^+	1	6	38

This simple result shows that data can be obtained which accurately reflects the surface structure and using static conditions no evidence of surface damage was detectable. It also provides a very sensitive method for monitoring the extent to which a surface is fully annealed after cleaning by ion bombardment. This method has been shown to be more reliable than the use of LEED diffraction patterns which are sensitive to the top few layers rather than the top 2 layers monitored by SSIMS.

5.2 Adsorption on single crystal metal surfaces - CO on Metals

Apart from the value of being able to use SSIMS to identify and monitor the structure and coverage of adsorbed molecules, the adsorbate state is a particularly delicate model to test the ability of SSIMS to monitor surface chemistry. The CO adsorbate state is very well characterised by many techniques of surface science, thus it was the obvious choice for study by SSIMS.

5.2.1 Molecular or dissociative adsorption ?

The earliest studies investigated the adsorption of CO on a series of polycrystalline metals. It was known from other studies using electron spectroscopy that the adsorption of CO at 300 K would be molecular or dissociative depending on the adsorption energy. Where the enthalpy of adsorption is less than 250 kJ mol^{-1}, as it is on Cu, Pd and Ni, the adsorption is molecular, but where it is greater, as on W, it is dissociative, while adsorption on Fe lies between and both forms are observed[52]. Fig. 5 shows that SSIMS can distinguish clearly between the two. Dissociative adsorption is characterized by MC^+, MO^+, M_2O^+ and M_2C^+ secondary ions, while the molecular adsorbate state is identified by MCO^+ and M_2CO^+ ions[53].

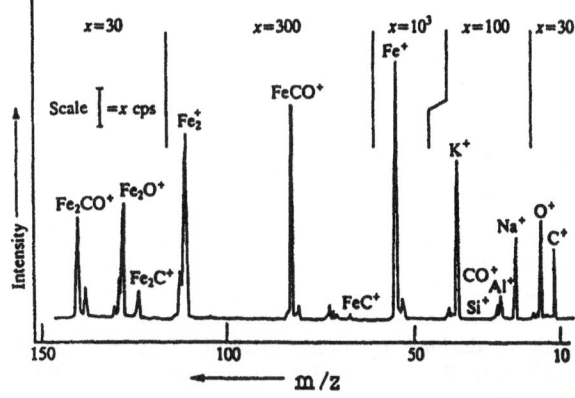

Fig. 5 SSIMS spectrum of iron foil exposed to 10^{-8} Torr CO.
Ip = 0.5 nA cm^{-2}, Ep = 3 keV

5.2.2 <u>Surface coverage measurements</u> The measurement of the surface coverage of adsorbates is an important requirement but is not always very straightforward.   Thermal desorption can give a measure of the total number of adsorbed molecules and an indication of the energetics of adsorption.   The adsorbate state can be characterized by electron and vibrational spectroscopies but the intensities of the resulting spectra are not necessarily related directly or simply to coverage.   It would therefore be of great benefit if the SSIMS cluster ions, MCO^+ and M_2CO^+, which identify the adsorbate state could also be related to the surface coverage of CO adsorbate species.   At first sight, this is complex.   The secondary-ion intensities are sensitive to the quantity of surface CO *and to the work-function change* that accompanies adsorption.   Thus as CO coverage increases on a surface, the secondary-ion intensities increase non-linearly as a consequence of both of these parameters.   At present there is no satisfactory theoretical model that would allow the two parameters to be unravelled.   An empirical approach has been found to be very successful.   It was argued that, while the M_xCO^+ species were sensitive to both coverage and work function, the M_x^+ ions also increased but only as a consequence of the work-function change[54].   It was suggested that the *ratio* M_xCO^+/M_x^+ should only be sensitive to coverage.   Since some CO is removed as MCO^+ and some as M_2CO^+, total coverage should be represented by

$$S(M_xCO^+/M_x^+) \; \propto \theta_{CO}. \qquad\qquad 10$$

Linear plots of this function against coverage have demonstrated that SSIMS data are sensitive to adsorbate coverage, see fig. 6[54,55].   There is a break at a relative coverage of about 0.5 in the plots for most of the surfaces.   The gradient change is correlated with changes in the structure of the adlayer, which are usually 'compressed' in this coverage region. It can be seen that the gradient of the linear relationship is surface-dependent.   There is evidence that it is a function of the enthalpy of adsorption.

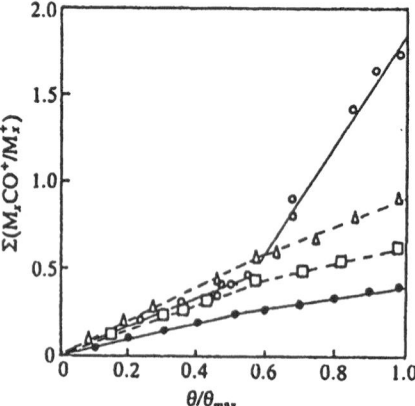

Fig. 6 Variation of sum of ion ratios $\Sigma(M_x CO^+ /M_x^+)$ as a function of CO coverage on, O Pd(111); □ Ni(111); △ Ru(0001); ● Pd particles

The ability to monitor coverage by SSIMS has also provided the possibility of measuring the enthalpy of adsorption by the isosteric heat method.   Thus the sum of ion ratios is monitored as a function of temperature for particular ambient pressures of CO.   From these plots, isosteric heats can be derived which are in very good agreement with those derived by other methods (see fig. 7)[54].   Other workers have subsequently confirmed that ion ratios are related to surface coverage for other adsorption systems, for example, H_2 on Ni, CO on Ni,[56] and NO and N_2 on Rh[57].   Furthermore it has been shown that the method is applicable to metal-on-metal adsorption. Thus the coverage of Cu on Ru(0001) has been monitored by the $RuCu^+/Ru^+$ ratio. In this case it has been cross-checked using thermal desorption spectrometry (TDS)[58].

5.2.3 <u>Studies of adsorbate structure</u> Mass spectrometry is best known for its ability to provide data related to the molecular chemical structure. In the same way a SSIMS fragmentation pattern of an adsorbed state can also be defined, which allows the distribution of the different types of adsorbed CO on a surface to be identified and quantified. On metals, CO can adsorb into linear, bridged, or triply bridged surface states. Thus by carefully selecting single-crystal faces that were known from vibrational spectroscopy to adsorb CO into particular adsorbate states, it was possible to show, by comparative studies utilizing vibrational spectroscopy and SSIMS on these metal surfaces, that the relative populations of the MCO^+ the M_2CO^+ and the M_3CO^+ ions can be used to identify the state of the CO on the surface, see Table 5.

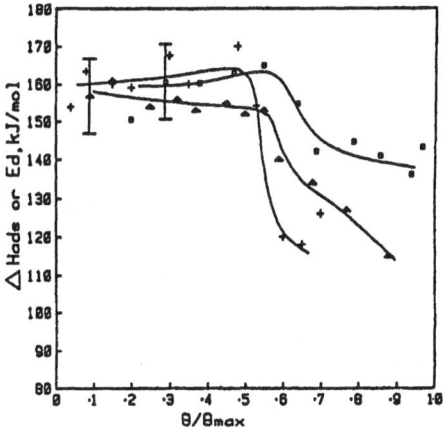

Fig. 7 Coverage dependence of the isosteric heat of adsorption ΔH_{ads} from the sum of the SIMS intensity ratios ($\triangle$), the energies of desorption, E_d ($\square$) and the literature (+)

In fig. 8 the variation in CO adsorbate geometry is monitored by SSIMS via the variation in the ratio of ion intensities $MCO^+/\Sigma M_xCO^+$ and by high-resolution electron energy loss spectroscopy (HREELS) by plotting the proportion of linear CO from the ratio of the intensities of vibrational bands due to linear CO (v') to the sum of the intensities for the bridged and linear structures, (v" + v'). The relative extinction coefficients for the C—O stretching vibrations of linear and bridge CO are not known so the latter ratio cannot be quantitative[55,57].

TABLE 5

Surface	SSIMS			CO structure
	MCO^+	M_2CO^+	M_3CO^+	IR/HREELS/LEED
Cu(100)	0.9	0.1	- -	linear only
Ru(0001)	0.9	0.1	- -	linear only
Ni(100)	0.8	0.2	- -	linear + bridge
Ni(111)	0.6	0.4	- -	bridge + linear
Pd(100)	0.3	0.6	0.1	bridge only
Pd(111)	0.3	0.4	0.3	triple bridge
Pt(100)	0.65	0.35	- -	linear and bridge
TABLE 5				

5.2.4 Adsorbate-adsorbate interactions   Another important parameter that characterizes the adsorbate state is the extent to which molecules interact with one another.   There are various modes of interaction: dipole-dipole and indirect chemical interactions via the substrate at low coverage, and direct repulsive interactions at high coverage.These interactions are usually detected by wave number shifts of the C—O stretching vibration as CO coverage varies.   There is good evidence that these effects may also be seen using the SSIMS M_2^+/M^+ ratio.   This can be rationalized as follows: The formation of the adsorbate-surface bond will have a statistical and chemical effect on the emission of substrate ions, M^+ and M_2^+: statistical in that, if a CO molecule adsorbs onto a particular M site, the increased probability of its emission as MCO^+ will reduce the probability of its emission as M^+ and its chances of being emitted in an M_2^+ cluster will be even further reduced.

There will also be strong chemical effects on the cohesion of M—M bonds in the metal surface resulting from the formation of an admolecule-surface bond. Thus it might be expected that a random distribution of CO on the surface would result in a steady fall of M_2^+/M^+ with θ_{CO}.   When island growth occurs, the fall will be much less steep or a plateau may occur until the islands coalesce, when a sharp fall may occur.   On hexagonal surfaces there is an exact correspondence between the fall in the SSIMS M_2^+/M^+ ratio and the rise in the vibrational band due to adsorbate-adsorbate interactions.

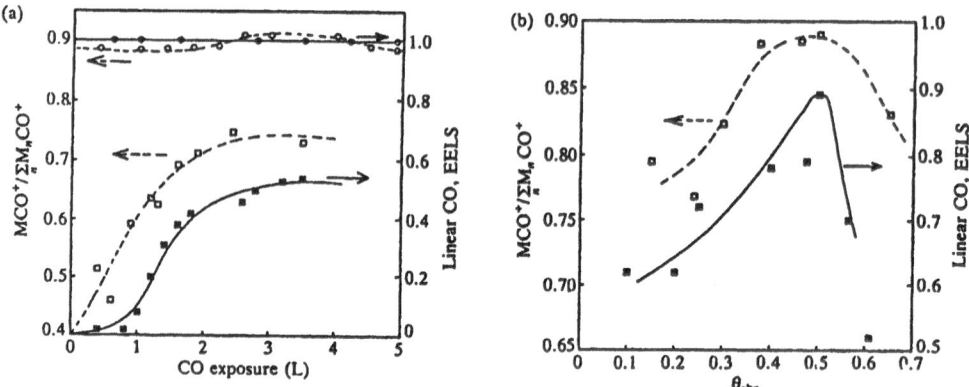

Fig. 8 Variation of the ion fractions $MCO^+/\Sigma(M_x CO)$ with coverage or exposure for CO adsorption on a) Ru(0001) at 300 K; compared with the proportion of linear CO calculated from EELS relative band intensities obtained from the literature a) O Ru(0001) SSIMS; ● Ru(001) EELS; □ Ni(111) SSIMS; ■ Ni(111) EELS and b)   Ni(100) □ SSIMS, ■ EELS

Fig. 9 compares the situation for CO adsorption on Pd(111) and Ru(0001) at 300 K with that for Ni(111) at 195 K.   On Ru (0001) for coverages below 0.33, disordered adsorption occurs.   In the region 0.20 to 0.33, dipolar coupling begins to occur as islands are generated of a ($\sqrt{3}$ x $\sqrt{3}$)R30° ordered overlayer, and as a consequence a plateau is observed.   Beyond a coverage of 0.33, there is a loss of order in the adlayer due to repulsive interactions and this is reflected in an increase in vibrational wavenumber and fall in the SSIMS ratio[60].   In the case of Pd(111) the constant frequency for the IR band at coverages up to 0.3 is good evidence for attractive interactions between admolecules and island formation.   This is fully supported by the constant Pd_2^+/Pd^+ ratio up to θ = 0.3.   The sharpness of the fall in Pd_2^+/Pd^+ and the rise in the vibrational frequency in the region beyond θ = 0.3. indicates a rapid loss of order as further admolecules give rise to increasing repulsive interactions and loss of the ($\sqrt{3}$ x $\sqrt{3}$)R30° domains.   On Ni(111) at 195 K a continuous fall in rise in the vibrational frequency indicates a statistical distribution of molecules in the adlayer. These measurements were made at 140 K. The SSIMS data obtained at 195 K suggest that below 1 L exposure some island formation occurs. It may be that at the lower temperature used for the vibrational studies there was insufficient energy to reorder the molecules.[55,59].

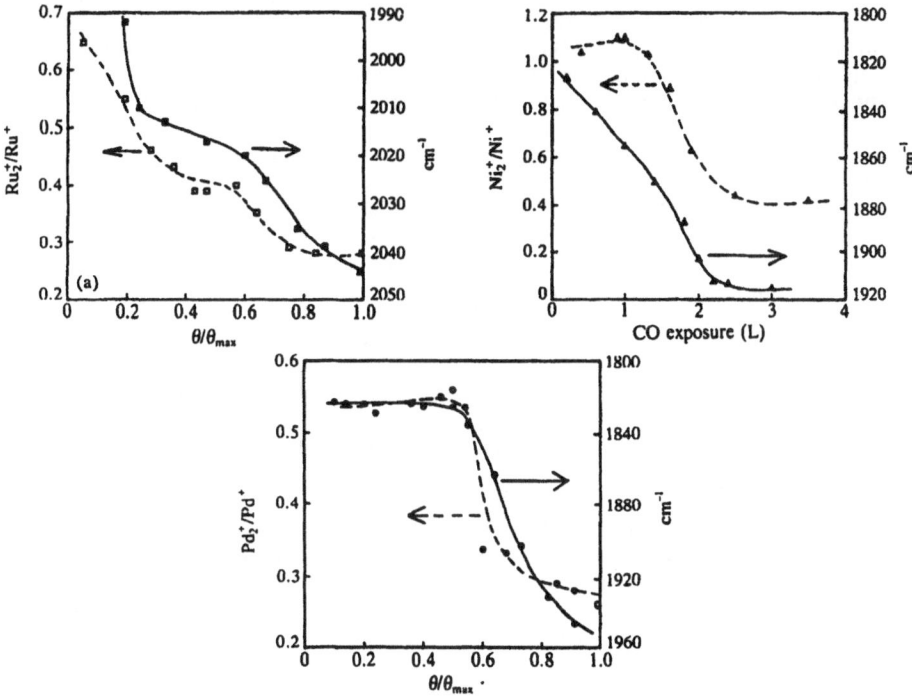

Fig. 9 Comparison of the exposure or coverage dependence of M_2^+/M^+ from SSIMS with CO vibrational frequency from IR and EELS for CO adsorption on (a) Ru(0001) at 300 K, (b) Ni(111) at 195 K (SSIMS) and 140 K (EELS) and (c) Pd(111) at 300 K

5.2.5 <u>Conclusions</u> In this study of perhaps the best understood of the adsorbate systems it can be seen that static SIMS is able to provide significant data which reflects not only the chemistry of the surface but enables the analyst to quantify the coverage and the energetics of adsorption. The fact that the analysis of such delicate surface structures is possible gives great confidence that chemically significant data should be obtainable from most material types. Thus although the technique is basically destructive it is the use of static analysis conditions combined with the fact that the ions which are detected are emitted relatively gently following an indirect collision process which makes true surface analysis possible.

6 APPLICATIONS OF SIMS IN ADSORPTION AND HETEROGENEOUS CATALYSIS

6.1 <u>Comparison of surface state of real and model catalysts</u>

Over the last 20 years many studies have been carried out using model single crystal surfaces to try to understand the surface mechanism by which real catalysts operate. The complex nature of most heterogeneous supported metal catalysts makes it very difficult to isolate the role of the different surface parameters in determining surface activity. Many of the most useful catalysts are bimetallic, for example the Pt/Re system supported on alumina is widely used as a reforming catalyst. The activity of the bimetallic combination is very different from the activity of pure platinum or pure rhenium. There are indications why this should be but it is difficult to use polycrystalline supported materials to study the problem. It is not easy to find techniques which enable us to investigate or to study processes occurring on very small particles (diameter ≈ 5 to 100 nm). These difficulties have led to the use of single crystal metal systems to model the catalyst. Obviously the ease of characterisation and study is attractive however an unsupported single crystal plane would seem to be far removed from a polycrystalline supported catalyst. Reaction studies on these highly regular model surfaces have frequently been found to show surprisingly similar catalytic activity to the practical catalyst. One real problem is that few of the methods of surface physics used to characterise the single crystal models yield useful data on small supported metal particles. Static SIMS is an exception as the following example will show.

A number of years ago Sinfelt showed that the combination of a 1B metal such as Cu with Ru in a silica supported catalyst changed the catalytic behaviour in hydrocarbon hydrogenolysis rather significantly whereas activity in hydrogenation was little affected[60]. The 1B metal/Ru system is interesting since these metals do not form alloys and it is therefore rather easier to define the surface composition. The reason for the observed behaviour was thought to be partly a consequence of Cu disturbing the chemical geometry of the Ru surface (the so-called ensemble effect) and partly an electronic effect of the Cu on the mode of adsorption of the reacting molecules. Vickerman et al initiated a study of three different catalytic systems which are intended to connect model single crystal studies to polycrystalline supported catalysts via model catalysts consisting of small metal particles supported on films of oxidised aluminium or silicon.

6.1.1 Preparation and surface characterisation

Preparation of the catalyst surface is very different for the three types of catalyst. The polycrystalline materials are prepared by impregnation of the support by a Ru and Cu salt. This is followed by drying and reduction of the metal before use. The single crystal Ru surfaces are cleaned by oxidation/reduction and ion etching cycles in UHV followed by annealing. The 1B metal is then evaporated or sputtered onto the surface to the required coverage. Finally the model supported catalyst is prepared first by careful oxidation of, for example, an aluminium surface by electrochemical or gas phase methods to form the support. Ru and Cu are then sputtered onto the surface to the required coverage.

The prepared surfaces should then be characterised. An active catalyst is usually about 1% by weight Ru. This concentration is too low to detect effectively by electron spectroscopies, although the single crystal can be easily analysed by Auger or X-ray photoelectron spectroscopy. SIMS however is able to provide useful data on all the surfaces. Fig. 10 shows the Ru^+/Al^+ and Cu^+/Ru^+ ratios for a polycrystalline catalyst series as a function of Cu concentration.

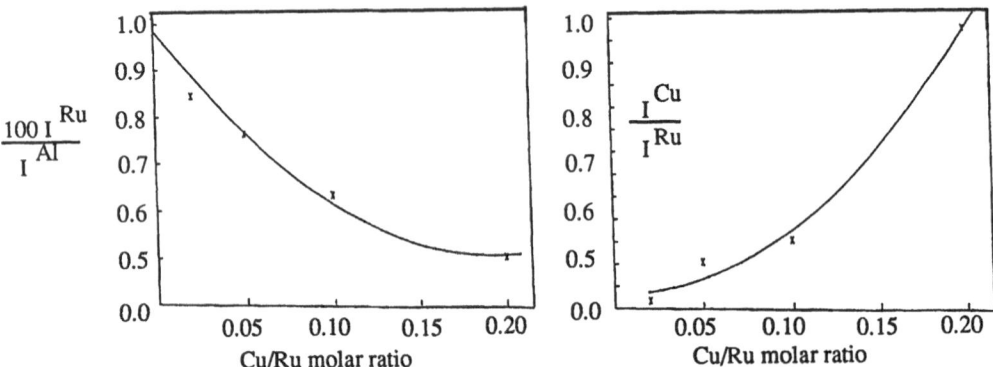

Fig. 10 a) Ru^+/Al^+ b) Cu^+/Ru^+ vs. Cu content of CuRu catalysts

It can be seen that as the Cu content of the catalyst increases the exposed Ru level falls indicating that the Cu covers the Ru surface. This is an important result because it confirms that the Cu does indeed associate with the small Ru crystallites which transmission electron microscopy has shown to be between 1 and 6 nm in diameter[61,62]. This conclusion has been confirmed by EXAFS[63]. Thus it is valid to model such a surface by depositing Cu or Au on a Ru(0001) surface. Extensive studies of the character of such surfaces with sub-monolayer quantities of the 1B metal deposited under different conditions have been carried out[64,65,66], and SSIMS has made a significant contribution to our understanding[58]. Two deposition conditions have been used, in one the Ru(0001) surface was held at 1100 K and in the other at 540 K whilst Cu or Au was evaporated onto the surface. The surface was then cooled down rapidly to room temperature.

Both SSIMS and electron spectroscopy showed that at 1100 K· smooth statistical deposition occurred which developed into layer by layer growth beyond 1 monolayer, whereas at 540 K island clustering occurred such that even at nominal 1 monolayer deposition as monitored by thermal desorption, the Ru surface is not completely covered. Gradient changes in the Auger plot of the intensity of the Cu $M_{2,3}M_4M_4$ transition as a function of the intensity of the Ru $M_5N_{4,5}N_{4,5}$ line for the high temperature surface demonstrate the formation of successive monolayers, whereas the absence of a definite break in this plot for the low temperature surface suggests 3-dimensional island growth. In fig. 11 the SSIMS data augments these conclusions. The plot of Cu^+/Ru^+ shows that beyond 1 nominal monolayer the gradient rises rapidly for the 1100 K surface whereas it does not for the 540 K surface. Clearly in the former case the surface is covered with Cu whereas in the latter it is not. The more even coverage of Cu for 1100 K surface is also reflected in the higher $RuCu^+/Ru^+$ cluster ratio. Indeed from the plot in fig 11(d) it is possible to estimate that 25% of the Ru surface is left uncovered on the lower temperature surface, when the high temperature surface is covered.

SSIMS was very helpful in similar studies on the model supported surface[67]. When Ru is sputtered onto an oxidised alumina surface it is highly dispersed, but it then has to be annealed so that the Ru can stabilise into crystallites. XPS and the Auger parameter have been used to characterise the dispersion and interaction of the Ru particles on the support. SSIMS data have helped to confirm some of the findings. The $Ru^+/AlOH^+$ ratio decreases from 10 to 3 after progressively annealing in vacuum to 400 °C and the Ru_2^+/Ru^+ ratio increases from 0.03 to 0.06 both indicating that the Ru is forming larger crystallites leaving some alumina surface uncovered, fig 12. It is interesting to note however that the Ru_2^+/Ru^+ ratio from a single crystal surface is much higher at about 0.4, the probability of removing a Ru_2^+ is much smaller from very small crystals.

SSIMS is thus capable of providing detailed information regarding the chemical state of the surface, coverage, extent of clustering etc. independent of the physical state of the material.

6.1.2 Adsorption on Catalysts

As we have seen, SSIMS can be used to monitor coverage and adsorption state. The above catalysts are to be used for the Fischer Tropsch reaction, thus it is of interest to study the adsorption of CO. The adsorption on the model supported catalysts was compared with that on single crystal Ru(0001). At saturation at 300 K $\Sigma(Ru_xCO^+/Ru^+)$ was 0.75 on Ru(0001) and 0.41 on a supported Ru surface before annealing and 0.32 after annealing.

This demonstrates that the high density of CO which can be accommodated on the smooth single crystal cannot be obtained on small crystallites. The values obtained on the supported Ru correspond to coverages of about 0.4 relative to Ru(0001). Furthermore as the surface is annealed and crystallites are formed and grow, the coverage relative to Ru falls. This development is reflected in the CO adsorbate state which can also be monitored by SSIMS[67]

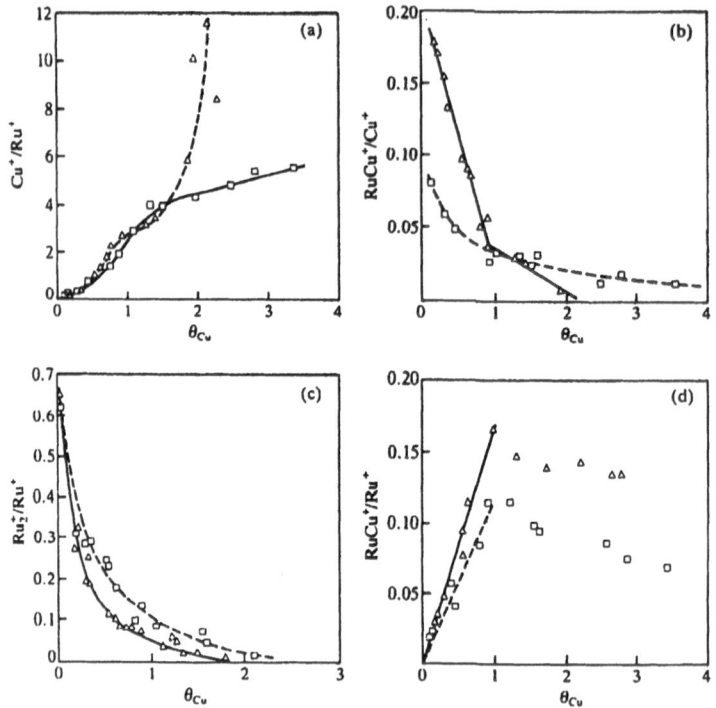

Fig. 11　Variation of SSIMS intensity ratio Cu^+/Ru^+ as a function of Cu coverage for the 540 K (□) and the 1080 K (△) series, b) Variation of $RuCu^+/Cu^+$ for 540 K (□) and 1080 K (△), c) Variation of Ru_2^+/Ru^+ for 540 K (□) and 1080 K (△), d) Plot of the secondary ion ratio .$RuCu^+/Ru^+$ as a function of Cu coverage for the 540 K (□) and 1080 K (△) series.

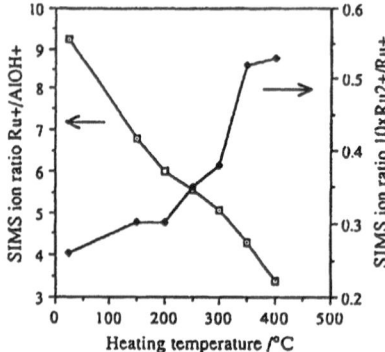

Fig. 12　Variation in SIMS intensities and ion ratios with increasing temperature of heat treatment in vacuo; heating time 15 mins

The RuCO⁺/(RuCO⁺ + Ru₂CO⁺) ratio reflects the proportion of CO which adsorbs into bridge sites (see earlier). A value of 0.9 as found on the single crystal (0001) surface suggests almost 100% linear bonding. This is the value observed just after the Ru surface has been sputtered onto the support. After annealing to 670 K the value is *even higher than 0.9* indicating that whatever the physical state the CO predominantly adsorbs in the linear form. However if Cu is added to the surface some CO is forced into bridge sites. The single crystal Ru(0001) data in fig 13 shows that the $RuCO^+/(RuCO^+ + Ru_2CO^+)$ falls to 0.7 at $\theta_{Cu} = 0.35$ suggesting that about 25% of the CO is located on bridge sites.

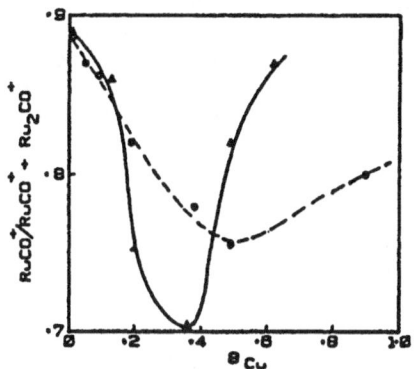

Fig. 13   Variation of $RuCO^+/(RuCO^+ + Ru_2CO^+)$ at $\emptyset/\emptyset$max Ru = 0.2
with Cu coverage for the 540 K (O) and 1080 K (Δ) series.

The fact that Ru invariably adsorbs CO into linear sites suggests that Cu is exerting an electronic effect. This is not unreasonable since it is indeed known that electron density is transferred to the Ru from the Cu. Concurrent with this shift to bridge sites there is also evidence of adsorption onto a mixed Ru-Cu site from the appearance of a RuCuCO⁺ cluster. Such an adsorbed species would be more weakly held. Thermal desorption measurements confirm the presence of an entity intermediate between the weakly adsorbed species on purely Cu sites and the strongly adsorbed CO on Ru itself, fig. 14. When Au is deposited on the Ru surface there is no evidence of CO being relocated to bridge sites nor of new mixed sites[68,69].

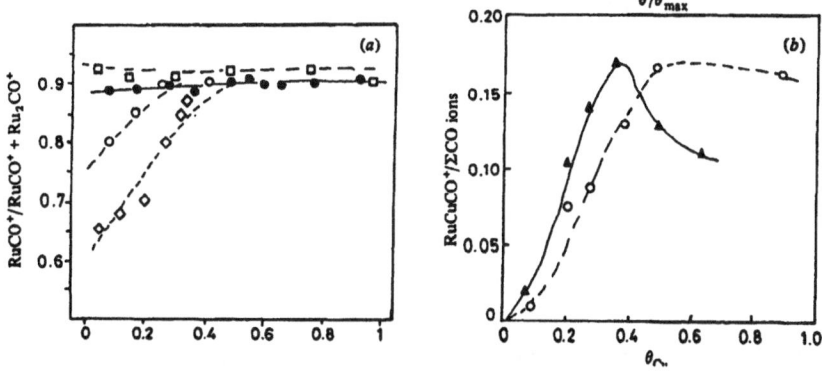

Fig.14   Variation of (a) $RuCO^+/(RuCO^+ + Ru_2CO^+)$ and (b) $RuCuCO^+/\Sigma\ CO^+$ ions as a
function of copper coverage at a relative CO coverage of 0.2. (a) Ru(0001),
□ $\theta_{Au} = 0.34$, O $\theta_{Cu} = 0.13$, ◇ $\theta_{Cu} = 0.36$, (b) 1100 K ▲ series, 540 K O series

SSIMS has the great advantage that the mass spectral form of the data means that the adsorption of more complex molecules can be relatively easily followed. Thus for example ethene adsorption on Ru(0001) has been studied in detail and the data correlated with vibrational information from electron energy loss spectroscopy[70]. Fig 15a shows the Ru isotope spectrum obtained for $RuC_2D_4^+$ when ethene is adsorbed at 130 K.

Heating the adsorbed species to 230 K results in some dehydrogenation to ethylidyne and the resulting complex spectrum. Further heating results in further loss of hydrogen and ultimately the formation of surface carbon. The spectra can be deconvoluted to enable the different surface species to be followed. Thus the m/z=134 is uniquely associated with the C_2D_3 ethylidyne species, and m/z=124 will indicate the presence of C_2D_2 species. Furthermore the ratio $I_{136}/^{104}Ru^+$ will indicate the coverage of adsorbed C_2D_4. The addition of Cu and Au to these surfaces inhibits the di-σ adsorption of ethene inducing π-bonding instead. This is less reactive and is just desorbed as the surface temperature is increased. Similar SSIMS studies of propene adsorption have been able to show that on Ru(0001) σ- bonding decomposes first propylidyne and then to ethylidyne as the surface temperature is raised from 130 K to 250 K.

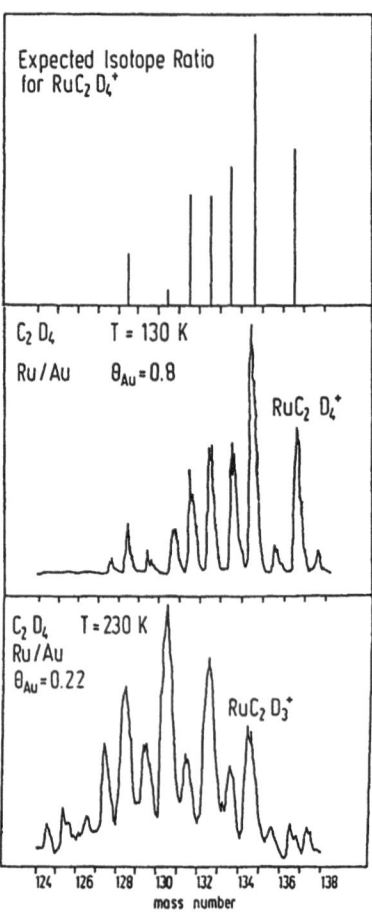

Fig. 15   (a) expected isotope ratio for $RuC_2D_4^+$; (b) $RuC_2D_4^+$ spectrum observed from a Au/Ru(0001) surface ($\emptyset_{Au}$= 0.8) after adsorption of d_4 ethene at 130 K; (c) C_2D_4 on $\emptyset_{Au}$ = 0.22 heated to 230 K

6.1.3   The State of the Catalyst after Use

Using high pressure catalysis cells attached to UHV surface analysis systems has enabled the catalytic behaviour of model catalysts to be compared with the 'real' polycrystalline materials. In the case of the present example it has been demonstrated that the Fischer Tropsch activity is very similar on the single crystal and polycrystalline Cu/Ru and Au/Ru catalysts. In the main the 1B metal reduces activity but shifts in selectivity occur. For example on a surface having small quantities of Au there is an increased selectivity to ethene. This is probably due to the reduced capacity for adsorbed hydrogen. SSIMS together with HREELS was used to investigate the surface of the single crystal catalyst after use, see fig 16.

The Ru(0001) surface was covered with extensive amounts of hydrogenated carbon as judged from the intense CH_x^+ and $RuCH_x^+$ SSIMS peaks and the CH stretch signals in the EELS. On the other hand a Au/Ru surface showed much reduced evidence of retained hydrocarbon, although adsorbed oxygen and water were now evident. Surface hydrogen starvation could explain the retention of adsorbed oxygen. Furthermore adsorbed oxygen would tend to accelerate the desorption of ethene[62]. It can be seen that SSIMS has made a significant contribution to the overall understanding of the behaviour of this catalyst system. Its ability to probe the detailed surface chemistry of almost any type of surface is its particular strength in such studies.

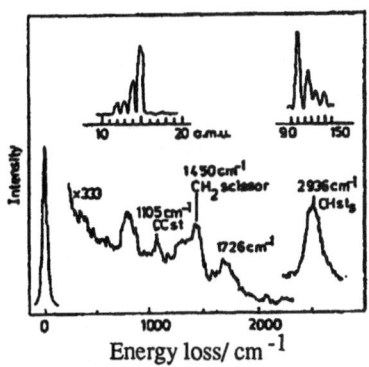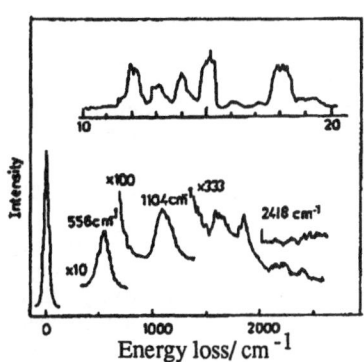

Fig. 16 (a) EELS and SIMS spectra of the Ru(0001) surface after reaction, (b) EELS spectra of the Au/Ru(0001) surface with $\emptyset_{Au}$ = 0.32 (540 K) after reaction.

6.2 Adsorption and reaction on small palladium particles

SSIMS also finds an application in the study of 'real' catalyst surface reactivity *during reaction*. The oxidation of CO over Pd has been studied in detail by a number of authors but SSIMS enables the reaction to be investigated over small metal particles as well as over the smooth single crystal surfaces of surface science. Gillet et al have studied CO oxidation over small palladium particles supported on mica (a model of a real supported catalyst), and more recently on alumina and magnesia[71,72,73]. The Pd particle sizes range from 2 to 12 nm. When CO is adsorbed the same linear relationship was observed for θ_{CO} versus $\Sigma_x(Pd_xCO^+/Pd_x^+)$ for the Pd particles as for the Pd single crystal. The particles above 6 nm in diameter are stable, they are triangular with their (111) plane parallel to the mica substrate. When they are exposed to CO the SSIMS analysis, monitored by $(PdCO/\Sigma Pd_xCO)$, indicates that the bonding is initially linear, but as coverage increases there is evidence for the development of bridge bonding, this is similar to the results obtained from the Pd(111) single crystal. The particles below 5 nm in diameter are unstable and tend to become spherical when exposed to gas and faceting may even occur under more severe conditions.

A SSIMS investigation of the oxidation of CO over the series of *alumina* supported Pd catalysts has shown that the small crystallites dissociate CO to form extensive coverages of surface carbon as indicated by the appearance and growth of Pd_2C^+ in the presence of CO at high temperature[73]. This does not occur to a significant extent on a monocrystal or on smooth films of supported Pd. Efficient oxidation only occurs as the carbon coverage is reduced and a reasonable oxygen coverage grows. Fig 17 shows what occurs on an alumina supported Pd surface as the temperature is increased in the presence of an oxygen/CO mixture. The CO coverage $(\Sigma\, Pd_xCO^+/Pd^+)$ is initially high and predominantly linearly coordinated $(PdCO^+/\Sigma\, Pd_xCO^+ = ca\ 0.9)$ and falls as temperature is increased. Above 300 K carbon coverage (Pd_2C^+) increases and maximises around 380 K. This intriguingly correlated with a plateau in CO_{ads} coverage and the movement of a proportion of the CO_{ads} into 2-fold coordination. As the carbon coverage falls the oxygen coverage increases (Pd_2O^+) and is accompanied by a rise in CO_2 production. In some studies the intermediate $PdCOO^+$ has been observed.

Where reaction begins to occur a significant proportion of the CO_{ads} is bridged but as oxygen coverage increases the CO_{ads} becomes entirely linear. This study shows that although the Langmuir-Hinshelwood reaction between linearly coordinated CO_{ads} and O_{ads} is the main reaction route, oxidation of surface carbon formed from the dissociation of CO is a significant alternative mechanism on small particles.

There is clear evidence that the electronic state of small particles is different from bulk metal. This factor and the many low coordination sites on such particles will be responsible for the different surface reactivity. In this example SSIMS has been able to obtain data on the surface processes occurring on small particles which is *inaccessible* by other surface physics techniques.

6.3 Surface activity of phthalocyanine based sensors

Metal phthalocyanines are organic semiconductors whose conductivity is sensitive to the adsorption of strongly electrophilic gases. The magnitude and reversibility of the conductivity changes is dependent on the central metal ion. The reasons for these differences were not known and XPS has not been able to distinguish any clear effects. A SSIMS study of oxygen and chlorine adsorption on H_2Pc, MgPc, NiPc, CoPc, and PbPc using ToF-SIMS showed rather contrasting behaviour. Treating films of the phthalocyanines in oxygen at 470 K had no significant effect on the SSIMS spectra, see fig. 18(a). Although treatment at 590 K left the H_2Pc, CoPc, and NiPc unchanged, in the case of MgPc and PbPc the molecular ion peaks $MPc^{\pm}$ were lost, and in PbPc, Pb_xO_y ion species were found up to high mass. Clearly, dissociation of the complex occurs in the surface layers. Exposure of the films to 200 p.p.m. Cl_2 at 440 K had little effect on H_2Pc and NiPc, some weak $(MPc + Cl)$ peaks appearing, however, the remaining samples all showed strong Cl^- yields. For MgPc and CoPc, ions of the general formula $MPcCl_x$ (x=1-8) were formed, see fig. 18(b).

Very significant effects were seen for PbPc where Pb_yCl_x ions were strongly evident, with weaker peaks corresponding to H_2PcCl_x, although there was no evidence for $MPcCl_x$. Chlorine interacts with these phthalocyanines in different ways. A mechanism for multiple chlorine adsorption appears to exist for CoPc and MgPc, which suggests that the chlorine is associated with the organic part of the molecule. In contrast, chlorine adsorption on PbPc seems to involve the metal atom. Although surface science studies of organic surfaces are still in their infancy, the use of SSIMS along with the electron and vibrational spectroscopies should ensure that the tools necessary to make significant progress are there.

6.4 Characterisation of surface treatments of zeolites[9]

This class of alumino-silicates form an important class of cracking catalysts. The surface chemical composition is important in influencing the catalytic behaviour of the zeolites. Such materials are frequently pretreated by steam treatments, alternatively the surface may be de-aluminated by treatment with

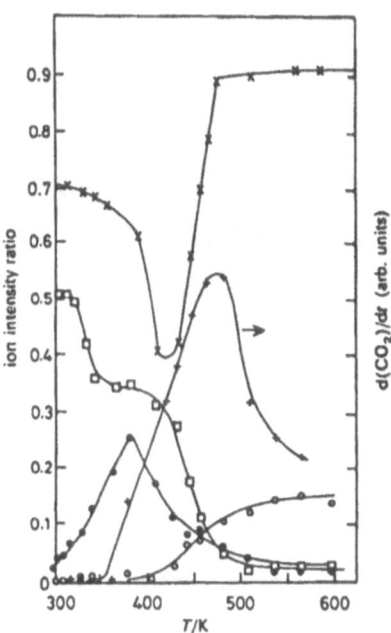

Fig. 17 Intensity of S_{CO} (□), S_O (○), S_C (●), $PdCO^+/\Sigma_n Pd_n CO^+$ (x) and P_{CO_2} (+) as a function of surface Temperature under catalytic conditions $P_{CO} = 2\times10^{-6}$ Pa $P_{O_2} = 4\times10^{-6}$ Pa, Pd particles on Al_2O_3

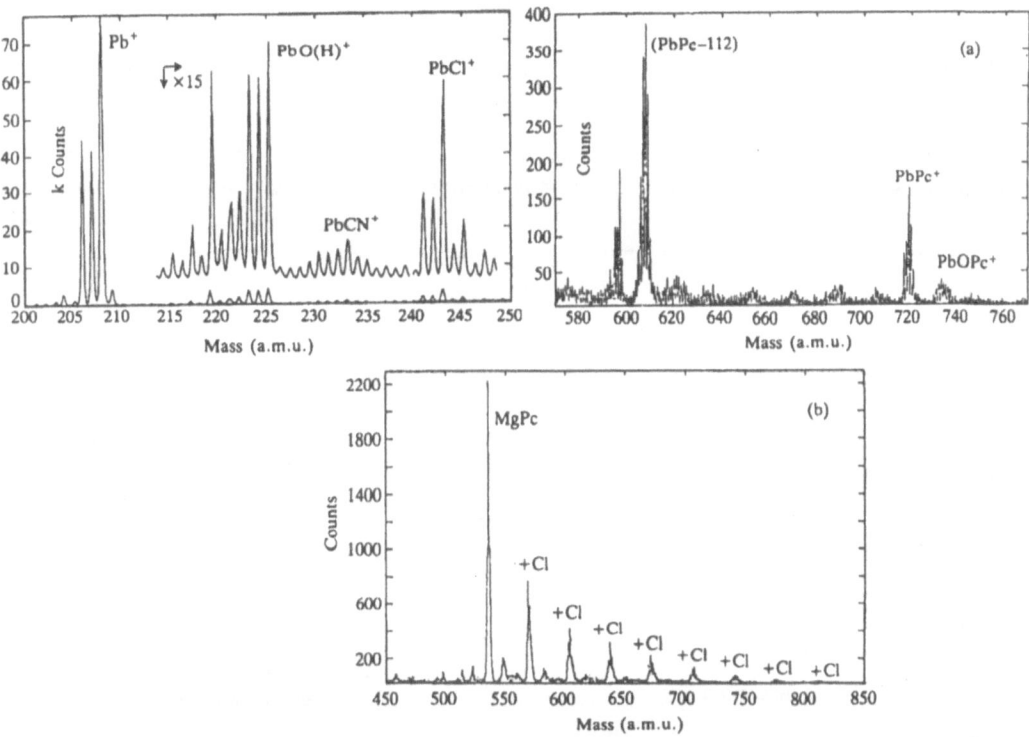

Fig.18 Part of the positive ToF SIMS spectrum of (a) Lead phthalocyanine obtained with a primary Ga^+ particvle dose of 6×10^9 (b) magnesium phthalocyanine showing multiple chlorine addition

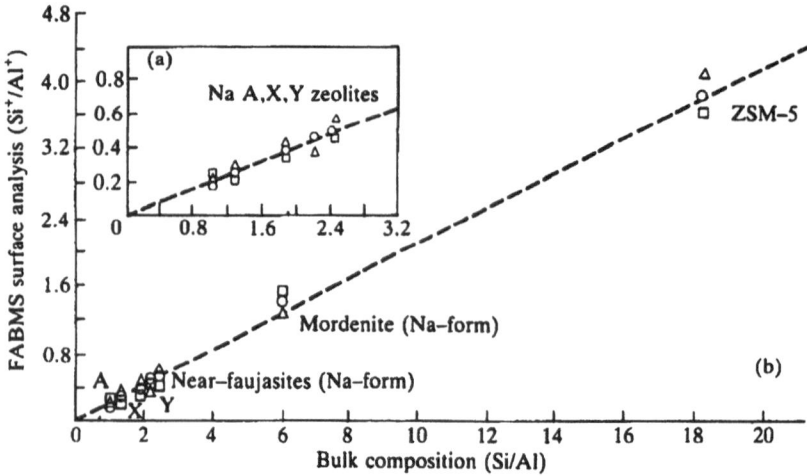

Fig. 19   Correlation between the secondary ion ratio Si^+/Al^+ and the bulk
composition Si/Al for zeolites; $\triangle$ = first layer; $\bigcirc$ = after first
bombardment; $\square$ = after second bombardment.

acids or $SiCl_4$.   SIMS can be used to investigate the effect of such treatments on
surface and sub-surface composition.   The catalyst materials were in powder form.
Each sample was pressed into a shallow, 1 cm diameter, depression in the sample
holder.   The sample was physically stable in this form.

It was first necessary to produce a calibration plot.   A series of untreated
zeolites ranging from faujastite X having a Si/AI of close to 1 to ZSM5 having an
Si/AI ≈ 18 were subjected to SIMS analysis using a fast atom beam. The observed
Si^+/Al^+ ratio (or SiO_2^-/AlO_2^-) was obtained after a 5 min. etch (~ 20 μA cm^{-2}
equivalent) and after a 20 min etch, the values obtained were in close agreement.
Fig. 19 shows the resulting calibration plot.   The gradient is close to 0.2. Using the
calibration it was possible to depth profile a number of treated zeolite samples.

The effect of steam treatment can be seen in fig. 20(a) which suggests that
in the presence of steam at 600 °C Al moves from the framework and the surface

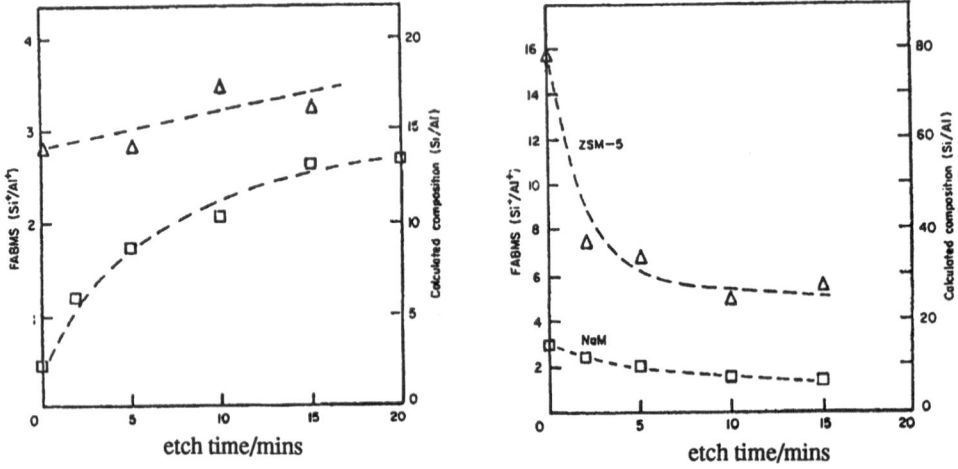

Fig. 20   Depth profiles of H-ZSM, $\triangle$ = untreated; $\square$ = heated in steam at 600 °C for 14 hrs. b)Depth
profiles of zeolites treated with HCl, $\triangle$ = ZSM-5 heated in 4 M HCl for 38 hrs at 50 °C. $\square$ =
sodium mordenite heated in 4 M HCl for 24 hrs at 25 °C

becomes rich in Al. Figure 20(b) shows the effect of acid treatment on a ZSM-5 and a mordenite sample. It can be seen that considerable dealumination of the surface layers occurs. This may be a significant process during ion exchange procedures, having an important effect on subsequent catalytic behaviour. Dealumination may be carried out by treatment with SiCl4 at 500 °C. This too results in a surface rich in Al and the bulk rich in Si. Subsequent leaching of Al with acid results in a bulk Si/Al ratio of 300.

6.5   Characterisation of an aluminium/TiO2 catalyst system[74]

This study of the surface composition of the catalyst should shed light on the uniformity of dispersion of Al in the solid. Fig. 21(a) shows the variation of the secondary ion ratio $Al^+/(Al^+ + Ti^+)$ as a function of composition, x, before and after a 3-min ion etch at 75 μA min^{-1} to yield a measure of the surface and bulk Al concentration. It should be remembered that the sensitivity to Al is approximately 100 times greater than to Ti. An almost linear rise in the ion ratio with x is observed up to x = 0.03 for the etched series indicating a uniform incorporation of Al up to this level but thereafter the concentration plateaus. Since SIMS only measures the top surface concentration, the observation that maintaining constant surface concentration, whilst the bulk concentration continues to increase, suggests that islands of α-Al2O3 are being formed at the surface whose depth is increasing more rapidly than their lateral size. There is clearly phase separation of α-Al2O3 occurring. Below x = 0.04 the steeper rise in the ion ratio as a function of x for the non-etched series indicates some segregation of Al towards the surface even in the solid solution regime.

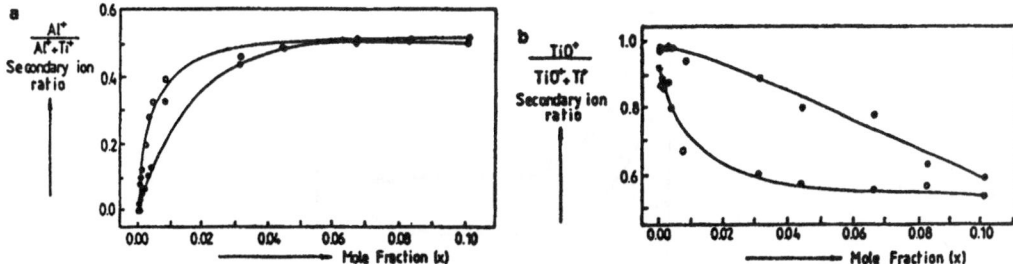

Fig. 21   Variation of secondary ion ratio (a) $Al^+/(Al^+ + Ti^+)$ and $TiO^+/(Ti^+ + TiO^+)$ as a function of Al incorporation in TiO2 before (O) and after (●) a 3 min Argon ion etch at 75 μA min

The effect of Al incorporation on bonding in the surface of the catalysts can be monitored via the TiO^+ secondary ion signal where emission is sensitively dependent on the metal-oxygen bond strength. In fig. 21(b) the ion ratio $TiO+/(TiO^+ + Ti^+)$ is plotted against x again for both ion-etched and non-etched surfaces. It is clear that the bonding of oxygen to Ti is weakened up to x = 0.01 in line with the higher concentration of Al in the surface. Beyond x = 0.03, as we indicated previously, the constant ion ratio is probably a consequence of phase segregation, further Al mainly going to form α-Al2O3. However, the variation after etching shows an approximately linear fall with increasing x all the way to x = 0.1. This demonstrates that although phase segregation does occur as the concentration of Al increases there is an increase of Al in the solid-solution phase throughout the range up to x = 0.1, as indeed would be expected if thermodynamic equilibrium of the Al dispersal was reached during preparation.

These catalysts were studied in collaboration with Professor Knozinger of Munich University. The details of preparation and characterization are given elsewhere. Two types of catalyst are described here, the first designated Ni3Mol2 Al in which a 12 wt % molybdenum oxide was added to a γ-Al$_2$O$_3$ support followed by 3 wt % NiO; the second was prepared in the reverse order and is designated Mol2Ni3 Al. Two further samples were studied as standards namely Ni3Al and Mo12Al.

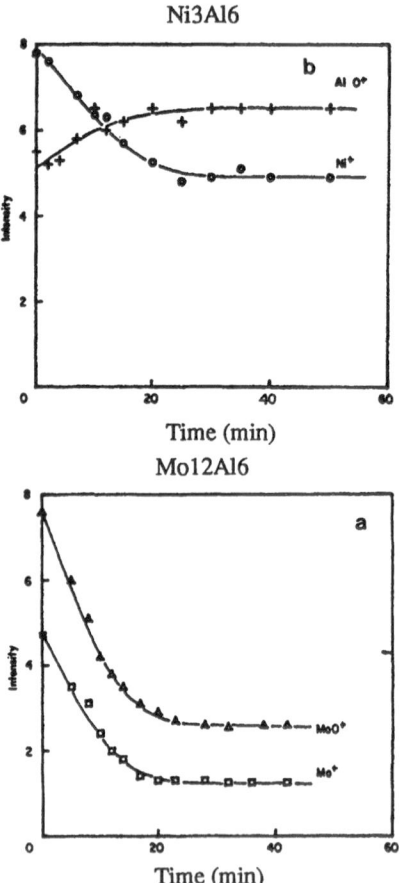

Fig. 22   Depth profiles of (a) Mo12Al and (b) Ni3Al

After the preparation procedures, which utilized impregnation with salt solutions, the samples were dried and calcined at 400 °C in air. It is clearly of considerable interest to know something about the composition, thickness and structure of the surface layer. For this type of catalyst it is important to have data on (a) surface composition, (b) the extent to which an overlayer is formed and (c) the state of the Mo and Ni in the surface. Fig. 22 shows the results of a FAB-SIMS depth profile study of the Mol2 Al and Ni3 Al precursor powders. The Mo$^+$ and MoO$^+$ secondary ions fall to constant level after 15-20 minutes of etching. This would correspond to the removal of approximately one monolayer. Since the sample is in powder form, new particles will be constantly uncovered so if the molybdenum oxide is in the form of a surface layer the observed signal will reach a 'steady state'. Table 6 shows that the Mo$^+$/Al$^+$ ratio falls by a factor of three. The observed behaviour is in good agreement with that expected for a one to two layer surface   structure.

TABLE 6

Secondary ion ratios observed (B)Before etching and (A)After etching

Sample		Ni^+/Al^+	Mo^+/Al^+	Ni^+/Mo^+	Ni^+/Al_2O^+	$Mo^+/Al_2O_3H^+$
Mo12Al	(B)	-	0.006	-	-	3.4
	(A)	-	0.002	-	-	1.3
Ni3Al	(B)	0.16	-	-	13	-
	(A)	0.13	-	-	7.5	-
Ni3Mo12Al	(B)	0.04	0.007	5.6	9.2	3.3
	(A)	0.09	0.001	10.0	8.9	0.9
Mo12Ni3Al	(B)	0.03	0.006	4.0	30	74
	(A)	0.02	0.003	9.0	8.4	13

The Ni3 Al sample also showed a fall in the Ni^+ signal but it was proportionately smaller. The steady state level is considerably higher than for Mo^+ suggesting the diffusion of Ni into the surface layers of alumina, in agreement with earlier studies[75]. The Ni^+/Al^+ ratio hardly falls, although there is evidence that the Ni density falls with depth from the increase in the Al_2O^+ signal.

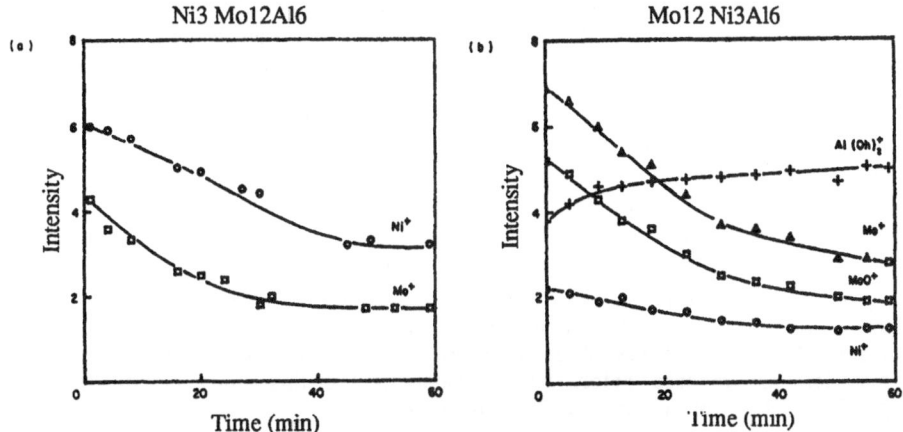

Fig. 23   Depth profiles of (a) Ni3Mo12Al and (b) Mo12Ni3Al catalyst precursors

The depth profile behaviour of the two Ni/Mo systems is shown in fig. 23. Here the etching rate is half that in fig. 22. When Mo is deposited on the Al_2O_3 before Ni (i.e. Ni3 Mol2 Al) the behaviour of the Ni^+ signal is similar to Ni3Al and the Mo^+ signal to Mol2Al. Table 6 indicates that the Ni/Al ratio actually rises with etching whilst the Mo/Al ratio falls by factor of ten. Mo forms a thin oxide layer on the Al_2O_3 and Ni ions diffuse through and into the Al_2O_3 carrier. The Ni^+/Al_2O^+ and $Mo^+/Al_2O_3^+$ ratios before etching suggest that quite large areas of the alumina carrier are exposed, thus the Mo layer is not continuous. The situation is again similar to Mol2Al. When Ni is deposited before the Mo the resulting surface is rather different. This is most clearly seen from Table 6. Whilst the changes in Ni^+/Al^+ and Mo^+/Al^+ with etching are similar to the other samples, the Ni^+/Al_2O^+ and $Mo^+/Al_2O_3^+$ ratios are very high at the start and fall very sharply with etching. A rather complete Ni/Mo overlayer must be formed on the Al_2O_3. Whether this is a consequence of depositing Ni first is difficult to say, although it is clear from fig. 23 and Table 6 that the Ni diffuses into the Mo layer as well as into the γ-Al_2O_3.

Differences in the chemical structure of the three Mo layers can be inferred from an examination of the cluster ions, especially those observed in the negative ion spectra. The oxygen containing clusters can give an indication of the state of oxidation of the molybdenum layer, so an examination of the MoO_3^-/MoO_2^- and the MoO_4^-/$Al_3O_6^-$ ratios is shown in Table 7 before and after the depth profile etch. It can be seen that the Ni3Mol2 Al sample behaves differently from the other two. The MoO_3^-/MoO_2^- ratio is significantly lower when Ni is incorporated on top of the Mo layers, whilst the MoO_4^-/$Al_3O_6^-$ ratio is very much lower.

TABLE 7   Negative secondary ion ratios observed (B)Before and (A)After etching Mo-containing catalyst precursors

Sample		MoO_3^-/MoO_2^-	MoO_4^-/$Al_3O_6^-$	$MoAlO_5^-$/$Al_3O_6^-$
Mo12Al	(B)	3.3	27	5.0
	(A)	3.1	13.5	1.7
Mo12Ni3Al	(B)	4.0	18	3.0
	(A)	3.3	13.5	2.0
Ni3Mo12Al	(B)	3.0	1.5	0.2
	(A)	2.5	4.5	0.2

It would seem therefore that the oxidation state of the Mo is lowered when Ni is incorporated second. Indeed the MoO_4^- intensity is very much less on this surface than on the other two. Without standard spectra as a function of Mo oxidation state it is not possible to define the precise state. Such spectra are being investigated in this laboratory for a wide range of inorganic components. An alternative explanation for the high MoO_4^-/$Al_3O_6^-$ ratio on the Mol2Ni3 Al surface could be the completeness of the molybedenum overlayer observed from the Mo^+/$Al_2O_3^+$ ratio. This however was not observed for the Mo12Al surface.

The chemical structure of the molybdenum layer-alumina interface may be suggested from the mixed cation cluster ions. $MoAlO_5^-$ and $MoAlO_4^-$ are observed, but they are of low intensity on the Ni3 Mo12 Al surface. The $MoAlO_5^-$/$Al_3O_6^-$ ratio is shown in table 7. It is clear that there is considerable chemical interaction between the Mo layer and the alumina except when Ni is incorporated second. These materials are reduced before use as catalysts but this study demonstrates the power of FAB-SIMS for applied catalyst analysis.

7 REFERENCES

1   J.J. Thomson, (1910), Phil. Mag., 20, 252
2   R.F.K. Herzog and F.P. Viebock, (1949), Phys. Rev., 76, 855L
3   R. E. Honig, (1958), J. Appl. Phys, 29, 549
4   A. Benninghoven, (1970), Z. Physik, 230, 403
5   P.D. Prewett and D.K. Jeffries, (1980), Inst. Physics Conf. Ser., 54, 316
6   F.M. Devienne, (1973), Vide, 167, 193
7   D.J. Surman and J.C. Vickerman, (1981), Appl. Surf. Sci., 9, 108
8   D.J. Surman, J.A. van den Berg, and J.C. Vickerman, (1982), Surf. Interface Anal., 4, 160
9   K. Wittmaack, (1975), Surf. Sci., 53, 626
10   A. Brown, J.A. van den Berg and J.C. Vickerman, (1985), Spectrochim. Acta, 40B, 871
11   D. Briggs and M.J. Hearn, (1986), Vacuum, 36, 1005
12   P. Williams, (1979), Surf. Sci., 90, 588
13   C.A. Anderson and J.R. Hinthorne, (1972), Science, 175, 853
14   G. Slodzian and J.F. Hennequin, (1966), Compt. Rend. (Paris), B263, 1246

15 A. Benninghoven, (1975), Surf. Sci., 53, 596

16 H. Oechsner, (1984) in Thin Film and Depth Profile Analysis, Topics in Current Physics, Vol. 37, Springer Verlag, Berlin and Heidelberg, p. 63

17 D. Briggs and M.J. Hearn, (1985), Int. J. Mass Spec. Ion. Proc., 67, 47

18 H.H. Anderson and H.L. Bay, (1981), in Sputtering by Particle Bombardment I, Springer Series in Topics in Applied Physics, Vol 47, (R. Behrisch ed.), Springer Verlag, Berlin and
Heidelberg, p.145

19 S.Y. Lai, D. Briggs and J.C. Vickerman, Surf. Interface Anal., submitted

20 A. Brown and J.C. Vickerman, (1986), Surf. Interface Anal., 8, 75

21 A. Brown, J.A. van den Berg and J.C. Vickerman, (1985), Spectrochimica Acta, 40B, 871.

22 A.R. Waugh, A.R. Bayly and K. Anderson, (1986), in Secondary Ion Mass Spec SIMS V, Springer Series in Chemical Physics, Vol. 44, (A. Benninghoven, R.J. Colton, D.S. Simmons and H.W. Werner, eds.) Springer Verlag, Berlin and Heidelberg, p.138

23 R. Levi-Setti, P.H. La Marche, K. Lam and Y.L. Wang, (1984), Soc. Photo-Opt. Instrumen. Eng., 471, 75

24 K. Wittmaak, (1982), Vacuum, 32, 65

25 P.H. Dawson and N.R. Whetten, (1969), Advances in Electronics and Electron Physics, Vol. 27, (L. Marton, ed.), Academic Presss, New York, p.61

26 N. Winograd, D.E. Harrison and B.J. Garrison, (1978), Phys. Rev. Lett. 41, 1120.

27 B.J. Garrison, N. Winograd and D.E. Harrison, (1978), Phys. Rev. B18, 6000.

28 B.J. Garrison, (1980), J. Am. Chem. Soc. 102, 6553.

29 Tandem Mass Spectrometry, (F.W. McLafferty, ed.), John Wiley & Sons (1983)

30 R.D. MacFarlane and D.F. Torgerson, (1976), Science, 191, 920

31 K.G. Standing, R. Beavis, W. Ens and B. Schuler, (1983), Int. J. Mass Spec. Ion Phys., 53, 125

32 W.P. Poschenrieder, (1972), Int. J. Mass Spec. Ion. Phys., 357.

33 B.A. Mamyrin, V.I. Karataev, D. Shmikk and V.A. Zagullin, (1973), Sov. Phys. JETP, 37, 45

34 P. Steffans, E. Niehuis, T. Freise, D. Greifendorf and A. Benninghoven, (1984), in Secondary Ion Mass Spectrometry SIMS IV, Springer Series in Chemical Physics, Vol 36, (A. Benninghoven, J. Okano, R. Shimizu and H.W. Werner, eds.), Springer Verlag, Berlin and Heidelberg, p.404

35 R. Kelly, (1984) in Ion Bombardment Modification of Surfaces: Fundamentals and Applications, (O. Auciello and R. Kelly, eds.), Elsevier Science Publishers, BV, Amsterdam, p.27

36 P. Sigmund, (1981), in Sputtering by Particle Bombardment I, Springer Series Topics in Applied Physics, Vol 47, (R. Behrisch, ed.), Springer Verlag, Berlin and Heidelberg, p.9

37 R. Kelly, (1979), Surf. Sci., 90,280

38 H. Overiejnder, A. Haring and A.E. de Vries, (1978), Rad. Effects, 37, 205

39 M.L. Knotek and P.J. Freibelman, (1978), Phys. Rev. Lett., 40, 964

40 N. Winograd, (1981), Prog. Solid State Chem., 13, 285

41 B.J. Garrison, (1983), Int. J. Mass Spec. Ion Phys., 53, 243

42 J.K. Norskov and B.I. Lundqvist, (1979), Phys. Rev., B19, 5661

43 K. Wittmaack, (1977), in Inelastic Ion-Surface Collisions,(N.H. Tolk, J.C. Tully, W. Heiland and C.W. White, eds.), Academic Press, New York, p.153

44 D.W. Moon and N. Winograd, (1983), Int. J. Mass Spectrom. Ion Phys. 51, 217.

45 J.C. Vickerman, in J.C. Vickerman, N.M. Reed and A. Brown (Eds), Secondary Ion Mass Spectrometry - Principles and Applications, Oxford University Press, Oxford 1989, p54.

46 M. Barber, J.C. Vickerman and J. Wolstenholme, (1977), Surf. Sci. 68, 130.

47 B.J. Garrisom, A.C. Diebold, J.H. Lin and Z. Sroubek, (1983), Surf. Sci., 124, 461

48 R.D. Macfarlane, (1985) in Desorption Mass Spectrometry, Am. Chem. Soc. Symp. Ser., 291, 56

49   W. Gerhard and C. Plog, (1983), Z. Phys. B. Cond. Matt., 54,59

50   R.G. Cooks and K.L. Busch, (1983), Int. J. Mass Spec. Ion Phys., 53,111

51   M. Barber, R. Bordoli, J. Wolstenholm and J.C. Vickerman, (1977) Proc. 7th Int. Conf. Vac. Sci. and 3rd Int. Conf. on Solid Surf., Vienna p.983

52   K. Kishi and M.W. Roberts, J. Chem. Soc., Faraday Trans. I, (1975), 71, 1715.

53   M. Barber, J.C. Vickerman and J. Wolstenholme, (1977), Surf. Sci. 68, 130.

54   R.S. Bordoli, J.C. Vickerman and J. Wolstenholme, (1979) Surf. Sci., 85, 244

55   A. Brown and J.C. Vickerman, (1982), Surf. Sci., 117, 154

56   K.E. Foley, N. Winograd and B.J. Garrison, (1984), J. Chem Phys. 80, 5254

57   L.A. Delouise and N. Winograd, (1985), Surf. Sci. 154, 79

58   A. Brown and J.C. Vickerman, (1984), Surf. Sci. 140, 261

59   A. Brown and J.C. Vickerman, (1985), Surf. Sci. 151, 319

60   J.H. Sinfelt, (1972), *Chem. Eng News*, 50, 18; (1977), Acc. Chem. Res. 10, 15

61   S.Y. Lai and J.C. Vickerman, (1984), J. Catalysis, 90, 337

62   N.I. Dunhill, B. Sakakini and J.C. Vickerman, (1988), Proc. 9th Int. Conf. Catalysis, Calgary, p 1166

63   J.H. Sinfelt and G.H. Via, (1980), J.Chem. Phys. 72, 4832

64   J.C. Vickerman, K. Christmann, G.Ertl, P. Heimann, F.J. Himpsle and D.E. Eastman, (1983), Surf. Sci., 134, 367

65   J.C. Vickerman, K. Christmann and G. Ertl, (1981), J Catalysis, 71, 175.

66   C. Harendt, K. Christmann, W. Hirschwald and J.C. Vickerman, (1986), Surf. Sci., 165, 413

67   N. Aas, PhD Thesis, UMIST 1990.

68   A. Brown, J.A. van den Berg and J.C. Vickerman, (1984), Proc . 8th Int Congr. Catal., (Dechema, West Berlin), vol IV, p35.

69   B. Sakakini, A.J. Swift, J.C. Vickerman, C. Harendt and K. Christmann, (1987), J. Chem. Soc., Faraday Trans. I , 83, 1975.

70   B. Sakakini, N. Dunhill, C. Harendt, B. Steeples and J.C. Vickerman, (1987), Surf. Sci., 189/190, 211

71   E. Gillet, S. Channakhone, V. Matolin and M. Gillet, (1985), Surf. Sci., 152/153, 603.

72   V. Matolin, E. Gillet and M.Gillet, Surf. Sci., 162, 354 (1985).

73   N.M. Reed, V.Matolin, E. Gillet and J.C. Vickerman, (1990) J. Chem. Soc. Faraday Trans., 86(15), 2749-2755

74   A. Boronicolos and J.C. Vickerman, (1986), J. Catal, 100, 59-68

75   H. Knozinger and H. Jeziorowski, (1980), Proc. 7th Int. Congr. Catalysis, Tokyo

STUDY OF THE $Pt_3Sn(100)$, (111) AND (110) SINGLE CRYSTAL SURFACES

BY LEISS AND LEED

Ugo Bardi, Luca Pedocchi, Gianfranco Rovida

Dipartimento di Chimica, Università di Firenze
50121 Firenze- Italy

Alexandra N. Haner and Philip N. Ross

Materials and Chemical Science Division
Lawrence Berkeley Laboratory
Berkeley CA 94720 - USA

ABSTRACT

The structure and the composition of the (111), (100) and (110) oriented surfaces of single crystal samples of the ordered Pt_3Sn compound were studied by low energy Ne^+ scattering and other surface sensitive techniques (mainly XPS and LEED). The LEISS results for all three low index surfaces show that the surface plane contains both Pt and Sn and indicate an outward displacement of Sn atoms from the plane of Pt atoms.

1. INTRODUCTION

Pt-Sn alloys have considerable technological importance for their uses as catalysts for hydrocarbon conversion[1,2] and in fuel cells[3,4]. In order to obtain data about the surface structure and properties of these alloys, we undertook a study on the low index surfaces of well characterized single crystal samples of the Pt_3Sn compound, which has the FCC "AuCu3" type structure. Several other FCC Pt-M alloys (also of interest for their catalytic properties) have been studied in single crystal form by techniques capable of determining the composition of the outermost surface plane[5,6,7]. These studies have shown that for M=Ti, Fe, Co and Ni, the outermost layer of the clean, low index surfaces is strongly enriched in Pt (with the only exception the (110) surface of $Pt_{0.5}Ni_{0.5}$[7]). In the present work, however, we will show that all the Pt_3Sn low index surfaces have a mixed Pt-Sn composition and that some evidence appears to indicate an enrichment in Sn with respect to the bulk composition. Furthermore, LEED indicates that the (110) and the (100) surfaces are reconstructed.

Fundamental Aspects of Heterogeneous Catalysis Studied by Particle Beams,
Edited by H.H. Brongersma and R.A. van Santen, Plenum Press, New York, 1991

2. EXPERIMENTAL

All experiments were performed in an UHV chamber equipped with a hemispherical analyser with multichannel detector used for both XPS and LEISS measurements. XPS was performed using a conventional Mg Kα X-ray source. For LEISS we used a primary beam of Ne$^+$ ions at energies variable in the range from 300 to 2000 eV. The ion current on the sample was approximately 7×10^{-8} A cm^{-2} (ca. 5×10^{11} ion s^{-1} cm^{-2}). The fixed scattering angle was 135°. Recording a complete LEISS spectrum took approximately 3 minutes. No appreciable variation in the surface composition was observed during repeated LEISS measurements under the same conditions. Angle-resolved LEISS measurements were performed rotating the sample along the azimuthal or the polar angle and keeping fixed the ion source and detector assembly. The vacuum system was also equipped with a 3 grids optics, used for LEED, and with facilities for clean gas introduction.

The samples were single crystal discs (all cut from the same Pt$_3$Sn rod) approximately 10 mm diameter, with one face cut and polished along the (100), (111) or (110) planes. The samples were cleaned in situ by a procedure of Ar$^+$ ion bombardment and annealing in vacuum at about 1000 K until no impurities could be detected by XPS. LEED examination after this treatment showed that the samples were homogeneous single crystals over the whole surface.

3. RESULTS

The LEISS examination of the three low index Pt$_3$Sn surfaces at non grazing incidence of the Ne$^+$ beam showed the presence of both Pt and Sn on the outermost surface plane. The results at 500 eV primary energy and at nearly normal incidence of the ion beam are shown in fig. 1. At all energies examined, the Pt/Sn ratio was found to be higher for the (111) surface. The reproducibility of the LEISS signal ratios was found to be better than 10% over repeated cycles of preparation of the surfaces. The surface composition was found to be constant for annealing in a range of temperatures of approximately 800-1000 K. For all three surfaces, we observed an increase of the Pt/Sn LEISS ratio as a function of increasing energy of the Ne$^+$ ions in the range of 300 to 1000 eV.

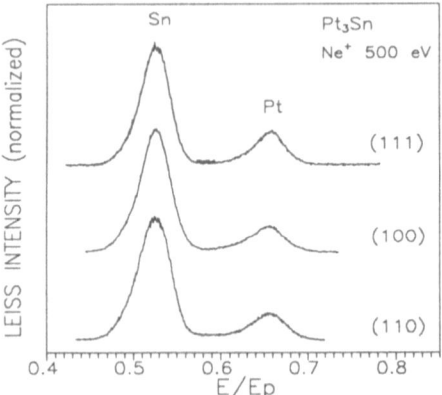

FIG. 1. LEISS spectra at near normal angle of incidence of the primary Ne$^+$ beam at 500 eV. The fixed scattering angle is 135°.

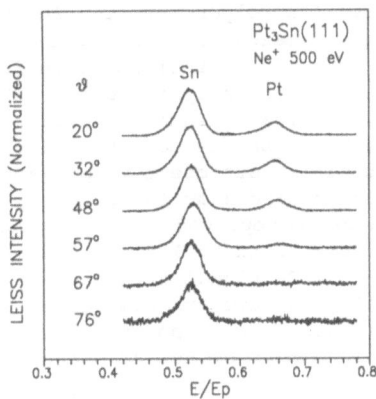

FIG. 2. LEISS spectra at variable polar angle of incidence of the y Ne+ beam for the Pt3Sn(111) surface. Angles are defined with respect to the surface normal. Ne+ beam energy: 500 eV.

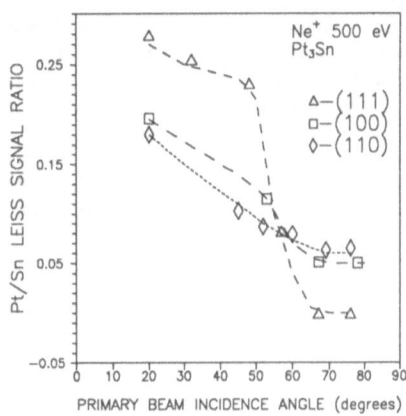

FIG. 3. Polar angle dependency of the Pt/Sn LEISS signal intensity ratio for the three surfaces examined. Angles are measured with respect to the surface normal. Ne+ beam energy: 500 eV.

In fig. 2 we show a series of LEISS spectra for the (111) sample for variable polar angles and in fig. 3 the variation of the Pt/Sn ratio for the three faces as a function of the polar angle of incidence of the Ne+ beam (polar angles are defined with respect to the normal to the surface). In all cases the Pt/Sn signal ratio is significantly lowered at incidence angles of the ion beam grazing to the surface; however the effect is more pronounced for the (111) face. The same qualitative angle dependency of the Pt/Sn ratio was observed with a 1 keV Ne+ beam. This behavior was found to be only weakly dependent of the azimuthal orientation of the sample with respect to the ion beam, at least for the (110) surface. Angle-resolved XPS measurements showed a weak variation of the Pt/Sn signal ratio as a function of the polar angle.

The LEED results for the low index Pt3Sn surfaces will be reported elsewhere[9]. Briefly, the (110) surface was found to show "extra" spots in the pattern which could be interpreted in terms of a rhombic unit mesh. Also the (100) surface showed characteristic "streaks" in the pattern which may be interpreted as due to a surface reconstruction. Extended annealing in vacuum at temperatures above 1000 K led to the weakening or to the disappearance of the "extra" features. Only the (111) surface showed a surface periodicity compatible with bulk truncation parameters under all conditions of annealing. However, also in this case, the parameters of the unit mesh alone cannot exclude the possibility of a structural reconstruction.

4. DISCUSSION

For the 3 Pt3Sn low index faces studied, the presence of both Pt and Sn in the outermost surface plane is in qualitative agreement with the literature data[10,11,12], as well as with theoretical predictions[17]. A quantitative analysis of our LEISS data at 2 keV was performed using experimental sensitivity factors for Pt and Sn at the same Ne+ energy and at nearly the same scattering angle[11]. In this way we measured a surface composition of 50% at. Sn for Pt3Sn(111). For the other two faces, the Sn surface atomic concentration may be estimated to be of the order of 70%.

These values appear consistent with the results for polycrystalline Pt$_3$Sn, where a surface composition of ca. 60% at. Sn was measured by LEISS and by CO chemisorption[11]. We remark also that a strong surface enrichment in Sn was observed for polycrystalline Ni$_3$Sn[11]. However, an analysis of our LEISS data based on different elemental sensitivity factors[18] leads to values of the Sn surface concentration closer to the bulk values. Hence, it is not clear that the accuracy of the LEISS measurement is sufficient to prove the presence of a large Sn enrichment in the Pt$_3$Sn low index surfaces.

The angle dependency of the Pt/Sn LEISS ratio reported in the present work is qualitatively similar to the results for the Sn/Pt(111) system[8]. In particular, the reduction and disapperance of the Pt signal at angles higher than ca 60° off normal for the Pt$_3$Sn(111) surface are similar to the data for the Pt(111) surface after low temperature deposition of Sn[8]. This result was interpreted[8] as due to a "shadowing" effect of the deposited Sn atoms. In the case of the formation of Pt-Sn "surface alloys" at high temperature, the similar - albeit weaker - angle dependency of the Pt/Sn ratio was interpreted in terms of upward "rippling" of the Sn atoms in the outermost surface plane[8]. This interpretation should be considered with caution, until a detailed model of the Ne$^+$ scattering process on this surface can be developed. However, it appears at present as a reasonable description of the available data. Preliminary calculations[19] permitted us to estimate the upward displacement of Sn atoms in the case of the Pt$_3$Sn(110) surface to be roughly between 1 and 2 Å.

The LEISS data can be examined in view of structural models for the Pt$_3$Sn surfaces. The simplest conceivable model corresponds to bulk truncation. In the Pt$_3$Sn lattice, the planes normal to the <100> and <110> directions are alternate "Sn rich" (50% at. Sn) and "Pt rich" (0% at. Sn) planes. Conversely, all planes normal to the <111> direction have the same composition (25% at. Sn). In this model, if we assume that both the (110) and the (100) surfaces terminate with the "Sn-rich" planes, the expected variation of the Sn concentration going from the (111) to the (100) and (110) surface is qualitatively reproduced by the experimental results. However, the LEISS mesurements appear to indicate a significantly larger Sn concentration. Furthermore, if the strong angular dependency of the Pt/Sn LEISS ratio is interpreted in terms of a shadowing effect, the upward Sn displacement may be considered too large to be explained simply in terms of bulk truncation models. The LEISS results may therefore indicate some kind of reconstruction of all low index Pt$_3$Sn surfaces, with the formation of a "Sn rich" surface layer.

A possible structural model, alternative to the bulk truncation ones and which takes into account both the LEISS and the LEED data may be based on the structure of the (0001) plane of the PtSn compound. In the PtSn structure (hexagonal "NiAs" type), planes of pure Pt and of pure Sn alternate along the [0001] direction. The apparent Sn enrichment and the shadowing effect may be therefore explained assuming that the "pure Sn" plane is the stable termination. This model is compatible with the LEED data for the Pt$_3$Sn(110) surface if we assume an in-plane distortion (5% contraction in one direction and 10% expansion in the other) of the unit mesh. Such a model, however, does not appear compatible with the LEED results for the Pt$_3$Sn(111) surface. At present, therefore, it is not yet possible to propose a single structural model accounting for the results of all three faces studied.

5. CONCLUSION

The presence in the surface outermost plane of the minority component (Sn) and its apparent outward displacement render the Pt$_3$Sn compound unique among the platinum alloys studied so far as single crystals. Furthermore, Pt$_3$Sn is so far the only *ordered* FCC "AuCu$_3$" type compound to show a surface reconstruction for the (110) and (100) planes. In all the other cases studied (Ni_3Al[13,14], Pt_3Ti[15] and $AuCu_3$[16]) the surface structure for all low index faces studied was found to correspond to bulk truncation. Work is in progress in order to study the effect of the surface structure and composition on the chemisorptive and catalytic properties of Pt_3Sn in comparison to other Pt alloys.

ACKNOWLEDGMENTS

This work was supported in part by the Assistant Secretary for Conservation and Renewable Energy, Deputy assistant Secretary for Utility Technologies, Office of Energy Management, Advanced Utility Concepts, Division of the U.S. Department of Energy under contract DE-ACO3-76SF00098 and by the Consiglio Nazionale delle Ricerche (Italy). The authors are indebted to Brunetto Cortigiani for assistence during the measurements.

REFERENCES

1. F. Dautzenberg, J. Helle, P. Biloen and W. Sachtler, J. Catal. 63 (1980)119.
2. Z. Karpinski and J. Clarke, J. Chem Soc., Faraday Trans. 1 71(1975)893.
3. K.J. Cathro, J. Electrochem Soc. 116 (1986) 1608.
4. B. McNicol, R. Short and A. Chapman, J. Electroanal. Chem. 72(1976) 2735.
5. U.Bardi, A. Atrei, P.N. Ross, E. Zanazzi and G. Rovida, Surface Science 211/212(1989)441.
6. J. Paul, S.D. Cameron, D.J. Dwyer and F.M. Hoffmann, Surface Science, 177(1986)121.
7. Y. Gauthier, R. Baudoing, M. Lundberg and J. Rundgren, Phys rev. B. 35(1987)7867.
8. M.T. Paffett and R.G. Windham, Surface Science, 208(1989)34.
9. H.N. Haner, P.N. Ross, and U. Bardi, in "Structure of Surfaces III", Proc. ICSOS-III Eds. M. Van Hove, K. Takayanagi and X. Xide, Springer-Verlag, Berlin, in press.
10. R. Bowman, L. Toneman and A. Holscher, Surface Sci. 35(1974)348.
11. P. Biloen, R. Bouwman, R. Van santen and H. Brongersma, Appl. Surf. Sci. 2(1979)532.
12. R. Bouwman, L. Toneman and A. Holscher, Surface Science 35(1973)8.
13. D. Sondericker, F. Jona and P.M. Marcus Phys. Rev. B 34 (1986) 6770.
14. D. Sondericker, F. Jona and P.M. Marcus Phys. Rev. B 33 (1986)900.
15. U. Bardi and P.N. Ross, Surface Science 146(1984)L555.
16. V.S. Sundaram, B. Farrell, R.S. Alben and W.D. Robertson, Phys. rev. Lett. 31(1973)1136
17. R. Van Santen and W. Sachtler, J. Catalysis 33(1974)202
18. W. Heiland and E. Taglauer, Nucl. Instr. Methods 132(1976)535
19. A. Haner, P.N. Ross and U. Bardi, LBL report 28074, 1990

STUDY OF THE SURFACE STRUCTURE OF $V_2O_5/$ γ-Al_2O_3 CATALYSTS BY LEIS

J.-P. Jacobs, G.C. van Leerdam, and H.H. Brongersma

Faculty of Physics and Schuit Institute of Catalysis
Eindhoven University of Technology
P.O. Box 513, 5600 MB Eindhoven, The Netherlands

ABSTRACT

LEIS has been used to study the surface structure of V_2O_5/γ-Al_2O_3 catalysts as a function of V-loading. At low loadings (< 7 wt%) the vanadium species is optimally dispersed over the support, while at medium loadings (< 11 wt%) multilayered structures are formed. Three-dimensional models for the surface structures are proposed.

.1. INTRODUCTION

During the last decade low-energy ion scattering (LEIS) has developed to a valuable technique for the analysis of the atomic composition and structure of solid surfaces. LEIS is based on the unique capability of probing only the topmost atomic layer. Depth information may be obtained by sputter etching. Recent studies showed that LEIS can provide a detailed picture of the surface structure of the support material γ-Al_2O_3 [1] and the location of molybdena on this support [2]. The use of absolute rather than relative LEIS intensities proved to be very useful and lead to a quite different model for the alumina supported molybdena catalysts [1,2]. Since both molybdenum and vanadium are classified in the category of monolayer-type catalysts [3,4], it seems interesting to investigate the surface structure of V_2O_5/γ-Al_2O_3 catalysts.

Supported vanadium oxides are widely used as oxidation catalysts and have, therefore, been studied extensively by various experimental techniques. In this paper only the interaction between γ-alumina and V_2O_5 will be considered. According to Murakami et al. [5,6], who used benzaldehyde-ammonia titration, NO-NH_3 rectangular pulse methods, X-ray diffraction, IR, ESR, and UV-reflectance spectrometry, V-species are optimally dispersed on the support at low loadings (< 4 wt%). At medium loadings (< 35 wt%) lamellae are formed and at even higher loadings the support is completely covered. Haber et al. [3] concluded from chemical analysis and reduction-oxidation studies that monolayer coverage is reached at a loading of 20 wt%. The formation of dimeric V-O species is inherent to the γ-alumina surface and not forced by too high concentration. This is confirmed by Nag and Massoth [7] by means of XPS and

Fundamental Aspects of Heterogeneous Catalysis Studied by Particle Beams,
Edited by H.H. Brongersma and R.A. van Santen, Plenum Press, New York, 1991

gravimetric reduction studies. Laser Raman spectroscopic experiments by Roozeboom et al. [8,9] showed that three different species may be present on the catalyst: vanadate tetrahedra, two-dimensional polymeric octahedral clusters and at higher loadings V_2O_5 crystallites. Solid state ^{51}V NMR structural studies on the catalyst by Eckert and Wachs [10] show a predominance of a tetrahedral oxygen coordination of the V atoms at low loadings. At higher loadings (up to monolayer coverage) a strong increase of octahedrally coordinated V-species is found. Average interatomic distances of the vanadium oxide supported on γ-alumina (9.8 wt%) have been given by Bergeret et al [11] using RED (radial electron distribution). These results indicate that the structure of the supported vanadium oxide is different from bulk vanadium pentoxide. In this paper the different models are compared and contrasted with the results of our LEIS experiments.

2. EXPERIMENTAL SECTION

2.1 Catalyst preparation

The materials used in the experiments were obtained from the Delft University of Technology, Chemical Engineering Department (Scholten/Kreemers). The supported vanadium catalysts were prepared by leading a NH_4VO_3 solution over the γ-Al_2O_3 support (250 m^2/g) at 343 K following the method described by Roozeboom et al. [12]. The catalysts were then calcinated in air at 673 K for 4 hrs. A monolayer coverage was estimated to correspond to 15.3 wt% V_2O_5 [13].

2.2 Surface analysis equipment

The surface analysis was performed with the LEIS apparatus NODUS, which basic design has been outlined earlier [14]. Mono-energetic ions are nowadays produced in a Leybold ion source and directed onto the target. A kind of CMA (cylindrical mirror analyzer) selects ions scattered at an angle of 142^0. The energy spectrum of the selected ions is recorded. The differential pumping system ensures that the background pressure increases only from $5*10^{-8}$ Pa to $1*10^{-6}$ Pa when the ion beam is used. This increase is mainly due to He. The catalysts studied were pressed in tantalum disks and then mounted onto the target holders. Before entering the UHV chamber the samples were treated with 10^3 Pa O_2 for 15 min. at 500 K in a reaction chamber. Then the reaction chamber was evacuated for 10 min. to limit contamination of the experimental UHV chamber when the samples are inserted via a loadlock onto the carousel. The carousel can hold twelve samples which ensures identical experimental conditions when analyzing a series of targets. Different series can be compared by means of a normalization sample. Typical experimental conditions were a beam current of 200 nA, a $^4He^+$ probe, a beam diameter between 0.5 and 5 mm depending on the desired sputtering rate, and a primary energy of 3 keV for the ions. Surface charging effects were eliminated by flooding the surface with low-energy electrons from a ring shaped neutralizing system which ensures flooding from all sides.

3. RESULTS AND DISCUSSION

A series of V_2O_5/γ-Al_2O_3 catalysts with a V_2O_5 content ranging from 0 to 14 wt% has been analyzed. Figure 1 shows a typical LEIS spectrum which is used in the analysis of the alumina supported vanadia catalysts. The peak height is taken as a measure of the scattered intensity. Identical results were obtained if area's were used instead. The V_2O_5 loading dependence of the peak intensities is shown in figure 2.

The behaviour of the LEIS signal as function of vanadium pentoxide loading can be divided into three regimes. At low loadings (<7 wt%) a linear increase of the V signal is found together with a linear decrease of the Al signal. This linearity points to a monolayer dispersion of V species, which shield part of the aluminium from detection. In the regime from one half to three quarters of a monolayer (7-11 wt%) both the V and Al signals have a constant value. The plateau of the V signal indicates that extra V atoms are deposited on the surface in such a way that they cover already present V atoms. This means that a multilayered structure is formed. The growth of multilayered species, which doesn't cause any additional shielding of the aluminium atoms, is consistent with the Al signal plateau. An increase of the vanadium signal and simultaneously a decrease of the aluminium signal is observed when the loading is increased up to monolayer coverage. This can be accounted for by the deposition of V species on locations where they shield aluminium. The fact that the slope of the second linear increase of the V signal is the same as in the first regime suggests not only a monolayer dispersion, but also a deposition of V species in such a way that they do not cover any already present V atoms. Note that there is no significant change in the O signal over the whole regime.

The insensitivity of the oxygen signal indicates a growth of V species on the γ-alumina surface which extends the oxygen lattice of the support. The structure of γ-Al$_2$O$_3$ is a defected spinel [15,16]. The preferential exposure of one of the cleavage planes has been discussed before but no conclusive answer has been reached yet. Recent results at our laboratory [1] point to the D-layer of the (110)-face as the predominant face of the γ-alumina. The D-layer, as shown is figure 3.a, distinguishes itself by the low aluminium content at the surface and the exposure of only octahedral interstices which are filled with aluminium atoms. In the following a model for the V$_2$O$_5$/γ-Al$_2$O$_3$ catalysts is proposed which assumes the D-layer to be the predominant surface structure of the support.

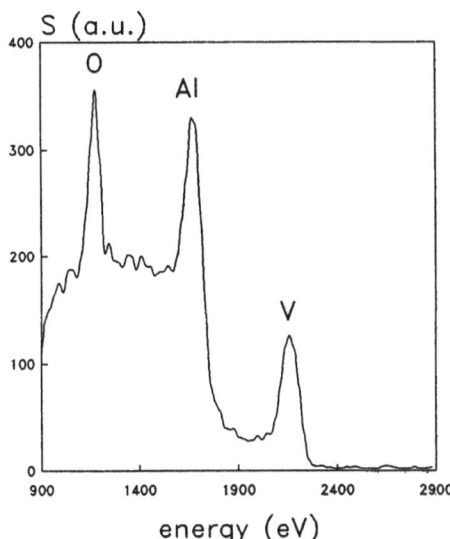

Fig. 1.
Typical LEIS spectrum of a
8.84 wt% V$_2$O$_5$/γ-Al$_2$O$_3$ catalyst
(⁴He$^+$, primary energy of 3000 eV)

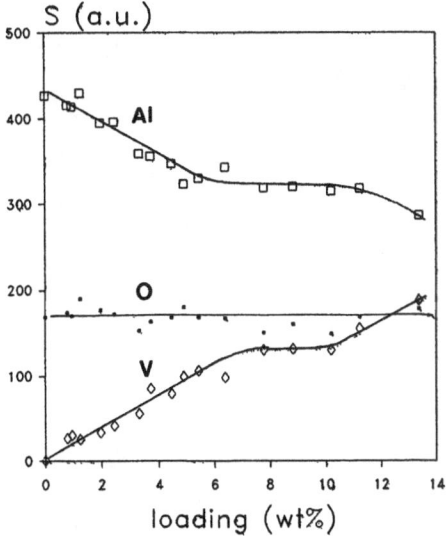

Fig. 2.
The peak intensities (S)
of Al, O, V as a function of
V$_2$O$_5$ content

Extension of the spinel structure starting from the D-layer will give the C-layer. This layer contains alternating rows of tetrahedral and octahedral interstices in the oxygen lattice. As stated in the introduction the vanadium is found tetrahedrally coordinated at low loadings which suggests a filling of only tetrahedral interstices in the C-layer. This procedure will preserve the oxygen lattice of the spinel structure and gives rise to a monolayer dispersion of vanadium on the surface as is shown in figure 3.b. The shielding of aluminium atoms, as observed in the LEIS experiments, is clearly visible in the figure. In the C-layer the tetrahedral interstices would be completely filled at 4.6 V atoms per nm^2 which corresponds to 14.6 wt% [13]. The end of the linear increase of the V signal however is reached at about 7 wt%. This indicates that only half of the tetrahedral interstices are filled. This behaviour could be accounted for by the fact that only one OH-group is available per octahedrally coordinated aluminium atom in the D-layer of the γ-Al_2O_3 [16]. The formation of dimers as observed by Haber et al. [3] at low coverage is supported by this model. In the next regime (7-11 wt%) the experiments show shielding of V atoms by additional V species without affecting the Al or O signal. The additional V atoms have an octahedral environment [9,10], which is consistent with the growth of a multilayered structure or lamellae as is shown in figure 3.c. The octahedral V species will shield the tetrahedrally coordinated V atoms from detection. From the size of the plateau in figure 2 it follows that only half the amount of vanadium which could be accommodated in the tetrahedral interstices can be deposited in the octahedral sites. This growth process supports observations by Inomata et al. [5], who detect the coverage of inactive tetrahedrally coordinated sites by active octahedral sites. When the loading is increased they detect further coverage of the surface until full coverage is reached. At loadings greater than 11 wt% the increase of the V signal and the simultaneous decrease of the Al signal can be explained by the coverage of the empty rows between the lamellae. Following the C-

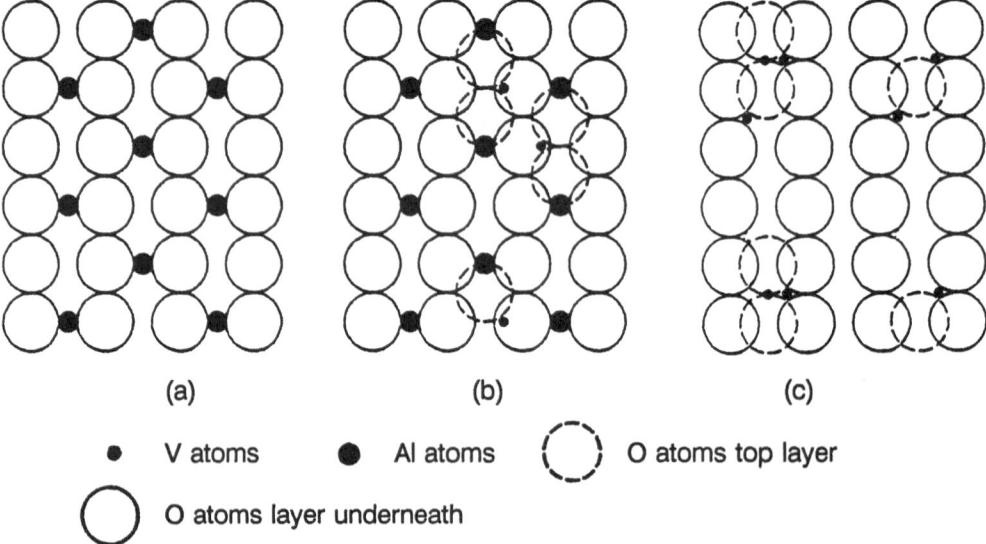

(a)   (b)   (c)

• V atoms   ● Al atoms   (◯) O atoms top layer

◯ O atoms layer underneath

Fig. 3. (a) D-layer of (110) plane, (b) continuation of the D-layer by V species, which occupy the tetrahedral sites of the complementary C-layer, (c) formation of a multilayered structure by deposition of octahedral V species on top of the tetrahedrally coordinated V atoms layer, which has left octahedrally favoured interstices, the oxygen lattice is preserved and a further increase of octahedral species will be the result. The proposed structure gives good agreement with the V-V distances presented by Bergeret et al. [11]. The difference is no more than a few hundredths of a nanometre.

4. CONCLUSION

The present investigations show how powerful a tool LEIS can be because of its ability of selective probing of the topmost atomic layer only. When combining these results with the results presented in literature one is able to build a consistent model for the surface structure of the V_2O_5/γ-Al_2O_3 catalyst when assuming the D-layer to be the predominant face of the support crystallites. Further investigations, which will consider the stability and thermodynamics of the proposed structure, should give a conclusive answer whether it is really possible to pinpoint the location of the atoms on the catalyst.

REFERENCES

[1]   G.C. van Leerdam, PhD. Thesis, T.U. Eindhoven (1991), and to be published
[2]   G.C. van Leerdam, PhD. Thesis, T.U. Eindhoven (1991), and to be published
[3]   Haber, A. Kozlowska, and R. Kozwolski, J. Catalysis 102(1986)52
[4]   J. Leyer, R. Margraf, E. Taglauer, and H. Knözinger, Surf. Sci. 201(1988)603
[5]   M. Inomata, K. Mori, A. Miyamoto, and Y. Murakami, J. Phys. Chem. 87(1983)761
[6]   M. Niwa, Y. Matsuoka, and Y. Murakami, J. Phys. Chem. 93(1989)3660
[7]   N.K. Nag, and F.E. Massoth, J. Catalysis 124(1990)127
[8]   F. Roozeboom, J. Medema, and P.J. Gellings, Z. Phys. Chem. N.F. 111(1978)215
[9]   F. Roozeboom, H.C. Mittelmeijer-Hazeleger, J.A. Moulijn, J. Medema, V.H.J. de Beer, and P.J. Gellings, J. Phys. Chem. 84(1980)2783
[10] H.Eckert,and I.E. Wachs, J. Phys. Chem. 93(1989)6796
[11] G. Bergeret, P. Gallezot, K.V.R. Chary, B. Rama Rao, V.S. Subrahmanyam, Appl. Cat. 40(1988)191
[12] F. Roozeboom, T. Fransen, P.Mars, and P.J. Gellings, Z. Anorg. All. Chemie 449(1979)25
[13] E. van der Heide, PhD. thesis, T.U. Delft (1990)
[14] H.H. Brongersma, N. Hazewindus, J.M. van Nieuwland, A.M.M. Otten, and A.J. Smets, Rev. Sci. Instrum. 49(6)(1978)707
[15] B.C. Lippens, and J.J. Steggerda, in: Physical and Chemical Aspects of Absorbents and Catalysts, B.G. Linsen, ed., Academic Press, New York, 1970, 171
[16] H. Knözinger, and P. Ratnasamy, Catal. Rev.-Sci. Eng. 17(1978)31

GROWTH OF PALLADIUM ON ZINC OXIDE SURFACES

C. A. Leighton, A.J. Swift and J.C.Vickerman

C.S.M.A., UMIST, P.O. Box 88, Manchester M60 1QD
United Kingdom

[1] INTRODUCTION

Metal oxide semiconductors, both pure and metal doped, play a key role in many of todays industries and, as such, it is desirable to seek a fuller understanding of the mechanisms governing the many processes in which these compounds are involved. Areas such as catalysis and gas sensing rely heavily on the use of metal oxide semiconductors, and it is the unusual surface properties which these compounds exhibit that allow them to be employed in such areas. Within these fields it has been found for several semiconductors that, through doping the surface of the semiconductor with a reactive metal, the sensitivity and selectivity of the system can be improved to a level greater than that of either of the two components. In order to improve our understanding of the processes occurring at the surface of the metal-semiconductor system, so called 'model systems' are employed. These involve the simulation of the real surface through the deposition of measured amounts of metal onto a single crystal surface of the semiconductor under Ultra High Vacuum conditions. In this way, surface sensitive spectroscopic techniques such as XPS, AES and SIMS can be used to examine surfaces prepared under conditions designed to minimise surface defects and contamination. The work described in this poster is an example of such a model system - that of palladium covered zinc oxide.

[2] EXPERIMENTAL

All experiments were carried out using an ESCA III (Vacuum Generators) comprising two chambers (analysis and preparation) and pumped by two hot oil diffusion

Fundamental Aspects of Heterogeneous Catalysis Studied by Particle Beams,
Edited by H.H. Brongersma and R.A. van Santen, Plenum Press, New York, 1991

pumps, an 8 ls^{-1} ion pump and a titanium sublimation pump [1]. After bakeout and subsequent filament degassing, base pressures in the analysis chamber of better than 5x10^{-9} mbar were attained. Samples were mounted on a tantalum sample stage, heating and cooling of the sample being effected by conduction from a filament and liquid nitrogen reservoir located within the probe tip. Temperature measurements on the sample were made via a chromel-alumel thermocouple junction also housed in the probe tip. All spectra were taken using 12kV Mg Ka radiation from a MK II source fitted with twin Mg/Al anodes and operated at 240W. The spectrometer was fitted with a concentric hemispherical analyser (150°) and constant analyser energy settings of 10 and 50 eV were employed for qualitative and quantitative work respectively. Calibration of the spectrometer was performed using the $2p_{3/2}$ peak from a clean zinc single crystal, giving a binding energy of 1021.45 eV, which agrees well with figures from reference [2], and a FWHM of 1.3 eV at a pass energy of 10 eV. Various energy windows were repeat scanned using an Apple IIe microcomputer to produce the final spectra.

For the palladium depositions, two different substrate surfaces were employed. Firstly, a zinc single crystal rod was cut in order to expose the basal (001) plane. The resulting discs were polished until features were <1/4 mm and then electrochemically cleaned in a mixture of methanol and concentrated nitric acid.

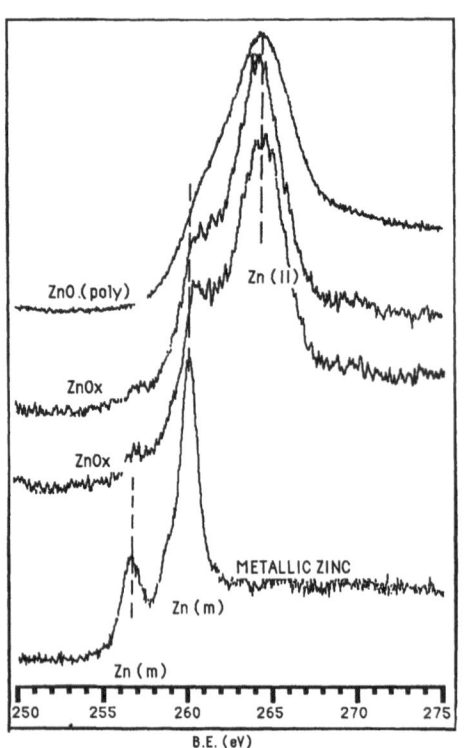

Figure 1. The zinc LMM Auger peak for Zn, ZnO and ZnOx.

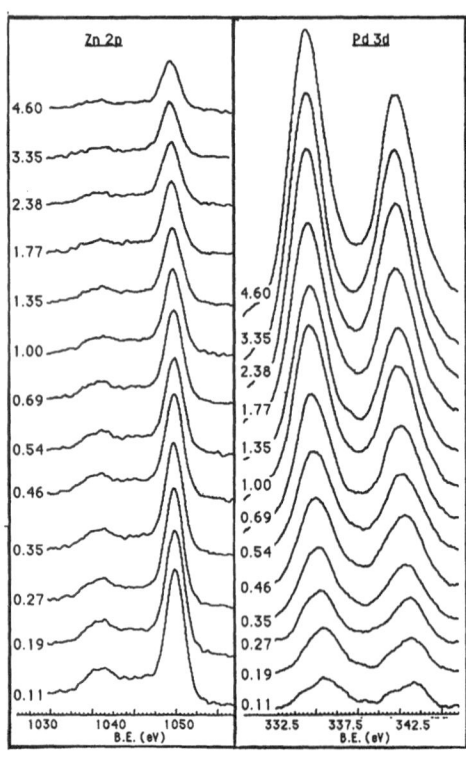

Figure 2. Zinc 2p and palladium 3d peaks for a number of coverages. Coverages in monolayer equivalents.

The resulting surface was placed in-vacuo and cleaned by repeated cycles of argon ion etching followed by annealing (423K). An oxidised zinc surface was then produced by heating the crystal to 573K in 10^{-4} mbar. O_2 for 30 mins. X-Ray induced Auger spectra confirmed that an oxide surface had been produced.

The second material employed as a substrate were discs of zinc oxide powder pressed under a pressure of 10 tons. The discs were placed in-vacuo and cleaned by cycles of argon ion etching and annealing in oxygen (10^{-5} mbar, 765K) to replace surface oxygen preferentially sputtered during etching.

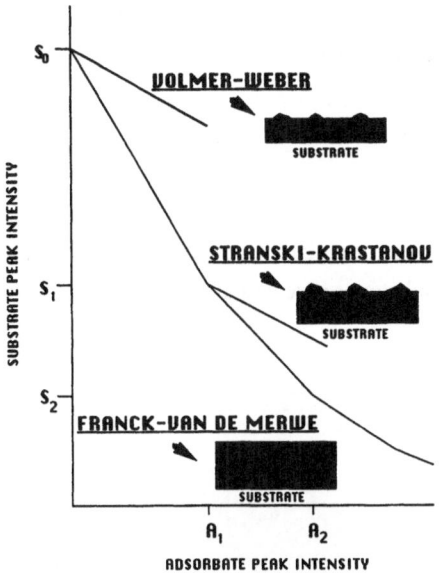

Figure 3. Model illustrating different growth systems.

[3] THE OXIDISED ZINC SURFACE

Figure 1 shows the zinc $L_3M_{4,5}M_{4,5}$ Auger transition for several oxide surfaces of zinc. The figure shows that oxide coverages have been generated on the zinc single crystal surfaces following the oxidation procedure described in reference [1]. The shoulders to the low binding energy side of the oxide peaks are attributed to metallic zinc lying at interstitial sites within the oxide lattice. Further oxidation produced no further decrease in the metallic shoulder and by taking angle dependent spectra, the concentration of metallic zinc interstitials was found to be constant throughout the sampling depth of Auger [3].

[4] DEPOSITION OF PALLADIUM

In the palladium doping experiments, the oxidised zinc surfaces, together with pressed polycrystalline discs, were employed as substrates and an investigation was undertaken to discover -

1. The reproducibility of the palladium source behaviour.
2. The growth characteristics of palladium on zinc supported zinc oxide thin films.
3. The effect of palladium coverages upon the chemical and electronic state of the surface.

For palladium doping of the sample surface a simple evaporation source was employed, based on a hot wire filament. The source was powered by a smoothed power supply capable of providing currents of up to 6A.

[5] RESULTS

Comparison of the results for palladium deposition on both the oxidised zinc and the polycrystalline zinc oxide substrates revealed differences in behaviour. Figure 2 shows

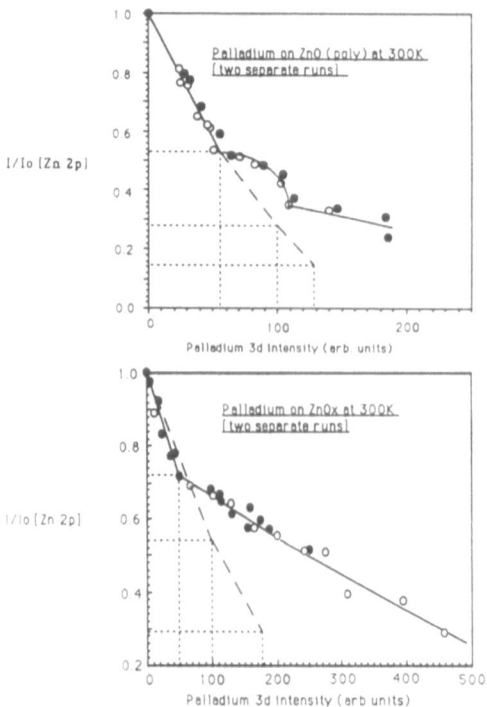

Figure 4. Attenuation of zinc signal with palladium coverage. Dotted lines represent model layer by layer growth.

the palladium 3d and zinc 2p$_{1/2}$ peaks for a gradation of palladium coverages on the oxidised zinc surface, and reveals a steady growth in the palladium peaks with exposure time. Also visible in the spectra is a negative shift in binding energy with increasing coverage which is discussed in section [6]. A corresponding shift in the zinc peak is not observed. Figure 3 illustrates how a plot of the substrate peak intensity against the adsorbate peak intensity can be utilised to attain information about the growth mechanism of the adsorbate. If the plot yields a straight line then the adsorbate grows on the surface as clusters. When a single break in gradient is observed, a Stranski-Krastanov growth mode is inferred (i.e. clusters forming on an initial monolayer) and more than one break in evidence indicates a layer by layer growth. Figure 4 plots the peak intensity for palladium against that of zinc for both oxide substrates. Clearly visible for polycrystalline zinc oxide is a break at Pd intensity of around 55, corresponding to completion of the first monolayer. When comparing the attenuation of the original zinc signal at this point with the theoretical value for a uniform covering of palladium as derived by Seah and Dench [4], good agreement is obtained (within 5%) indicating a simple monolayer with negligible clustering. The broken line in the figure indicates the theoretical behaviour of the plot after completion of the first monolayer using the same model as referenced above. Clearly, for a given zinc signal, the palladium signal for the experimental points is greater than that predicted by the model, suggesting that either palladium is diffusing into the oxide or that clustering of the palladium is occurring i.e. the deposition mode is Stranski-Krastanov. Previous studies of palladium on zinc oxide single crystal surfaces have differed in their conclusions regarding the growth mechanism of palladium [5,6].

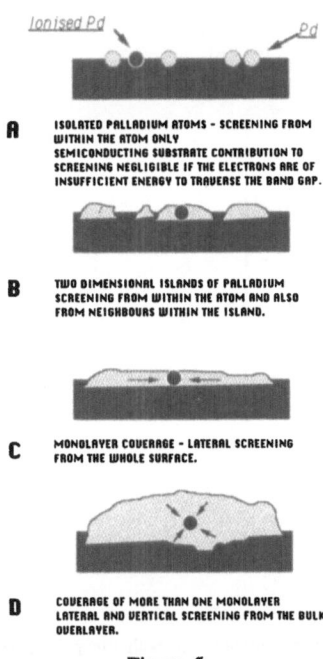

Figure 5.

For the ZnOx surface, the zinc signal is seen to drop linearly with coverage until the first monolayer is half completed, whereupon a sharp drop in gradient is observed indicating that large scale clustering is occurring.

[6] SURFACE REACTION

The possible surface reaction of the palladium and zinc oxide has also been addressed. As seen in section [5] (figure 2) a shift to lower binding energy of the palladium 3d peaks is observed with increasing palladium coverage, from a value at low coverages characteristic of oxidised palladium to a metallic-like position at higher coverages. However in addition to the chemical effects occurring on the surface, we must also consider surface electronic relaxation effects [7]. XPS and Auger processes, by their very nature, leave the atom momentarily in a positively charged state. Thus for an isolated atom, the positive hole(s) left by the photoemission process can only be shielded from the photoelectron by electrons from within the atom itself. A diagram to illustrate this feature is shown in figure 5.

With the completion of the first monolayer, subsequently deposited palladium atoms will have the benefit of screening not only laterally but from above and below as well. The kinetic energies of the substrate photoelectrons is predicted, therefore, to continue to fall until almost all the atoms contributing to the palladium signal are in a bulk metal environment.

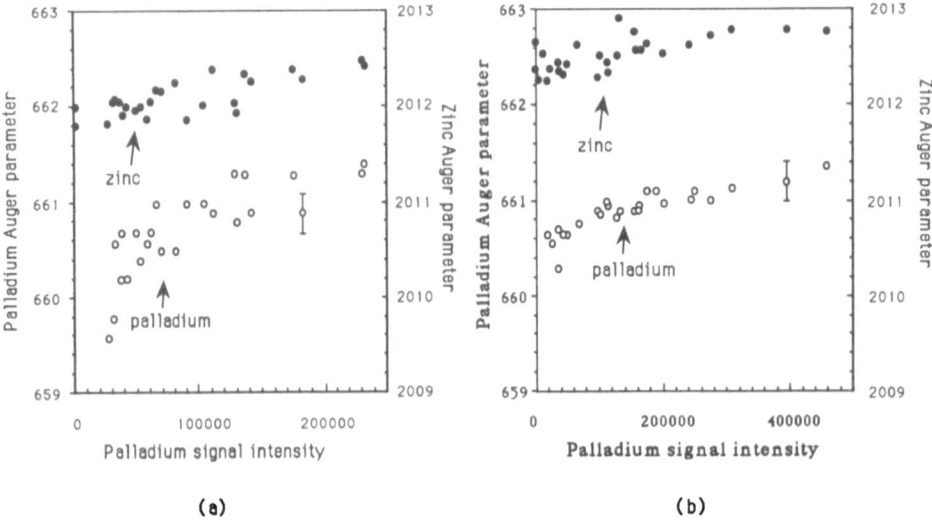

Figure 6. Palladium and zinc auger parameters for both the polycrystalline (a) and ZnOx surfaces (b).

For semiconductors and insulators, shifts in peak binding energy can occur due to charge build up on the surface of the sample. By utilising the Auger parameter [8], the difference in energy between the most intense XPS and Auger peaks), changes observed will be independent of any static charging. Static charging can be a problem when dealing with insulators and semiconductors. Charge formed through the photoemission process is unable to leak away and thus builds up on the sample surface. Because this charge is spread across the whole surface and not localised in any particular atom, all departing photoelectrons will be retarded (and thus peaks shifted) by the same amount. By taking the difference between two peaks, this contribution to the shift can be eliminated. Figure 6. shows the Auger parameter for both zinc and palladium as plotted against palladium signal. Similar shifts are seen for both palladium and zinc on both surfaces, although the phenomena producing the shifts are different. For palladium, the shift is due to electronic effects such as relaxation. However for zinc, the shift is due to reduction of the oxide surface.

[7] EVIDENCE OF ALLOYING

As mentioned in section [3], for zinc oxide there exists a shoulder on the zinc $L_3M_{4,5}M_{4,5}$ Auger peak due to metallic zinc present in interstices. By differentiating the peak and taking the ratio of peak to peak heights, the relative contribution of the metallic and oxidic zinc can be ascertained. For the deposition of 1ML of palladium, the metallic contribution was seen to increase by 20% relative to the oxidic contribution for both types of substrate, indicating that the metallic underlayer on the oxidised zinc substrate does not effect this behaviour. The fact that an increase in metallic zinc is observed could be a consequence of the reduction of the oxide by palladium, however no evidence for the oxidation of palladium (in the form of core level broadening/splitting or Auger peak shape changes) was found. Alternatively, a migration of zinc into the palladium overlayer seems likely, with alloy formation occurring. Indeed this is supported by the fact that at almost saturation palladium coverages, the zinc signal that persisted was entirely metallic. Deposition of palladium onto a zinc (001) surface resulted in splitting of the palladium peaks, indicating that some degree of alloying was occurring. Evidence of alloy formation has been presented for the deposition of palladium on tin oxide [9,10] although the phenomenon has previously not been seen for the palladium-zinc oxide system.

[8] CONCLUSION

Palladium has been deposited onto both polycrystalline and oxidised zinc surfaces and has shown differing physical and similar chemical and electronic properties. Growth mode has been found to be Stranski-Krastanov, with a greater degree of clustering occurring on the ZnOx surface. A shift in the palladium core level peak with increasing palladium coverage has been observed as predicted. No evidence for the oxidation of the deposited palladium has been found, however it is thought that some degree of alloying is occurring between the substrate and adsorbate.

[9] REFERENCES

1. A.J. Swift, Ph. D. thesis (1988) UMIST.
2. K.S. Kim et al., Chem. Phys. Lett., **41** (1976) p503.
3. C.A. Leighton, A.J. Swift and J.C. Vickerman, unpublished results.
4. M.P. Seah and W.A. Dench, Surf. Int. Anal., **1** (1979) p2.
5. H. Jacobs et al., Surf. Sci. **160** (1985) p217.
6. W. Gaebler, K. Jacobi and W. Ranke, Surf. Sci. **75** (1978) p355.
7. S. Kohiki, Appl. Surf. Sci. **25** (1986) p81.
8. C.D. Wagner, Trans. Faraday Soc. **60** (1975) p291.
9. R. Hück, D. Kohl and G. Heiland, International Symposium on Trends and New Applications in Thin Films, Strasbourg, 1987.
10. T.B. Fryberger, J.W. Erickson and S. Semancik, Surf. Int. Anal. **14** (1989) p83.

ION SCATTERING ANALYSIS OF Rh
ADLAYERS ON ALUMINA FILMS

Christian Linsmeier[1,2], Edmund Taglauer[1], Helmut Knözinger[2]

[1] Max-Planck-Institut für Plasmaphysik, EURATOM-Association
D-8046 Garching/München, Fed. Rep. of Germany

[2] Institut für Physikalische Chemie der Universität München
Sophienstraße 11, D-8000 München 2, Fed. Rep. of Germany

INTRODUCTION

Rhodium shows catalytic activities for a variety of chemical reactions. Its flexibility, as regards reactivity in CO-hydrogenation, makes it very interesting for the examination of catalytic mechanisms [1]. Moreover, rhodium and other group VIII metals, have been found to show strong interactions between the metal and typically reducible oxidic supports such as TiO_2 (SMSI, strong metal-support interactions) [2]. Similar effects have also been reported with non-reducible supports (Al_2O_3) [3,4].

To study systems with surface-sensitive methods using charged particles, it is convenient or even necessary that the samples show sufficient electrical conductivity. In this work, thin oxide films prepared by anodic oxidation of aluminium exhibited no disturbing charging effects. Model systems for Rh/Al_2O_3 catalysts with metal loadings of one monolayer were prepared and characterized by ion scattering spectroscopy (ISS) and Rutherford backscattering spectroscopy (RBS).

EXPERIMENTAL SECTION

The alumina films were prepared by anodic oxidation of high purity (99.999 %) Al foils of dimensions 10×14 mm^2 and thickness 1 mm. The metal was polished using diamond polish, the grain size of which was decreased from $45\mu m$ to $1\mu m$, followed by an additional treatment with 0.05 μm γ-Al_2O_3 on pellon discs. An aqueous solution of 3% ammoniumcitrate was used as electrolyte, in which the formed alumina is insoluble. The anodic oxidation was carried out at a constant current density of approximately 8 mA/cm^2, until the barrier voltage reached 10 V. According to earlier results, the formed barrier layer is expected to have a thickness of about 137 Å and is amorphous or of γ-Al_2O_3 structure [5,6,7]. The films were dried at 100 °C for 4.5 hours, some samples were calcined in air at 550 °C for 3.5 hours. The thermal treatment is carried out in order to get a conversion from amorphous to γ-Al_2O_3, which takes place at 400–500 °C [8].

The rhodium (MARZ grade, 99.99 %) was evaporated by electron heating in a uhv chamber with a base pressure of $3 \cdot 10^{-8}$ hPa. The growth rate was measured with a quartz balance next to the specimens with a value of 0.04–0.05 Å$/s^{-1}$. In 60 s 2.4–3.0 Å were deposited, which corresponds to a thickness of one monolayer of metal.

Fundamental Aspects of Heterogeneous Catalysis Studied by Particle Beams,
Edited by H.H. Brongersma and R.A. van Santen, Plenum Press, New York, 1991

The measurements for ion scattering spectroscopy (ISS) were carried out in a uhv chamber with a base pressure of $1 \cdot 10^{-11}$ hPa equipped with a cylindrical mirror analyzer (scattering angle 137°) with an integral ion gun [9]. He^+ was used at three different energies (500 eV, 1 and 2 keV) and Ne^+ at 1 keV. The pressure of noble gas during the measurements was $7 \cdot 10^{-7}$ hPa, the current density 3–7 $\mu A/cm^2$. The electrical conductivity of the thin alumina films was sufficently high that no charging effects could be observed. The different energies and projectile masses caused a different sputtering yield and surface sensitivity, so depth profiles could be measured. No cleaning steps were carried out prior to analysis in order to not destroy the thin metal film.

Rutherford backscattering spectroscopy (RBS) was performed at a van de Graaf accelerator using $^4He^+$ with a primary energy of 1 MeV, perpendicular incidence and scattering angle of 165°. The detector resolution was measured using several elemental targets (Si, Si with a Au adlayer of 1000 Å thickness, Rh) and was 12.8 keV. Due to the detection of both ions and neutrals, RBS is a quantitative method without the need of standards. In taking into account the scattering cross-sections and the experimental parameters, the absolute coverage of rhodium and the amount of oxygen up to a depth of several 1000 Å can be calculated from the RBS spectra.

RESULTS AND DISCUSSION

Ion Scattering Spectroscopy

ISS is a technique which is sensitive only for the outermost atomic layer, especially, if noble gas ions are used as projectiles [9,10]. Therefore, the chemical composition of the surface can be determined. The examined samples showed only the expected elements — oxygen, aluminium and rhodium — in the spectra with helium at 500 and 1000 eV. With a primary energy of 2000 eV, carbon could be detected, which was found to be a contamination in AES studies [11]. No sodium or nitrogen was measured, so no contamination is introduced during the sample preparation. The carbon signal is caused by hydrocarbons, which are unavoidable during the sample transport in air. These hydrocarbons also explain the increase in the intensity of all ISS signals in the first minutes, which can be seen in all spectra. Due to the mass ratio, there is a good energy transfer from the projectile (He^+) to hydrogen, so that the main part of the primary energy in the beginning of a depth profile is transferred to hydrogen atoms. No backscattering signal is possible for $He^+ \rightarrow H$, and a detection of hydrogen recoil-ions is also impossible in the applied scattering geometry. During the sputtering of the contaminants, the intensities of all signals increase.

Fig. 1 shows a series of spectra of a depth profile, taken with a primary energy of 500 eV. With increasing fluence, a different development of the signals can be seen. While the aluminium and oxygen peaks increase, the rhodium signal reaches a maximum at $8.7 \cdot 10^{16}$ He^+/cm^2 and then decreases. This indicates that the surface layer of the metal is sputtered away, while the amount of detectable alumina increases. Fig. 2a shows the depth profile corresponding to fig. 1. The specimen is not anodically oxidized, the detected oxygen results from oxidation by air. This film is relatively thin and the measured composition changes gradually. A fluence of $27 \cdot 10^{16}$ He^+/cm^2 corresponds to 12–13 monolayers of γ-Al_2O_3, calculated from experimental sputtering yields [12]. Since the neutralisation terms for Al and O in alumina are unknown, it is impossible to calculate a stoichiometry for this layer. In fig. 2b the intensity ratio of O and Al is shown for two different positions on the sample. The two lines are close together, there is hardly any difference in the composition over the sample. It is also obvious that the composition of the oxide, formed by oxidation in air, varies over the whole depth examined. The strong decrease in the beginning, up to a fluence of about $3–4 \cdot 10^{16}$ He^+/cm^2, is caused by the sputtering of the contaminants. The high oxygen content can be explained by adsorbed O-containing species from the air. Beginning at this fluence, the changes in the ratio of the intensities reflect the alteration in the stoichiometry of the alumina.

Additional to the changes in stoichiometry, two other effects can be observed, which are caused by the ion beam. Fig. 3 shows the depth profile (3a) and the ratio of intensities O/Al (3b) for an anodically oxidized film without a rhodium overlayer. The spectra are taken with helium at 1000 eV. The first effect is the roughening of the sample surface, induced by the impinging ions [13]. This roughening leads to a decrease in the intensity. In the depth profile, a decrease of the oxygen signal starts at a fluence of $11 \cdot 10^{16}$ He$^+$/cm^2, followed by the aluminium signal at $20 \cdot 10^{16}$ He$^+$/cm^2. Even though both intensities decrease, the unoxidized aluminium is not yet reached and a change in the composition towards metallic aluminium can be excluded. The decrease in the ratio of the O and Al intensities rather shows the influence of preferential sputtering, although the difference of the masses of the two elements is small [14].

With a higher primary energy, a greater depth of the sample can be examined, because the sputtering yield increases with energy. Using helium at 2000 eV, calcined and uncalcined films with an overlayer of rhodium give different results (fig. 4). The uncalcined film (4a) is much thinner, the depth profile of oxygen has a maximum at $20 \cdot 10^{16}$ He$^+$/cm^2. The slopes of the aluminium and oxygen profiles change at $35 \cdot 10^{16}$ He$^+$/cm^2, which indicates the end of the alumina layer and the beginning of the metallic material. This value corresponds to a thickness of about 32 monolayers, taking into account the hydrocarbon contamination. In contrast to this, the calcined sample (4b) shows no change in the aluminium profile. Only the oxygen signals decrease slightly, which demonstrates the influence of preferential sputtering. The difference in the alumina signals between the calcined and the uncalcined samples suggests the additional formation of aluminium oxide during the calcination procedure. This explanation is supported by the RBS results (see below), in which the calcined samples show about twice the oxygen content than the uncalcined ones. In both cases, the rhodium signal has a maximum at $7 \cdot 10^{16}$ He$^+$/cm^2 and shows an exponential decay. This is explained by three possible

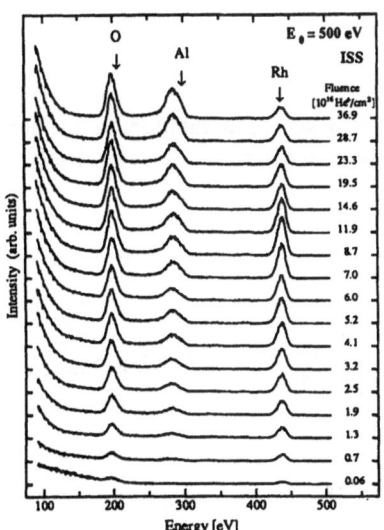

Figure 1. ISS spectra of an alumina film, formed by oxidation in air, with an overlayer of rhodium. To the right, the fluence at the beginning of the spectra is noted. The arrows indicate the peak positions, calculated with the binary collision model.

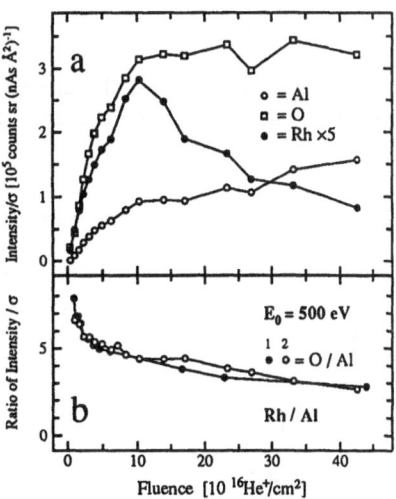

Figure 2. ISS depth profile of an air-oxidized alumina film with a rhodium overlayer of one monolayer (a) and the corresponding intensity ratios of oxygen and aluminium at two different positions on the sample (b).

415

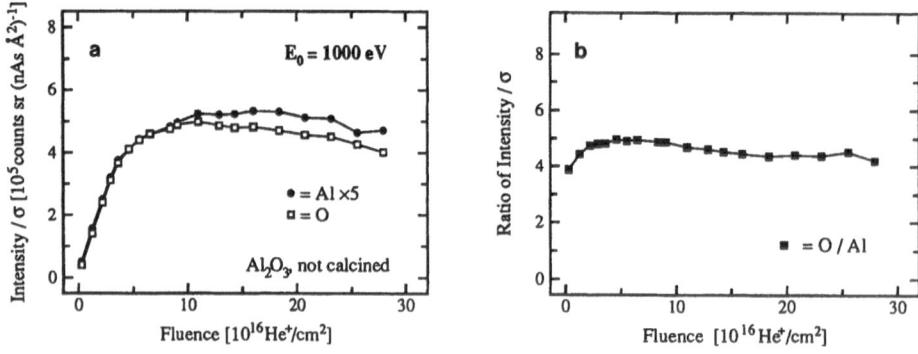

Figure 3. ISS depth profile (a) and ratio of O and Al intensities (b) for an anodically oxidized alumina film.

mechanisms. The first possibility for rhodium signals from deeper layers than the first one are implanted metal atoms by the impinging ions (recoil implantation). Secondly, the crater effect, caused by the Gaussian profile of the ion beam, always leaves some higher layers at the edge of the beam spot, while deeper layers in the centre contribute to the signal. The third mechanism is redeposition of sputtered atoms. Mainly the crater effect is responsible for the limitation of the depth resolution in this case.

The depth profiles with helium at the three applied energies show no slope changes in the aluminium and oxygen curves after the removal of rhodium. Therefore, the real area of alumina is much greater than the geometric one, for which the amount of rhodium was calculated to form one monolayer. The detected amount of metal (see below) does not cover the oxide surface completely, so the support signals are independent of the overlayer signals. This shows, that, despite the polishing of the aluminium surface, the oxide surface increases drastically during anodic oxidation.

Rutherford Backscattering Spectroscopy

From the RBS spectra, the absolute amounts of oxygen and rhodium in the surface region of the samples can be determined. It is not possible to determine the layer depths from

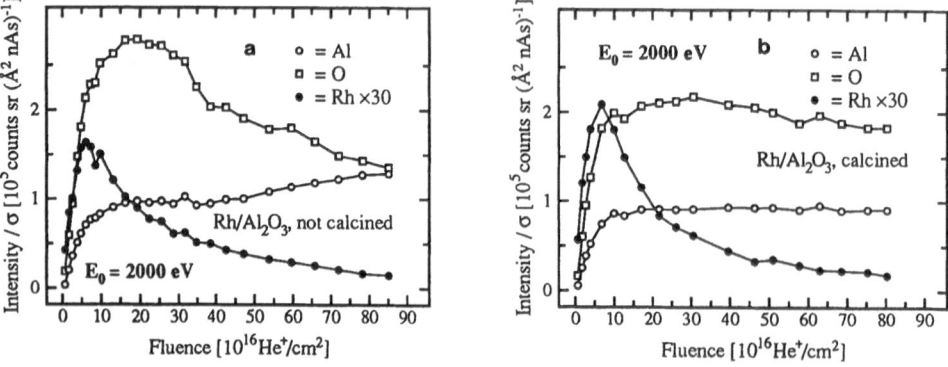

Figure 4. ISS depth profiles of an uncalcined (a) and a calcined (b) sample of alumina films with a rhodium overlayer. Maximum fluence corresponds to a depth of approximately 80 monolayers, calculated for γ-alumina.

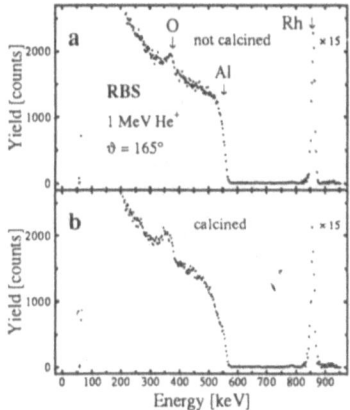

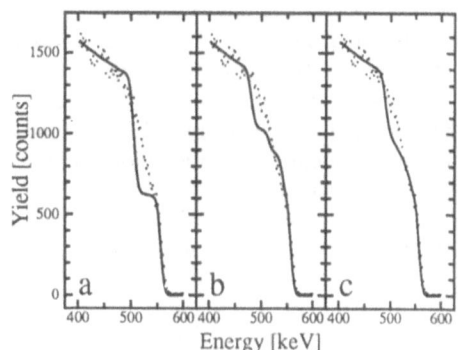

Figure 5. RBS spectra of an uncalcined (a) and a calcined (b) sample of Rh/Al_2O_3 model catalysts with an adlayer of 1 ML rhodium. The rhodium signal is enhanced.

Figure 6. Al edges of the calcined sample. The dots are the the measured spectrum, the line is calculated by a simulation using different oxide layer compositions (see text).

the energy dispersion, even for the alumina, because the energy resolution of the analyzer is too coarse to resolve the energy differences caused by these small distribution depths. For the rhodium layer, a thickness between 2.1 and 2.4 Å was calculated using the metallic density, a value of almost one monolayer (2.7 Å). Inside the beam spots produced by ion bombardment during ISS measurements, the depth varies between 0.3 and 0.7 Å. It is impossible to examine separate positions inside the spots, because the beam size for ISS is comparable to that of the RBS beam. As already mentioned, the oxygen amount of the calcined and uncalcined samples varies by a factor of 2. Calculated with the density for γ-Al_2O_3, the thicknesses are 600 Å and 300 Å, respectively.

Fig. 5 shows the spectra for an uncalcined (5a) and a calcined (5b) sample. They differ in the shape of the aluminium edge, which is influenced by the oxidic layer. In the uncalcined case, the signal rises without a visible step, only with a slight change in the slope at nearly half the height, which corresponds to a stoichiometric Al_2O_3 layer. In contrast to that, the calcined sample shows a steep rise until the Al_2O_3 height, which is in agreement with the detector resolution, a small step and then a continuous decrease in slope until the top of the Al signal.

Although the distribution depth of oxygen is too small to be ascertained by the energy dispersion, an approach is possible via the slope of the elemental edge of aluminium in the energy spectrum. Fig. 6 shows the spectrum of the calcined sample (dots), compared to three simulated edges. The simulations take into account all experimental parameters, so no fit to the measured data is necessary. In the three calculated curves, the total amount of oxygen is held constant to a value which was determined from the oxygen peak in the measured spectrum (fig. 5). The calculated edges therefore differ only in the distribution of the oxygen over the alumina layers. In fig. 6a, a homogeneous Al_2O_3 layer is assumed, which leads to a steep rise of the aluminium edge (557 keV), followed by a step corresponding to an oxygen mole fraction of $x_O = 0.6$. In fig. 6b and 6c, the oxygen is distributed over 3 and 6 layers with compositions of $x_O = 0.6, 0.4, 0.3$ and $x_O = 0.6, 0.6, 0.5, 0.4, 0.38, 0.33$, respectively. As it is shown by this series of simulations, the measured edge is approximated best by a continuous oxygen gradient. The minimum amount of atoms per layer is restricted in the simulation programme, so no further refinement was possible.

CONCLUSIONS

It could be shown that alumina layers up to some hundred Å thickness show a sufficiently high electrical conductivity to be examined with ion beam methods. The surface of the anodically prepared oxide films is much higher than the geometric area of the specimen. Their thickness can be increased by calcination of the samples. The distribution of oxygen in the calcined alumina layer is inhomogeneous, the stoichiometry changing continuously from Al_2O_3 to metallic aluminium with increasing depth. The metal is uniformly distributed over the samples, no indication of clustering or island formation was found. Due to the large oxide surface, the rhodium does not cover the support completely.

ACKNOWLEDGEMENTS

The support of B.M.U. Scherzer and R. Siegele is gratefully acknowledged. The authors are also grateful to John Davies for his advise to improve the RBS part of the manuscript. This work was carried out under the framework of Sonderforschungsbereich 338 of the Deutsche Forschungsgemeinschaft.

REFERENCES

[1] Knözinger H., Proc. 5th Int. Symp. on Rel. between Homogeneous and Heterogeneous Catalysis,
    Novosibirsk (1986)
[2] Stevenson S.A., Dumesic J.A., Baker R.T.K., Ruckenstein E., '*Metal-Support Interactions in
    Catalysis, Sintering and Redispersion*', Van Nostrand Reinhold's Catalysis Series, New York (1987)
[3] Cairns J.A., Baglin J.E.E., Clark G.J., Ziegler J.F., J. Catal. **83**, 301 (1983)
[4] Den Otter G.J., Dautzenberg F.M., J. Catal. **53**, 116 (1978)
[5] Cocke D. L., Johnson E. D., Merrill R. P., Catal. Rev.—Sci. Eng. **26(2)**, 163 (1984)
[6] Tajima S., Adv. Corros. Sci. Technol. (USA), **1**, 229 (1970)
[7] Walkenhorst W., Naturwiss. **12**, 373 (1947)
[8] Hönicke D., Appl. Catal. **5**, 179 (1983)
[9] Taglauer E., Heiland W., in: Applied Surface Analysis, eds. T.L. Barr, L.E. Davis, ASTM STP
    669, Philadelphia (1980), p. 111
[10] Taglauer E., Appl. Phys. **A38**, 161 (1985)
[11] Linsmeier Ch., Taglauer E., Knözinger H., to be published
[12] Roth J., Bohdansky J., Ottenberger W., IPP-Report 9/26 (1979)
[13] Margraf R., Knözinger H., Taglauer E., Surf. Sci. **211/212**, 1083 (1989)
[14] Taglauer E., Heiland W., in: Proc. Int. Symp. on Sputtering, eds. P. Varga, G. Betz, F.P. Vieh-
     böck, Inst. Allg. Phys. TU Wien (1980)

ENERGY-DEPENDENT EFFECTS IN LOW ENERGY
ION SCATTERING FROM TiO_2, Ti, and H_2O ICE

B. L. Maschhoff, J-M. Pan, R. A. Baragiola[*], and T. E. Madey

Dept. of Physics and Astronomy, Rutgers, The State University of New Jersey
Piscataway, NJ 08855 USA

Abstract

Low energy He^+ ion scattering (LEIS) studies have been performed on TiO_2(110), polycrystalline Ti, and Ti with oxide and water overlayers. It is found that the energy dependence of the O/Ti LEIS intensity ratio for TiO_2 is due to ion neutralization and/or reionization and is not the result of shadowing effects. The studies of water covered surfaces indicate attenuation of LEIS intensities due to the presence of hydrogen atoms.

Introduction

Low energy ion scattering (LEIS) is a useful technique for the determination of surface composition [1]. It is often of particular utility in catalytic studies, since the structure of the topmost layer is critical in determining the chemical activity of the surface. A number of LEIS studies have been performed on oxide systems, due in part to their widespread use as catalysts and catalyst support media [2,3]. Quantitative interpretation of the data is often difficult because the relative contributions of structural and electronic effects in LEIS are not well known [4,5].

An important structural effect in LEIS is the shadowing of incident ion intensity from one near-surface atom by another. For near normal ion incidence and large scattering angles, this effect is most significant when the "shadowing" atom lies above the plane containing the atom of interest [1]. The extent of the shadowing is also determined by the strength of the repulsive potential between the incident ion and the shadowing atom, and is predicted to decrease as the incident ion kinetic energy is increased. Indeed, several workers have observed that the O/metal He^+ scattering yield ratio for a variety of oxides is strongly dependent on the incident ion energy; a decrease is commonly observed with increasing energy [6-8]. The conclusion of the majority of the studies into this effect is that structural factors are the likely cause. Similar energy-dependent behavior has also been observed by Taglauer and Heiland for adsorbed S and O on Ni(111) [9]. Van Leerdam and Brongersma [10], however, have recently shown that, at least for SiO_2, the energy dependence of the yield ratio more likely results from energy-dependent differences in the relative ion survival probabilities for scattering from the respective atoms (i.e. the trend is characteristic of the isolated atoms) rather than from surface structural effects. Their conclusion is based on a comparison of absolute scattering yields for Si and O in different matrices.

In this work, structural effects in low energy He^+ ion scattering from TiO_2 are investigated. Comparisons are made between LEIS results for TiO_2(110), polycrystalline Ti, and Ti exposed to oxygen or water vapor. It is found that while the O/Ti scattering yield ratios for TiO_2 and oxygen-exposed Ti are strongly energy dependent, the trend is not explained by an energy-dependent shadowing effect. We also observe that the absolute He^+ LEIS yields from O atoms in submonolayer to multilayer

Fundamental Aspects of Heterogeneous Catalysis Studied by Particle Beams,
Edited by H.H. Brongersma and R.A. van Santen, Plenum Press, New York, 1991

419

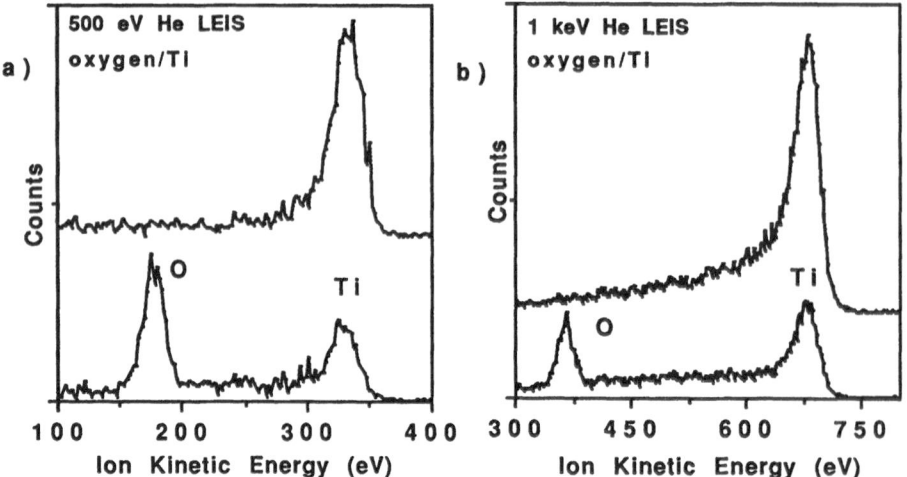

Fig 1 He+ LEIS for clean Ti metal (polycrystalline) and following 40L O$_2$ exposure
a) 500 eV incident ion energy, b) 1000 eV. The upper spectra are displaced vertically
for clarity

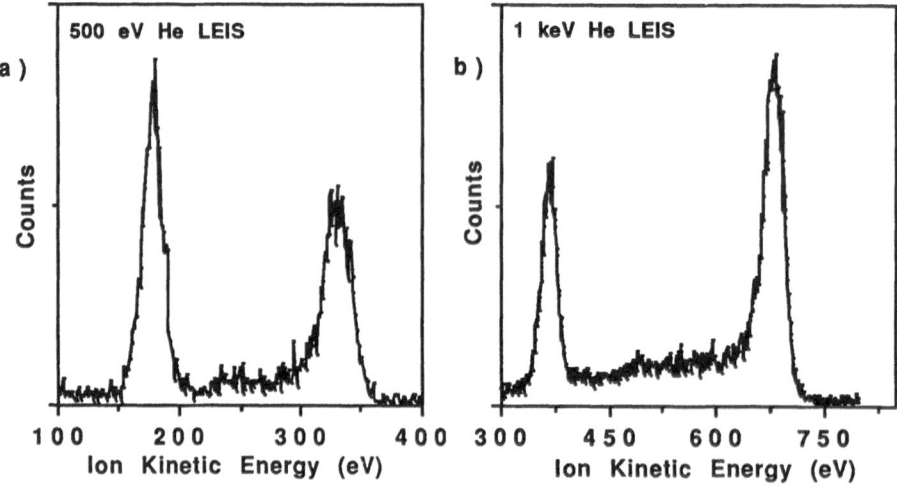

Fig. 2 He+ LEIS for TiO$_2$(110) obtained at a) 500 eV incident energy, b) 1000 eV.

films of H_2O on Ti are rather low, indicating that hydrogen in the vicinity of a target atom has an unexpectedly large effect on the backscattering yield.

Experimental

These studies are performed in an ultrahigh vacuum apparatus with a base pressure of 2 x 10^{-10} Torr. XPS and LEED are used for sample characterization. A Perkin Elmer model 04-161 ion gun and a Vacuum Science Workshop HA-100 hemispherical energy analyzer (fixed retard ratio=5) are used for the low energy ion scattering measurements. The typical He$^+$ ion flux used is 100 nA/cm^2 with sample bombardment times of 1-5 min. during analysis. Beam damage is not appreciable during analysis. The fixed scattering angle for these measurements is 140°. The angle of incidence is approximately 20° from the normal direction.

$TiO_2(110)$ surfaces (Commercial Crystal Laboratories) were prepared by Ar$^+$ ion etching and annealing in oxygen to 1000K, resulting in a Ti2p XPS spectrum characteristic of the fully oxidized surface. Clean Ti metal surfaces were prepared by evaporating Ti onto annealed polycrystalline Ti foil at room temperature. O_2 and H_2O were introduced into the vacuum chamber using a leak valve, and partial pressures measured using an uncalibrated ion gauge. H_2O was purified using several freeze/pump/thaw cycles prior to use.

The spectra in figs. 1-3 are uncorrected data. For the calculation of O/Ti ratios, the raw data are corrected for analyzer transmission and differential scattering cross section. The transmission efficiency of the hemispherical analyzer for the FRR mode is assumed to be proportional to kinetic energy. Scattering cross section values for O and Ti are obtained using the Moliere potential with a reduced screening length [1].

Results and Discussion
LEIS from Ti and Ti/O_2

Shown in fig. 1a are the results obtained for 500 eV He+ scattering from clean Ti metal (polycrystalline) and for the same surface exposed to 40 L O_2 (1L = 1 x 10^{-6} Torr·sec). Shown in fig. 1b are the results obtained for 1000 eV incident energy. The asymmetry of the Ti peak for the clean metal is more pronounced than for the O_2-exposed surface although additional structure between the Ti and O peaks for the oxide spectrum might partially obscure this effect. The exposure of the metal surface to oxygen causes strong attenuation of the Ti peak. The resulting O/Ti peak ratio is similar to that obtained for $TiO_2(110)$, as shown in fig. 2. Similar energy-dependent behavior is also observed. Interestingly, the percentage attenuation of the Ti peak intensity is nearly identical for the two different energies in fig 1. Evidently, the shadowing (attenuation) effect of oxygen on titanium in the surface layer depends little on the incident ion energy. This result is consistent with the observations of van Leerdam and Brongersma for SiO_2; the O/Ti ratio energy dependence observed here is apparently unrelated to geometric effects.

LEIS from H_2O overlayers on Ti

As demonstrated previously [10], it is useful to compare the energy-dependent behavior of LEIS from TiO_2 with that for Ti and O atoms independently . The data for Ti are readily obtained from the clean metal; similar data for oxygen atoms were obtained from H_2O thin films formed on Ti. Shown in fig. 3 are LEIS results at 1 keV for clean Ti, Ti exposed to 100 L H_2O at 300K, and 10 L H_2O exposure at < 150 K. At room temperature, H_2O dissociates on clean Ti to produce adsorbed H and OH [11]. At lower temperatures, H_2O multilayers can be formed on the surface. It is estimated that 5-10 layers are formed on the surface from the 10 L exposure. For the 300K exposure, the extent of attenuation of the Ti peak is similar to that for the 40 L O_2 exposure. The resulting O peak intensity for water exposure, however, is substantially less than for O_2. One possible explanation for this effect is that the coverage of OH on the surface (for H_2O/Ti) is less than O (for O_2/Ti). Additional attenuation of the Ti intensity could be due to H adsorbed on Ti. We have found that H_2 exposure to clean Ti does result in slight attenuation of the Ti LEIS peak, but the effect is small [12]. It has been suggested that the H adsorption site on Ti(001) is is either a threefold hollow or subsurface[13], thus shadowing is not expected to be large. A more likely explanation is that the O is partially shielded by H, since the O-H bond distance is reasonably short (0.1 nm) and the H is expected to lie above (although not necessarily directly atop) the O. ESDIAD measurements for H_2O on single crystal Ti surfaces should clarify the geometric arrangement [14].

The LEIS data for multilayer H_2O (fig. 3) are consistent with this explanation, in that the shielding of O in molecular H_2O is expected to be larger due to the proximity of two hydrogens. For a direct comparison of intensities, it is necessary to consider differences in the surface density of O between TiO_2 and H_2O. The TiO_2(110) surface (the most stable face) has a O atom density of ca. 1.5×10^{15} atoms/cm^2 whereas the surface O atom density in H_2O is ca. 1.04×10^{15} atoms/cm^2. Assuming that scattering from second layer atoms is not observed, a 33% decrease in the yield could be expected. The results in fig. 2 suggest a decrease greater than 50%. This additional decrease is thought to be due to shielding by H.

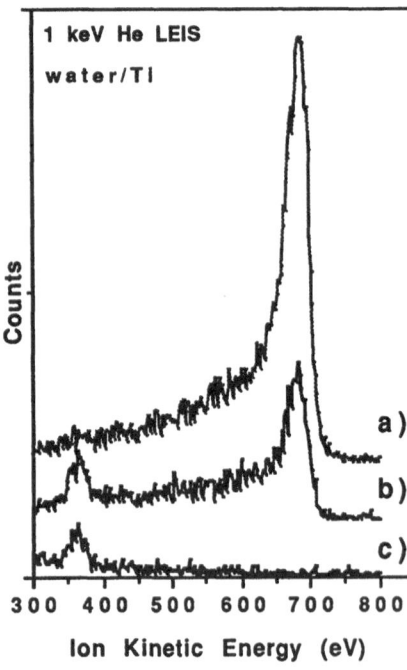

Fig. 3 1keV LEIS results for: a)Ti metal, b) 100L H_2O@300K, c) 10L H_2O@150K

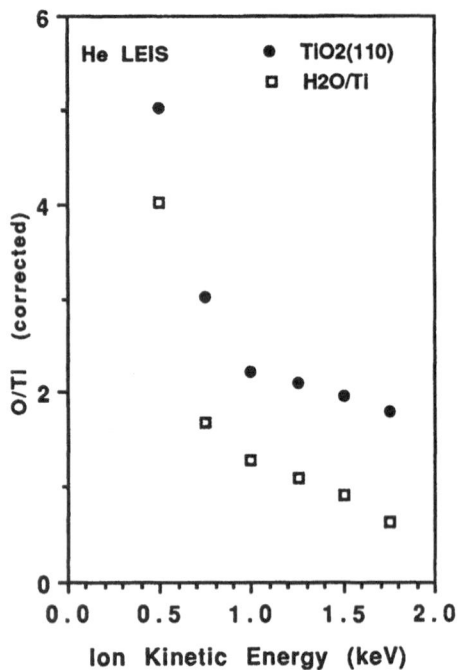

Fig. 4 Corrected O/Ti LEIS intensity ratios from TiO_2(110) and from O (H_2O overlayer) vs. Ti (metal).

Energy dependence of O/Ti LEIS ratio

The incident He^+ energy dependence of the O/Ti scattering yield ratio was measured for TiO_2(110) in the range 0.5 - 1.75 keV. The data were corrected for the analyzer transmission function and the differential scattering cross section as described in the experimental section. The results are shown in fig. 4 with the solid circles. Energy dependent scattering yields for Ti and O were also obtained from the clean Ti and H_2O multilayer surfaces respectively; results are shown by the open squares in fig. 4. These data were also corrected as described above, but no further normalization was performed. The surface area and experimental orientation of the two surfaces was identical such that direct comparison is meaningful. Both sets of data clearly indicate a strong decrease in the O/Ti ratio with incident He^+ energy. The lower O/Ti ratios for the Ti/H_2O could be a result of a reduced surface O density, shadowing by H atoms, or electronic effects (ion neutralization probability). It should also be noted that integration of the Ti peak, particularly for the clean metal surface, is rather problematic.

The similarity of the two sets of data indicates that geometric factors do not account for the energy dependence of the yield ratios. Thus, differences in the ion survival probabilities (and their energy dependences) are indicated. Neutralization occurs either by a resonant mechanism or an Auger process [15]. The data in fig. 1 show the the LEIS Ti peak intensity increases approximately by the amount expected upon increasing the surface density of Ti from that in TiO_2 to that for Ti metal. This indicates that the neutralization probability of He^+ incident on Ti is not enhanced by increasing the local valence electron density. We therefore believe that an increase in the relative probabilities for reionization upon scattering from Ti vs. O, as the incident energy is increased, is the most likely explanation for the energy dependent yields for TiO_2.

Conclusions

Low energy He^+ ion scattering studies of TiO_2, Ti, and H_2O overlayers have been performed. Two experiments have demonstrated that the energy dependent O/Ti scattering yields for TiO_2 are most likely characteristic of the nature of the He^+-O and He^+-Ti collisions and not the result of shadowing effects. Studies of H_2O overlayers on Ti have identified an attenuation of the LEIS yield from O atoms by hydrogen. This phenomenon represents an additional complication in the determination of the composition of hydroxylated oxide surfaces, and requires further investigation.

Acknowledgements

Partial support for this work was obtained from the National Science Foundation, Materials Research Division, Grant No. DMR-8907553

References

*   Present address: Dept. Nuclear Engineering and Engineering Physics, University of Virginia, Charlottesville, VA 22901 USA.
1)   D. G. Armour, Vacuum 31 (1981) 10.
2)   B. A. Horrell and D. L. Cocke, Catal. Rev.-Sci. Eng., 29 (1987) 447.
3)   R. Margraf, H. Knozinger, and E. Taglauer, Surface Sci. 211/212 (1989) 1083.
4)   D. P. Woodruff and T. A. Delchar, Modern Techniques of Surface Science,Cambridge University Press, Cambridge UK (1986) pp.220-239.
5)   C. A. Moyer and K. Orvek, Surface Sci. 121 (1982) 138.
6)   G. C. Nelson, J. Vac. Sci. Technol. 15 (1978) 702.
7)   W. L. Baun, Phys. Rev. A17 (1978) 849.
8)   R. C. McCune, J. Vac. Sci. Technol. 16 (1979), 1569.
9)   E. Taglauer and W. Heiland, Surface Sci. 47 (1975) 234.
10)   G. C. van Leerdam and H. H. Brongersma, Proc. Sympos. Surf. Sci., La Plagne, 1990.
11)   R. Stockbauer, D. M. Hanson, S. A. Flodstrom, and T. E. Madey, Phys Rev. B26 (1982) 1885
12)   B. L. Maschhoff, J-M. Pan, and T. E. Madey, to be published.
13)   P. J. Feibelman, D. R. Hamann, and F. J. Himpsel, Phys. Rev. B22 (1980) 1734.
14)   T. E. Madey, Science 234 (1986) 316.
15)   D. P. Woodruff, Nucl. Instrum Methods 194 (1982) 639.

FACTORS WHICH AFFECT PEAK SHAPES AND AREAS IN ION SCATTERING

SPECTROSCOPY

Orlando Melendez and Gar B. Hoflund

Department of Chemical Engineering
University of Florida
Gainesville, Florida 32611, USA

Richard E. Gilbert

Department of Chemical Engineering
University of Nebraska
Lincoln, Nebraska 68588, USA

Vaneica Y. Young

Department of Chemistry
University of Florida
Gainesville, Florida 32611, USA

ABSTRACT

This study is a further attempt to understand how experimental factors influence signal intensities and peak shapes in ion scattering spectroscopy (ISS). Different inert gases and mixtures have been scattered off polycrystalline Pt and Cu using primary beam energies ranging from 500 to 2000 eV. The ISS signal intensities decreases significantly as the primary beam energy decreases or as the mass of the scattering ion increases. This fact is consistent with the assertion that the neutralization probability increases as the period of time that an ion spends near a metallic surface increases. The influence of isotopic distribution of the scattering gas on ISS peak shape is also examined using mixtures of $^{20}Ne^+$ and $^{22}Ne^+$.

INTRODUCTION

In recent studies by Young and Hoflund [1] and Young et al. [2], factors which affect resolution in ion scattering spectroscopy (ISS) have been investigated. These factors include mass and isotopic distributions of the elements at the solid surface, width of the scattering angle and spread in primary ion beam energy. Since the peak shapes vary with isotopic distribution, Young and Hoflund [1] argue that peak areas give a better measure of intensities than peak heights. In this present study the factors which affect ISS peak shapes and areas are examined further. Specifically, the influence of primary beam energy on peak area and the influence of isotopic distribution of the scattering ions on peak shapes are considered.

Fundamental Aspects of Heterogeneous Catalysis Studied by Particle Beams,
Edited by H.H. Brongersma and R.A. van Santen, Plenum Press, New York, 1991

EXPERIMENTAL

The ISS data were collected using a Perkin-Elmer PHI 15-255GAR double-pass cylindrical mirror analyzer with the 12^0 slotted aperture. ISS was performed in the nonretarding mode using primary beam energies of 500, 1000, 1500 and 2000 eV and a scattering angle of 151.3 ± 6^0. The isotopic gases used were obtained from Isotec (Miamisburg, OH) and had the following compositions (atomic %): ^{3}He - 99.995, ^{20}Ne - 99.95, ^{22}Ne - 99.9, ^{22}Ne - 76/^{20}Ne - 24 and ^{20}Ne - 90/^{22}Ne - 10. The samples used in this study consisted of polycrystalline Pt and Cu foils. The Pt was cleaned by annealing repeatedly at 1200 ^{0}C in O_2 and then H_2 before sputtering with 2 keV Ar^+. The Cu was cleaned simply by sputtering with 2 keV Ar^+.

RESULTS AND DISCUSSION

Table 1 gives the ISS Pt peak areas obtained using various primary inert gas ions at energies ranging from 500 to 2000 eV. For each gas the pressure was adjusted so that the total current to the positively biased sample was the same value while each spectrum was taken. The values given in table 1 were obtained by subtracting the inelastic background from the peak using the procedure described by Nelson (3). He claims that scattered ion spectra can be divided into two portions; a peak due to elastically scattered ions which can be represented by a combined Gaussian and Lorentzian function

$$I_i(E) = \frac{A \exp\left\{-(1-R) \times 2.772559 \left[\frac{E - E_0}{[FWHM]}\right]^2\right\}}{R[E-E_0]^2 + \left[\frac{[FWHM]}{2}\right]^2} \tag{1}$$

and a background tail due to inelastically scattered ions

$$I_B(E) = B\{\pi - 2 \times \tan^{-1}[2(E-E_0)/[FWHM]]\}, \tag{2}$$

A is the peak amplitude, R is the Gaussian/Lorentzian factor (R = 1 is only Lorentzian; R = 0 is only Gaussian), [FWHM] is the full width at half maximum, E_0 is the energy at the peak maximum, and B is the intensity of the tail. Furthermore, the intensity of the tail decays by the factor

$$I_T(E) = \exp\left[\frac{-k}{\sqrt{E}}\right] \tag{3}$$

where k is the neutralization decay rate. Then the total spectral intensity as a function of energy is given by $I_i(E) + E_B(E)I_T(E)$.

The data in table 1 demonstrates two strong trends. The first is a decrease in signal strength as the primary beam energy decreases for a given scattering ion, and the second is a decrease in signal strength as the mass of the scattering increases at a given primary beam energy. Both of these trends are related to the neutralization process. If an ion remains near a metallic surface for a longer period of time, it has a higher probability of being neutralized. Consistent with the data in table 1 and based on this fact, the neutralization probability should increase as the mass increases at a constant energy or as the kinetic energy decreases for a given mass. This process can be quantified by assuming that equation 3 also applies to the intensity of ISS peaks as well as to the inelastic background. This is reasonable since the $1/\sqrt{E}$ term is proportional to the period

Table 1.   ISS

Peak areas obtained by scattering different probe ions of
varying energy off polycrystalline Pt

gas	2 keV	1.5 keV	1.0 keV	0.5 keV
^{3}He	1,120,000	879,000	450,000	93,000
^{4}He	650,000	502,000	295,000	62,000
^{20}Ne $-$ 99.95%	515,000	270,000	100,000	22,000
^{22}Ne $-$ 99.9%	360,000	180,000	50,000	16,000

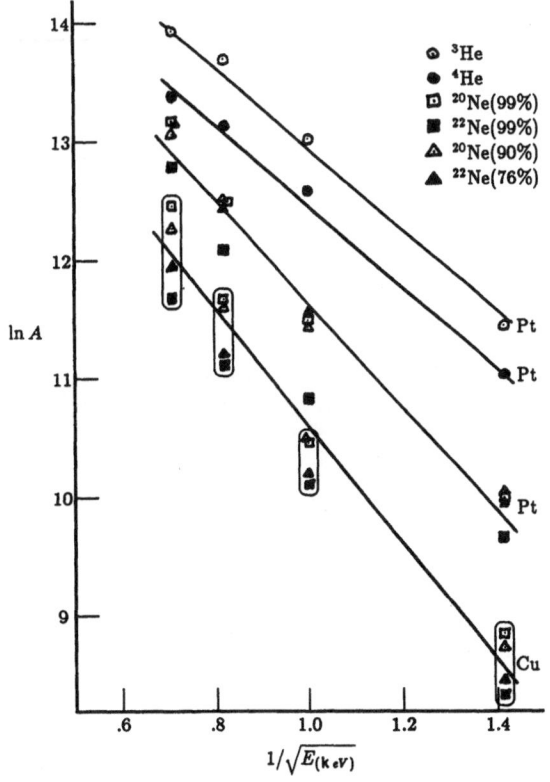

PEAK AREA vs. BEAM ENERGY

Figure 1.   Ln of the ISS peak areas given in table 1 versus $1/\sqrt{E}$ for various inert gas
ions scattering off polycrystalline Pt and Cu. The slope is the neutralization
decay rate according to equation 3.

of time that an ion spends in the vicinity of the metal surface. A plot of the ln of the peak areas versus $1/\sqrt{E}$ is shown in figure 1 for different inert gas ions scattered off polycrystalline Pt and Cu. Straight lines are obtained indicating the validity of equation 3, and the neutralization decay rate k is equal to the slopes of these lines. The parameters used to fit the experimental $^3He^+$ and $^4He^+$ spectra to Nelson's model are given in table 2.

Table 2. parameters used for background removal from $^3He^+$ and $^4He^+$ spectra

ion	2 keV	1.5 keV	1.0 keV
$^3He^+$			
E_0	1905 eV	1420	940
FWHM	50 eV	50	50
B	0.2×10^9	0.8×10^9	1.4×10^{10}
k	600	600	600
$^4He^+$			
E_0	1890 eV	1395	935
FWHM	50 eV	50	50
B	0.7×10^8	0.4×10^9	0.9×10^{10}
k	590	590	590

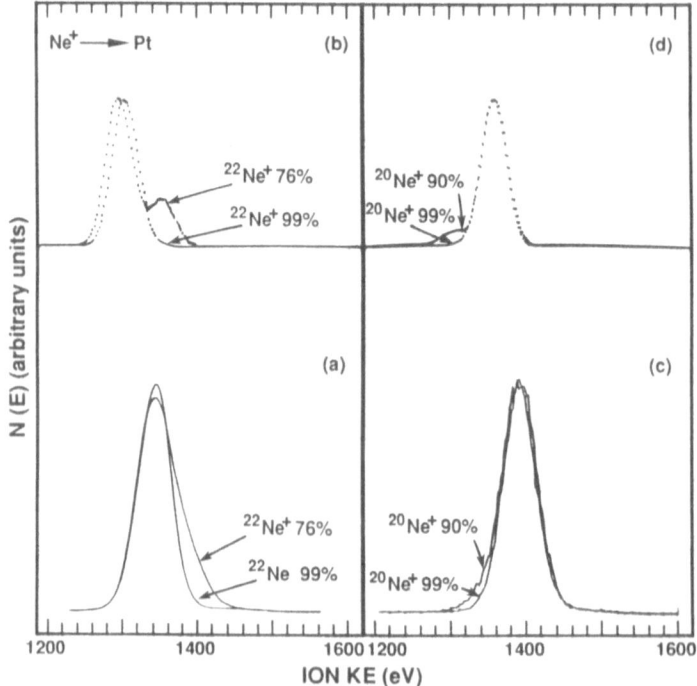

Figure 2. Experimental ISS spectra obtained by scattering 2 keV $^{20}Ne^+$ − 99.95%, $^{20}Ne^+$ − 24%/$^{22}Ne^+$ − 76%, $^{20}Ne^+$ − 90%/$^{22}Ne^+$ − 10% and $^{22}Ne^+$ − 99.9% primary ion beams off polycrystalline Pt. Corresponding calculated spectra using the model described by Young et al. (2) and a 12^0−wide scattering angle and a 10 eV spread in the primary beam energy distribution are also shown.

Two ISS spectra taken from a polycrystalline Pt sample using a 2 keV $^{20}Ne^+$ – 24%/$^{22}Ne^+$ – 76% and a $^{20}Ne^+$ – 0.1%/$^{22}Ne^+$ – 99.9% primary ion beam are shown in figure 2a. The isotopic mixture yields a peak which is considerably broader than that obtained using $^{22}Ne^+$ – 99.9%. Calculated spectra obtained using the method described by Young et al. (2) are shown in figure 2b. These calculations were made using a 12^0 spread in scattering angle and a 10 eV spread in primary beam energy. The calculated spectra are in qualitative agreement with the experimental spectra except that the shoulder due to the presence of $^{20}Ne^+$ is much better resolved in the calculated spectra. This is probably due to an experimental spread in primary beam energy which is larger than the 10 eV spread assumed for the calculations. Similar experimental and calculated spectra for $^{20}Ne^+$ – 90%/$^{22}Ne^+$ – 10% and $^{20}Ne^+$ – 99.95% are shown in figure 2c and d respectively. The contribution of the $^{22}Ne^+$ appears as a small shoulder on the low energy side of the ISS peak. Again these spectra are consistent and illustrate the influence of isotopic distribution of the scattering ions on ISS peak shape. The calculated peaks based on the binary scattering equation appear at energies which differ from the experimental peaks indicating that the binary scattering equation is only approximate.

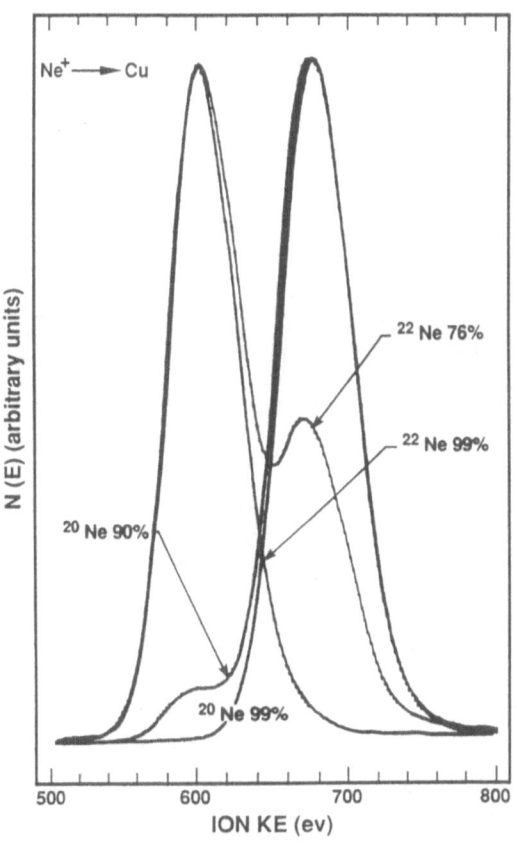

Figure 3. ISS spectra obtained by scattering 2 keV $^{20}Ne^+$ – 99.95%, $^{20}Ne^+$ – 24%/$^{22}Ne^+$ – 76%, $^{20}Ne^+$ – 90%/$^{22}Ne^+$ – 10%/$^{22}Ne^+$ – 99.95% off polycrystalline Cu.

The isotopic effects on the spectra in figure 2a and 2c are poorly resolved because the mass of the target atoms is large resulting in low resolution of spectral features (4). Better resolution of the ISS peaks due to scattering of $^{20}Ne^+$ and $^{22}Ne^+$ is obtained by using Cu as the target. ISS features obtained by scattering 2 keV $^{22}Ne^+ - 99.9\%$, $^{20}Ne^+ - 24\%/^{22}Ne^+ - 76\%$, $^{20}Ne^+ - 90\%/^{22}Ne^+ - 10\%$ and $^{20}Ne^+ - 99.95\%$ off polycrystalline Cu are shown in figure 3. The features due to the two different Ne isotopes in these spectra are clearly resolved. The total spectra obtained from scattering the four Ne isotopic mixtures off Cu are shown in figure 4a to d. Features due to the presence of both Ne isotopes as Ne^+, Ne^{++} and Ne^{+++} ions are present in these spectra close to the energies predicted by the elastic binary collision model. As expected, the intensities of the ISS features decrease as charge increases. This appears to be the first observation of ISS features due to triply charged inert gas ions produced by a sputter ion gun.

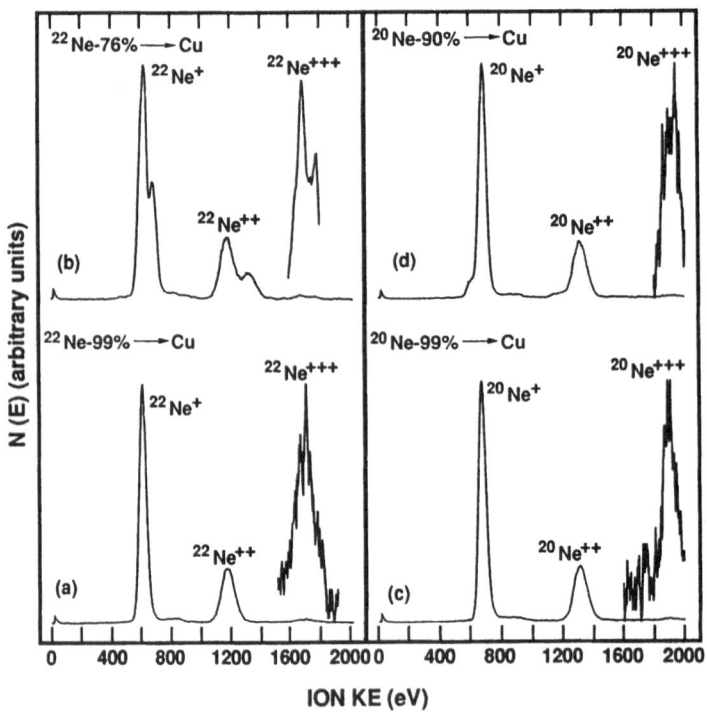

Figure 4.   ISS spectrum obtained by scattering 2 keV $^{20}Ne^+ - 99.95\%$, $^{20}Ne^+ - 24\%/^{22}Ne^+ - 76\%$, $^{20}Ne^+ - 90\%/^{22}Ne^+ - 10\%$ and $^{22}Ne^+ - 99.95\%$ off polycrystalline Cu showing contributions from Ne^+, Ne^{++} and Ne^{+++}.

REFERENCES

1. V.Y. Young and G.B. Hoflund, Anal. Chem 60(1988) 269.
2. V.Y. Young, G.B. Hoflund and A.C. Miller, Surface Sci. 235 (1990) 60.
3. G.C. Nelson, J. Vac. Sci. Technol. A2(1986)1567.
4. T.M. Buck in Methods of Surface Analysis edited by A.C. Czanderna, Elsevier, Amsterdam 75(1975)75.

THE MEASUREMENT OF ABSOLUTE SURFACE COVERAGES USING NUCLEAR REACTION ANALYSIS

I. V. Mitchell, W. N. Lennard, G. R. Massoumi

Department of Physics

and K. Griffiths

Department of Chemistry

The University of Western Ontario
London, Ontario N6A 3K7 Canada

ABSTRACT

The use of ion beam methods for determining the adsorbate coverage of single crystal surfaces can lead to absolute values with uncertainties less than 0.05 ML (1 ML = 1 monolayer $\approx 10^{15}$ atoms cm^{-2}) in favourable cases.   This accuracy arises from the availability of reliable nuclear reaction cross section data for light ions incident at MeV energies on low atomic number atoms, specifically H(D), C and O isotopes. Examples of such coverage measurements are given for CO/Pt(111) at 185K assayed separately for C and O, O_2/Ni(100) at 300K determined for both chemisorbed ^{16}O and ^{18}O, and C_2D_4/Pt(111) at 100K determined through measurements of both C and D.

1.   INTRODUCTION

The measurement of absolute adsorbate coverages remains an important aspect of experimental studies of surface phase stability / phase transitions and surface reaction mechanisms on catalytically active single crystal surfaces. The conventional approach relies upon a measurement of gas exposure (i.e. pressure x time);   provided the sticking coefficient, S, is known or can be inferred, the coverage can be calculated. This approach is of limited use in the high pressure regime where S may not be well known, or where the use of a directed flux (e.g. from a capillary array beam doser) makes it difficult to measure the pressure over the surface.

It is possible to determine coverages through XPS peak intensity measurements, but calibration is non-trivial, and the method is not applicable to hydrogen isotopes. Similarly, while static secondary ion mass spectrometry

Fundamental Aspects of Heterogeneous Catalysis Studied by Particle Beams,
Edited by H.H. Brongersma and R.A. van Santen, Plenum Press, New York, 1991

(SSIMS) has high sensitivity, the derivation of absolute coverage values requires calibration. The variation of ion yield with simultaneous changes in surface chemistry is also problematic. Finally, thermal desorption spectroscopy (TDS) in combination with quadrupole mass spectroscopy identifies the desorbing molecular species via its mass to charge ratio, but again requires calibration to yield absolute values. To date, ion beam methods have been used in only a limited way. The advantage of methods such as nuclear reaction analysis (NRA) at MeV ion energies and heavy ion induced recoil scattering (ERDA) at a few tens of MeV incident energy is that the reaction or collision product yield is directly proportional to the product $N\sigma$, where N is the areal density of target atoms and σ is the relevant cross section. The yield and thus the value for N is made absolute by measuring: (i) the total number of beam particles impinging on the target during the experiment (beam fluence), and (ii) the detector solid angle, and (iii) the detector efficiency (usually unity for charged particle detection). In many cases, the cross sections are already known, as is the case for elastic scattering, or they can be measured separately. The precision for cross section values is often of the order of a few percent. With respect to experimental requirements, the vacuum conditions needed for UHV operation can be maintained on-line to accelerator facilities through the use of differentially pumped apertures [1] and UHV-compatible (i.e. bakeable) solid state detectors for charged particles are readily available. In this paper, we illustrate with three specific examples the use of NRA techniques performed using a 2.5 MV Van de Graaff accelerator facility to measure the saturation coverages of common gases (CO, O_2, C_2H_4) on well known single crystal surfaces. At the ion beam energies used, the reaction product groups can be easily recognized in the charged particle energy spectrum and are free of spectral background problems.

2. EXPERIMENTS

(A) CO / Pt(111)

The saturation coverage of CO on the Pt(111) surface at 185K has been reported previously [1,2]. The *atomic* coverages were determined by measurements of the $^{12}C(d,p)^{13}C$ and $^{16}O(d,p_1)^{17}O$ reaction yields, both at a deuteron beam energy of 970 keV. At the laboratory detection angle of 150°, the values used for the two cross sections were 25.5 ± 0.8 mb sr^{-1} and 13.3 ± 0.4 mb sr^{-1}, respectively, leading to areal densities of 0.79 ± 0.03 atoms cm^{-2} (C) and 0.69 ± 0.05 atoms cm^{-2} (O).

The value for the $^{16}O(d,p_1)^{17}O$ cross section had been measured in a separate experiment using a thin Ta_2O_5 target of known areal density. A new measurement of this reaction cross section [3] is in excellent agreement (within 2.2 %). However, inconsistencies between the cross sections for $^{12}C(d,p)^{13}C$ and $^{12}C(\alpha,\alpha)^{12}C$ [4] have prompted a re-examination of the carbon reaction. A study has been undertaken therefore in our laboratory using condensed gas targets of CO, CO_2 and CH_3OH at T $\approx$ 60K. Corrections were made for target thickness using excitation functions measured with thin carbon and thin oxide targets. Yields have been extrapolated back to zero beam fluence equivalent for the CO and CO_2 data - the two cases where ion beam erosion of the target was observable. Details of these experiments are reported elsewhere [5].

The extracted reaction cross section ratio $\sigma\{^{12}C(d,p)^{13}C\}$ / $\sigma\{^{16}O(d,p1)^{17}O\}$ for an incident deuteron energy of 972 keV is listed in Table 1 for each target species. The internal agreement is excellent and leads to a value for the carbon reaction cross section of 27.7 ± 1.1 mb sr^{-1} at 150°, higher than the value reported in ref. 1, viz. 25.5 ± 0.8 mb sr^{-1}.

Application of the new cross section value leads to revised C and O values for CO/Pt(111), viz. 0.48 ± 0.02 and 0.46 ± 0.03 ML, respectively (cf. 0.53 ± 0.02 reported in ref. 1 for C), where 1 ML corresponds to a coverage of 1.5×10^{15} atoms cm^{-2}. The new coverage values are consistent with a C:O ratio of unity, the expected stoichiometry. We believe the new value for the cross section should be adopted; the authority rests on the condensed gas target data [5] with the implication that the coverage reported for CO on Pt(111) has a mean value $\theta = 0.47 \pm 0.03$ ML.

TABLE 1. Measured values for the cross section ratio $\sigma\{^{12}C(d,p)^{13}C\}$ / $\sigma\{^{16}O(d,p1)^{17}O\}$ for a laboratory angle of 150° and a deuteron energy of 972 keV, derived from experiments using condensed gas targets at $T \approx 60K$.

Target Species	Cross Section Ratio
CO	2.05 ± 0.06
CO_2	2.08 ± 0.06
CH_3OH	2.09 ± 0.06

(B) O_2 / Ni(100)

Ni single crystals are known to form passive oxide layers following exposure to O_2 at room temperature and below. The saturation coverage of an oxide layer formed on a clean Ni(100) surface at room temperature has been measured using a combination of LEED, work function, AES and NRA techniques [6,7]. In a series of three exposures, separated by standard cleaning procedures, the Ni(100) surface was saturated in: (i) $^{16}O_2$, (ii) 97 % enriched $^{18}O_2$, and (iii) 10 L of $^{18}O_2$ followed by ~ 200 L of $^{16}O_2$ (1 L = 10^{-6} torr-s). Oxygen coverages were determined absolutely through use of the $^{16}O(d,p1)^{17}O$ and $^{18}O(p,\alpha)^{15}N$ reactions at a deuteron energy of 972 keV and a proton energy of 750 keV, respectively, i.e. at energies where the relevant cross sections are known to within 4 % . Details of these experiments can be found in ref. 7.

We found total oxygen coverage values of: (i) 2.45 ± 0.10 ML for the $^{16}O_2$ exposure; (ii) 2.49 ± 0.08 ML (= 0.14 ± 0.02 ^{16}O + 2.35 ± 0.08 ML ^{18}O) for the $^{18}O_2$ exposure; and (iii) 2.42 ± 0.10 ML (= 1.89 ± 0.09 ^{16}O + 0.53 ± 0.04 ML ^{18}O) for the composite oxide layer. Agreement is very good between the three and may be compared with an earlier measurement of 2.4 ± 0.2 ML [8]. These results should perhaps be viewed as a test of the consistency of the nuclear reaction cross sections used. The $^{16}O_2$ content in experiment (ii) above was attributed to contamination of the $^{18}O_2$ gas either at the source or in the transfer lines. It is worth noting in these room temperature measurements that *no* evidence for change in oxygen content was observed under prolonged beam bombardment, contrary to observations for the condensed gas experiments described in (A).

(C) Ethylene / Pt(111)

The adsorption of ethylene (C2H4) onto the Pt(111) surface is one the most extensively studied of all catalytic systems; however, there is still a persistent discrepancy in the literature concerning the saturation coverage values over a range of temperatures [9-15]. We have studied the ethylene on Pt(111) system using LEED, AES, TDS and ion beam methods for fully deuterated ethylene, i.e. C_2D_4. This substitution permitted *independent* measurements of both the carbon and deuterium coverages (cf. C and O in the CO case cited in

(A) above) via the $^{12}C(d,p)^{13}C$ and $D(^{3}He,p)^{4}He$ nuclear reactions, respectively, for incident projectile energies of 972 keV (d) and 630 keV (^{3}He) at a detection angle of $150°$ with respect to the beam direction. (Contrary to the CO case, we used two *different* projectiles in these experiments.) A full description of the experiment will be given in a forthcoming publication [16] but we give here only the data obtained at T = 100K. Application of the appropriate cross section values (see refs. 1 and 17 for the cross section for the D + ^{3}He reaction) to the measured reaction yields results in the areal densities listed in Table 2. It can be seen that the ratio of D:C is 1.90 ± 0.16, i.e. consistent with the value 2:1.

TABLE 2. Measured saturated surface coverage of deuterated ethylene (C_2D_4) on Pt(111) at 100K using the yields from the nuclear reactions $^{12}C(d,p)^{13}C$ and $D(^{3}He,p)^{4}He$.

Target Species	Coverage [10^{15} atoms cm^{-2}]
^{12}C	0.70±0.05
D	1.33±0.06

This observation supports the commonly held view that ethylene is adsorbed stoichiometrically at this temperature [9,11]. Thus, we derive a *molecular* coverage for the ethylene species of 0.23 ± 0.02 ML based on these data. No changes in coverage were observed for extended ion beam bombardment of the surface, i.e. we are inclined to believe that the saturation coverage value θ_{sat} ~ 1/4 ML is unperturbed by the analyzing beam.

This result is to be compared with the value 0.5 ML reported by Yu and Gustafsson [9] based on the use of high energy heavy ion induced recoil methods. Our TDS data are fully consistent with those reported by other investigators [13,14,18-21], i.e. for all reported cases, some molecular ethylene was observed to desorb at T ≈ 290K following adsorption at 100K with subsequent heating. No TDS data were recorded in the experiments of ref. 9 and the constancy of the carbon signal intensity over a temperature range where C_2H_4 is known to desorb is therefore puzzling. The use of low energy (keV) ion beams has been illustrated by Masson et al. [10], also for the case of C_2H_4/Pt(111), but for adsorption temperatures of 340K and above, i.e. outside the range of our ion beam measurements.

3. DISCUSSION

The usefulness of the ion beam method in determining absolute surface coverage is traceable to the direct relationship between the coverage and the reaction product or recoil product yield. The spectrometry is straightforward and the necessary cross sections usually can be predicted or measured independently. Existence of a suitable nuclear nuclear reaction for ^{13}C (viz. the $^{13}C(d,p)^{14}C$ reaction) and an elastic recoil process, $H(^{4}He,p)^{4}He$, applicable to protium opens the way for studying reaction mechanisms and kinetics via isotopic labelling for all three species - H, C and O. Contrary to the case for MeV energy light ions and fast heavy ions, the use of low energy (keV) ions may require a consideration of shadowing and focussing effects. (For a discussion of recoil scattering at low incident ion energy, see the paper by J.W. Rabalais in these proceedings.)

For all ion beam measurements, it is important to test for beam induced desorption. A variety of measurements for ion induced erosion of surface films has been recognized (see, for example, the works of Schou [22] and Reimann et al. [23]). Much more work needs to be done to clarify the processes responsible for desorption of the saturated layers, as distinct from extended condensate/ice layers. The separate issue of thermal desorption or decomposition due to bulk heating by the beam can be eliminated through control of the current intensity. In the work reported here, we believe the accuracy of the results is not influenced by either effect.

One advantage of the ion beam method that has not been discussed in this paper is the possibility for determining not only the coverage but also the lattice sites occupied by the adsorbate species on a crystal surface. This extension has been shown in the elegant studies by the Århus group, using transmission channeling in thin crystals [24].

Finally, it is worth pointing out that with further development, MeV energy ion beam methods could be implemented at higher target chamber pressures, i.e. closer to the operating conditions required for many catalysts. This area of research falls in a regime where the conventional surface science techniques are not applicable.

4. SUMMARY

Three examples of the use of MeV energy light ion beams to measure adsorbate coverages on catalytically-active single crystal surfaces have been discussed. The measurement uncertainties were less than ten percent at monolayer levels. The coupling of an ion beam facility in situ, in an UHV environment with a target chamber equipped with modern surface science analytical tools is therefore seen to be a valuable aid to research on adsorbate behaviour for these materials.

5. ACKNOWLEDGEMENTS

The authors would like to acknowledge the many valuable contributions to this work provided by Prof. P. R. Norton, Dr. P. F. A. Alkemade, S. J. Bushby, S. Y. Tong, J. W. Huppertz and the technical support staff of Interface Science Western at the University of Western Ontario.

REFERENCES

[1] J.A. Davies and P.R. Norton, Nucl. Instr. and Meth. 168 (1980) 661.
[2] P.R. Norton, J.A. Davies and T.E. Jackman, Surf. Sci. 122 (1982) L593.
[3] W.N. Lennard, S.Y. Tong, I.V. Mitchell and G.R. Massoumi, Nucl. Instr. and Meth. in Phys. Res. B43 (1989) 187.
[4] S.Y. Tong, W.N. Lennard, P.F.A. Alkemade and I.V. Mitchell, Nucl. Instr. and Meth. in Phys. Res. B45 (1990) 30.
[5] W.N. Lennard, G.R. Massoumi, P.F.A. Alkemade, I.V. Mitchell and S.Y.Tong, to be published.
[6] S.J. Bushby, T.D. Pope, B.W. Callen, K. Griffiths, C.-S.Zhang and P. R. Norton, J. Vac. Sci. Technol. (1990, in press).
[7] I.V. Mitchell, G.R. Massoumi, W.N. Lennard, S.Y. Tong, P.F.A. Alkemade, K. Griffiths, S.J. Bushby and P.R. Norton, Nucl. Instr. and Meth. in Phys. Res. B45 (1990) 107.
[8] P.F.A. Alkemade, S. Deckers, F.H.P.M. Habraken and W.F. Van der Weg, Surf. Sci. 189/190 (1987) 161.
[9] R. Yu and T. Gustafsson, Surf. Sci. 182 (1987) L234.
[10] F. Masson, C.S. Sass, O. Grizzi and J.W. Rabalais, Surf. Sci. 221 (1989) 299.
[11] N. Freyer, G. Pirug and H.P. Bonzel, Surf. Sci. 125 (1983) 327.
[12] N. Freyer, G. Pirug and H.P. Bonzel, Surf. Sci. 126 (1983) 487.
[13] J.R. Creighton and J.M. White, Surf. Si. 129 (1983) 327.

[14] M. Abon, J. Billy and J.C. Bertolini, Surf. Sci. 171 (1986) L387.
[15] S.M. Davis, F. Zaera and G.A. Somorjai, J. Catal. 77 (1982) 439.
[16] I.V. Mitchell, W.N. Lennard, K. Griffiths, G.R. Massoumi and J.W. Huppertz, Surf. Sci. (1990, to be published).
[17] W. Möller and F. Besenbacher, Nucl. Instr. and Meth. 168 (1980) 111.
[18] M. Salmeron and G.A. Somorjai, J. Chem. Phys. 86 (1982) 341.
[19] A.M. Baro and H. Ibach, J. Chem. Phys. 74 (1981) 4194.
[20] J.R. Creighton, K.M. Ogle and J.M. White, Surf. Sci. 138 (1984) L137.
[21] S.M. Davis, F. Zebra, B.E. Gordon and G.A. Somorjai, J. Catal. 92 (1985) 240.
[22] J. Schou, Nucl. Instr. and Meth. in Phys. Res. B27 (1987) 188.
[23] C.T. Reimann, R.E. Johnson and W.L. Brown, Phys. Rev. Lett. 53 (1984) 600.
[24] I. Stensgaard, Nucl. Instr. and Meth. in Phys. Res. B15 (1986) 300.

ON THE NEUTRALIZATION MODELS IN LEIS

G. Verbist°, H.H. Brongersma

Eindhoven University of Technology, Department of Physics
P.O.Box 513, 5300 MB Eindhoven, The Netherlands

and J.T. Devreese*

University of Antwerp (UIA), Department of Physics
Universiteitsplein 1, B-2610 Antwerp, Belgium

ABSTRACT

Three different neutralization models are discussed. In the continuum limit their connection is clarified for ion trajectories with unspecified time-dependence. The unifying model is fitted to experiment for $He^+ \rightarrow MgO$ scattering. It is pointed out how difficulties, related to the use of relative ion fractions, can be circumvented. Therefore the results of DISC-experiments, which are *absolute* intensities, are compared with calculations on the basis of the Muda-Newns model.

INTRODUCTION

Low-Energy Ion Scattering (LEIS) has become an important tool for the analysis of the atomic composition and structure of solid surfaces [1]. Since only charged particles are detected in LEIS measurements, neutralization is a key process. This is especially important for inert gas ions such as He^+, where the high neutralization rate generally reduces the information depth to only one atomic layer.

For quantitative structure analysis by means of LEIS, neutralization theories are needed. In this contribution, different neutralization models are discussed. For a recent model [2], which was introduced by the present authors, the calculations are extended to obtain the time-dependent neutralization rate. Special attention is devoted to the continuum limit, in which the connection between the existing models of Hagstrum [3] and Godfrey and Woodruff [4] is clarified. Assuming asymptotic straight-line trajectories, the LEIS-experiments of Souda et al. [5] on MgO(001) are fitted to this model. These results are compared to previous calculations on the NaCl(001)-surface [2,6]. Finally, some general remarks are made concerning the fitting of these models in particular with respect to *absolute* intensities.

Fundamental Aspects of Heterogeneous Catalysis Studied by Particle Beams,
Edited by H.H. Brongersma and R.A. van Santen, Plenum Press, New York, 1991

The different Neutralization Models

The fraction P^+ of low-energy inert gas ions which are _not_ neutralized (after scattering at a solid surface), is conventionally written in terms of a neutralization rate $R(t)$ as

$$P^+ = \exp\left[-\int R(t)\,dt\right] \quad . \tag{1}$$

We shall discuss three different forms of the neutralization rate, as well as the underlying models: a) the Hagstrum model [3], b) the Godfrey-Woodruff model [4], and c) the unifying model, which was introduced by the present authors in ref. [2].

In the _Hagstrum model_ one considers the neutralization at metallic surfaces. In this case, the electron density can be considered homogeneous and, therefore, one assumes that $R(t)$ only depends on the perpendicular distance $z(t)$ of the ion above the surface. Hagstrum parametrized

$$R_m(t) = A_m e^{-a_m z(t)} \quad , \tag{2}$$

where A_m and a_m are the so-called _neutralization constants_ of the metal. (An index m is used to denote quantities referring to a metal.)

Godfrey and Woodruff [4] considered a local neutralization model, which can also be used to explain the neutralization by adsorbed atoms. Therefore, the distance between the ion and the (adsorbed) atom is the relevant quantity, in contrast to the perpendicular distance above the surface in the Hagstrum model. An analogeous parametrization leads to

$$R_{at} = A_{at} e^{-a_{at}|\vec{r}(t)-\vec{r}_{at}|} \quad , \tag{3}$$

where A_{at} and a_{at} are again characteristic constants, but now for the (adsorbed) atom. (the subscript at denotes atomic neutralization constants.)

In ref. [2] a _unifying model_ was proposed, which starts from the Godfrey-Woodruff parametrization at one surface atom. The contributions of all atoms in the surface-lattice were then summed. A so-called asymptotic (straight line) trajectory was used for the ion motion. In the present paper, this condition will be relaxed. The present calculation gives the neutralization rate without an assumption of the explicit form of the ion motion. This allows us to examine the time-dependence of the neutralization rate $R(t)$ in the continuum limit, and to compare it with the Hagstrum parametrization.

Calculation of the Time-dependent Neutralization Rate

Let us consider a square lattice (the surface) with lattice constant a and with N atoms in the unit cell at positions $\vec{u}_j$ ($j = 1, \ldots, N$). The ion fraction, eq. (1), takes the form

$$P^+ = \exp\left(-\sum_{j=1}^{N} \int dt\, R_j(t)\right) \quad , \tag{4}$$

where the neutralization rates $R_j(t)$ are given by the formula

$$R_j(t) = A_j \sum_{h,k=-\infty}^{+\infty} e^{-a_j|\vec{r}(t)-\vec{R}_{h,k}-\vec{u}_j|} \quad . \tag{5}$$

Eq. (5) represents a lattice sum of Godfrey-Woodruff neutralization rates for the j-th atom.

We introduced the following notations: $\vec{R}_{h,k} = a(h\hat{x} + k\hat{y})$ for a lattice vector, A_j and a_j as the neutralization constants for the j-th atom. The lattice periodicity is best exploited in reciprocal space. Therefore, we employ the Fourier transform

$$e^{-b|\vec{r}|} = \int \frac{d^3q}{(2\pi)^3} \frac{8\pi b}{(q^2+b^2)^2} e^{i\vec{q}\cdot\vec{r}} \quad , \tag{6}$$

by which we can transform $R_j(t)$ into a summation over reciprocal lattice vectors $\vec{g}_{h,k} = (2\pi/a)(h\hat{x} + k\hat{y})$ as

$$R_j(t) = \frac{A_j}{a^2} \sum_{h,k=-\infty}^{+\infty} \int \frac{dq_z}{2\pi} \frac{8\pi a_j}{(a_j^2 + q_z^2 + g_{h,k}^2)^2} e^{i(\vec{g}_{h,k}+q_z\hat{z})\cdot(\vec{r}(t)-\vec{u}_j)} \quad . \tag{7}$$

Performing the q_z-integral, the following result is obtained for the neutralization rate

$$\begin{aligned}
R_j(t) = A_j \frac{4\pi a_j^2}{a^2} \sum_{h,k=-\infty}^{+\infty} \left(\frac{a_j^2}{a_j^2 + g_{h,k}^2}\right)^{3/2} e^{i\vec{g}_{h,k}\cdot(\vec{r}(t)-\vec{u}_j)} \\
\times \tfrac{1}{2} \left(1 + \sqrt{a_j^2 + g_{h,k}^2}\, z(t)\right) e^{-\sqrt{a_j^2+g_{h,k}^2}\, z(t)} \quad .
\end{aligned} \tag{8}$$

¿From this formula, we can conclude that
- if the ion is at an appreciable distance ($z(t) \gg 1/a_j$) above the surface, only the term $h = k = 0$ is important, due to the exponential factor. As a consequence, a Hagstrum-like neutralization rate is found.
- if the ion is closer to the surface orientational effects become important. The ion will actually *see* the different atoms. The main contribution stems from terms for which the phase-factors in eq. (8) are unity, i.e. for which $\vec{g}_{h,k} \parallel (\vec{r}(t) - \vec{u}_j)$.

In the *continuum limit*, the lattice constant a tends to zero at fixed atomic density n. (For finite values of a: $n = a^{-2}$.) Hence the following neutralization rate is obtained

$$R_{j,\text{cont.}} = A_j \frac{4\pi n}{a_j^2} e^{-a_j z(t)} \left(\frac{1 + a_j z(t)}{2}\right) \quad . \tag{9}$$

This result can be related to the Hagstrum parametrization by introducing *metallic* neutralization constants: $A_m = A_j(4\pi n/a_j^2)$ and $a_m = a_j$. The functional dependence of $R(t)$ on $z(t)$, however, is different. This model predicts a linear polynomial, while the Hagstrum parameterization has a constant. Hagstrum [3] remarked that in general a polynomial is expected, which he chose constant for simplicity.

APPLICATION TO MgO

In the experiments of Souda *et al.* [5], 1 keV He$^+$-ions were scattered on the (001)-surface of different rocksalt structures (NaCl, MgO, NiO and TiC). In ref. [6], the NaCl-surface was studied by means of the unifying model. In this section we present our results for MgO. The intensity ratio Mg/O was estimated to be 5.6 [7]. Since it is impossible to obtain *absolute* intensities experimentally, both the experimental and theoretical results were normalized to the ($\beta = 60°, \psi = 45°$)-intensity of Mg. Numerically it was found that there are a magnitude of "best fits".

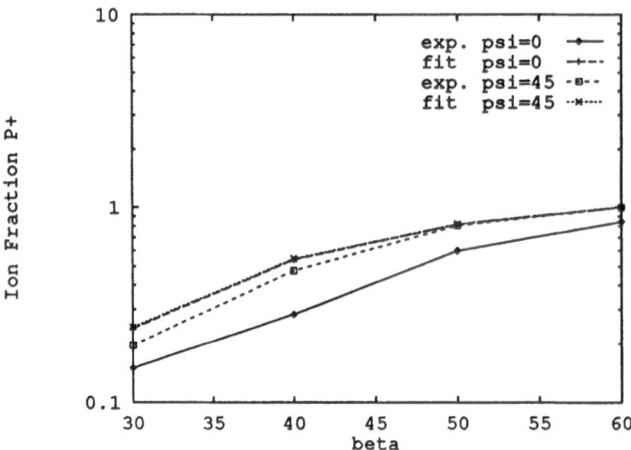

Fig. 1 The experimental and fitted (relative) ion fractions are compared for scattering at Mg as a function of the polar angle β for two different azimuths $\psi = 0°$ and $\psi = 45°$. (All results are normalized to the $(\beta = 60°, \psi = 0°)$ value.)

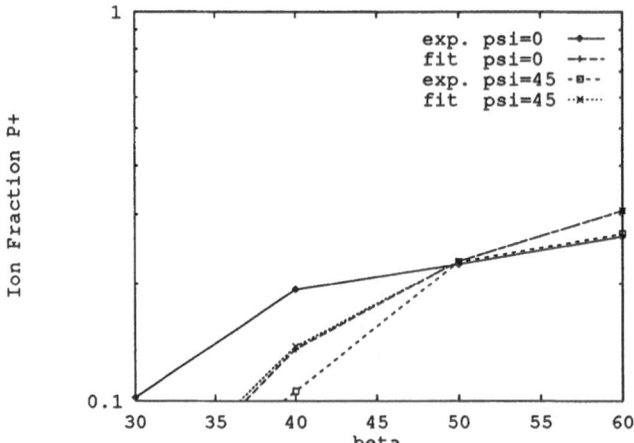

Fig. 2 Identical to fig. 1 for scattering at O.

Most of them, however, have to be rejected because they do not result in physical meaningful absolute intensities. This a direct consequence of the fact that *relative* intensities are fitted. The neutralization constants, which lead to the best fit, are presented in Table 1. In figs. 1 and 2, the fitted results are compared with experiment for MgO. These results lead to ion fractions of the order of 0.7–3% for Mg and 0.1–0.8% for O.

Although our results predict the qualititative behavior of the ion fractions as a function of both the azimuth ψ and the polar angle β, they are, not as accurate for O as for Mg.

Table 1 The results of our fit for $He^+ \rightarrow MgO(001)$ scattering at $E = 1$ keV.

quantity	Mg	O	units
a_{at}	4.70	1.00	Å^{-1}
A_{at}	2.05×10^{16}	2.70×10^{15}	s^{-1}
γ_{at}	0.318	1.49	$(-)$
ρ_{at}	0.426	2.00	Å
v_{at}	4.37	2.70	$10^5 \ m/s$

In particular, it must be noted that our model is unable to fit the sudden change in the oxygen signal as can be observed from fig. 2. In the NaCl-case, the same problem was noted from Cl. In ref. [6], an extra term was included in the neutralization rate to account for the long-range neutralization at the most electronegative element (Cl for NaCl). Such a calculation, gives a rather good agreement with experiment by the inclusion of two extra fitting parameters. The problem related to the absolute intensity, however, persists. In the next section, preliminary results are presented to circumvent this.

A REMARK ON ABSOLUTE INTENSITIES

Recent experimental work [8] used $^4\text{He}^+$, as well as $^3\text{He}^+$, for the scattering at silicon. After correcting for different beam intensities and kinetic cross sections [9], the unnormalized ion fractions were obtained. If the inverse square root of the incident energy E is plotted versus the logarithm of P^+, an excellent linear behavior is obtained in the low-energy region, as predicted by the Hagstrum model. Extrapolation of these lines to infinite energy ($E^{-1/2} \rightarrow 0$) leads, for both $^4He^+$ and $^3He^+$, to the same value N_0 (within the experimental error) for the unnormalised ion fraction. Since at infinite energy, the normalized ion fraction is expected to be unity, N_0 is the (previously unknown) normalization factor. In fig. 3 the normalized ion fraction is plotted as a function of the energy E. Also indicated is a theoretical calculation [10] that we performed along the lines of the model of Y. Muda and D.M. Newns [11]. The agreement seems rather promising and it is hoped that

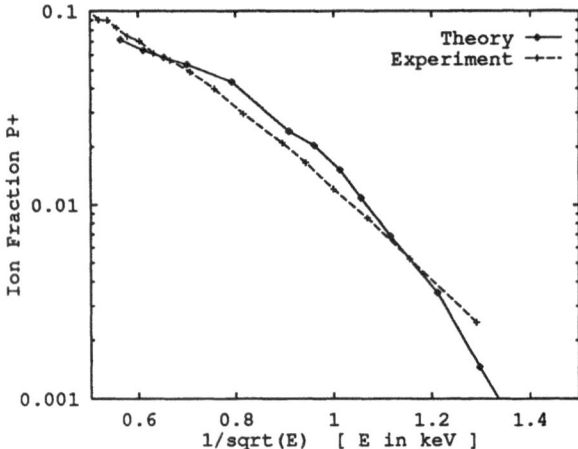

Fig. 3 The theoretical and experimental ion fraction P^+ is plotted as a function of the incident energy E for $He^+ \rightarrow Si$ scattering. Unlike in figs. 1 and 2, absolute ion fractions are presented here.

such a procedure would help to obtain absolute, i.e. normalized, values for the ion fraction P^+.

CONCLUSION

Three different neutralization models were reviewed. For the unifying model the time-dependent neutralization rate was calculated in order to clarify the connection to the Hagstrum model in the continuum limit. Fitting results were presented for MgO, which was experimentally studied by Souda et al. [5] and numerical difficulties related to the use of relative intensities were pointed out. Finally, a procedure to circumvent this difficulty was indicated. A satisfactory agreement between experiment and theory was found for silicon.

AKNOWLEDGEMENT

One of us (GV) thanks the Belgian National Science foundation for a travel grant and NATO for the admission grant to the conference.

REFERENCES

° Permanent address: University of Antwerp (UIA), Department of Physics, Universiteitsplein 1, B-2610 Antwerp, Belgium.

* Also at: University of Antwerp (RUCA), Groenenborgerlaan 171, B-2020 Antwerp, Belgium; and Eindhoven University of Technology, P.O.Box 513, 5300 MB Eindhoven, The Netherlands.

[1] H.H. Brongersma in: *"Surface Analysis of High Temperature Materials: Chemistry and Topography"*, Ed. G. Kemeny (Elsevier, Barking 1984) pp. 131-140.

[2] G. Verbist, J.T. Devreese and H.H. Brongersma, Surf. Sci. **233** (1990) 323.

[3] H.D. Hagstrum, Phys. Rev. **96** (1954) 336; and in: *"Inelastic Ion-Surface Collisions"*, Eds. N.H. Tully, W. Heiland and C.W. White (Academic Press, New York, 1977) pp. 1-15.

[4] D.J. Godfrey and D.P. Woodruff, Surf. Sci. **105** (1981) 438 and *ibid.* 459.

[5] R. Souda, M. Aono, C. Oshima, S. Otani, and Y. Ishizawa, Nucl. Instr. Meth. Phys. Res. **B15** (1986) 138.

[6] J.-M. Beuken, E. Pierson and P. Bertrand, Surf. Sci. **223** (1989) 221.

[7] O. van Kessel and H.H. Brongersma (unpublished).

[8] G.A.C. van Leerdam and H.H. Brongersma (to be published).

[9] The Moliére potential was used to calculated the cross sections, as is discussed in e.g. H.H. Brongersma and T.M. Buck, Nucl. Instr. **24** (1974) 437.

[10] G. Verbist, H.H. Brongersma and J.T. Devreese (to be published).

[11] Y. Muda and D.M. Newns, Nucl. Instr. Meth. Phys. Res. **33** (1988) 388.

DEUTERIUM-OXYGEN REACTION ON Pt(111)

CATALYSIS BY DEFECTS

Laurens K. Verheij, Markus B. Hugenschmidt,
Martin Freitag, Bene Poelsema and George Comsa

Institut für Grenzflächenforschung und Vakuumphysik
der KFA Jülich GmbH, Postfach 1913, D-5170 Jülich
Federal Republic of Germany

INTRODUCTION

Because of the simplicity of the reactants and products, the hydrogen-oxygen reaction on Pt(111) was believed to be an ideal model system for investigating chemical reactions on single crystal surfaces. However, in spite of the large number of studies [1-6] (and references there in), a clear picture of the mechanism of the reaction has not yet emerged. The main reason for this appears to be that the role of defects has been ignored. We have shown previously that the kinetics of the hydrogen-oxygen reaction is sensitive to the presence of atomic steps [3]. However, this effect could still not explain the discrepencies reported in the literature. More recently, we found that the reaction is also sensitive to the presence of other defects, even at concentrations of 10^{-3} or below [5]. We propose that these defects, which are very efficient in catalyzing the hydrogen-oxygen reaction, are kinks in atomic steps. Here we present additional evidence for this reaction mechanism and we will show that the observed kinetics of the reaction can be explained, at least qualtitatively, by assuming that the reaction takes place exclusively at these reactive kink sites for surface temperatures around 400 K.

EXPERIMENTAL

In order to reduce background effects, we have chosen to investigate the reaction between deuterium and oxygen which shows very similar behaviour to the hydrogen-oxygen reaction [1,2]. The reaction is studied by re-

Fundamental Aspects of Heterogeneous Catalysis Studied by Particle Beams,
Edited by H.H. Brongersma and R.A. van Santen, Plenum Press, New York, 1991

active scattering [1,3]: A modulated supersonic beam containing one of the reactants is directed onto the surface. A small fraction of the beam molecules adsorbs and reacts there with the other reactant which is adsorbed from the ambient gas. The product molecules (D_2O) are monitored with a quadrupole mass analyser. The demodulation of the D_2O signal supplies information on the reaction rate [1]. More details about the experimental system are given in refs [3,4].

The reaction is studied on the Pt(111) surface which was subject to three different preparation procedures [5]. The starting point, common for the three procedures is a Pt(111) surface, which we consider "ideal" at the present state of the art: Residual impurities are no more detectable by AES and the high reflectivity for thermal He scattering indicates that both impurities and defects are reduced to a minimum. The step density as determined by He interference measurements is below $3 \cdot 10^{-3}$.

Each of the three surface preparation procedures outlined below is applied on such an "ideal" surface:

1) No further treatment is applied. The surface is in the "ideal" state called in the following S_1.

2) Heating for 5 minutes at 550 K in oxygen ($P_{O_2} = 2 \cdot 10^{-5}$ mbar) and subsequent removal of the chemisorbed oxygen by heating in hydrogen at T_s < 550 K. As shown below, the reactivity of the resulting surface state S_2, differs dramatically from that of state S_1. In spite of this, states S_1 and S_2 looked identical when analysed with the commom surface-analytical techniques (LEED, AES, He-scattering), i.e. the oxygen coverage is below the detection level of our AES system ($5 \cdot 10^{-3}$ ML [5]) and no increase in the number of steps could be detected.

3) Argon ion bombardment at 600 K and subsequent cooling at the reaction temperature. State S_3 is characterized by randomly distributed steps preferentially oriented along <110> directions [7]. The step density of state S_3 determined in situ was $\Theta_{step} \approx 2 \cdot 10^{-2}$.

RESULTS AND DISCUSSION

The experimental results, which are presented, will be discussed in terms of a reaction model in which it is assumed that the reaction takes place exclusively at special reactive sites, presumably kinks in atomic steps. Even on the "ideal" surface S_1, some atomic steps with kinks are present. By the oxygen treatment (surface S_2) additional reactive sites are created, which seem to be kinks in preexistant almost kinkless step rows [5]. (Note that oxygen is known to react actively with steps [8]). On the bombarded surface (S_3), additional kinks will be found in the newly created atomic steps. So on both surfaces S_2 and S_3 more reactive sites are present than on the ideal surface; the distribution of these sites, however, will be very different.

In our reaction model, the overall reaction is thought to be divided into three more basic processes: adsorption of the reactants, diffusion of these towards the reactive kink sites and the actual reaction in which the product D_2O is formed. In a separate adsorption study [6], we have found equal sticking probabilities of the reactants on surface S_1 and S_2. The observed difference in reactivity of these surface is therefore not due to the adsorption process. In order to explain the major phenomena observed, it seems reasonable, therefore, to assume that adsorption takes place homogenously on the surface. Of course this assumption is too simple if details of the measurements are considered: a kink is a part of an atomic step and it is known that the sticking probability of both oxygen and hydrogen at step sites is much higher than on the flat Pt(111) terraces. Within our model one may distinguish between two extreme cases: 1) the reaction itself is rate limiting and 2) the diffusion of oxygen towards the kinks is rate limiting. Hydrogen diffusion is supposed to be fast, so we assume that this process is never rate limiting for the reaction conditions considered here. Below we discuss these two extreme cases, i.e. we will investigate how the overall "reaction rate" depends on the concentration and the distribution of the kink sites. The reaction rate is defined here as the reaction rate for an individual atom, i.e. as the inverse of the average time spent by a minority reactant atom on the surface between adsorption and desorption in the form of D_2O. A qualitative measure for this quantity is the slope of the $I_{D_2O}(t)$ curve, i.e. if the reaction is fast, I_{D_2O} will reach its steady state value almost immediately whereas for a small reaction rate $I_{D_2O}(t)$ will increase slowly.

At high oxygen coverage, all kink sites are occupied with oxygen and the reaction itself will be rate limiting. Experiments will reveal in this case information about the details of the actual reaction. (See e.g. ref. [3], $\Theta_0 > 0.08$. The clean surface in that study is in fact the surface S_2

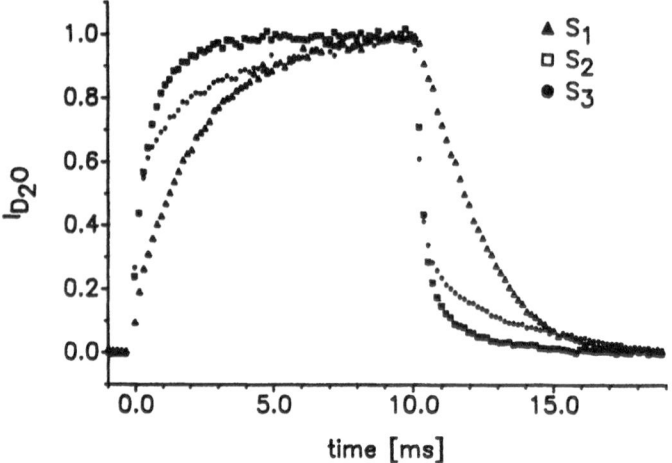

Fig. 1. D_2O-product waveform measured on Surfaces S_1, S_2 and S_3 (resp. "ideal", "oxygen modified" and ion bombarded, see text). Surface temperature: 400 K. Beam: D_2 modulated with 48.8. Hz (square wave). Oxygen excess.

in this work). Relevant for this work is that one expects the reaction rate to be proportional to the concentration of kinks and independent of their distribution, i.e. for both surfaces S_2 and S_3 we predict a clear increase of the reaction rate with respect to surface S_1. Experiments at an oxygen coverage $\Theta_0 \approx 0.1$ and using a modulated D_2 (minority reactant) – beam show exactly this behaviour (Fig. 1). The D_2O intensity spectrum measured on surface S_2 is only slightly demodulated. We infer that most of the adsorbed deuterium atoms react within 0.5 ms with oxygen to form D_2O. In contrast the signal measured on surface S_1 is clearly demodulated. The slow rise and decay of I_{D_2O} shows that the time which elapses between the adsorption of the D_2 molecules and the formation of D_2O is much longer (2–4 ms). The reaction rate on surface S_2 is therefore larger than on surface S_1. The reaction on the stepped surface S_3 appears to be more complicated. It seems to involve two different processes: A fast one leading to a rapid rise and decay of the I_{D_2O} intensity, practically identical to that on surface S_2, and a much slower process. The origin of the latter has been uncovered recently [3]. It is due to the temporary storage of D–atoms at step sites; the D–atoms being unreactive as long as they are stored; the water–reaction is slowed down by the higher binding energy of the D–atoms at step sites. At the surface temperature considered here (400 K), this process competes with the direct reaction. Important for the present discussion is that we observe a clear increase of the reaction rate when increasing the concentration of reactive sites regardless of the distribution along the surface.

At zero oxygen coverage, the kink sites are not occupied with oxygen and upon adsorbing this reactant, the O-atoms have to diffuse towards the reactive kink sites before the reaction can take place. It can be shown that the mean diffusion time τ is determined primarily by the distance which the atoms have to diffuse to reach a reactive site and depends only weakly on the density of these sites [9-11]. In a qualitative way this can be understood by noting that the efficiency of reactive sites to remove diffusing O-atoms from the surface becomes smaller the more they are clustering together, since neighbouring reactive sites will cause a decrease of the local oxygen concentration. If the mean distance between kinks along a step is smaller than the mean distance between steps, one has effectively a clustering of kinks, i.e. the mean diffusion distance is primarily determined by the step density and to a much lesser extent by the kink density. Thus, if this condition is fulfilled on surface S_1, i.e. $\Theta_{kink} > 10^{-5}$, one will have about the same mean diffusion distance on surfaces S_1 and S_2 determined by Θ_{step}. Consequently we expect the reaction rates on these surfaces to be similar (but not equal) if oxygen diffusion is rate limiting. On the bombarded surface (S_3), Θ_{step} is much larger and one expects therefore that the reaction is much faster on this surface.

Experiments conducted in conditions of deuterium excess, performed with a modulated O_2 (minority reactant) beam agree with this prediction as shown in Fig. 2. The reaction on all three surfaces is much slower than in the case of oxygen excess (see fig. 1) and appears to be independent of the deuterium pressure, i.e. of the deuterium coverage. This is to be expected when oxygen diffusion is rate limiting. On surfaces S_1 and S_2 the mean residence time of the O atoms before reaction takes place is of the order of 0.1 to 0.25 s and very similar as predicted by our model. Note, however that the shape of the I_{D_2O} curves for surfaces S_1 and S_2 exhibit some differences. It appears that the details of the reaction still depend on the concentration of reactive sites, but our present model is too simple to explain the difference. On surfaces S_3 the reaction is much faster than on surfaces S_1 and S_2, the mean residence time of the O atoms before the reaction takes place is of the order of 5 ms (determined in a separate experiment).

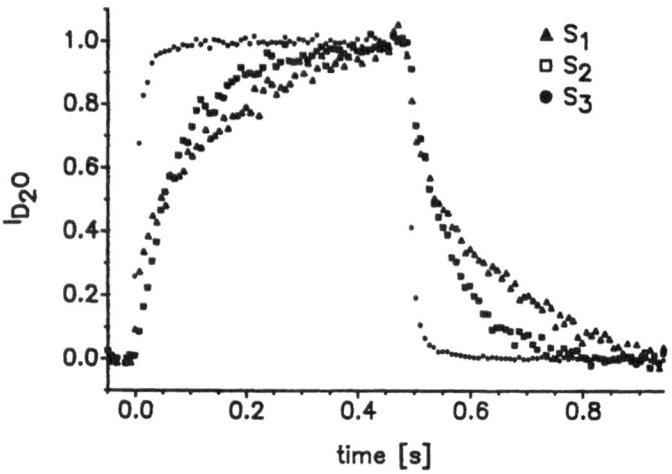

Fig. 2. D_2O-product waveform measured on surfaces S_1, S_2 and S_3. Surface temperature: 400 K. Beam: O_2, modulated with 1 Hz (square wave) Hydrogen excess.

To summarize, we conclude that the experiments presented here give further support to our previous conclusion that the water reaction on Pt(111) surfaces is catalyzed by reactive sites related to atomic steps, being probably kinks in step rows. Moreover, we found that this process is determining the reaction not only in the case of oxygen excess but also in the case of hydrogen excess, where the reaction proceeds much slower. In this latter case, oxygen diffusion appears to be rate-limiting for the reaction.

REFERENCES

1. G.E. Gdowski and R.J. Madix, Surf. Sci. 119 (1982) 184
2. J.L. Gland, G.B. Fisher and E.B. Kollin, J. Catal. 77 (1982) 263
3. L.K. Verheij, M.B. Hugenschmidt, L. Cölln, B. Poelsema and G. Comsa, Chem. Phys. Lett. 166 (1990) 523
4. L.K. Verheij, M.B. Hugenschmidt, B. Poelsema and G. Comsa, Surf. Sci. 233 (1990) 209
5. L.K. Verheij, M.B. Hugenschmidt, B. Poelsema and G. Comsa, submitted for publication
6. M.B. Hugenschmidt, L.K. Verheij, M. Freitag, B. Poelsema and G. Comsa, submitted for publication.
7. B. Poelsema and G. Comsa, Springer Tracts in Modern Physics, Vol. 115, Springer-Verlag, Berlin, Heidelberg 1989
8. G. Comsa, G. Mechtersheimer and B. Poelsema, Surf. Sci. 119 (1982) 159
9. L.A. Girifalco and D.R. Behrendt, Phys. Rev. 124 (1961) 420
10. M.J. Stowell, Phil. Mag. 21 (1970) 125
11. L.K. Verheij, M.B. Hugenschmidt, B. Poelsema and G. Comsa, to be published.

Directors and Lecturers

Directors

Prof. H.H. Brongersma
Eindhoven University of Technology
Department of Physics
P.O. Box 513
5600 MB Eindhoven
The Netherlands

Prof. A. Gras–Marti
Dept. Fisica Aplicada Apt. 99
E–03080 Alacant
Spain

Lecturers

Prof. J.A. Davies
Institute for Materials Research
McMaster University
1280 Main street West
Hamilton, Ontario L8S 4M1
Canada

Dr. H. Jobic
Institut de Catalyse
CNRS
2 Avenue A. Einstein
F–69626 Villeurbanne Cedex
France

Prof. R.W. Joyner
Leverhulme Centre for Innovative
Catalysis
Dept. of Chemistry
University of Liverpool
P.O. Box 147
Liverpool L69 3BX
United Kingdom

Dr. A. Kaldor
Corporate Research Laboratory
Exxon Research & Engineering Co.
RT22E, Clinton TWSP
Annandale, NJ 08801
USA

Prof. H. Knözinger
Institut für Physikalische Chemie
Universität München
Sophienstrasse 11
8000 München 2
Germany

Dr. R. Miranda
Dept. de Fisica de la Materia
Condensada, C–III
Universidad Autonoma de Madrid
Cantoblanco
28049 Madrid
Spain

Prof. H. Pfeiffer
Sektion Physik
Universität Leipzig
Linneausstrasse 5
Leipzig 7010
Germany

Prof. J.W. Rabalais
Dept. of Chemistry
University of Houston
Houston
Texas 77004
USA

Prof.dr. R.A. van Santen
Faculteit Scheikunde SH 7.10
Eindhoven University of Technology
P.O. Box 513
5600 MB Eindhoven
The Netherlands

Dr. E. Taglauer
Max–Planck–Institut für
Plasmaphysik
Boltzmannstrasse 2
D–8046 Garching bei München
Germany

Dr. J.C. Vickerman
Centre for Surface and Materials
Analysis
Dept. of Chemistry
UMIST
P.O. Box 88
Manchester M60 1QD
United Kingdom

Prof. J.T. Yates
Dept. of Chemistry
University of Pittsburgh
PA 15260
USA

Other participants

Dr. N. Aas
Department of Chemistry
UMIST
P.O. Box 88
Manchester M60 IQD
England

Dr. I. Abril
Departament de Fisica Aplicada
Fac. de Ciencies
Universitat d'Alacant, Apt. 99
E–03080 Alacant
Spain

Mr. N.G. Alameddin
2145 Sheridan Road
Technological Institute
Dept. of Chemistry
Northwestern University
Evanston, Illinois 60208
USA

Mr. R. Alvarez
Instituto de Ciencias de Materiales
Centro Mixto CSIC–Univ. Sevilla
P.O. Box 1115
41071 Sevilla
Spain

Dr. Suna Balci
Middle East Technical University
Chemical Engineering Department
06531 Ankara
Turkey

Mr. T.H. Ballinger
Box 46 Chevron Science Center
University of Pittsburgh
Pittsburgh, PA 15260
USA

Dr. R.O. Barrachina
Departament de Fisica Aplicada
Fac. de Ciencies
Universitat d'Alacant, Apt. 99
E–03080 Alacant
Spain

Dr. J. Beltramini
Department of Chemical Engineering
The University of Queensland
St. Lucia QLD 4067
Australia

Mr. J.M. Beuken
1, Place Croid du Sud, Lab PCPM
1348 Louvain–La–Neuve
Belgium

Mrs. J.W. Bijsterbosch
Department of Chemical Engineering
Universiteit van Amsterdam
Nieuwe Achtergracht 166
1018 WV Amsterdam
The Netherlands

Ir. H.J. Borg
Lab. of Inorganic Chemistry and
Catalysis, SH 7.10
Eindhoven University of Technology
P.O. Box 513
5600 MB Eindhoven
The Netherlands

Prof. C. Bruno
Dip. di Meccanica
University of Rome
Via Eudossiana 18
00184 Roma
Italy

Mr. F.J. Castella
Departament de Fisica Aplicada
Fac. de Ciencies
Universitat d'Alacant, Apt. 99
E–03080 Alacant
Spain

Dr. A.R. Cholach
Institute of Catalysis of the Siberian
Branch of the USSR
Academy of Sciences
630090, Novosibirsk, 90
Prospekt Akademika Lavrentieva, 5
USSR

Dr. G.W. Coulston
Physics Department
University of Kaiserslautern
Postfach 3049
D6750 Kaiserslautern
Germany

Prof. T.N. Durlu
Science Faculty of Ankara University
Besevler
06100 Ankara
Turkey

Prof. J.M. Feliu
Department Quimica Fisica
Universitat d'Alacant, Apt. 99,
03080 Alacant
Spain

Mr. M.L. Forcada
Departament de Fisica Aplicada
Fac. de Ciencies
Universitat d'Alacant, Apt. 99
E–03080 Alacant
Spain

Dr. R. Garcia–Molina
Departamento de Fisica
Universidad de Murcia
E–30071 Murcia
Spain

Dr. J.J.C. Geerlings
Koninklijke Shell Laboratorium
Amsterdam
Postbus 3003
1003 AA Amsterdam
The Netherlands

Prof. J.W. Geus
Analytisch Chemisch Laboratorium
Croesestraat 77A
3522 AD Utrecht
The Netherlands

Prof. W. Heiland
F.B. Physik
University of Osnabruck
D–4500 Osnabruck
Germany

Mr. O. Heintz
CNRS
B.P. 138
21004 Dijon Cedex
France

Mr. A. Henderson
Ctr. Surface and Materials Analysis
(CSMA)
Chemistry Dept.
UMIST
P.O. Box 88, Sackville Street
Manchester M60 1QD
United Kingdom

Mr. J.P. Holgado
Instituto de Ciencias de Materiales
Centro Mixto CSIC–Univ. Sevilla
P.O. Box 1115
41071 Sevilla
Spain

Dr. R. Imbihl
Fritz–Haber–Institut
Faradayweg 4–6
1000 Berlin 33
Germany

Mr. J.P. Jacobs
Dept. of Physics
Eindhoven University of Technology
P.O. Box 513
5600 MB Eindhoven
The Netherlands

Dr. M.M. Jakas
Departament de Fisica Aplicada
Fac. de Ciencies
Universitat d'Alacant, Apt. 99
E–03080 Alacant
Spain

Dr. A.P.J. Jansen
Faculteit der Scheikundige
Technologie, SH 6.80
Technische Universiteit Eindhoven
Postbus 513
5600 MB Eindhoven
The Netherlands

Dr. H.P.C.E. Kuipers
Kon./Shell Lab., afd. CGP–1
Postbus 3003
1003 AA Amsterdam
The Netherlands

Dr. J. Laszlo
Max Planck Institut für Plasmaphysik
Boltzmannstrasse 2
D–8046 Garching bei München
Germany

Dr.Ir. L. Lefferts
DSM Research
P.O. Box 18
6160 MD Geleen
The Netherlands

Mr. C. Leighton
Department of Chemistry
UMIST
P.O. Box 88, Sackville Street
Manchester M60 1QD
United Kingdom

Mr. C. Linsmeier
Max Planck Institut für Plasmaphysik
Boltzmannstrasse 2
D–8046 Garching bei München
Germany

Mr. J. Llorca–Santos
Departament de Quimica Fisica
Universitat d'Alacant, Apt. 99,
E–03080 Alacant
Spain

Dr. M.–J. Luys
DSM Research B.V.
Industrial Catalysis
P.O. Box 18
6160 MD Geleen
The Netherlands

Dr. B. Maschhoff
Department of Physics
Rutgers University
Piscataway NJ 08855
USA

Mr. O. Melendez
University of Florida
Chem. Eng. Dept
Gainesville, FL 32611
USA

Prof. I.V. Mitchell
Dept. of Physics
University of Western Ontario
London, Ontario N6A 3K7
Canada

Mr. J.C. Morena–Marin
Departament de Fisica Aplicada
Fac. de Ciencies
Universitat d'Alacant, Apt. 99
E–03080 Alacant
Spain

Dr. M.A. Morris
Catalysis Res. Ctr.
ICI, Billingham
Cleveland, TS23 ILB
England

Dipl. Ing. P. Novacek
Institut für Allgemeine Physik
Technische Universität Wien
Wiedner Hauptstrasse 8–10/134
A–1040 Wien
Austria

Mr. J.M. Orts
Departament de Quimica Fisica
Universitat d'Alacant
Apartat 99, E–03080 Alacant
Spain

Mr. L. Pedocchi
Department of Chemistry
v. G. Capponi, 9
50121 Firenze
Italy

Dr. N.M. Reed
Dept. of Chemistry
UMIST
P.O. Box 88
Manchester M60 1QD
United Kingdom

Mr. A. Rodes–Garcia
LEI du CNRS
1, Place Aristide Briand
92195 Meudon Cedex
France

Dr. U. van Slooten
FOM–Inst. for Atomic and Molecular
Physics
Kruislaan 407
1098 SJ Amsterdam
The Netherlands

Dr. H.M. Urbassek
Institut für Theoretische Physik, TU
Postfach 3329
D3300 Braunschweig
Germany

Dr. P. Varga
Institut für Allgemeine Physik
Technische Universität Wien
Wiedner Hauptstrasse 8–10/134
A–1040 Wien
Austria

Ir. H.R.J. ter Veen
DSM Research B.V.
FA–GF
P.O. Box 18
6160 MD Geleen
The Netherlands

Dr. A. Fdez de la Vega
Dept. Quimica–Fisica
Universitat d'Alacant
03080 Alicante
Spain

Mr. G. Verbist
Dept. of Physics UIA
Universiteit Antwerpen
Universiteitsplein 1
B–2610 Antwerpen
Belgium

Dr. L.K. Verheij
EGV/KFA Jülich
D–5170 Jülich
Germany

Mr. M. Vicanek
Departament de Fisica Aplicada
Fac. de Ciencies
Universitat d'Alacant, Apt. 99
E–03080 Alacant
Spain

Mr. R.–J. Vreeburg
Debye Institute
Padualaan 8
3584 CH Utrecht
The Netherlands

FUNDAMENTAL ASPECTS OF HETEROGENOUS CATALYSIS
STUDIED BY PARTICLE BEAMS
September 3-14, 1990
Alicante, Spain

INDEX

ESDIAD (Electron Stimulated
        Desorption Ion Angular
        Distribution), 238
EXAFS (Extended X-ray
        Absorption Fine
        Structure), 191
  ab-initio, 201
  data analysis, 201
  theory of, 192

FAB-SIMS (Fast Atom
        Bombardment-
        Secondary Ion Mass
        Spectrometry),
        358, 388
Facetting, 125
Fast atom bombardment,
        see FAB
Faujasite, 11
Force field calculations, 265
Formic acid,
  electrooxidation, 63
Fourier,
  transform method, 203
  transform Raman spectroscopy,
        173
Frequency modulation
        spectroscopy, 172
FTIR (Fourier Transform
        Infrared Reflectance),
        146, 221

GaAs(110), 43
Glucose,
  electrooxidation, 63
Godfrey-Woodruff model, 437
Group orbital, 83
Growth mechanism, 216, 409

Hagstrum model, 437
HEIS, see RBS
High-energy ion scattering,
        see RBS
HREELS (High Resolution
        Electron Energy Loss
        Spectroscopy),
        249, 375
Hydrocarbon on metals, 274
Hydrodesulphurization, 18, 204
Hydrogen bonding, 176
Hydrogenation, 249
Hydrotreating, 17
Hydroxyl group, 9
  on alumina, 174
  on silica, 173
Hypersonics, 69

Impregnation, 38

Impregnation (continued)
  incipient wetness, 38
Inelastic neutron scattering,
        see INS
Infrared
  emission spectroscopy, 170
  transmission-absorption
        spectroscopy, 168
Infrequent event, 138
INS (Inelastic Neutron
        Scattering), 255
Interaction,
  adsorbate-adsorbate, 376
Interfacial reaction, 41
Ion
  beam analysis, 431
  energy, 425
  fraction,   437
  scattering, 113, 301, 419
    energy dependence, 290
    helium, 419
    high-energy, see RBS
    low-energy, see LEIS, ISS
    spectrometry, 313
        see LEIS,ISS
Ionization, 367
Isotope,
  exchange, 15
  ^{18}O, 177
  distribution, 425
ISS (Ion Scattering
        Spectrometry),
        301, 425
        see also LEIS

Junction,
  metal-semiconductor, 45

Kinetic oscillation, 125
Kink, 445
Kubelka-Munk function, 169

La_2O_3 on alumina,
  surface structure, 283, 293
Langmuir-Hinshelwood, 126
LEED (Low Energy Electron
        Diffraction), 393
LEIS, 41, 283, 393, 399, 419,
LEIS (continued),   437
  quantification of, 288, 298
        see also ISS
LEISS, see LEIS
Lewis acidity, 10, 179
Local density of states, 83

Magnesia, 9
MAS (Magic Angle Spinning)
        NMR, 159